Mechanochemistry and Emerging Technologies for Sustainable Chemical Manufacturing

This unique volume describes advances in the field of mechanochemistry, in particular the scaling up of mechanochemical processes. Scalable techniques employed to carry out solvent-free synthesis are evaluated. Comparability to continuous flow chemistry, the current industrial benchmark for continuous efficient chemical synthesis, is presented. The book concludes that mechanochemical synthesis can be scaled up into a continuous, sustainable process. It demonstrates that large-scale mechanochemistry can meet industrial demands, especially in the pharmaceutical industry.

Features:

- Mechanochemistry is rapidly developing as a multidisciplinary science on the borderline between chemistry, materials science and environmental science
- This unique text focuses on mechanochemistry with the ability to scale up and illustrates how mechanochemical synthesis is no longer an obstacle
- This timely book highlights recent advancements describing what can be achieved in chemical synthesis
- Mechanochemistry enables the synthesis of multiple polymorphic crystalline forms in the production of drugs in the form of tablets or granules in capsules

I0131964

Mechanochemistry and Emerging Technologies for Sustainable Chemical Manufacturing

Edited By
Evelina Colacino and Felipe García

CRC Press
Taylor & Francis Group
Boca Raton London New York

CRC Press is an imprint of the
Taylor & Francis Group, an **informa** business

Contents

Part 1   Mechanochemistry – A General Introduction: Mechanochemical-Based Technologies

Part 2   Solvent-Free Sustainable Technologies at Large Scale

Part 3   Case Studies/Perspective

Part 4   Reduced Solvent Sustainable Technologies

Foreword

Mechanochemistry: A Brief Historical Introduction

Over the past decades, the world has witnessed a rapid growth of *mechanochemistry*, a field studying chemical reactions initiated or accelerated by the direct absorption of mechanical energy. Mechanically induced chemistry has a long history and continues to be of high importance. In this *Foreword* to the book on *Mechanochemistry and Emerging Technologies for Sustainable Chemical Manufacturing*, which by no means exhausts the history of mechanochemistry, I present a short survey of *milestones* from the last about 50 years, which have influenced the more recent development of the field.

As described in Boldyrev's detailed accounts of the history of mechanochemistry in Siberia [1], the establishment of the first research *Group of Mechanochemistry* in 1968 at the Siberian Branch of the Academy of Sciences of the former Union of Soviet Socialist Republics (USSR) in Akademgorodok (Novosibirsk) provided a key stimulus for concentrated and systematic mechanochemical studies [2]. Since then, research in mechanochemistry was no longer the endeavour of a few disconnected researchers but was the common mission of a community of scientists pursuing similar objectives. Consequently, in 1972, the *first publication* documenting international collaboration in the field of mechanochemistry appeared in the literature [3]. Several new research groups began to work in the field, especially in the former USSR, Eastern Europe, Israel, and Japan. The development of mechanochemistry in this period is associated with the work of well-known scientists from Russia (V. V. Boldyrev, E. G. Avvakumov and P. Yu. Butyagin), Germany (*P. A. Thiessen, G. Heinicke* and *H.-P. Hegen*), Slovakia (*K. Tkáčová*), Hungary (*Z. A. Juhász* and *L. Opoczky*), Israel (*I. J. Lin*), and Japan (*T. Kubo, G. Jimbo* and *M. Senna*). Within this context, it should be mentioned that *mechanical alloying*, which is a distinct branch within the field of mechanochemistry, developed independently from "classical" mechanochemistry, mostly in the USA and in Western Europe [4].

An important milestone in the history of mechanochemistry is the foundation of the *International Mechanochemical Association (IMA)* [5] in 1988 at Tatranská Lomnica (Slovakia). Shortly after, IMA became one of the associated organizations of the *International Union of Pure and Applied Chemistry (IUPAC)*. The founders of IMA are portrayed in the historical photograph (Figure 0.1), toasting in honour of the newly born association. In its early stage, IMA played a crucial role in unifying the community, as well as promoting the exchange of ideas among scientists working in *mechanical alloying* and *mechanochemistry*. Note that there were no cross-references between the two areas for more than 20 years! IMA brought together previously unfamiliar scientists working in these two clearly related areas under one umbrella in 1993 when IMA members initiated the organization of the first dedicated, fully international meeting, the *International Conference on Mechanochemistry and Mechanical Alloying (INCOME)*. Since then, *INCOME* represents the most important convention dedicated to the science of physical and chemical transformations initiated and driven by mechanical forces. In almost 30 years, ten *INCOME* editions have taken place and the last one occurred in 2022 at Cagliari (Italy) led by Evelina Colacino (France) and Francesco Delogu (Italy).

Mechanochemistry is currently undergoing rapid expansion in a wide range of areas, and has been identified by IUPAC as *one of ten world-changing technologies* [6]. This field, in addition to its fascinating history, promises exciting results in a variety of applications in the future [7-9]. In this context, it is a paradox that this once obscure discipline is becoming a "game-changer" and increasingly mainstream. This is particularly true now when the impact of the European Programme *COST Action CA18112 'Mechanochemistry for Sustainable Industry' (MechSustInd)* [10–11] in expanding and coordinating the ever-growing mechanochemical community has become evident and with the several new initiatives worldwide aiming at further promoting mechanochemistry and its full potential for science and technology, for instance, the *NSF Center for the Mechanical Control of Chemistry (NSF CMCC)* [12]. In this context, IMA appears as the crucial actor that can catalyse the establishment of the authoritative

FIGURE 0.1   Foundation of International Mechanochemical Association (IMA) in 1988 at Tatranská Lomnica, Slovakia. From left to right: A. P. Purga (Russia), V. Jesenák (Slovakia), I. Hocmanová (Slovakia), L. G. Austin (USA), P. Baláž (Slovakia), M. Senna (Japan), interpreter, E. G. Avvakumov (Russia), L. Opoczky, (Hungary), K. Tkáčová (Slovakia), N. Z. Lyachov (Russia), V. V. Boldyrev (Russia), H.-P. Hennig (Germany), N. Števulová (Slovakia), H.-P. Heegn (Germany) and P. Yu. Butyagin (Russia). Author of photography: V. Šepelák (Slovakia).

international cooperation needed to strengthen and harmonize researchers' efforts in the field of mechanochemistry all around the globe in a time when mechanochemical processes are investigated at both laboratory and large scales, targeting potential applications in different industrial market sectors.

I hope that the present book will provide a comprehensive approach to the field of *mechanochemistry for chemical manufacturing* in a way that captures its breadth, realized impact and vast potential.

Vladimír Šepelák
President of the International Mechanochemical Association
December 2022

REFERENCES

1. V. V. Boldyrev, Mechanochemistry in Siberia. *Herald Russ. Acad. Sci.* 88(2), 142–150 (2018).
2. (a) V. V. Boldyrev, and E. G. Avvakumov, Mechanochemistry of inorganic solids. *Russ. Chem. Rev.* 40(10), 847–859 (1971); (b) V. V. Boldyrev, Kinetic factors that determine the specifics of mechanochemical processes in inorganic systems. *Kinet. Katal.* 13(6), 1411–1421 (1972).
3. (a) V. V. Boldyrew, E. G. Awwakumov, G. Heinicke, and H. Harenz, Zur triebochemischen Zersetzung von Alkali-Bromaten und Nitraten. *Z. Anorg. Allg. Chem.* 2, 152–158 (1972); (b) *Festkörperchemie: Beitrage aus Forschung und Praxis*, Eds. W. Boldyrev, and K. Meyer, VEB Deutscher Verlag für Grundstoffindustrie, Leipzig, 1973.
4. J. S. Benjamin, *New Materials by Mechanical Alloying Techniques*, Eds. E. Artz, and L. Schultz, DGM Informationsgesellschaft, Oberursel, p. 3, 1989.
5. For more details, please see http://imamechanochemical.com/.
6. F. Gomollón-Bel, Ten chemical innovations that will change our world: IUPAC identifies emerging technologies in chemistry with potential to make our planet more sustainable. *Chem. Int.* 41(2), 12–17 (2019).

7. X. Z. Lim, Interlocking screws crank out pharmaceuticals. *Chem. Eng. News* 98(38) (2020).
8. F. Gomollón-Bel, Mechanochemists want to shake up industrial chemistry. *ACS Cent. Sci.* 8(11), 1474–1476 (2022).
9. E. Colacino, V. Isoni, D. Crawford, and F. García, Upscaling mechanochemistry: Challenges and perspectives. *J. Trends Chem.* 3(5), 335–339 (2021).
10. For more details, please visit www.mechsustind.eu.
11. J. Hernández, I. Halasz, D. E. Crawford, M. Krupička, M. Baláž, V. André, L. Vella-Zarb, A. Niidu, F. García, L. Maini, and E. Colacino, European research in focus: Mechanochemistry for sustainable industry (COST action *MechSustInd*). *Eur. J. Org. Chem.* 2020(1), 8–9 (2020).
12. For more details, please visit http://www.chem.tamu.edu/cmcc/.

Preface

Mechanochemistry generally refers to the use of mechanical force to perform chemical reactions, instead of relying on traditional methods based on the combination of heat and the use of solvents. Mechanochemical approaches offers several advantages, including the ability to perform reactions under milder conditions, reduced waste production, and more precise control over reaction outcomes.

At industrial level, technologies that incorporate mechanochemistry into chemical manufacturing processes are expected to play a key role in reducing the environmental footprint of the chemical industry as well as help achieving both United Nations' Sustainable Development goals (SGDs) and the European Green Deal objectives. For example, the use of mechanochemistry in pharmaceutical manufacturing has been shown to drastically reduce the amount of waste produced, while improving the quality and purity of active pharmaceutical ingredients (API). Similarly, mechanochemical methods have shown great promise in the production of energy storage materials, catalysts, and other products (fine chemicals, agrochemicals, etc.) that are critical for the current transition to a more sustainable economy.

This book focusing on mechanochemistry and other enabling technologies encapsulates the enthusiasm and motivation of a wide range of world leading experts to discover and adopt chemical manufacturing methods that are cleaner, safer, and more eco-friendly. The book, which is complemented by a brief historical introduction and a terminology and nomenclature section, is divided into four distinctive yet complementary parts.

The first part, "Mechanochemistry – A General Introduction", includes an overview of mechanochemistry with a focus on *in-situ* monitoring methodologies, followed by an insight on kinetics and thermodynamics, which discuss both computational and experimental approaches (Chapters 2 and 3). The part is accompanied by a discussion on life cycle analysis (Chapter 4) and current aspects of intellectual property (Chapter 5).

The introductory segment of the book is followed by a section covering "Solvent-Free Sustainable Technologies at Large Scale". This section discusses process intensification and scaling up using continuous mechanochemical methodologies (Chapter 6) and batch processes (Chapter 7), also including solvent-free acoustic synthesis and Resonant Acoustic Mixing (RAM) (Chapter 8).

The third section of the book, "Case Studies/Perspective", focuses on mechanochemical technologies for the large-scale sustainable synthesis of APIs and biologically active compounds (Chapter 9), or the use of industrial eccentric vibratory mills on energy related materials (Chapter 10). This section is completed by a chapter on scalable solutions for continuous manufacturing on organic synthesis (Chapter 11).

To ensure the book provides a comprehensive coverage of the topic, it includes two final chapters that discuss other enabling technologies that use reduced or benign solvents. Chapter 12 focuses on flow chemistry at large scale, while Chapter 13 highlights sustainable industrial chemical manufacturing using water as a solvent.

Presenting a comprehensive overview of *Mechanochemistry and Emerging Technologies for Sustainable Chemical Manufacturing* is challenging, but creating a concise, engaging, and well-structured textbook is even more difficult. However, the assembled team, including both authors and editors, has devoted their efforts to addressing this challenge and producing a book that we hope will engage students, educators, enthusiasts, and experts alike.

Evelina Colacino and Felipe García

Acknowledgements

This book is based upon work from COST Action CA18112, supported by COST (European Cooperation in Science and Technology).

COST (European Cooperation in Science and Technology) is a funding agency for research and innovation networks. Our Actions help connect research initiatives across Europe and enable scientists to grow their ideas by sharing them with their peers. This boosts their research, career and innovation.
http://www.cost.eu/

COST Action CA18112 – Mechanochemistry for Sustainable Industry (MechSustInd, www.mechsustind.eu) is a European Union networking initiative aimed to nurture and catalyse interactions between European and overseas researchers in Mechanochemistry.

COST Action CA18112 is supported for 4 years (2019-2023) by European COST funds and gather more than 140 scientists from 38 Countries within Europe. The international dimension of COST Action CA18112 goes out the borders of Europe and counts members from Canada, China, Japan, Mexico, Russia, South Africa, South Korea and USA.

COST Action CA18112 aims to nucleate a critical mass of actors from EU research Institutions, enterprises and industries, bringing together different areas of expertise and application. This in turn will create the necessary synergy to establishing a vigorous multi-disciplinary network of European scientists, engineers, technologists, entrepreneurs, industrialists and investors addressing the exploitation of mechanical activation in the production of chemicals through sustainable and economically convenient practices on the medium and large scales. Among the objectives, there is the implementation of mechanochemistry as a common practice at Undergraduate Level, as a way to contribute to the development of the *sustainable thinking* of the future generations of scientists.

For more information www.mechsustind.eu and the Memorandum of Understanding (MoU).

About the Editors

Evelina Colacino received her double Ph.D. (with European Label) in 2002 at the University of Montpellier II (France) and at the University of Calabria (Italy). She was appointed Research Fellow at the Catholic University of Louvain (Belgium, 2003), working on the preparation of new hydantoin scaffolds as antibacterial agents. Research Scientist at Sigma-Tau Pharmaceuticals (Italy, 2004), Post-Doctoral Fellow University of Montpellier II (France, until 2007), she was hired as Assistant Professor in 2008. Associate Professor of Organic and Green Chemistry since 2013, at the University of Montpellier, France, her main research activities concern the development of eco-friendly methodologies for the preparation of biomolecules, heterocyclic compounds and hybrid materials by mechanochemistry, with a main focus on Active Pharmaceutical Ingredients (*medicinal mechanochemistry*). She also investigates sustainable approaches to homogeneous or heterogeneous metal-catalysed processes by combining enabling technologies with non-conventional media. She is also a promoter of sustainability in Higher Education by integrating green chemistry at the undergraduate level in organic chemistry courses, teaching laboratories and across the sub-disciplines of chemistry, with a special focus on the fundamentals and the practice of mechanochemistry. She is a member of the Advisory Board of the Green Chemistry Commitment (GCC, www.beyondbenign.org) and the International Mechanochemical Association (IMA, http://imamechanochemical.com). She leads the European Programme COST Action CA18112 (MechSustInd, 2019-2023) – 'Mechanochemistry for Sustainable Industry' (www.mechsustind.eu and https://www.cost.eu) and the EU Horizon Project IMPACTIVE (**I**nnovative **M**echanochemical **P**rocesses to synthesise green **ACTIVE** pharmaceutical ingredients, 2022–2026, www.mechanochemistry.eu).

Felipe García is originally from the coastal town of Gijón (Spain) and gained both his B.Sc. and M.Sc. degrees in Chemistry at the local Oviedo University (Spain). In 2001, he moved to the University of Cambridge (UK) to carry out his graduate studies on main group imides and phosphides as a Cambridge European Trust and Newton Trust Scholar under the supervision of Prof. Dominic Wright. He then gained Junior Research Fellowship at Wolfson College (UK, 2005) and was appointed College Lecturer in Inorganic Chemistry at Newnham and Trinity Colleges (UK, 2006). In March 2011, he moved to Nanyang Technological University (Singapore) as Assistant Professor, where he developed different aspects of main group chemistry. In 2022, he moved back to his *alma mater* (*i.e.*, University of Oviedo) as a Margarita Salas Senior researcher (funded by FICYT). The Margarita Salas Senior programme is a new scheme from the *Foundation for the Promotion in Asturias of Applied Scientific Research and Technology* (*FICYT*) to attract highly qualified well-established researchers to R&D and higher education institutions in the Principality of Asturias (Spain). In late 2023, he will move to Monash University in Australia continue his research career. Felipe has published over 90 papers on Main Group Chemistry and maintains a strong interest in the synthesis of novel compounds for industrial and biological applications.

Contributors

Marcela Achimovičová
Institute of Geotechnics
Slovak Academy of Sciences
Košice, Slovakia

Peter Baláz
Institute of Geotechnics
Slovak Academy of Sciences
Košice, Slovakia

Matej Baláž
Institute of Geotechnics
Slovak Academy of Sciences
Košice, Slovakia

Tanja Bendele
LL.M., RUHR-IP Patent Attorneys
Essen, Germany

Wessel Bonnet
Department of Chemistry
Faculty of Natural and Agricultural Sciences
University of Pretoria
Pretoria, South Africa

Duncan Browne
University College of London (UK)

Charles E. Diesendruck
Grand Technion Energy Program
Technion – Israel Institute of Technology
Haifa, Israel
and
Faculty of Chemistry
Technion – Israel Institute of Technology
Haifa, Israel

Erika Dutková
Institute of Geotechnics
Slovak Academy of Sciences
Košice, Slovakia

Michael Felderhoff
Department of Heterogeneous Catalysis
Max-Planck-Institut für Kohlenforschung
Mülheim, Germany

Tomislav Friščić
Ruđer Bošković Institute,
Zagreb, Croati

Or Galant
Faculty of Civil and Environmental Engineering
Technion – Israel Institute of Technology
Haifa, Israel

Fabrice Gallou
Chemical and Analytical Development Novartis Pharma AG
Basel, Switzerland

Ivan Halasz
Ruđer Bošković Institute
Zagreb, Croatia

Valerio Isoni
The Institute of Sustainability for Chemicals, Energy and Environment
Agency for Science, Technology and Research (A*STAR)
Singapore

Jaimee Jugmohan
Council for Scientific and Industrial Research (CSIR)
Future Production: Manufacturing Cluster
Pretoria, South Africa

Jamie A. Leitch
University College of London, UK

Stipe Lukin
Ruđer Bošković Institute
Zagreb, Croatia

Feliu Maseras
ICIQ – Institut Catala d'Investigacio Quimica
Tarragona, Spain

Adam Michalchuk
Federal Institute for Materials Research and Testing (BAM)
Berlin, Germany
and
School of Chemistry
University of Birmingham
Birmingham, UK

Manisha Mishra
Department of Chemistry
University of Massachusetts Boston Boston, MA

Nicole Neyt
Council for Scientific and Industrial Research (CSIR)
Future Production: Manufacturing Cluster
Pretoria, South Africa

Jenny-Lee Panayides
Council for Scientific and Industrial Research (CSIR)
Future Production: Manufacturing Cluster
Pretoria, South Africa

Bruna S. Pladevall
ICIQ – Institut Catala d'Investigacio Quimica
Tarragona, Spain

Darren Riley
Department of Chemistry
Faculty of Natural and Agricultural Sciences
University of Pretoria
Pretoria, South Africa

Steffen Reichle
Department of Heterogeneous Catalysis
Max-Planck-Institut für Kohlenforschung
Mülheim, Germany

Maria Rivas-Velazco
Johnson Matthey Technology Centre
Reading, UK

Sabrina Spatari
Faculty of Civil and Environmental Engineering
Technion – Israel Institute of Technology,
Haifa, Israel
and
Grand Technion Energy Program Technion – Israel Institute of Technology
Haifa, Israel

Bela Torok
Department of Chemistry
University of Massachusetts Boston
Boston, MA

Matthew T. J. Williams
University College of London, UK

Ning Ye
Chemical and Analytical Development
Suzhou Novartis Technical Development Co., Ltd
Jiangsu, China

Part 1

Mechanochemistry – A General Introduction

Mechanochemical-Based Technologies

1

In Situ Monitoring of Mechanochemical Ball-Milling Reactions

Ivan Halasz, Stipe Lukin, and Tomislav Friščić

CONTENTS

1.1 Introduction

Mechanochemistry by ball milling,[1–6] being conducted in rapidly moving reaction chambers (also known as milling jars or vessels) made of thick, hard, and often non-transparent material, has for long been treated as a black-box technique where the milled materials undergo poorly understood processes. This has severely hindered reaction optimization, limiting it to laborious, time-consuming step-wise *ex situ* analysis where the sample is periodically extracted from the milling chamber to be analyzed.[7] Such an approach requires either periodic interruption of the milling process for sampling or repeating

DOI: 10.1201/9781003178187-2

the reaction from the same starting point and letting it run for increasingly longer times.[8] Initially, the interest in understanding the processes taking place during ball milling has led to the development of methods for temperature and pressure monitoring.[9, 10] These were accomplished through modifications to the milling apparatus, notably the attachment of temperature sensors close to the inner surface of the reaction chamber,[11] or installing valves that could connect the inside if the reaction chamber to a pressure sensor.[9]

Temperature and pressure monitoring provide valuable information but usually offer only an indirect insight into the chemical reaction taking place during ball milling. Recently, a new field of in situ reaction monitoring has started developing based on methods that can directly probe the composition of the reaction mixture.[12] These are based primarily on synchrotron powder X-ray diffraction (PXRD)[13] and Raman spectroscopy,[14] as well as their use in tandem.[15, 16] Initially developed in 2013, in situ reaction monitoring by powder X-ray diffraction enabled the first insight into the evolution of crystalline species: depletion of reactants, product formation, and notably, formation of intermediate phases. PXRD monitoring was followed in 2014 by in situ Raman spectroscopy monitoring providing a complementary insight since Raman spectroscopy is not limited by the crystallinity of the reaction mixture and enables monitoring the evolution also of amorphous or liquid mixture components. Such complementarity stipulates their use in tandem since Raman spectroscopy may not always be the method of choice in transformations of crystalline materials, and PXRD is less informative when it comes to transformations of non-crystalline reaction mixture components, which may constitute a significant portion of the mixture.[17]

The need for in situ reaction monitoring has further stimulated the application of other analytical techniques and has led to the use of NMR spectroscopy,[18, 19] X-ray absorption,[20] and luminescence[21] for monitoring of transformations during milling. Currently, however, in situ PXRD and Raman spectroscopy remain the two most established techniques for in situ monitoring of mechanochemistry, particularly since protocols for each have been described in detail.[22, 23] In situ monitoring techniques are applicable to the mechanosynthesis of all types and classes of compounds,[24] including industrially relevant materials[13] and active pharmaceutical ingredients.[25–27]

Overall, in situ monitoring has revealed a highly dynamic milling environment comprising fast reactions, a plethora of intermediates, which are sometimes short-lived, has enabled the recognition of various parameters influencing ball-milling reactions, including the understanding of the catalytic effect of additives, which are becoming ubiquitous in ball milling. The development of in situ monitoring techniques for ball-milling mechanochemistry has been tremendous.[28–30] Considering that in situ PXRD is a synchrotron-based technique,[22] more prospects for a wider use can be expected for in situ Raman spectroscopy,[23] which is an every-day laboratory technique based on the use of small and affordable Raman spectrometers.

1.2 Methods

1.2.1 In Situ Temperature Monitoring

When performing a ball-milling reaction, one readily notices that the reaction chamber heats up during processing. The increase in temperature is more substantial in a planetary ball mill where, with the use of a large number of heavy milling balls, temperatures above the boiling point of water can be reached. On a vibratory ball mill, the temperature increase is usually 10–15 °C so temperatures above 40 °C are hardly ever surpassed. The temperature rise will largely depend on the type of material from which the jar is made, for example, switching between steel and poly(methylmetacrylate) (PMMA), as well as on the size, type, and number of milling balls, and the milling frequency. In particular, the milling frequency as a parameter of the milling process will determine the energy input, and consequently, the rise in the temperature through friction heating.[31, 32] It is important to bear in mind that energy input is also determined by the mill geometry. Thus, milling at the same frequency on two different ball mills can still lead to different energy inputs. Noteworthy, some ball mills may contribute to the heating of the reaction vessel by blowing warm air from under the mill's hood and over the reaction vessels.[23]

Temperature monitoring during milling is difficult to achieve since placing a temperature sensor in direct contact with the reaction mixture is not possible because it would be exposed to the violent impacts of the milling media. Circumventing this obstacle can be attempted by drilling holes in the wall of the reaction chamber where the temperature sensor (a thermocouple, for example) could be placed closer to the inner surface of the reaction chamber. If the jar material was stainless steel, and therefore a good heat conductor, such an approach seems more accurate than, for example, if the jar material is plastic or ceramic, such as zirconia or tungsten carbide.

Currently the most precise and direct approach to temperature monitoring during a ball-milling reaction was achieved by using a custom-built PMMA reaction chamber with a small aluminum pin embedded in the chamber wall and serving as the heat bridge between the reaction mixture and the temperature sensor.[33] This enabled temperature monitoring with a precision better than ± 0.05 °C, leading to highly reliable measurements of small temperature changes. Simpler implementations of temperature monitoring involve the use of thermal cameras which, however, are limited as they record the temperature of the outside wall of the reaction vessel.[34] Thermal cameras may also be a good choice for immediate temperature measurement after stopping the milling process and opening the jar.[35]

Precise and reliable temperature profiles are essential to understand heat flow and heat generation during milling processes. Numerical simulation of heat flow during milling, performed to reproduce experimentally observed temperature profiles, revealed that the dominant factor contributing to temperature increase is friction arising from inelastic collisions of the milling media. Friction is strongly dependent on the material that is being milled, and the corresponding characteristics of inelastic ball collisions will change with changes to the milled material associated with chemical or structural transformations such as interconversion of polymorphs, amorphization, but also particle size distribution. With the overwhelming contribution of friction to the observed temperature profiles, any heat generation arising from the chemical change will most often be negligible, unless the reaction is extremely fast and highly exothermic or endothermic, such as in mechanochemical self-sustained reactions (MSR) characterized by a high ratio of reaction enthalpy to the heat capacity of the milled material.[36] In such self-sustained reactions, usually the whole sample undergoes a sudden transformation, resulting in large energy release or absorption. Such energy-intense events are readily observed in the form of a sharp peak in the temperature profile. In comparison, chemical transformations that are not accompanied by large enthalpy changes, or that are gradual, will have a small contribution to temperature variations. The temperature profile will then be dominated by friction.

Unless milling introduces such significant physicochemical changes to the material, the process of milling will generally represent a stable heat source, leading to an increase in the temperature of the whole milling assembly. Once the temperature of the milled system rises above the ambient temperature, it begins to dissipate heat to the surroundings. Such heat dissipation will be more pronounced at higher temperatures of the milling assembly and, since energy input due to the milling process is constant, at some point the sensor will detect a constant temperature, indicating that the rates of mechanically induced heating and heat dissipation are equal. Any change in heat generation will lead to a different steady-state temperature, which can be a consequence of a change in the chemical composition or overall structure of the milled material.[33] The steady-state temperature is characteristic not only of the milled material but also of the amount of the milled material and will be sensitive to any changes in the surroundings of the mill, which may influence heat transfer from the milling assembly, such as a change in ambient temperature or in air flow around the mill.

1.2.2 In Situ Pressure Monitoring

Currently, a distinct field is developing where the atmosphere inside a reacting chamber is being controlled and where gases can be sampled in situ for a spectroscopic analysis that is conducted in tandem with the in situ PXRD monitoring.[37] A remarkable example of reactions with gases during milling is the recent ammonia synthesis from elements at room temperature[38] and in general catalyzed reactions of gases that require mechanochemical processing of the catalyst.[39, 40]

Using pressure to monitor mechanochemical reactions has a long history.[9, 10, 41, 42] Besides specific technical modifications of the milling vessel, the main obstacle to the general use of pressure as a handle for reaction monitoring is its limitation to reactions involving gaseous reaction participants.[43–46] However, with an increasing number of mechanochemical reactions involving gases, it can become very useful also for the optimization of milling parameters.[47] In situ pressure monitoring was employed in the manometric study of the formation of zeolitic imidazolate frameworks (ZIFs)[48] and when molybdenum hexacarbonyl ($Mo(CO)_6$) was used as an in situ source of CO for palladium-catalyzed amidation and esterification reactions.[49] Indeed, in the latter case, in situ pressure monitoring was essential to determine that there is no significant release of CO during milling proving that metal carbonyls may be a more convenient source of CO that avoid the necessity of handling this poisonous gas.

Most examples of using a change in pressure of the reaction vessel due to the release, expansion, or sorption of gas have focused on reactions of inorganic materials involving gases, for example, nitrides (involving NH_3, N_2), hydrides (involving H_2), or carbonates (involving CO_2). Manometric monitoring was also applied to reactions that exhibit strongly exothermic behavior. Real-time monitoring of the gas pressure within a milling jar was recently used by the Borchardt team, in studies of mechanochemical syntheses of nanostructured carbon-based materials. These studies revealed MSR-like behavior, i.e. the appearance of an induction period followed by a strongly exothermic process, across a wide range of different systems. Some of the examples include the milling of calcium carbide with either hexachlorobenzene (C_6Cl_6) or trichlorotriazine ($C_3Cl_3N_3$) to form layered graphitic materials,[50, 51] the mechanochemical Scholl reaction achieved through ball milling of dendritic oligophenylenes to form nanographenes,[47] or the mechanochemical Friedel–Crafts arylation of cyanuric chloride.[52]

Early reports of thermal monitoring of mechanochemical reactions focused on highly exothermic mechanically induced self-propagating reactions, an area that has previously been extensively reviewed.[53] One notable example of monitoring such reactions is the reduction of the oxide Fe_3O_4 with zirconium metal. Monitoring this reaction with a K-type thermocouple revealed a slow, gentle heating for approximately the first 100 seconds of the reaction. This gentle increase is then suddenly followed by an almost immediate increase of ca. 70 °C. Following the temperature spike, the system was found to rapidly cool and reach an equilibrium temperature of ca. 60 °C after approximately 7 minutes. Very recently, the Weidenthaler group demonstrated a combination of real-time pressure monitoring with in situ synchrotron PXRD. The tandem methodology was applied to the mechanochemical reaction of metallic zinc with sulfur (S8), forming zinc sulfide (ZnS). Analysis by real-time and in situ synchrotron PXRD revealed the very fast formation of the hexagonal form of ZnS (wurtzite), which was followed by conversion to the thermodynamically more stable cubic polymorph, sphalerite.[54] Importantly, the appearance of the wurtzite phase was accompanied by a marked jump in pressure, which was interpreted as evidence that the mechanochemical process involves impact-induced vaporization of S_8.

In the context of metal-organic framework (MOF) synthesis, Brekalo et al. have used a manometric technique to follow the mechanochemical formation of ZIFs by planetary milling of imidazole or 2-methylimidazole (to form the well-known and commercially relevant ZIF-8 framework) with basic zinc carbonate.[48] Following of changes in the pressure of the gas inside the milling vessel revealed an unexpected feedback process, in which the CO_2 by-product reacts further with the generated ZIF-8 material, yielding a zinc imidazolate carbonate. The formation of this material was also found to be avoidable by using an excess of the imidazole linker precursor.

1.2.3 In Situ PXRD Monitoring

Despite providing an overall insight into the reaction course during uninterrupted milling, pressure and temperature monitoring[9, 10] do not provide direct information on the changes in the composition and/or the structure of the reaction mixture. The first approach for real-time and in situ mechanochemical reaction monitoring was introduced with the application of synchrotron powder X-ray diffraction (PXRD) to monitor the evolution of crystalline species during milling.[13, 22] With time resolution in seconds, such in situ PXRD analysis enabled simple monitoring of the depletion of reactants and the formation of intermediates and products, as well as following the evolution of particulate properties,[13, 55] such as average particle size, strain,[13, 55] and amorphization.[26]

In its first implementation, in situ PXRD monitoring of mechanochemical reaction on a vibratory ball mill required certain modifications to a usual milling setup (Figure 1.1).[22] In order for a standard ball

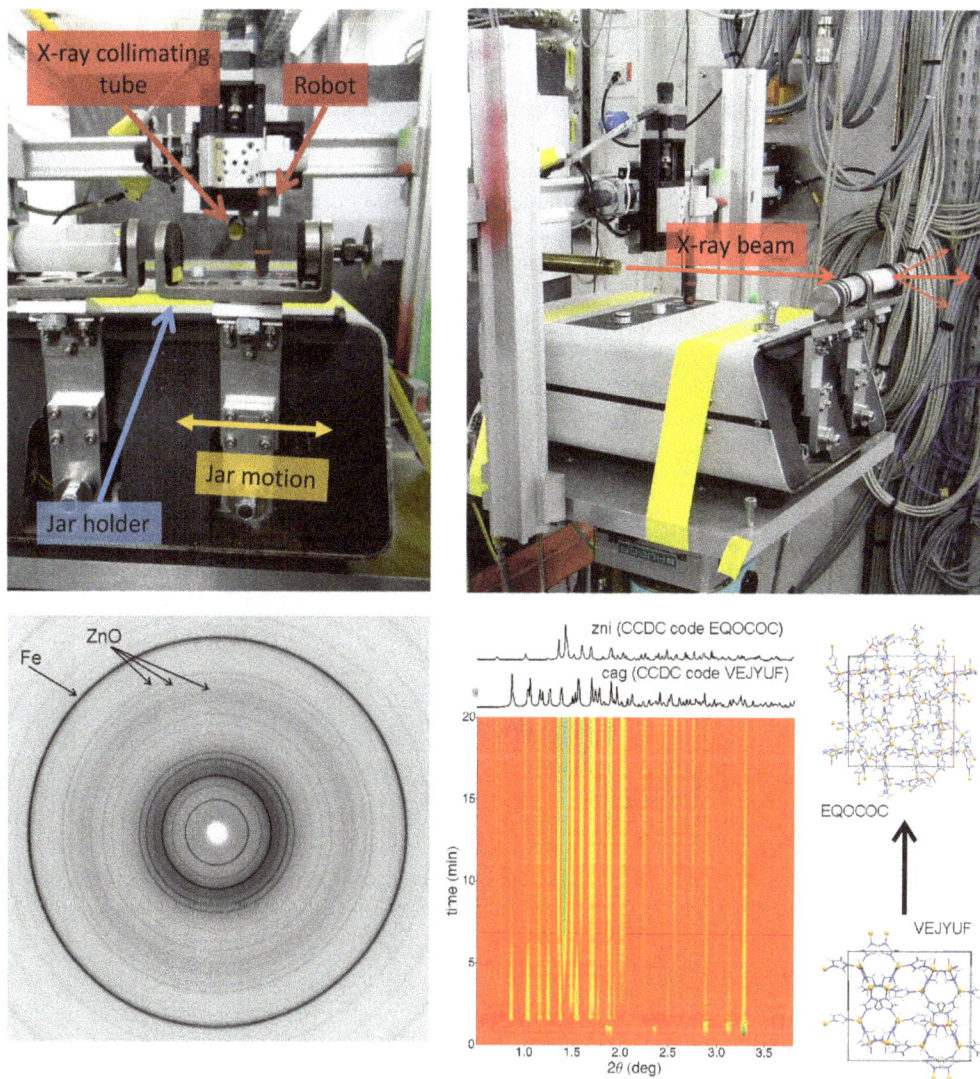

FIGURE 1.1   The first implementation of in situ monitoring using PXRD. (top row) The modified ball mill in the synchrotron experimental hutch. (bottom, left) A typical raw diffraction image with Debye–Scherrer rings. One such image is collected every few seconds. The Fe ring is coming from the diffraction by the milling ball. The inner rings belong to the product ZIF. (bottom, right) A series of images is integrated and plotted in a two-dimensional diffractogram. Contributions of the intermediate and the product phases are given above the 2D plot. The unusually small values for the 2β diffraction range are due to a very low radiation wavelength (0.14 Å). Adapted with permission. Copyright 2012 Nature Publishing Group.

mill to remain in the upright position while allowing access to the X-ray beam to the reaction jar, the holding hands were lifted above the ball mill, maintaining the oscillation of the reaction jar in the horizontal plane.[13] A setup using a type of vibratory ball mill where the jar oscillation occurs in the vertical plane is in this sense simpler since it does not require modifications of the ball mill and allows immediate access of the X-ray beam to the jar.[15, 56]

   Further essential modification to the usual mechanochemical setup was the replacement of the standard reaction jar made from stainless steel or some ceramic material, with a jar made from an X-ray amorphous and light-atom PMMA plastic. Such a reaction jar does not contribute with a diffraction signal, but rather with an amorphous halo. Though not impossible, testing using jars made from polycrystalline materials, such as stainless steel or aluminum, demonstrated a strong diffraction signal from the jar and a weak, but an observable signal from the sample. A synchrotron source is also essential for the success of this approach since an X-ray beam of sufficient intensity and photon energy is not available in a laboratory.

In addition, synchrotron sources are usually equipped with a large area detector necessary to immediately collect data across the entire diffraction pattern, enabling access to the required time resolution in seconds or even less. Since the safety procedure for closing and locking the experimental hutch at a synchrotron beamline takes 20–30 seconds, the final modification in the setup included the control of the mill from the outside of the hutch by using a robotic finger to press the start and stop buttons of the ball mill to enable the start of data collection along with the initiation of milling. Following the success of this in situ method, further modifications were introduced to control the mill remotely, enabling the software-controlled start of milling together with the initiation of data collection.[16] A particular ball mill may also provide enough room for the X-ray collimator to pass through the mill and come very close to the reaction jar, thus minimizing air scattering. The extensions to lift the milling arms still need to be in place but do not need to lift the jars above the hood of the mill (Figure 1.2). Such a setup exerts less strain on the mill and thus leads to lower vibrations during operation. The result is a more precise positioning of the beam at the bottom part of the reaction jar, which is essential to ensure that X-ray diffraction signals are not split.

If all the crystalline phases in the reaction mixture are structurally characterized, Rietveld refinement on the sequence of time-resolved diffraction patterns enables the construction of a reaction profile. However, as will often be the case in mechanochemical processes, the diffraction signal coming from the crystalline mixture components is accompanied by the scattering signal from amorphous or liquid

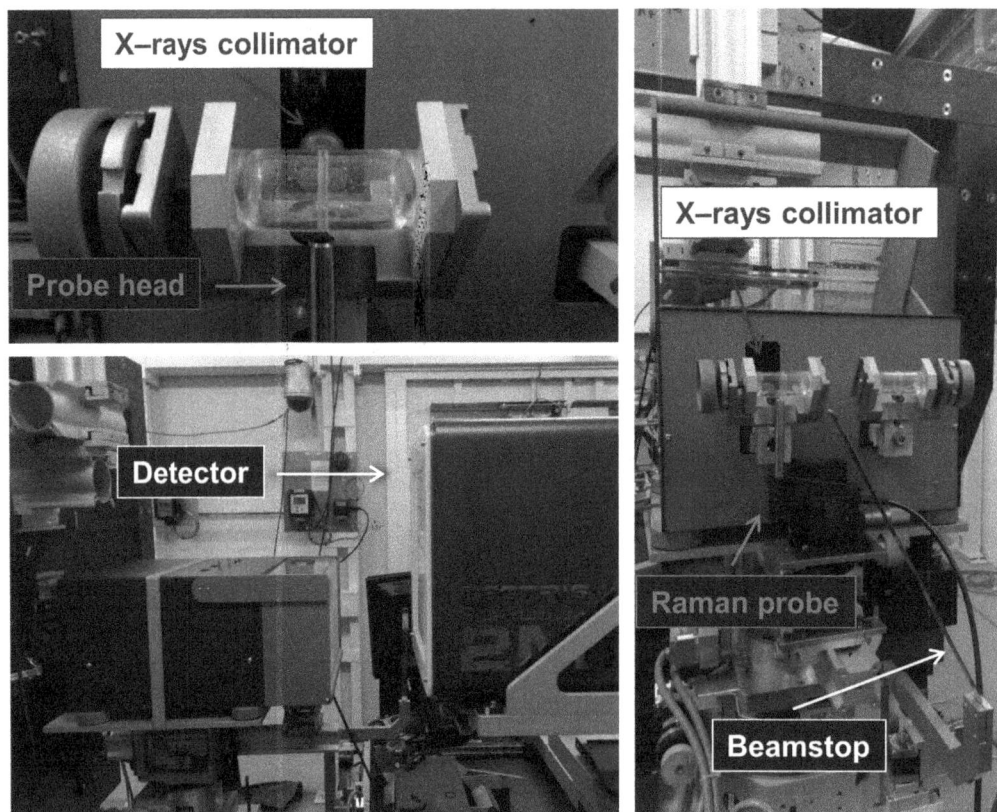

FIGURE 1.2   The mill setup at the ID15A beamline of the ESRF – the European synchrotron in Grenoble ready for tandem in situ monitoring.[12] The X-ray beam and the Raman laser focus are positioned to target the same part of the sample inside the jar. The X-ray detector is moved further away for standard X-ray diffraction monitoring experiments. Adapted in part with permission from reference [12]. Copyright 2021 American Chemical Society.

components, which remain buried in the high background caused by the scattering of the PMMA reaction vessel. This is partially remedied by using an internal X-ray scattering standard, which will usually be a highly crystalline inorganic solid such as silicon or a specialized NIST (National Institute of Standards and Technology) standard.[17] In this case, reaction profiles, derived from Rietveld-refined scale factors of individual crystalline phases and the Howard–Hill formula:[57]

$$w_p = \frac{S_p (ZMV)_p}{\sum_i S_i (ZMV)_i}, \tag{1.1}$$

will contain also the amounts of the amorphous phase but with, at best, only indirect information on the chemical composition of the amorphous fraction.

The in situ methodology was initially applied to the synthesis of zeolitic imidazolate frameworks (ZIFs),[58–60] a subclass of metal-organic frameworks (MOFs), and revealed a surprisingly dynamic reaction environment where the products could be formed within only a few minutes of milling. The in situ methodology enabled the study of the influence of additives and their amounts on mechanochemical reactivity. Thus, a fast formation of ZIF-8 required the use of a catalytic salt additive, while the use of different types and amounts of organic liquid additives enabled the templated formation of topologically distinct ZIFs based on the 2-ethylimidazole linker, adopting RHO-, ANA-, or qtz-topologies.

In this case of porous materials, the structure of only the ordered framework is not sufficient to describe the observed intensities since the pore content, which usually does not abide by the periodicity of the framework and the crystal, contributes to X-ray scattering. For this reason, Rietveld refinement is in general not applicable to porous materials and one needs to resort to individual peak fitting or a Pawley or Le Bail type of fitting to extract reaction profiles. Extracting the intensity of, usually, the most intense peak provides a means to compare different reaction profiles.[13] A more advanced approach will use a Pawley refinement to determine the relative intensities of reflections of a porous phase and then scale them all together in fitting the observed diffraction patterns.[61] As the amount of this phase increases so will the scale for this set of reflections and the variation of the scale in time provides the reaction profile. If the series of experiments is designed to always have the same amounts of reactants, different profiles based on the variation of this scale can be directly compared and any differences interpreted.

In a remarkable demonstration of its power, the in situ PXRD monitoring was applied to quantitative reaction monitoring of ZIF-8 synthesis where it enabled the discovery and subsequent characterization of a novel-topology polymorph of ZIF-8.[62] While studying the amorphization of the parent ZIF-8, a set of new diffraction signals emerged from a featureless diffraction image, which is characteristic of an amorphous reaction mixture. Insight into the reaction course from in situ monitoring was decisive for determining the appropriate moment when the milling process must be stopped in order to isolate the new phase. The subsequent crystal structure solution of the new ZIF from ex situ collected laboratory powder X-ray diffraction data revealed a complex, novel-topology, tetragonal crystal structure having four independent zinc(II) ions and four 2-methylimidazolate anions in the asymmetric unit and was named katsenite. Using crystalline silicon as an internal scattering standard demonstrated the recrystallization of the reaction mixture in full and the subsequent transformation of the katsenite material to the dense, diamondoid-topology polymorph of ZIF-8 as the final milling product (Figure 1.3).

In situ PXRD monitoring is now a well-established technique. Being initially developed at the ESRF – The European synchrotron in Grenoble,[63] it has now been applied at other synchrotron beamlines. At DESY in Hamburg, there is a dedicated beamline setup for in situ monitoring that also uses hard X-rays. At PSI in Villigen, a new and innovative ball mill design enables the acquisition of, comparatively, high-resolution data but has some drawbacks, and the milling apparatus is not always suited for all reaction conditions used in mechanochemistry.[64] BESYII in Berlin uses soft X-rays with lower penetrating power, which can be helped by using smaller reaction jars with thinner walls.[56] The use of high-energy X-rays (photon energy above 50 keV), as compared to softer X-rays, has another advantage in terms of low absorption, which is negligible for most materials and in particular for light-atom organic compounds and mixtures. Low absorption avoids the effect of micro-absorption, which would hamper quantitative Rietveld analysis.[65]

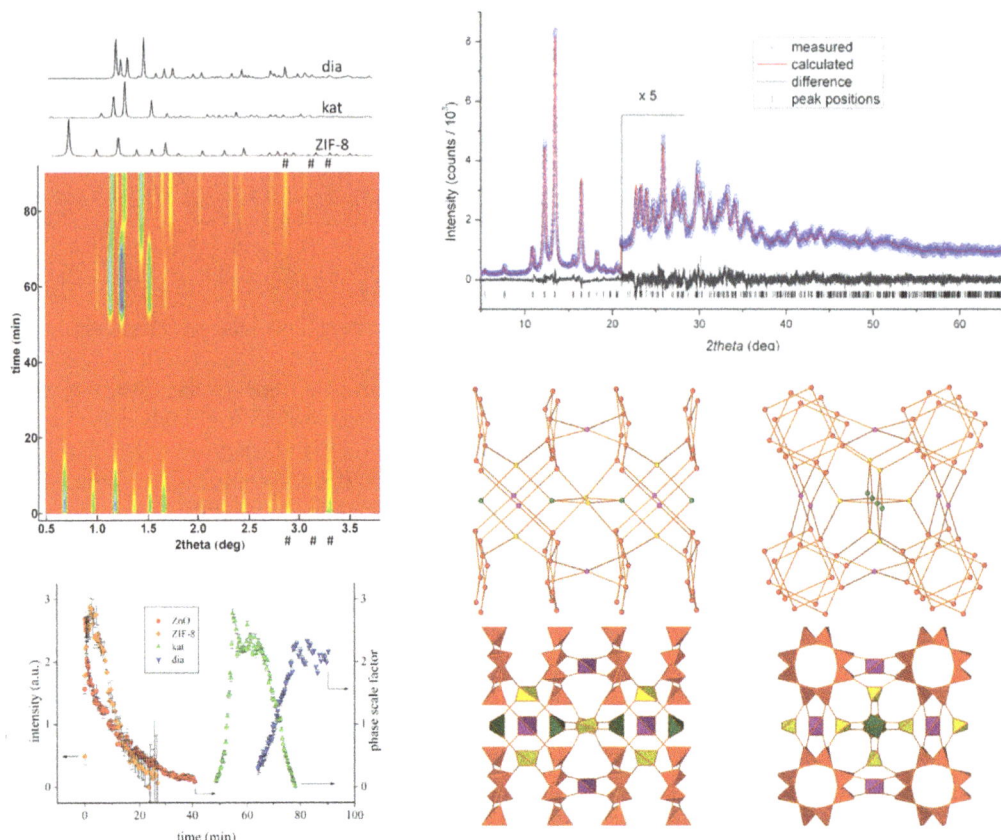

FIGURE 1.3　Formation of the katsenite framework, its characterization via PXRD, and its topology net.[62] Adapted with permission. Copyright 2015, Nature Publishing Group.

In our experience, and the experience of other groups,[66] the least squares-derived uncertainties on weight fractions seem to be underestimated and a Rietveld refinement may often attribute a small weight fraction to a phase that would appear not to be present. For example, the weight fraction of a reactant phase may, according to Rietveld refinement, persist in a small amount (e.g., 2–3 %) to the end of milling, particularly if its particle size is allowed to decrease to unreasonably low values. To prevent such an unrealistic situation, the widths of the calculated diffraction peaks should not be allowed to increase to values that lead to high correlations with the parameters used to describe the background function. It is therefore highly beneficial to attempt to experimentally reduce peak widths, which is currently possible using a dedicated mill, with a specialized compartment for the sampling of the reaction mixture, where the size of the diffracting sample is comparable to a sample contained in a larger capillary (Figure 1.4).[64] Another option could be to implement surface refinement,[67, 68] instead of sequential or independent Rietveld refinement, on the entire in situ data set and to enforce the time variation of scale factors with an appropriate kinetic equation.

Milling was traditionally devoted to particle comminution, a process intended to reduce the average particle size. As a consequence of crystal breaking, strain in the sample particles is also expected to increase. PXRD is suited also to monitor the evolution in particulate properties, such as average particle size and strain,[13, 55, 56] in a standard approach where the instrument contribution is measured independently and subtracted from the measured peak broadening. Remarkably, while some samples undergo full amorphization if subjected to prolonged milling with sufficient energy input,[69, 70] some samples exhibit steady-state particle size and strain,[71, 72] provided they are not being depleted in a chemical reaction or some other transformation. Liquid-assisted grinding (LAG) reaction conditions seem to be particularly beneficial for obtaining samples of better and more controlled crystallinity.[6, 8, 73]

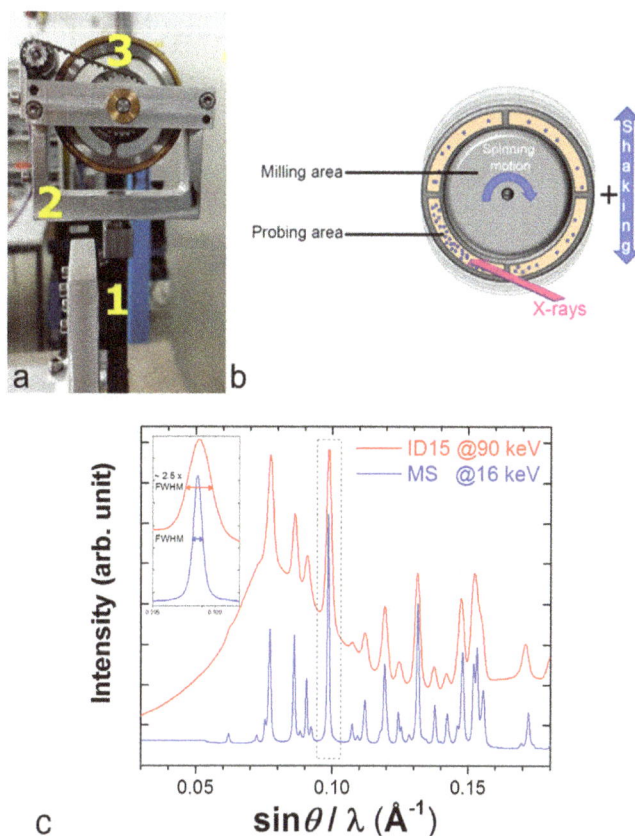

FIGURE 1.4 (a and b) In situ monitoring using a specialized mill with a sampling chamber enabling high-resolution of diffraction patterns and the use of a low-energy photon X-ray beam. (c) Comparison of peak widths obtained using a dedicated ball mill (MS, material science beamline at the Swiss Light Source) and the conventional in situ monitoring setup for a vibratory ball mill (ID15 beamline at the ESRF – the European synchrotron). Photon energy given in the top-right corner in c). Reprinted with permission from reference [64]. Copyright 2017 American Chemical Society.

1.2.4 In Situ Raman Spectroscopy Monitoring

The introduction of a synchrotron PXRD technique for in situ and real-time monitoring of mechanochemical ball-milling reactions was soon followed by the demonstration of a complementary technique, based on in situ Raman spectroscopy.[14] In a way similar to synchrotron PXRD, monitoring of milling reactions using Raman spectroscopy often involves reactions being performed in PMMA jars, but for a different reason. Notably, PMMA jars are translucent and this is essential because Raman scattering is based on the interaction of laser light with the sample, where a small portion of incident radiation undergoes inelastic collisions with samples leading to scattered light with a changed frequency. The PMMA itself produces a Raman scattering signal, which can usually be subtracted,[23] but it is clearly beneficial to use a jar material that does not interfere with Raman scattering from the sample, such as sapphire.[74]

The frequency of incident photons can be changed by energy quanta corresponding to a change in the vibration states of molecules in the sample. In a solid, molecular vibrations are influenced by inter- and intramolecular interactions, meaning their energy is under the influence of their local chemical environment. During physical and chemical transformations of a solid sample, the composition of the reaction mixture and the local molecular environment change. As a consequence, the Raman bands in the spectra change too: bands belonging to reactants diminish, while bands of the nascent species, intermediates, and products emerge. These changes will be most prominent when there is a breakage and formation of new, preferably covalent bonds, but they also occur in case of changing the intermolecular interactions, like when new hydrogen or halogen bonds form or break.

Polymorphism is specific. Two polymorphs can exhibit similar molecular environments and have the same pattern of intermolecular interactions. These similarities often result in both polymorphs having an almost identical Raman spectrum. However, in the solid state, vibrations of the entire crystal lattice, phonons, are located in the area of up to 250 cm^{-1}. Phonons strongly depend on the crystal packing and usually exhibit enough differences in Raman spectra of polymorphs for their unambiguous identification. Still, sometimes, a spectrometer will not collect the low-frequency part of the Raman spectrum posing a potential problem in identifying polymorphism.

1.2.4.1 Advantages and Limitations

When compared to the PXRD, the main advantage of Raman spectroscopy is its ability to provide useful information independent of the crystallinity of the sample. Even if the sample is amorphous, bands from the reactants and products are observed. Moreover, Raman spectroscopy, together with the other vibrational spectroscopy method – infrared spectroscopy – has long been used for the identification of molecular structure and specifically for the identification of various functional groups. In that context, changes in specific Raman bands during milling can be used to explain or corroborate a potential molecular reaction mechanism. Raman spectroscopy offers another possibility to directly monitor the interactions of additives during milling, assuming that the additive is present in sufficient amounts to become detectable.

On the other hand, irradiating a sample with the laser can lead to luminescence that negatively affects the Raman spectrum. In most cases, a strong luminescent signal from the sample can cover or distort the whole Raman signal resulting in the loss of information from Raman scattering and severely limiting the applicability of in situ Raman spectroscopy. Nevertheless, changes in the reaction mixture composition may alter the luminescence allowing, in some cases, monitoring of the reaction progress. Still, even in the absence of luminescence, some materials can display a very weak Raman signal. Sometimes, choosing a laser source with a different excitation wavelength can help to alleviate the problem with luminescence. Some limitations are common to both PXRD and Raman spectroscopy. Since they both probe only a small portion of the inside of the jar, they rely on the reaction mixture to be uniformly distributed and of uniform composition. It is not unlikely for the sample to become stuck and compacted on one side of the jar, effectively preventing the interaction with the X-ray beam or the laser light and resulting in partial or complete signal loss.

Raman spectroscopy is a contactless method allowing the positioning of the Raman probe in the vicinity of an oscillating jar and focusing the laser light at the bottom of the inside of the jar (Figure 1.5). Nowadays, Raman spectroscopy is an indispensable processing analytical technique (PAT).[75, 76] It is used as an in-line monitoring tool in manufacturing and synthetic processes in the pharmaceutical, chemical, and materials industry, as a field analysis tool, and also as a general technique in life science. For its widespread use, crucial was the development of accessible, small, portable, modular, and affordable Raman systems that include a spectrometer, a laser source, and a Raman probe, making Raman spectroscopy easy to implement in a laboratory,[77] in contrast to in situ PXRD that requires access to a synchrotron beamline.

A modular design allows various combinations of components, and thus, application-oriented use. For instance, laser sources come with different excitation wavelengths from the 500 nm range up to the near-infrared region (NIR) at around 1024 nm. Some laser sources even come with two distinct but very close exciting wavelengths. For example, a shifted-excitation Raman difference spectroscopy (SERDS) exploits the fact that for some minor shift in excitation wavelength (less than 2 nm), a fluorescent background remains almost the same, but the Raman signals slightly shift, since they depend on the excitation wavelength.[78] Therefore, collecting Raman spectra with two different wavelengths can be used to eliminate the fluorescent signal from the spectrum. Fluorescence in the Raman spectrum can be mitigated by using a laser source of lower energy, e.g., in near infrared (1024 nm), but red (785 nm) and green (532 nm) lasers are still the most commonly used ones. When it comes to monitoring milling, the choice of the laser source will depend mostly on the chemical systems studied. Organometallic, and

FIGURE 1.5 A setup for laboratory in situ monitoring of mechanochemical reactions on a vibratory ball mill using Raman spectroscopy, and with the Raman probe approaching the reaction vessel from below.

some metal-coordinated compounds, as well as aromatic molecules, may have a tendency to manifest a large fluorescent signal when excited with the red laser, but in general, the red laser will be adequate.

The spectrometer detects the Raman scattered photons collected by the Raman probe. Raman scattering is a low-probability event, and it happens, on average, once in a million collisions of photons and a molecule, meaning that most of the incident laser light scatters elastically. This constitutes Rayleigh scattering which must be filtered. The Raman scattered photons may be on both energy sides relative to the incident photons, characterized as Stokes or anti-Stokes scattering. Another vital parameter to consider is the resolution of a spectrometer. Again, the wanted resolution depends on the application, and having a better resolution usually trades with acquiring a lower-intensity signal. The resolution depends on the diffraction grating but also on the width of the spectrometer slit opening. Additionally, some manufacturers can include CCD detector cooling, which lowers the inherent detector noise and improves the signal-to-noise ratio. Finally, the choice of the Raman probe can also be important. Some probes focus on the laser light, and the laser spot size can be micrometers in diameter at the focal point. Others tend to maximize the laser spot size area to probe the larger sample area and reduce the effects of possible heterogeneity of the sample.

1.2.4.2 Basics of Raman Spectroscopy

Raman spectrum contains a plethora of qualitative and quantitative information, making Raman spectroscopy one of the most important analytical techniques. Raman spectrum can help corroborate the molecular structure of the material as well as identify and study the molecular interactions of a material in solids, liquids, slurries, emulsions, pellets, etc. However, the physical phenomenon of scattering of light on molecules offers an additional basis for the quantitative use of Raman spectroscopy using the Raman signal intensity, band position, and band shape.

The equation to describe the intensity of the Raman scattered light $I(\tilde{\nu})$ from a sample is:[79]

$$I(\tilde{\nu}) = \frac{h}{8\varepsilon_0^2 c} \frac{I_L N (\tilde{\nu}_0 - \tilde{\nu})^4}{\tilde{\nu}(1 - e^{-hc\tilde{\nu}/kT})} \frac{\left(45(a')^2 + 7(\gamma')^2\right)}{45}, \tag{1.2}$$

where c – speed of light, h – Planck's constant, I_L – laser excitation intensity, N – number of scattering molecules, $\tilde{\nu}$ – molecular vibration wavenumber, $\tilde{\nu}_0$ – laser excitation wavenumber, ε_0 – vacuum permittivity, k – Boltzmann constant, T – absolute temperature, a' – mean value invariant of the polarizability tensor, and γ' – anisotropy invariant of the polarizability tensor.

The intensity $I(\tilde{v})$ refers to the area under the width of a Raman band and it is proportional to the number of scattering molecules N. This is because the Raman scattered cross-sections are very small, thus making the probability of annihilation of a Raman scattered photon due to another Raman scattering event practically zero. That justifies the principle of linear superposition – a Raman spectrum of a mixture can be represented as the linear combination of pure components. Coefficients in this linear combination are determined by the Raman cross-sections and by quantities of the components. Still, chemical interactions, like hydrogen or halogen bonds, between components can influence and change the Raman spectrum of a mixture of components, making it different from just a weighted sum of pure components. Finally, in a case when the absorption of one or several components interacts with the transmission of Raman scattered photons, effects like the inner filter effect in fluorescence spectroscopy can arise, resulting in breaking the principle of linear superposition. Moreover, fluorescence can pose a limit to the ability to reliably extract quantitative information from the Raman spectrum.

For a reliable quantification of the composition mixture by Raman spectroscopy, one wants to increase the Raman signal and reduce the noise. Random fluctuations are always present and some sources of noise are unavoidable, like photon shot noise, whose intensity is proportional to the square root of detected photons. If photon shot noise is the predominant noise source, then the only way of increasing the signal-to-noise ratio (SNR) is by collecting more photons (by increasing exposure/collection time). The sample itself can be a source of the noise. The position and shape of the Raman band can change with temperature, or the sample can be heterogeneous. In milling experiments, a large temperature oscillation could occur at the point of the ball impact, but this affects only a small volume of the milled material and the temperature change of the bulk is usually gradual and, for vibratory ball mills, limited to less than 20 K increase from the ambient temperature.[33, 34, 80] Exceptions can be found with self-propagating reactions[81] or some other observable exothermic event that occurs over the bulk of the sample.[82]

The heterogeneity of the sample mixture may present a more serious problem since during milling, and if using a focused Raman probe, only a portion of the sample is illuminated. If this is the case, then the probed portion of the sample will not represent the composition of the entire sample. In some cases, such as when using liquid additives, a sample can stick to the milling balls or to the inner jar walls and thus inhibit proper mixing and redistribution of the reaction mixture in the milling vessel. Still, probably the largest concern comes from fluorescence that can lead to the total loss of Raman signal in some cases. In that case, changing to a laser source with lower energy could be beneficial. The stability of the laser source can also introduce noise or result in a drift of the wavenumber scale or the intensity in the spectra. In the long run, that could impact the reliability of measurements. Calibration helps to alleviate the drift problem by measuring the well-known standard material and appropriately modifying the excitation wavelength so that the Raman shift matches the known position of the selected Raman band of the standard. For solids, a common standard is a pure silicon that has a band at 520.5 cm^{-1}.

Noise in the spectra comes also from the instrument itself, and some spectrometers include an additional cooling system on their CCD detectors to reduce the thermal noise. Also, faulty detector pixels or cosmic rays passing through the detector can create spikes of intensities at a particular wavenumber. These two can be distinguished by observing where and when spikes occur. The cosmic rays result in the random appearance of spikes in the spectra, while faulty pixels are fixed and present in all spectra. Finally, certain common operations like subtracting the dark spectrum also introduce error, or bias, especially when subtracting the same dark spectrum from multiple experimental spectra. Ideally, each collected Raman spectrum should be subtracted by an independently collected dark spectrum.

1.2.4.3 Data Pre-processing

The purpose of data pre-processing is to remove the potential sources of errors and reduce noise to improve SNR, accuracy, and reliability of the quantitative and qualitative analysis. When analyzing spectral data, only certain parts of the spectra might be of interest. Segmentation of a spectrum and considering only parts that contain the relevant Raman signal is usually the first step in pre-processing. In a set of experimental spectra, identifying the spectral regions of interest could easily be done by plotting the spectra and analyzing the differences between starting and subsequent spectra. For time-resolved data, a two-dimensional plot is quite helpful because it can reveal the occurrence of intermediate phases. However, plotting raw spectra containing large baselines or even worse, a changing baseline, can severely limit the identification of spectral regions of interest.

The baseline in the Raman spectrum can occur as a consequence of several phenomena, but it is mostly present due to the fluorescence of a sample or impurities upon irradiation with the laser light. In a spectrum, a baseline is shown as a wide curve with slowly varying intensity. Modeling baseline relies on its additive nature – it is implied that the Raman spectrum is the sum of a signal, noise, and baseline. Then, the baseline is modeled by trying to fit some curve that should represent a true baseline. While in simulated data we can know exactly how a baseline looks like, it is impossible to know the exact nature of a baseline in real data, and therefore in modeling a baseline we will always introduce some error. Many algorithms have been developed for the removal of a baseline in Raman spectra. Some are based on polynomial or spline fitting mostly in an iterative fashion on a smaller spectral window.[83–86] Others are using the least-squares approach used in smoothing algorithms with additional restraints for the asymmetry of a baseline.[87, 88] Most of these algorithms have parameters that require human input and their optimal values differ between experiments and chemical species. For this reason, many adaptive algorithms are in development requiring no need for human input and optimization.[89, 90] Additionally, high-pass filters can also be used for the removal of baseline curves. Removing low-frequency signal from the Fourier transform (FT) of the spectral signal will eliminate wide curves like baseline from the original spectrum.[91] However, FT influences the entire spectrum, including baseline, noise, and the Raman signal. To improve on this, a different base can be used, like in wavelet transformation.[92] Finally, using the first or second derivatives, mostly accompanied by some kind of smoothing, like in the Savitzky–Golay derivation, is an alternative way of removing linear and different small baselines.[93]

1.2.4.4 Noise and Spike Removal and Smoothing

Averaging spectra will improve SNR at the cost of losing time resolution. For in situ Raman spectroscopy, this means that finding the most optimal conditions for spectrum acquisition is crucial to have a good time resolution. However, if the Raman signal is weak, or if the spectrometer is equipped with a narrow slit, photon collection will need to be longer to improve the signal-to-noise ratio. To obtain better data, several spectra should be averaged, which is equivalent to increasing the data collection time. Slow mechanochemical reactions will have no problem with a longer time resolution, but fast LAG or ion- and liquid-assisted grinding (ILAG) reactions, which can even be completed within minutes, will usually need to be individually optimized to obtain high-quality data with sufficient time resolution. Sometimes a collected noisy spectrum could be pre-processed using some denoising algorithm. Denoising attempts to remove or reduce noise from the spectra and sometimes is synonymous with smoothing. Low-pass filters remove high frequencies from the FT of a spectrum, which is characteristic of noise.

A different approach is smoothing. A moving window average is the simplest approach to smooth the spectra but often gives poor results. The much better Savitzky–Golay smoothing is based on fitting a polynomial on a small window of several points in the spectrum and iterating the procedure over the entire spectrum,[93] while Whittaker smoothing uses an entire spectrum to fit a new curve through the spectral points subjected to restraints.[94] The well-known effect of smoothing is peak broadening and the reduction of intensity at the peak maximum, especially when higher polynomials are used in the Savitzky–Golay, or with heavy restraints in Whitaker smoothing. Denoising could be performed before the baseline correction step, but only if deemed necessary.

Spike removal should be performed at the beginning of spectra pre-processing. If there are only a few spikes, they can be removed by hand with a moving average using a three- or five-point window. However, automated spike removal algorithms exist, while some of them still require human input.[95–97] The best practice is to test several approaches and use the one that performs best.

1.2.4.5 Subtraction of the Jar Signal

In situ measurements are mostly performed in a translucent PMMA jar. Like most materials, PMMA has a Raman spectrum. Thus, the collected in situ spectrum comprises Raman signal originating from the PMMA, reactants, products, and additives. The problem is that PMMA has a spectrum very "rich" in peaks covering mostly the entire region from 200 to 1750 cm^{-1}. Observing the changes during milling coming from the reactants and products thus can easily be hindered. PMMA is mostly inert towards the wide range of organic and inorganic solids, and considering the principle of linear superposition, its signal contribution can be removed by subtracting the properly scaled Raman spectrum of an empty jar.[23] To do this properly, a band in the experimental spectrum that belongs to the PMMA and does not overlap with any of the Raman signals from reactants or products should be regressed onto the same band from the spectrum of the empty PMMA jar. A simple least-squares procedure yields a scaling factor for the empty jar, and after multiplying the Raman spectrum with the scale factor, a simple subtraction from the experimental spectrum gives the spectrum with a removed PMMA signal. Instead of the single band, several bands could be introduced into the least-squares procedure and the mean of several scale factors could be taken as the correction factor. Ultimately, the whole spectrum of an empty PMMA jar could be regressed onto the experimental spectrum, but this could introduce errors of varying magnitude depending on the strength of the signal from reactants and products and it should be performed only when it is not possible to identify a single PMMA band without significant signal contribution from reactants and products. Also, using a jar material that does not contribute to Raman scattering is highly beneficial and sapphire has proven to be excellent for such purposes.[74]

1.2.4.6 Normalization, MSC

Finally, prior to the analysis of spectra, the final step is usually normalization. The purpose of this step is to reduce variations in absolute spectral intensity between spectra. Such variations originate from the uneven amount of the sample present in the laser beam path during the collection of individual spectra. There is no clear rule as to which norm to use for normalization and the most often used norms are the ℓ^1 (Taxicab) and the ℓ^2 (Euclidian) norms.[98] Optionally, if there is a band that does not change in intensity, shape, and position during the reaction, then its intensity can be used for normalization. Additionally, standard normal variate can be performed or even multiplicative scattering correction (MSC).[99] MSC is originally derived for correcting NIR spectra, as it relies on regressing the experimental spectrum on the average spectrum of all experimental spectra. It can perform well for NIR spectra, since the baseline in such spectra can be the result of scattering, thus the name. However, it was derived to correct the baseline of several experimental spectra of the same or similar samples, and it will not usually work well for Raman spectra obtained in situ because the idea of an average spectrum being the representative spectrum of the sample is inappropriate when there is a chemical reaction occurring and spectra change as the reaction mixture changes in composition.

1.2.4.7 Qualitative and Quantitative Analysis – Single Band and the Whole-Spectrum Approaches

As the number of in situ collected spectra can be in the range from dozens to thousands, and our interest lies in changes in peaks' intensity, shape, and positions in time, one of the efficient representations of time-resolved spectra is through two-dimensional plots. Here, the *x*-axis corresponds to wavenumbers, the *y*-axis to time, and the value of the intensity of each spectral point is represented with color. It is possible to plot time-resolved spectra as a waterfall plot, or even as a three-dimensional plot, but the former is effective when comparing only several spectra and the latter is not recommended, ever. However, even in two-dimensional plots, the choice of the color scheme, or rather its gradient, for representing intensity is important because changes on some low-intensity peaks can remain poorly resolved in some color gradients, especially when plotting the whole spectral range.

On a two-dimensional plot, it is usually rather easy to detect intermediates and identify Raman bands that change as the reaction proceeds.[16] Looking at the time axis, it is straightforward to estimate how much time is required for certain events or for how long an intermediate persisted in the mixture. Such information can be used to appropriately adjust the reaction time for a specific purpose and to compare different reaction conditions. Analysis of such plots can thus directly help to identify the time at which the target intermediate was abundant in the reaction mixture. Comparing the pure intermediate spectrum with real-time in situ collected Raman spectra, for example, in a repeated experiment, can facilitate in deciding when to stop the reaction in order to isolate the intermediate for further ex situ analysis and characterization.[100, 101] Therefore, just a qualitative insight into the milling process opens a plethora of possibilities, from reaction optimization, intermediate identification, and isolation up to deducing molecular changes and interactions by identifying molecular vibrations belonging to bands that change the most in intensity.

While qualitative information is valuable, quantifying the amount of reactants and products enables the derivation of reaction profiles that can be subjected to kinetic analysis.[102] It is difficult not to emphasize the importance of being able to extract reaction profiles, not just from Raman spectroscopy, but from other in situ methods for studying milling reactions, because underlying molecular mechanisms of mechanochemical reactions are still unknown and have been proposed in the past based on limited ex situ data. By amassing in situ kinetic data, a complex theoretical framework describing the physical and chemical nature of milling reactions can start to emerge and needs to be validated. Already, in situ studies from the last several years observed that mechanochemical reactions can be much faster than anticipated, showing some remarkable similarities with the reactions performed in solutions, but also some distinct features inherent to the solid-state reaction environment.

There are several possibilities for extracting reaction profiles from in situ Raman spectra and we may first distinguish them based on a single peak or the whole-spectrum approach. When we can identify all observable chemical species over the reaction course, and attribute to each specie its distinct Raman band, then the intensities of these bands will be directly proportional to their amounts.[103] There are several ways of extracting the intensity of a band: (i) the intensity value at the peak maximum, (ii) integrating the area under the peak, and (iii) fitting a peak with some analytical function, often Lorentzian or Gaussian, and estimating the area under the peak.

The problem with using just the values at the maximum of the signal peak is that it can be incorrect if there is a change in the peak shape, which often occurs due to chemical interactions in the reaction mixture. This is particularly relevant if the Raman band corresponds to the vibration of a functional group that can become engaged in a hydrogen or a halogen bonding, or if the spectral resolution is low (poorer than 4 cm^{-1}). This also means that a change in the intensity of one peak does not need to correlate with the change in intensity of other peaks, and thus its intensity does not need to accurately represent changes in the amount of a given species. Since the intensity of the Raman band is the area under the curve without the baseline, integrating the peak area is a better approach. Still, the problem can be in accurately determining the proper limits of integration and the baseline. On the other hand, peak fitting can give accurate peak position, area, and full width at half maximum (FWHM). It should be used with care, however, because analytical functions like Lorentzian or Gaussian can't take into account the peak asymmetry or other artifacts that can occur. Comparing absolute values for the extracted intensities for individual species is possible only if the cross section for each species is known, which is usually not the case. Thus, such reaction profiles do not represent the evolution of the amount of each species in absolute terms, but only the change in the Raman intensity of their respective peaks.

We can still, however, obtain the estimation of amounts of reactants and products without explicitly knowing their Raman cross-sections. If we are able to collect, independently, the Raman spectrum of the reactant(s) or the product, we then might use this as their representative spectrum. We will further elaborate what representative means. Mechanochemical reactions are usually performed in closed and relatively small jars (at least when performed on vibratory ball mills). For most solid-state reactions, we can safely assume that in situ spectrum, collected over some period of time, corresponds to the true sample spectrum. This, of course, is not true in cases if the milled material becomes sticky, or if it experiences drastic changes in rheology. However, when that is not the case, we can calculate the theoretical yield for a situation when all reactants transform into the product (or an intermediate). Collecting the Raman spectrum of pure reactants or products when the jar is charged with their theoretical maximum amounts permits us to obtain their representative spectra under the target experimental conditions. This means that these spectra correspond to the spectra we would expect to observe if all of the samples within the jar transforms to this one participating species. We can thus estimate the upper limit for the Raman intensities of each species. Then, by dividing the extracted intensities of single peaks with those upper limits, we can obtain the estimation of molar fractions for each species. By definition, the sum of molar fractions of all species is equal to 1, and we can use this to check how the sum of all derived molar fractions is close to 1. If the deviations are small, this is an indication that our estimation is valid. Otherwise, a different approach, where the whole spectrum is considered, might be a better solution.

1.2.4.8 The Whole-Spectrum Approach

We can obtain the reaction profile using the information over the whole spectral area. Again, the conditions are that all reactants and products in the reaction are known and being able to be prepared as isolated species. Then, we can use their representative spectra to estimate their molar fractions using the least-squares procedure.[16] Experimental spectrum can be decomposed as the sum of representative spectra with coefficients in that linear combination being estimates of molar fractions:

$$S_{exp} = \sum_{i=1}^{p} x_i S_{rep,i}, \tag{1.3}$$

where S_{exp} is the experimental Raman spectrum, $S_{rep,i}$ is the i-th representative spectrum, and x_i are molar fractions. To further ensure the physical meaning of x_i as molar fraction, we subject Equation 1.3 to the restriction that the sum of all x_i is equal to 1 (definition of the molar fraction). We can simultaneously solve these equations for all experimental spectra:

$$\mathbf{S}_{exp} = \mathbf{X}\mathbf{S}_{rep}, \tag{1.4}$$

and the solution of such a matrix system is a matrix $\hat{\mathbf{X}}$ of estimated coefficients x_i :

$$\hat{\mathbf{X}} = \mathbf{S}_{exp}\mathbf{S}_{rep}^{T}\left(\mathbf{S}_{rep}\mathbf{S}_{rep}^{T}\right)^{-1} \tag{1.5}$$

subjected to the previously mentioned condition: $\sum_{i}^{n} x_i = 1$.

We can estimate the errors for the molar fractions from each experimental spectrum using the covariance matrix $\mathbf{C}$:

$$\mathbf{C} = \sigma^2 \left(\mathbf{S}_{rep}\mathbf{S}_{rep}^{T}\right)^{-1} \tag{1.6}$$

where σ^2 is the variance of the residuals for a single experimental spectrum expressed as:

$$\sigma^2 = \frac{\mathbf{e}^T \mathbf{e}}{n-p} \tag{1.7}$$

where $\mathbf{e}$ is the vector of residuals, p is the number of independent species, and n is the length of a spectrum (number of wavelength points).

Restricted classical least squares (RCLS) works well when the covariance between representative spectra is small, or to put it simpler, when the representative spectra are not similar. To ensure smaller covariance between spectra, we can consider including only those parts of the spectrum where the correlation between species is small.

What if we observe some transient chemical species that we cannot characterize, isolate, or prepare independently? Since we are unable to obtain the representative spectrum of such species, we cannot derive the reaction profile using RCLS. In such a case, we can still estimate the reaction profile using other data analysis methods, like multivariate curve resolution alternating least-squares (MCR-ALS).[104]

In MCR, the goal is to estimate the correct number of chemical species in the reaction and then resolve the experimental spectrum into the linear combination of their estimated pure spectra directly from the experimental spectra. This method can be a powerful method to obtain reaction profiles in case when we observe intermediates in the reaction that we cannot isolate and characterize, or identify. MCR relies on the bilinear decomposition:

$$\mathbf{D} = \mathbf{CS}^{\mathrm{T}} + \mathbf{E} \qquad (1.8)$$

where $\mathbf{D}$ is the matrix of experimental spectra, $\mathbf{C}$ is the matrix of molar fractions, and $\mathbf{S}$ is the matrix of pure spectral responses or pure spectra of components. In the optimization procedure both $\mathbf{C}$ and $\mathbf{S}$ are estimated in alternating fashion, hence the name – alternating least squares. Additional set of constraints need to be applied to reduce the ambiguities and ensure physically meaningful decomposition $\mathbf{CS}^{\mathrm{T}}$. Usually, these constraints include non-negativity for spectra and mass balance or the previously mentioned definition of molar fraction for the $\mathbf{C}$ matrix.

In mechanochemistry, MCR-ALS was successfully applied in the mechanistic investigation of C–H bond activation in the unsymmetrical azobenzene substrate by various Pd^{II} catalysts,[105] in reaction monitoring of APIs silver sulfadiazine and dantrolene[25] as well as in a Knoevenagel condensation reaction.[106]

1.2.5 Tandem In Situ Monitoring and Other Techniques

As mentioned above, PXRD and Raman spectroscopy are independent but complementary in the sense that one is primarily suited for monitoring crystalline species, while the other is sensitive to all types of materials regardless of their crystallinity or their aggregate state.[15, 16] These two techniques can be used in tandem and also combined with in situ temperature and pressure monitoring.[107] Combining techniques provides better opportunities to understand the mechanochemical reactions, and moreover, temperature and pressure in situ monitoring require a parallel chemically sensitive technique to understand the observed profiles. Recently, in situ PXRD was coupled with in situ monitoring of gases in the reaction vessel offering an immense potential to study in operando mechanochemical reactions or mechanocatalysis involving gases, as well as vaporization and sublimation in the reaction mixture during milling.[37, 54] This combination will be the method of choice to study in situ the recently reported mechanochemical ammonia production.[38]

Besides Raman spectroscopy and PXRD monitoring, which are now the most established in situ monitoring methods, recent developments have extended in situ monitoring to other techniques such as NMR spectroscopy,[18] chemiluminescence,[21] and X-ray absorption spectroscopy.[20] These techniques provide a complementary insight into the standard PXRD and Raman monitoring, though care has to be taken and any differences in milling conditions should be taken into account.

1.2.6 Kinetic Analysis of In Situ-Derived Reaction Profiles

The reaction profiles derived from in situ monitoring present a unique opportunity for the kinetic analysis of a mechanochemical process.[16, 102, 108] Traditional models of solid-state reactivity, such as equations related to reaction order,[17, 109] the autocatalytic Prout–Tompkins equation[110, 111] or the nucleation-based JMAEK (Johnson–Mehl–Avrami–Erofeyev–Kolmogorov) model (also known as KJMA (Kolmogorov–Johnson–Mehl–Avrami), JMA (Johnson–Mehl–Avrami), the Avrami equation, or the Avrami–Erofeyev model),[112, 113] have been applied to fit the mechanochemical kinetic curves. However, using the mechanochemistry-specific, semi-empirical models that take into account the stochastic nature of motions and nucleation processes taking place in milling should be preferred.[102, 108, 114, 115]

Kinetic studies are in general sensitive to various changes in the environment or the sample. This can significantly affect the reproducibility, and all the parameters such as ambient temperature and humidity, the way the mill operates, sample preparation, and even seasonal changes can have a more or less pronounced influence on mechanochemical reaction kinetics.

The mechanochemical reaction will be either induced or sustained by ball impacts, and the characteristics of ball impacts, such as ball velocity upon impact and the frequency of ball impacts, determine the rates of mechanochemical reactions.[116] These parameters are in part available experimentally, either by the use of piezo-based sensors or by using a high-speed camera,[114, 117] but can also be accessed by theoretical modeling of the milling process. Next to the parameters that can be adjusted on the milling device, the amount of the milled materials in the reaction jar is also of great importance in determining ball velocity, and transfer of energy from the milling media to the milled material. With less of the material in the reaction vessel, ball collisions will be more elastic, while with more material, the collisions will approach fully inelastic collisions.[116] This can be visualized by imagining a ball drop to a surface lined with a very thin layer of materials where it will bounce back and lined with a thick layer of material where it will dig into the materials and will not bounce back. In the latter case, the ball has, upon impact, transferred all of its energy to the material.

Kinetic modeling of mechanochemical reactions should thus be based on assumptions pertinent to mechanochemical processing. Assuming that mechanochemical reactivity is occurring upon the impact of the milling media in the portion of the powder being compressed between the vial wall and the milling ball, this small portion (of the fraction) of the milled powder experiences the so-called critical loading conditions (CLC). Considering the characteristics of milling processing, kinetic equations can be derived that are able to model the simplest reaction profiles such as the exponential decay but also more complex profiles and even describe profiles of intermediate phases.[102, 116] In essence, one CLC leads to exponential decay, while, if more than one CLC is necessary to drive the reaction, more complex equations arise which can describe various shapes of experimental kinetic curves.

1.3 Mechanistic Understanding of Mechanochemical Reactions

In situ monitoring has revealed surprising aspects of mechanochemical reactivity such as fast reactions,[13] plethora of intermediates,[26] reactions obeying[118] or, rarely, violating Ostwald's rule of stages,[119] reaction selectivity,[118] and autocatalysis.[80, 120, 121] Despite that many of these aspects were partially known or were hinted at from previous ex situ studies,[122] their true scope became apparent only after the application of real-time in situ monitoring. For example, the mechanochemical reaction of zinc and sulfur to provide ZnS is a so-called, self-sustaining reaction characterized by a long induction period before a very sudden, and strongly exothermic reaction and the formation of the product.[36, 53] The reaction enthalpy released is manifested by a sudden increase in temperature, which is easily detected even with a thermocouple attached to the outside wall of the milling vessel. Except for an increase in temperature and a possible subsequent analysis of the reaction mixture, little more information could be acquired without in situ monitoring techniques.

Recently, Weidenthaler and co-workers have studied this reaction using in situ X-ray diffraction and have observed the anticipated sudden transformation.[54] ZnS forms abruptly and initially provides the metastable wurtzite that almost immediately engages in a slower transformation to the stable sphalerite (Figure 1.6). The sequence of transformations from the less stable to the most stable phase is in accordance with Ostwald's rule of stages, which thus seems to be applicable not only to soft organic and metal-organic systems but also to harder inorganic materials.[24] Such sudden transformations are not limited to inorganic systems as a similarly abrupt transformation, accompanied by rheological and color changes of the reaction mixture, has been observed in the mechanochemical benzil–benzilic acid rearrangement[82] and the related phenytoin synthesis[123] studied by in situ Raman spectroscopy.

1.3.1 Selectivity in Cocrystal Formation

The majority of mechanochemical reactions, however, are not abrupt and exhibit smooth reaction profiles in their transformations from reactants to products, usually via intermediates. Despite the wealth of information gathered by in situ monitoring and other techniques in the last decade, it cannot be said that mechanochemical reactions, and their mechanisms, are well understood and predictable. Nevertheless, some advances are evident and it seems that transformations of phases in cocrystal formation, in both hydrogen-bonded [118] and halogen-bonded cocrystals,[124] can be predicted based on Ostwald's rule and stabilities of potential reaction mixture compositions, which are accessible theoretically, if not available experimentally.

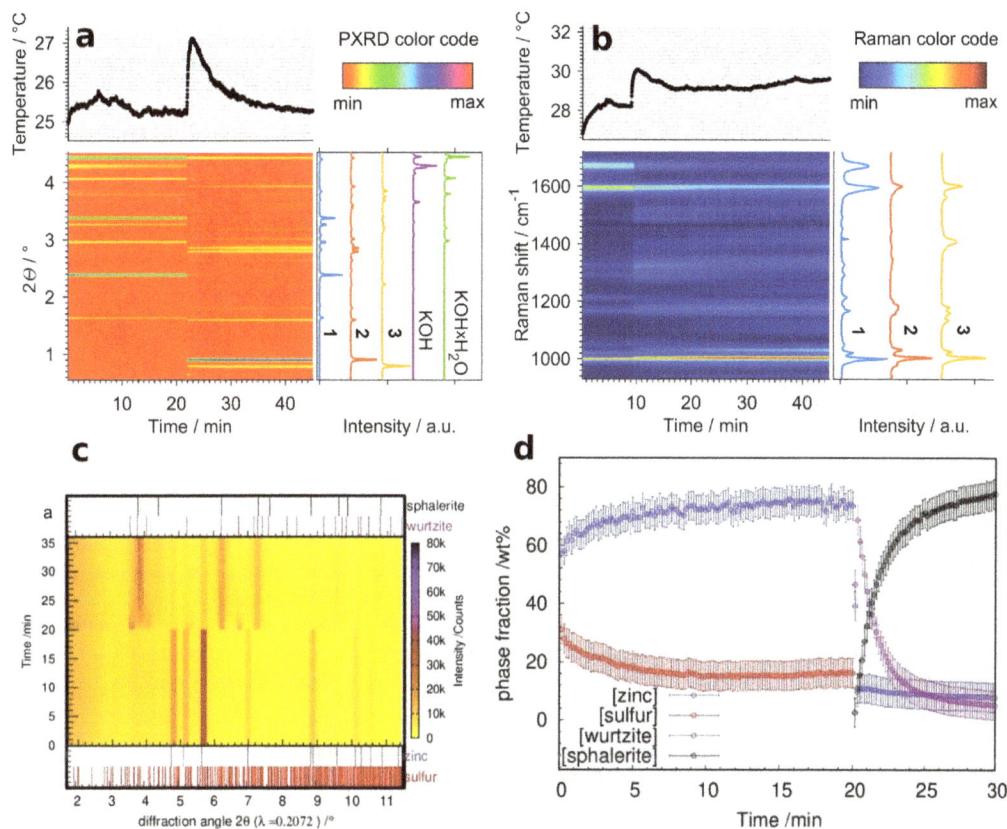

FIGURE 1.6 In situ temperature profiles collected in tandem with (a) time-resolved diffractograms or (b) Raman spectra for the benzil–benzilic acid rearrangement.[87] The sudden reaction is accompanied by a temperature increase. (c) Time-resolved diffractogram for the self-sustained reaction between elemental zinc and sulfur and (d) the corresponding reaction profile obtained by the Rietveld refinement.[54] Adapted with permission. Copyright 2020 and 2021 Wiley-VCH GmbH.

In the cocrystal selectivity study comprising nicotinamide (na), anthranilic acid (ana), and salicylic acid (sal), hydrogen-bonded cocrystals among all the three coformers are possible (Figure 1.7).[118] If the stabilities of all mixture compositions are evaluated using periodic dispersion-corrected DFT, we can anticipate the formation of each of the cocrystal, together with the remaining coformer in the reaction mixture. The mixture with three pure phases of coformers is highest in energy and each of the cocrystals: naana, nasal, salana can be formed, leaving the third conformer unreacted. As predicted by theoretical calculations, the mixture obtained by milling equimolar amounts of na, sal, and ana provides the nasal cocrystal, together with ana in the mixture, as the most stable mixture composition.

The difference in stabilities is the cause for selectivity also. In cocrystallization competing experiments, inspired by a pioneering work by Caira,[125] when sal is offered to the already prepared naana, we should expect the formation of nasal and separation of ana as a pure phase. Indeed, this is the end result after 30 minutes of milling at 30 Hz on a shaker mill, but the reaction path visits first the more stable salana cocrystal, which is formed as an intermediate. That is, sal reacts with naana to provide simultaneously salana and nasal. When these parallel reactions fully deplete pure sal, there is still naana present in the mixture. The salana cocrystal then becomes the source of sal and its amount is diminished to finally provide the mixture containing only nasal and ana, which represents the most stable composition.

Similar behavior was found with halogen-bonded cocrystals of the popular DABCO (diazabicyclooctane) and 12TFIB (1,2-diiodotetrafluorobenzene), which may form in different stoichiometric ratios of the coformers: 1:1, 1:2, and 2:1.[124] The experimentally established conversions among these halogen-bonded cocrystals could be a priori predicted from DFT calculated energies of the reaction mixtures.

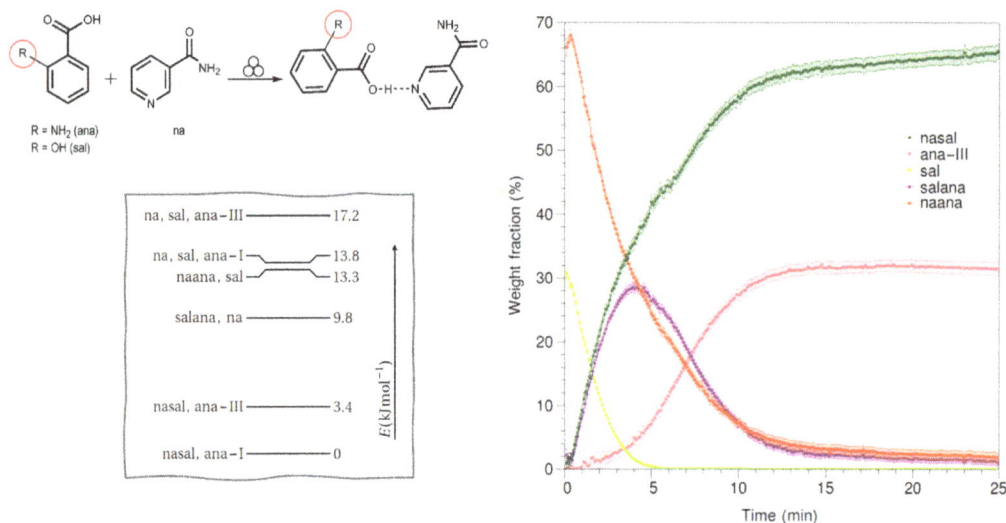

FIGURE 1.7 Formation of binary cocrystals from the three-component mixture of nicotinamide, anthranilic acid, and salicylic acid.[118] Energies of reaction mixture compositions determine the experimental reaction profiles. According to Ostwald's rule of stages, the initially formed less table mixture compositions are being replaced with increasingly more stable mixtures. Anthranilic acid is known to exist in three polymorphs, which are here designated as ana-I and ana-III for polymorphs I and III.[126, 127] Adapted with permission from reference [118]. Copyright 2018 American Chemical Society.

Moreover, this worked as a benchmark to evaluate different theoretical approaches to predict the correct (that is, experimentally verified) order of stabilities of different cocrystal and conformer mixtures and the course of mechanochemical transformations in various starting mixture compositions.

These two examples were found to be well in accordance with Ostwald's rule of stages, and most of the reaction systems can be expected to abide by the rule. Exceptions, however, can be found when a change in the input of mechanical energy changed selectivity. Using different balls and different materials for the reaction vessel led to the isolation of a less stable polymorph of the cocrystal of adipic acid and nico-tinamide where harsher milling conditions led to the selective formation of the less stable polymorph.[119] The reason for the switch in stability can potentially be found in different particulate properties, such as particle size distribution and accumulated imperfections or strain. That the particle surface, along with its surface energy, is vital in switching selectivity between polymorphs was well established in comprehensive studies by Belenguer and co-workers for liquid-assisted polymorphic transformations.[8, 128] It is, therefore, possible to use milling, under appropriate conditions, to prepare the less stable phase from a more stable one. This was, for example, demonstrated in a polymorphic transformation of the stable β-polymorph of D-mannitol to the metastable β-polymorph, which took over several hours of milling but retained the crystalline nature of the milled polymorph mixture throughout.[129]

1.3.2 The Influence of Temperature and Autocatalysis

The lack of mechanistic understanding of mechanochemical reactions is best illustrated by the lack of understanding of temperature on reaction kinetics. What is considered to be well understood in solution chemistry is, at present, largely unknown for the solid-state mechanochemical alternative, although there have been significant advancement recently.[130] The influence of temperature on reaction kinetics was first studied in situ only in 2016,[80] revealing a surprisingly strong influence of even a mild temperature increase. This was in contrast to expectations that could be made based on the models of mechanochemical reactivity that assume a sudden and very-short, but extremely intense increase in temperature and pressure at the small portions of the sample residing in the collision zone. Namely, if these mechanisms were at play, an increase of bulk temperature from room temperature to ca. 50 °C should not cause even a measurable, let alone a significant, reaction rate acceleration and a change in mechanism (Figure 1.8). In

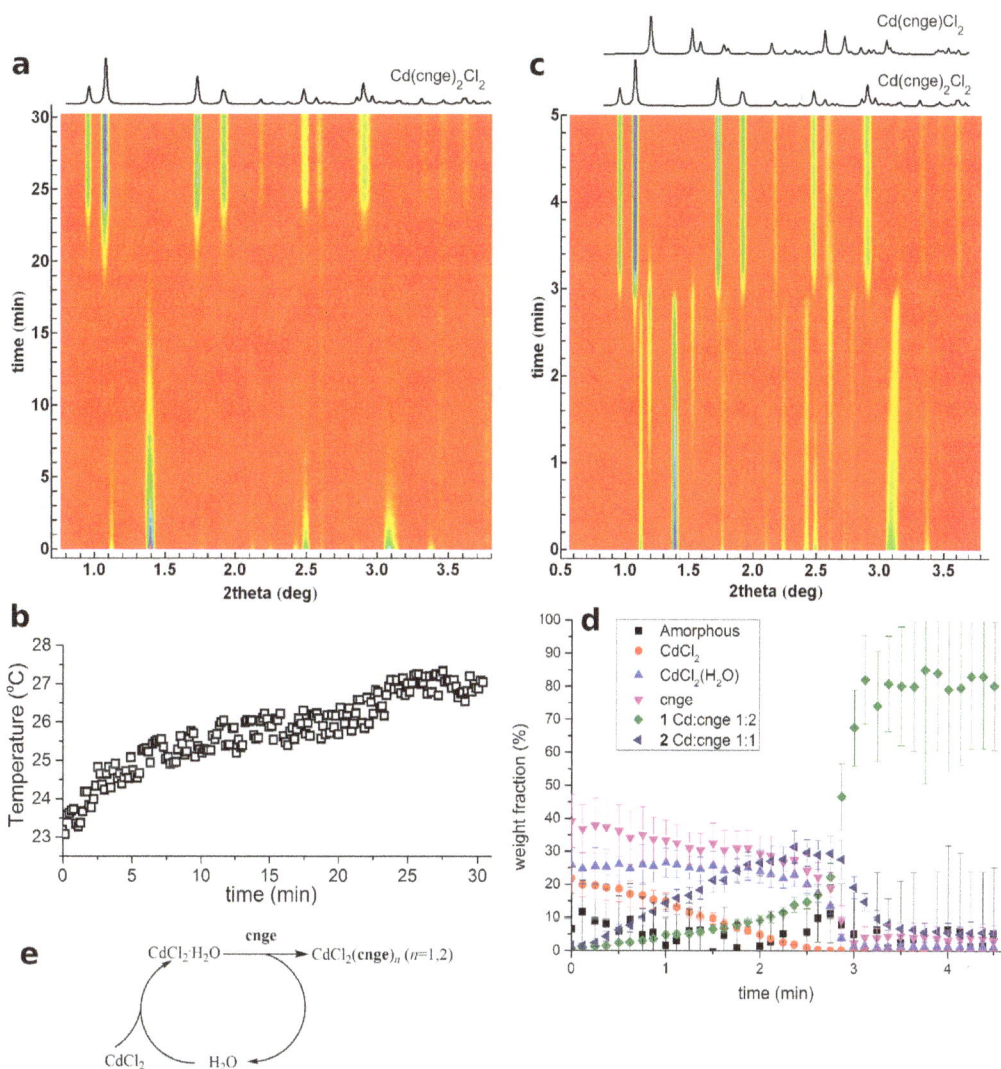

FIGURE 1.8 Time-resolved 2D diffractogram of the formation of the coordination polymers based on Cd(II) and CNGE conducted at (a) room temperature with (b) in situ temperature monitoring using a thermocouple and (c) at an elevated temperature. (d) The reaction profile for the 2D diffractogram in (c) showing a slow and steady depletion of $CdCl_2$ and an abrupt consumption of $CdCl_2(H_2O)$ after $CdCl_2$ is consumed. (e) A loop reaction mechanism explaining the unexpected reactivity of $CdCl_2$ and $CdCl_2(H_2O)$. Adapted with permission from reference [80]. Copyright 2016 American Chemical Society.

this study, the mechanochemical reaction of the formation of coordination polymers based on Cd(II) and cyanoguanidine (CNGE) was first conducted at room temperature followed by the reaction conducted at approximately 50, 60, and 70 βC. At room temperature, milling initially led to partial amorphization and direct formation of the product, while at elevated temperature the reaction mixture was significantly more crystalline, up to four times faster, and afforded an intermediate.

This study has revealed another set of perplexing observations. Namely, the initial reaction mixture of $CdCl_2$ and CNGE had inadvertently, during sample preparation, absorbed some moisture from air, leading to $CdCl_2$ being partially converted to its monohydrate $CdCl_2(H_2O)$. Extracting the reaction profile using Rietveld refinement revealed a steady and slow transformation of $CdCl_2$ into the coordination polymer, while its monohydrate displayed a stable weight fraction (Figure 1.8d). However, upon complete consumption of $CdCl_2$, the reaction experienced an abrupt acceleration. The lingering question was why would the monohydrate engage in the reaction only after the anhydrous $CdCl_2$ was consumed and not before, if it was more reactive? This was explained by a reaction loop where it is the monohydrate that reacts with CNGE and releases its one crystallization water molecule, which is however, immediately captured by the highly hygroscopic anhydrous $CdCl_2$, replenishing the consumed $CdCl_2(H_2O)$ and rendering its weight fraction stable. When the anhydrous $CdCl_2$ is fully consumed, the released water is not captured and enters the reaction mixture, leading to an autocatalytic liquid-assisted grinding reaction (Figure 1.8e).

While in the just described case the discovery of autocatalysis was serendipitous, many reactions have been deliberately started from hydrated (or solvated) reagents to exploit increased reactivity due to the in situ transformation of a neat grinding process into liquid-assisted grinding.[121, 131] Autocatalysis by water was recently described for the mechanochemical synthesis of calcium-based urea fertilizers where the calcium source was $CaCO_3$, which was poorly reactive, or $Cd(OH)_2$, which was highly reactive. The addition of a small amount of water significantly accelerated the reaction starting from $CaCO_3$.

1.3.3 Additives in Mechanochemical Reactions

Milling with additives, which enhance mechanochemical reactivity, be they liquids in liquid-assisted grinding (LAG),[132] salts and liquids in ion- and liquid-assisted grinding (ILAG),[133] or polymers in polymer-assisted grinding (POLAG),[73] was the subject of numerous studies. While the overall effect of these additives on reaction acceleration and templating was recognized based on ex situ work, only the in situ monitoring techniques offer a prospect to better understand the underlying mechanisms of action of these various additives. Understanding how to use additives in controlling a mechanochemical reaction allows one to direct the formation of the target product among several possible reaction outcomes. These target products may be, for example, different polymorphs,[127] different extended frameworks,[58] or different molecules.[134]

In the case of the formation of porous metal-organic frameworks (MOFs) or covalent-organic frameworks (COFs), the liquid can readily template the formation of a specific-topology product.[58, 62] In the recently studied COF formation,[135] the liquid additive enabled the formation of an intermediate that was essential for the construction of the target COF network. In the formation of zeolitic imidazolate frameworks (ZIFs),[59] the amount of the added liquid stabilized the more porous structures,[13] while in the formation of HKUST-1,[136] the chemical properties of the liquid, such as its proticity, determined its active participation in the MOF construction, while mere filling of the pores of the nascent MOF seemed irrelevant.[61] In POLAG, it was found that the amount of added polymer is not important and that the polymer can be added in small amounts, which will be highly relevant for any industrial application of POLAG and purification of the product from the polymer additive.[137]

The profound effect of ammonium and other salts on enhancing the reactivity of ZnO as the green source of Zn(II), in the formation of the archetypal ZIF-8, was extraordinarily demonstrated by in situ PXRD where the ILAG reaction conditions afforded an order of magnitude faster reaction than the corresponding LAG (Figure 1.9).[13] The sensitivity of mechanochemical reaction to the chemical nature of the additive was demonstrated by switching the salt and liquid additives in ZIF-8 synthesis with aqueous acetic acid. While the target ZIF-8 was quickly formed, it, however, became unstable under the very same reaction conditions and gradually turned amorphous. The amorphous material, surprisingly, recrystallized and afforded katsenite – a novel-topology polymorph of ZIF-8.[62]

Additives in mechanochemical reactions, and in particular liquid additives, have tremendously broadened the scope of mechanochemical reactions in terms of product formation and templating the formation of specific products. This was widely applied in selective polymorph preparation.[8, 101, 127, 132, 135] The underlying mechanism of action of liquid additives has mostly remained elusive: it is nevertheless known that the liquid additive may facilitate reactivity by partially dissolving reactants,[138] by participating in the rate-determining step of the reaction,[139] or by selectively stabilizing the surface of milled nanoparticles.[8] The choice of a liquid additive may also inhibit and even reverse multi-component crystal form formation.[140] The liquid additive will often participate by facilitating crystal growth during milling, providing products of better crystallinity exhibiting sharper peaks in their PXRD patterns.

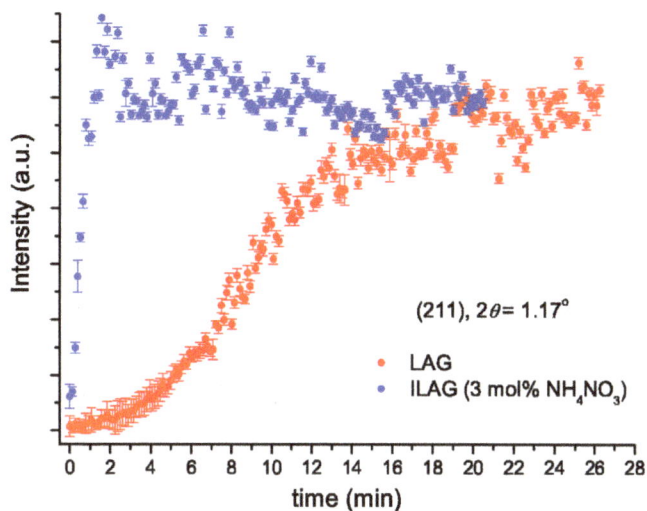

FIGURE 1.9 Comparison of reaction profiles in ILAG (with dimethylformamide [DMF] and NH$_4$NO$_3$) and LAG (with DMF) for the formation of ZIF-8 from ZnO and methylimidazole, based on the intensity change of the (211) reflection of ZIF-8.[13] Adapted with permission. Copyright 2012 Nature Publishing Group.

1.3.4 Intermediates in Mechanosynthesis

Mapping of the reaction course by in situ monitoring enables easier isolation of intermediates by knowing when to interrupt the milling process.[100, 101] Some intermediates isolated in this way are remarkable, for example, the cocrystal of barbituric acid and vanillin, which is formed prior to the target Knoevenagel condensation reaction.[100] This unique cocrystal intermediate positions the reacting molecules with their reacting centers in close proximity and highlights the importance of in situ monitoring, since it was discovered in a reaction that was previously studied in great detail.[106, 141, 142] The lifetime of the cocrystal intermediate could also be manipulated by the choice of the liquid additive. Similarly, a cocrystal of the substrate and the catalyst was isolated as an intermediate before the targeted rhodium insertion in the C–H bond.[143] In situ monitoring and mechanochemistry enabled the preparation and isolation of intermediates that were previously deemed too unstable to be isolated. This was, however, limited to the solution chemistry of aryl N-thiocarbamoylbenzotriazoles, while conducting the reaction in the solid state allowed the isolation of these intermediates as bench-stable solids.[144] A particularly rich set of intermediates was found in the LAG mechanosynthesis of MOF-74 which was studied in situ,[145] but also in a detailed ex situ study.[146]

The synthesis of lead halide organic–inorganic hybrid materials by Wilke and Casati was another reaction system revealing a rich and complex set of intermediates in the synthesis of four guanidinium lead halides differing in stoichiometry.[66] Remarkably, the formation of the final product was controlled by the initial ratio of reactants, but its formation was preceded by intermediates of various different compositions, demonstrating how the reacting systems visit various minima on the energy landscape before settling in the stable product (Figure 1.10). Similarly, the mechanochemical formation of polyoxometalates, studied by the same duo, shows stoichiometry control and intermediates on route to the final milling product.[117, 147] In other cases, expected intermediates may serve as competitors rather than intermediates on the reaction path towards the target product as was the case for ternary phases where the binary-phase intermediates were competitors and slowed down the reactions, while the overall reaction path was highly sensitive to reaction conditions.[148]

The plethora of intermediates in milling offers a possibility to selectively target various possible products, if specific reaction conditions to stabilize each are found. Also, drawing inspiration from facile intermediate formation in milling reactions, the solid state as the reaction medium allows to exploit different solids forms of a reacting molecule to manipulate mechanochemical reactivity. Using the previously studied Knoevenagel condensation of barbituric acid and vanillin, the starting solid form of

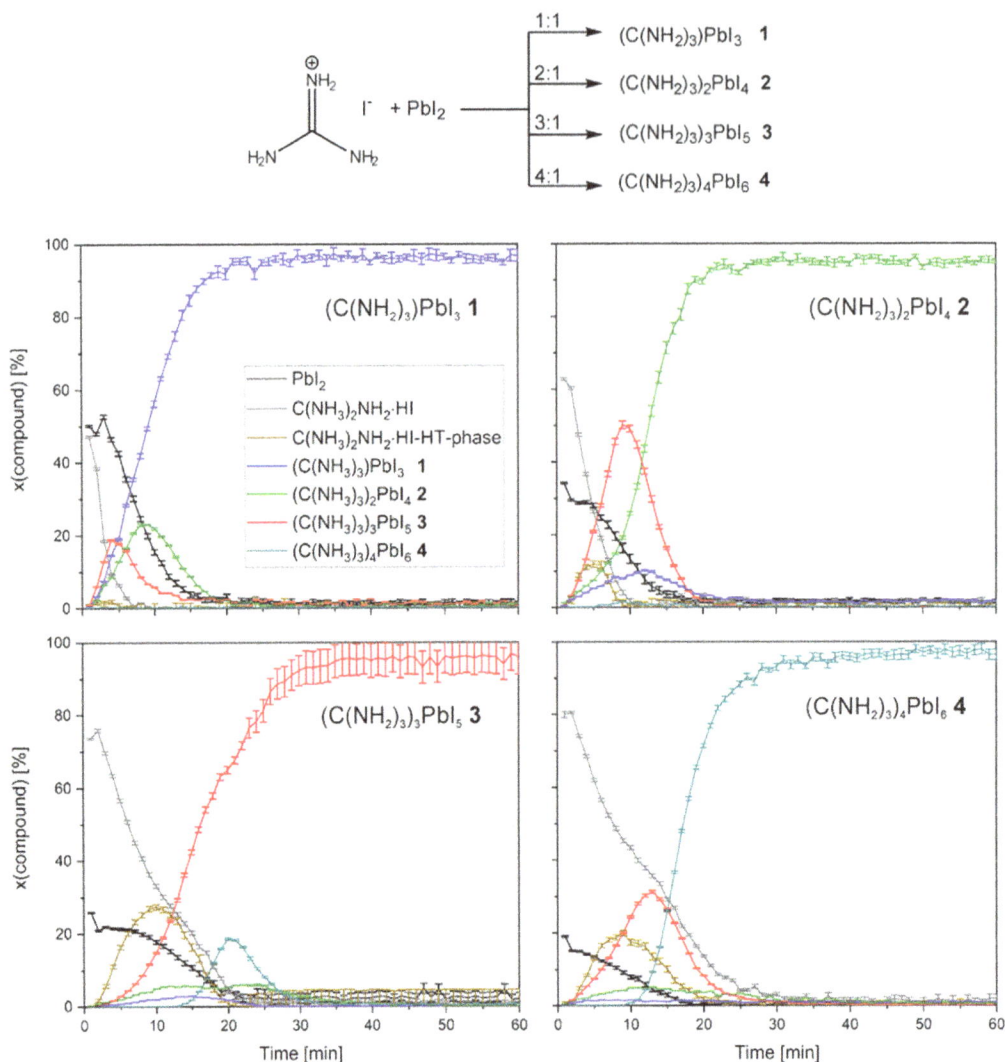

FIGURE 1.10   Reaction profiles in the targeted preparation of hybrid organic–inorganic lead halides visit various compounds en route to the target product determined by the initial stoichiometry of the reactants.[66] Adapted with permission. Copyright 2018 Wiley-VCH GmbH.

barbituric acid was modified by preparing its enol desmotrope, cocrystals, and salts and reacted these phases with a pure solid van.[149] The reaction rate of the formation of the Knoevenagel condensation product could thus be altered by an order of magnitude to become either slower or faster (Figure 1.11), demonstrating the vast potential of using different solid forms in controlling mechanochemical reactivity, an option which is less accessible in the corresponding solution reactions.

In complex reaction mechanisms that are, in solution, expected to proceed in more than a single step and provide intermediates, these are often observed also in analogous mechanochemical processing. For example, the formation of palladacyclic compounds of azobenzenes revealed numerous intermediate species along the reaction path to the formation of either monocyclopalladated or dicyclopalladated azobenzenes (Figure 1.12).[105, 150] These reactions were primarily studied in the context of C–H bond activation starting from the first mechanochemical C–H bond activation.[151] Later on, using different palladium precursors revealed the first palladium adduct intermediate formed before the insertion of the palladium atom in the C–H bond.[105] In situ monitoring enabled the ranking of different palladium precursors for mechanochemical C–H bond activation.

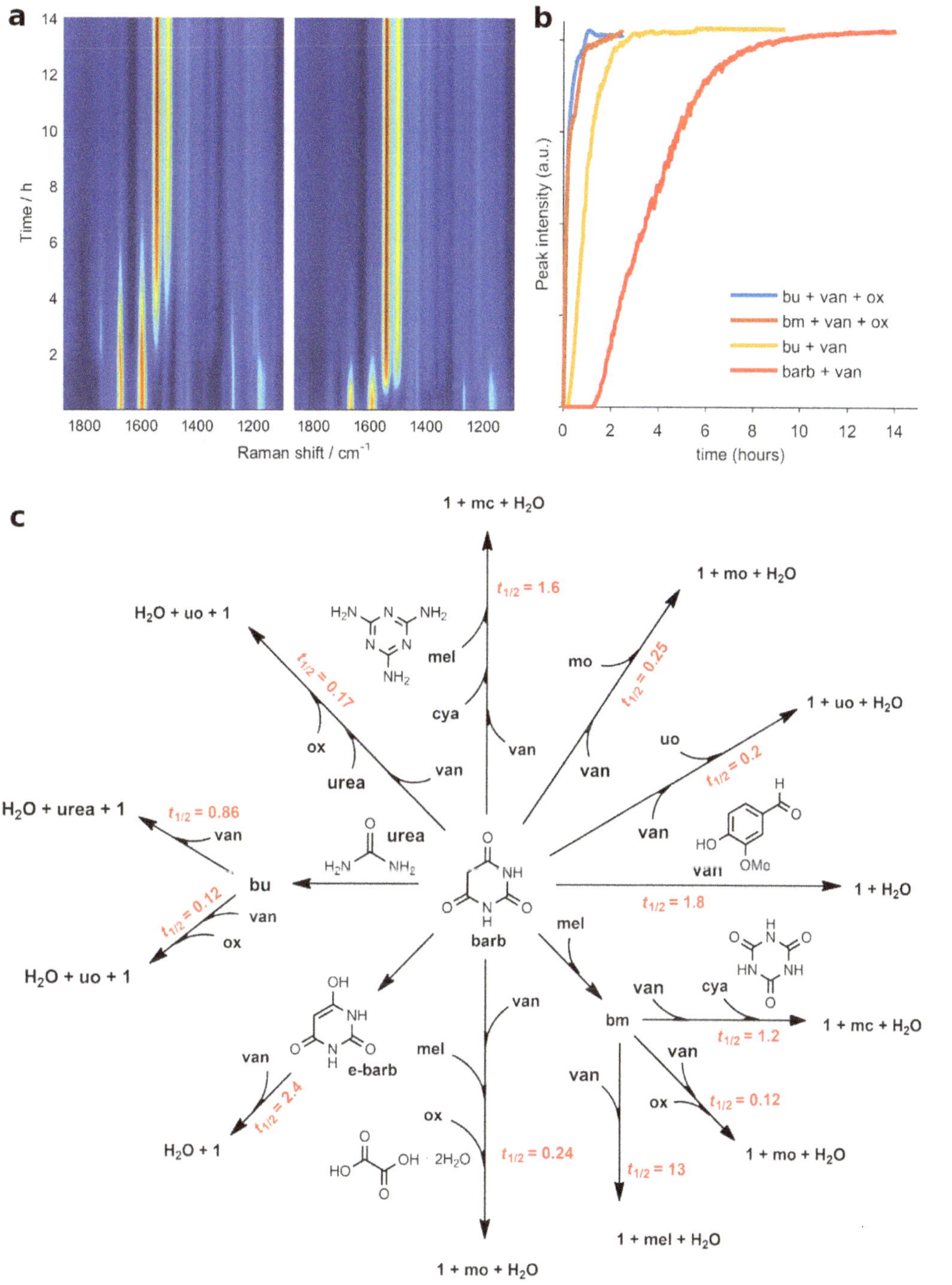

FIGURE 1.11 Modification of solid forms of the starting reagents enables modification and manipulation of mechano-chemical reactivity.[149] (a) Comparison of the parent reaction of barb with van (left) and the faster reaction of the barb:urea cocrystal (bu) with van (right). (b) Reaction profiles obtained by extracting the intensity of the strongest band belonging to the Knoevenagel condensation product showing reaction acceleration relative to the parent reaction. (c) Various modifications of the starting barb: e-barb – the enol tautomer of barb, bm – salt of barb and melamine, bu – cocrystal of barb and urea. Reaction half times ($t_{1/2}$) for each reaction are given in red. Adapted with permission from reference [149]. Copyright 2021 American Chemical Society.

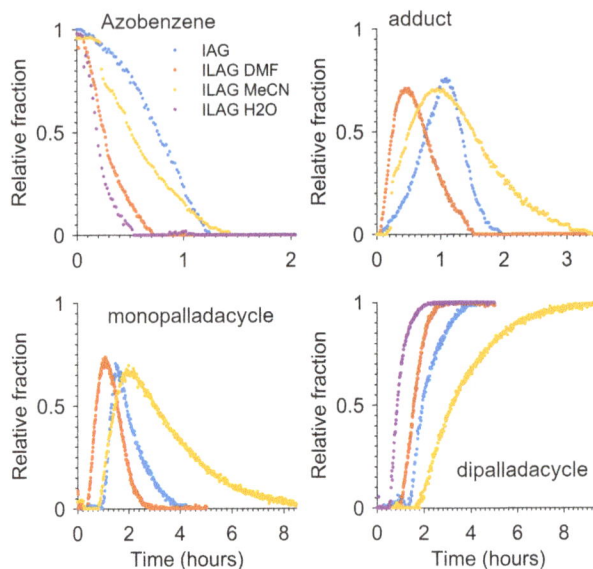

FIGURE 1.12   Reaction profiles for individual species under different reaction conditions obtained by MCR analysis of the in situ collected Raman spectra. The palladium adduct intermediate is formed as the first intermediate in the formation of a palladacycle. Adapted with permission from reference [105]. Copyright 2018 Wiley-VCH GmbH.

1.3.5 The Effect of Milling Conditions

The study of the influence of milling media and other milling parameters on mechanochemical reactivity was significantly limited before the introduction of in situ monitoring. While it is currently not possible to directly compare mechanochemical processing among different types of mills (such as planetary and vibratory), in situ monitoring on vibratory ball mills allows the study of variations in reactivity caused by a change in the size, type, and number of milling balls, or the shape and material of the milling jar,[152] as well as milling frequency.[32, 153–155] Studies of the influence of milling conditions should consider that the reaction occurring in a ball mill may arise from several factors, one of which is the chemical transformation resulting directly from the absorption of mechanical energy imparted to the sample in each impact. Other factors result from mass redistribution due to movement of the milling media, fracture, and growth of particles during mechanochemical processing,[72] and may also include changes in the rheological properties of the sample.[82, 133, 156] Overall dynamics of the milled sample is thus complex, and separating individual contributions to the experimentally observed bulk kinetics is, to say the least, challenging. It is also well known that solid regents, once "activated" by milling, may continue to react,[2] even without experiencing further collisions.

Many of the described issues can be circumvented by choosing a mechanochemical reaction that will fulfill five particular conditions: (i) it should have one reactant to remove the effect of mixing, (ii) the reaction should stop upon the cessation of milling, and (iii) the reactant should be a stable solid at room temperature. Further, the reaction should: (iv) be amenable to in situ reaction monitoring to obtain representative kinetic profiles, which implies that the milled material should be a free-flowing powder that remains uniformly distributed in the reaction jar, in that way ensuring that each of the time-resolved Raman spectra or PXRD patterns is representative of the average mixture composition. Finally, (v) the reaction must not be too fast in order to be able to collect a meaningful number of time-resolved spectra or diffraction patterns to reconstruct the reaction profile. With all these stipulations, finding a suitable mechanochemical reaction for the study of milling conditions is challenging. One such system can be found in the trimerization of the monomeric, brown nickel(II) dibenzoylmethanate complex, $Ni(dbm)_2$, that was previously found to afford the green trimer $[Ni(dbm)_2]_3$ upon heating (Figure 1.13).[157] The trimerization reaction, however, proceeds also under mechanochemical conditions at room temperature. A systematic study, where one milling ball of four different materials and two radii was used to monitor

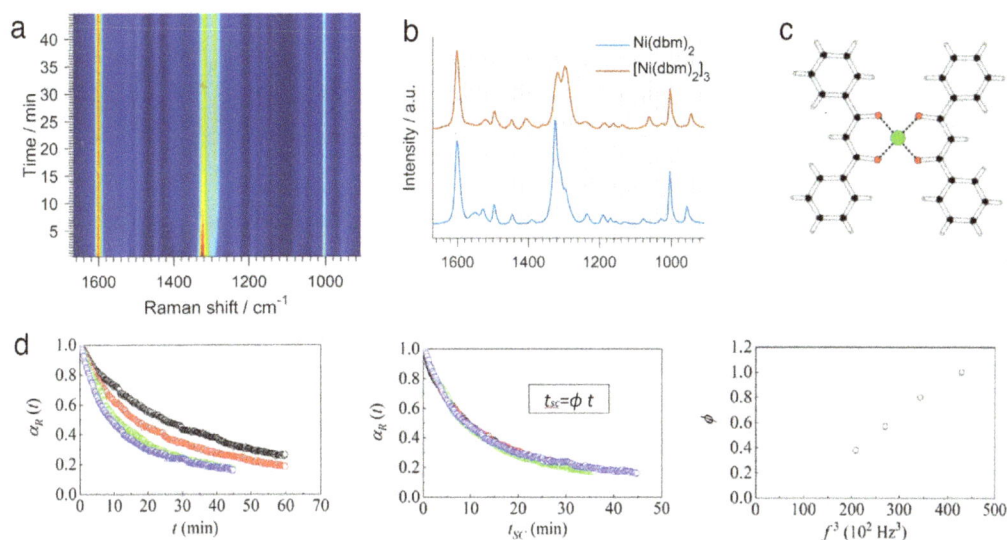

FIGURE 1.13 Purely mechanically activated trimerization of Ni(dbm)$_2$ to [Ni(dbm)$_2$]$_3$. (a) Time-resolved Raman spectra of the trimerization reaction. (b) Spectra of the reactant and the product used in least-squares decomposition of time-resolved spectra. (c) Molecular structure of the planar starting Ni(dbm)$_2$. (d) Reaction profiles derived from time-resolved spectra and the kinetic analysis showing linear scaling of reaction rates with the third powder of the milling frequency. Adapted with permission from reference [100]. Copyright 2018 Royal Society of Chemistry.

the reaction at four frequencies on a vibratory ball mill, revealed that the reaction kinetics scales linearly with impact energy.[158] The impact energy in turn, scales with the third power of frequency, and linearly with the ball mass. Thus, for example, energy input increases by ca. 30 % when frequency increases by 10 %.

1.3.6 The Dynamics of Milled Solids

Milled solids are far from being static and undergo more processes than particle comminution and, possibly, crystal growth: milled particles exchange atoms and molecules. Using isotope-labeled solids,[71] it was demonstrated that hydrogen atoms are not only exchanged between reactant molecules but also between molecules of solids that do not react (Figure 1.14). The exchange of whole molecules was also demonstrated experimentally[71] and corroborated by theoretical simulations of the collision of two crystalline particles.[159] The fact that the milled sample quickly adopts a statistical distribution of hydrogen atoms suggests that the slow solid-state diffusion is overcome by continuous fracturing and growth of milled particles leading to the whole of the sample becoming exposed to the surface over the course of a milling experiment. This observation is in accordance with recognized similarities between a solution and a mechanochemical environment.[109, 160]

Moreover, the exchange of hydrogen atoms extends to the interaction of the milled solids and the liquid additive demonstrating the direct participation of the liquid additive in mechanochemical reactions. The exchange of hydrogen atoms also has been employed to deuterate solids at the exchangeable hydrogen atoms.[71] The solid-state approach achieves full deuteration using a minimal amount of heavy water and will usually be more efficient and more economical than the corresponding solution approach.

1.3.7 Future Prospects

Currently, in situ monitoring is primarily applied to ball milling on vibratory (also known as mixer or shaker) ball mills, while no reports are available for in situ monitoring, either by PXRD or Raman, on planetary ball mills or the Spex-type mills. The reason is practical: the motion of the jar on a vibratory ball mill is simpler, with the jar oscillation amplitude being smaller than the jar itself and the jar

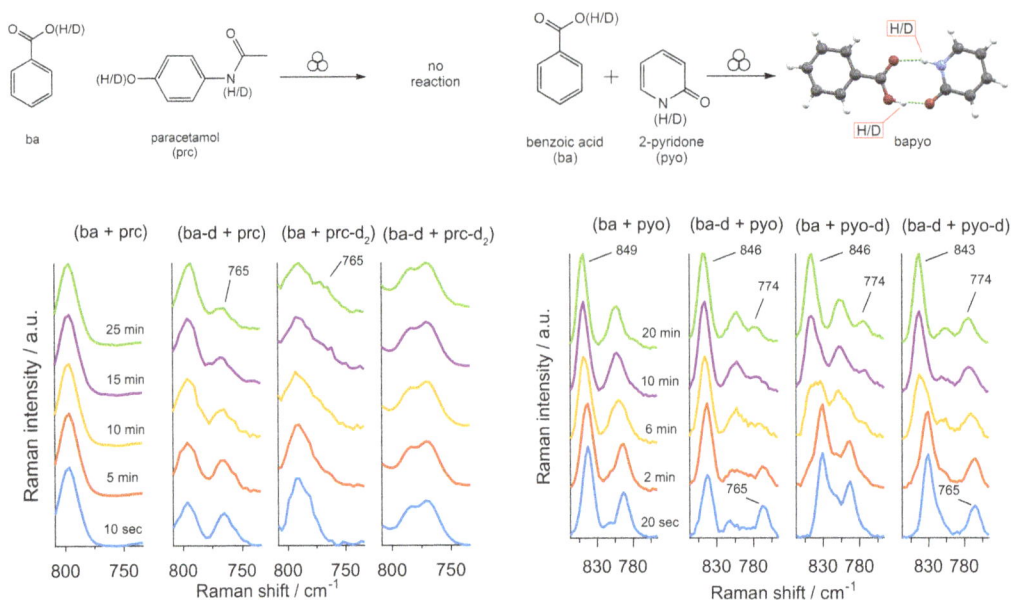

FIGURE 1.14 Hydrogen atom exchange between milled particles occurs without (left) and with new phase formation. Regardless of the initially deuterated partner, at the end of milling a uniform distribution of deuterium atoms is established, as evidenced by in situ collected Raman spectra. Adapted with permission from reference [71]. Copyright 2019 American Chemical Society.

moving in either the horizontal (more common) or the vertical plane. In this way, the fixed Raman probe or the X-ray beam is always probing the inside of the reaction jar during oscillation. This is currently not possible on a planetary ball mill, where the jar movement is wider and the rotating jar visits a point on the circumference once per revolution, or a Spex-type mill where the reaction jar moves along a curve shaped like the number "8" and also does not have a fixed point where the reaction mixture could be targeted either by an X-ray beam or by the laser light. Next to implementation on other types of mills, the introduction of new techniques for in situ monitoring will be welcome for each can provide, to a certain level, a complementary insight and therefore a better understanding of the milling processes.

1.4 Conclusion

Despite the enormous advances in the last 10 years, that have recently been summarized in review articles[12, 28–30] and published protocols,[22, 23] understanding the mechanistic framework of mechanochemical reactions, essential for their optimization as well as their systematic and targeted use in research laboratories and manufacturing industries, is still in its infancy.[161, 162] Techniques for in situ monitoring techniques have revealed a highly dynamic mechanochemical reaction environment, rich in intermediates and complex reaction paths, which are not only challenging to describe and understand, but demonstrated that mechanochemical reactions are highly sensitive to slight variations in reaction conditions. These challenges in understanding and controlling mechanochemistry are now being tackled with new in situ monitoring techniques adapted for milling,[18] by the application of kinetic modeling to in situ-derived reaction profiles,[102, 108, 114] and by the introduction of theoretical modeling to mechanochemical collisions.[159, 160] Prospects for future application of mechanochemistry in the context of green industrial manufacturing, and the full exploitation of its sustainability potential,[163] will depend on a deeper understanding of milling reactions and how to control not only the formation of a specific product but also its particulate properties.

REFERENCES

1. Do, J.-L.; Friščić, T. Mechanochemistry: A Force of Synthesis. *ACS Central Science*, 2017, *3*(1), 13–19. https://doi.org/10.1021/acscentsci.6b00271.

2. Friščić, T.; Mottillo, C.; Titi, H. M. Mechanochemistry for Synthesis. *Angewandte Chemie International Edition*, 2020, *59*(3), 1018–1029. https://doi.org/10.1002/anie.201906755.

3. Andersen, J.; Mack, J. Mechanochemistry and Organic Synthesis: From Mystical to Practical. *Green Chemistry*, 2018, *20*(7), 1435–1443. https://doi.org/10.1039/C7GC03797J.

4. Howard, J. L.; Cao, Q.; Browne, D. L. Mechanochemistry as an Emerging Tool for Molecular Synthesis: What Can It Offer? *Chemical Science*, 2018, *9*(12), 3080–3094. https://doi.org/10.1039/C7SC05371A.

5. Amrute, A. P.; De Bellis, J.; Felderhoff, M.; Schüth, F. Mechanochemical Synthesis of Catalytic Materials. *Chemistry – A European Journal*, 2021, *27*(23), 6819–6841. https://doi.org/10.1002/chem.202004583.

6. Friščić, T.; Halasz, I.; Štrukil, V.; Eckert-Maksić, M.; Dinnebier, R. E. Clean and Efficient Synthesis Using Mechanochemistry: Coordination Polymers, Metal-Organic Frameworks and Metallodrugs. *Croatica Chemica Acta*, 2012, *85*(3), 367–378. https://doi.org/10.5562/cca2014.

7. Julien, P. A.; Friščić, T. Methods for Monitoring Milling Reactions and Mechanistic Studies of Mechanochemistry: A Primer. *Crystal Growth & Design* 2022, *22*(9), 5726–5754. https://doi.org/10.1021/acs.cgd.2c00587.

8. Belenguer, A. M.; Lampronti, G. I.; Cruz-Cabeza, A. J.; Hunter, C. A.; Sanders, J. K. M. Solvation and Surface Effects on Polymorph Stabilities at the Nanoscale. *Chemical Science*, 2016, *7*(11), 6617–6621. https://doi.org/10.1039/C6SC03457H.

9. Doppiu, S.; Schultz, L.; Gutfleisch, O. In Situ Pressure and Temperature Monitoring during the Conversion of Mg into MgH_2 by High-Pressure Reactive Ball Milling. *Journal of Alloys & Compounds*, 2007, *427*(1), 204–208. https://doi.org/10.1016/j.jallcom.2006.02.045.

10. Colbe, J. M. B. von; Felderhoff, M.; Bogdanović, B.; Schüth, F.; Weidenthaler, C. One-Step Direct Synthesis of a Ti-Doped Sodium Alanate Hydrogen Storage Material. *Chemical Communications*, 2005, *37*(37), 4732–4734. https://doi.org/10.1039/B506502J.

11. Atzmon, M. In Situ Thermal Observation of Explosive Compound-Formation Reaction during Mechanical Alloying. *Physical Review Letters*, 1990, *64*(4), 487–490. https://doi.org/10.1103/PhysRevLett.64.481.

12. Lukin, S.; Germann, L. S.; Friščić, T., Halasz, I. Toward Mechanistic Understanding of Mechanochemical Reactions Using Real-Time *In Situ* Monitoring. *Accounts of Chemical Research*, 2022, *55*(9), 1262–1271. https://doi.org/10.1021/acs.accounts.2c00062.

13. Friščić, T.; Halasz, I.; Beldon, P. J.; Belenguer, A. M.; Adams, F.; Kimber, S. A. J.; Honkimäki, V.; Dinnebier, R. E. Real-Time and In Situ Monitoring of Mechanochemical Milling Reactions. *Nature Chemistry*, 2013, *5*(1), 66–73. https://doi.org/10.1038/nchem.1505.

14. Gracin, D.; Štrukil, V.; Friščic, T.; Halasz, I.; Užarevic, K. Laboratory Real-Time and In Situ Monitoring of Mechanochemical Milling Reactions by Raman Spectroscopy. *Angewandte Chemie - International Edition*, 2014, *53*(24), 6193–6191. https://doi.org/10.1002/anie.201402334.

15. Batzdorf, L.; Fischer, F.; Wilke, M.; Wenzel, K.-J.; Emmerling, F. Direct In Situ Investigation of Milling Reactions Using Combined X-ray Diffraction and Raman Spectroscopy. *Angewandte Chemie International Edition*, 2015, *54*(6), 1799–1802. https://doi.org/10.1002/anie.201409834.

16. Lukin, S.; Stolar, T.; Tireli, M.; Blanco, M. V.; Babić, D.; Friščić, T.; Užarević, K.; Halasz, I. Tandem In Situ Monitoring for Quantitative Assessment of Mechanochemical Reactions Involving Structurally Unknown Phases. *Chemistry - A European Journal*, 2017, *23*(56), 13941–13949. https://doi.org/10.1002/chem.201702489.

17. Halasz, I.; Friščić, T.; Kimber, S. A. J.; Užarević, K.; Puškarić, A.; Mottillo, C.; Julien, P.; Štrukil, V.; Honkimäki, V.; Dinnebier, R. E. Quantitative In Situ and Real-Time Monitoring of Mechanochemical Reactions. *Faraday Discussions*, 2014, *170*, 203–221. https://doi.org/10.1039/c4fd00013g.

18. Schiffmann, J. G.; Emmerling, F.; Martins, I. C. B.; Van Wüllen, L. In-Situ Reaction Monitoring of a Mechanochemical Ball Mill Reaction with Solid State NMR. *Solid State Nuclear Magnetic Resonance*, 2020, *109*. https://doi.org/10.1016/j.ssnmr.2020.101681.

19. Leger, M.; Guo, J.; MacMillan, B.; Titi, H.; Friscic, T.; Blight, B.; Balcom, B. Relaxation Time Correlation NMR for Mechanochemical In-Situ Reaction Monitoring of Metal-Organic Frameworks, 2021. https://doi.org/10.33774/chemrxiv-2021-rbj0t.

20. Oliveira, P. F. M. de; Michalchuk, A. A. L.; Buzanich, A. G.; Bienert, R.; Torresi, R. M.; Camargo, P. H. C.; Emmerling, F. Tandem X-ray Absorption Spectroscopy and Scattering for In Situ Time-Resolved Monitoring of Gold Nanoparticle Mechanosynthesis. *Chemical Communications*, 2020, *56*(71), 10329–10332. https://doi.org/10.1039/D0CC03862H.

21. Julien, P. A.; Arhangelskis, M.; Germann, L. S.; Etter, M.; Dinnebier, R. E.; Morris, A. J.; Friscic, T. Illuminating Mechanochemical Reactions by Combining Real-Time Fluorescence Emission Monitoring and Periodic Time-Dependent Density-Functional Calculations, 2021. https://doi.org/10.33774/chem-rxiv-2021-lw8sm.

22. Halasz, I.; Kimber, S. A. J.; Beldon, P. J.; Belenguer, A. M.; Adams, F.; Honkimäki, V.; Nightingale, R. C.; Dinnebier, R. E.; Friščić, T. In Situ and Real-Time Monitoring of Mechanochemical Milling Reactions Using Synchrotron X-ray Diffraction. *Nature Protocols*, 2013, *8*(9), 1718–1729. https://doi.org/10.1038/nprot.2013.100.

23. Lukin, S.; Užarević, K.; Halasz, I. Raman Spectroscopy for Real-Time and In Situ Monitoring of Mechanochemical Milling Reactions. *Nature Protocols*, 2021, *16*, 3492–3521. https://doi.org/10.1038/s41596-021-00545-x.

24. Lukin, S.; Stolar, T.; Lončarić, I.; Milanović, I.; Biliškov, N.; Di Michiel, M.; Friščić, T.; Halasz, I. Mechanochemical Metathesis between $AgNO_3$ and NaX (X = Cl, Br, I) and $Ag2XNO3$ Double-Salt Formation. *Inorganic Chemistry*, 2020, *59*(17), 12200–12208. https://doi.org/10.1021/acs.inorgchem.0c01196.

25. Sović, I.; Lukin, S.; Meštrović, E.; Halasz, I.; Porcheddu, A.; Delogu, F.; Ricci, P. C.; Caron, F.; Perilli, T.; Dogan, A. et al. Mechanochemical Preparation of Active Pharmaceutical Ingredients Monitored by *In Situ* Raman Spectroscopy. *ACS Omega*, 2020, *5*(44), 28663–28672. https://doi.org/10.1021/acsomega.0c03756.

26. Halasz, I.; Puškarić, A.; Kimber, S. A. J.; Beldon, P. J.; Belenguer, A. M.; Adams, F.; Honkimäki, V.; Dinnebier, R. E.; Patel, B.; Jones, W. et al. Real-Time In Situ Powder X-ray Diffraction Monitoring of Mechanochemical Synthesis of Pharmaceutical Cocrystals. *Angewandte Chemie - International Edition*, 2013, *52*(44), 11538–11541. https://doi.org/10.1002/anie.201305928.

27. Tan, D.; Loots, L.; Friščić, T. Towards Medicinal Mechanochemistry: Evolution of Milling from Pharmaceutical Solid Form Screening to the Synthesis of Active Pharmaceutical Ingredients (APIs). *Chemical Communications*, 2016, *52*(50), 7760–7781. https://doi.org/10.1039/C6CC02015A.

28. Užarević, K.; Halasz, I.; Friščić, T. Real-Time and In Situ Monitoring of Mechanochemical Reactions: A New Playground for All Chemists. *Journal of Physical Chemistry Letters*, 2015, *6*(20), 4129–4140. https://doi.org/10.1021/acs.jpclett.5b01831.

29. Michalchuk, A. A. L.; Emmerling, F. Time-Resolved In Situ Monitoring of Mechanochemical Reactions. *Angewandte Chemie International Edition*. https://doi.org/10.1002/anie.202117270.

30. Weidenthaler, C. In Situ Analytical Methods for the Characterization of Mechanochemical Reactions. *Crystals*, 2022, *12*(3), 345. https://doi.org/10.3390/cryst12030345.

31. Fang, Y.; Salamé, N.; Woo, S.; Bohle, D. S.; Friščić, T.; Cuccia, L. A. Rapid and Facile Solvent-Free Mechanosynthesis in a Cell Lysis Mill: Preparation and Mechanochemical Complexation of Aminobenzoquinones. *CrystEngComm*, 2014, *16*(31), 7180. https://doi.org/10.1039/C4CE00328D.

32. Julien, P. A.; Malvestiti, I.; Friščić, T. The Effect of Milling Frequency on a Mechanochemical Organic Reaction Monitored by In Situ Raman Spectroscopy. *Beilstein Journal of Organic Chemistry*, 2017, *13*(1), 2160–2168. https://doi.org/10.3762/bjoc.13.216.

33. Užarević, K.; Ferdelji, N.; Mrla, T.; Julien, P. A.; Halasz, B.; Friščić, T.; Halasz, I. Enthalpy vs. Friction: Heat Flow Modelling of Unexpected Temperature Profiles in Mechanochemistry of Metal-Organic Frameworks. *Chemical Science*, 2018, *9*(9), 2525–2532. https://doi.org/10.1039/c7sc05312f.

34. Kulla, H.; Wilke, M.; Fischer, F.; Röllig, M.; Maierhofer, C.; Emmerling, F. Warming up for Mechanosynthesis – Temperature Development in Ball Mills during Synthesis. *Chemical Communications*, 2017, *53*(10), 1664–1661. https://doi.org/10.1039/C6CC08950J.

35. Kubota, K.; Takahashi, R.; Uesugi, M.; Ito, H. A Glove-Box- and Schlenk-Line-Free Protocol for Solid-State C–N Cross-Coupling Reactions Using Mechanochemistry. *ACS Sustainable Chemistry & Engineering*, 2020, *8*(44), 16577–16582. https://doi.org/10.1021/acssuschemeng.0c05834.

36. Takacs, L. Self-Sustaining Reactions as a Tool to Study Mechanochemical Activation. *Faraday Discussions*, 2014, *170*, 251–265. https://doi.org/10.1039/C3FD00133D.

37. Rathmann, T.; Petersen, H.; Reichle, S.; Schmidt, W.; Amrute, A. P.; Etter, M.; Weidenthaler, C. In Situ Synchrotron X-ray Diffraction Studies Monitoring Mechanochemical Reactions of Hard Materials: Challenges and Limitations. *Review of Scientific Instruments*, 2021, *92*(11), 114102. https://doi.org/10.1063/5.0068621.

38. Reichle, S.; Felderhoff, M.; Schüth, F. Mechanocatalytic Room-Temperature Synthesis of Ammonia from Its Elements Down to Atmospheric Pressure. *Angewandte Chemie International Edition*, 2021, *60*(50), 26385–26389. https://doi.org/10.1002/anie.202112095.

39. Mori, S.; Xu, W.-C.; Ishidzuki, T.; Ogasawara, N.; Imai, J.; Kobayashi, K. Mechanochemical Activation of Catalysts for CO_2 Methanation. *Applied Catalysis A: General*, 1996, *137*(2), 255–268. https://doi.org/10.1016/0926-860X(95)00319-3.

40. Immohr, S.; Felderhoff, M.; Weidenthaler, C.; Schüth, F. An Orders-of-Magnitude Increase in the Rate of the Solid-Catalyzed CO Oxidation by In Situ Ball Milling. *Angewandte Chemie International Edition*, 2013, *52*(48), 12688–12691. https://doi.org/10.1002/anie.201305992.

41. Castro, F. J.; Fuster, V.; Urretavizcaya, G. MgH_2 Synthesis during Reactive Mechanical Alloying Studied by In-Situ Pressure Monitoring. *International Journal of Hydrogen Energy*, 2012, *37*(22), 16844–16851. https://doi.org/10.1016/j.ijhydene.2012.08.089.

42. Kozma, G.; Kukovecz, Á.; Kónya, Z. Spectroscopic Studies on the Formation Kinetics of SnO_2 Nanoparticles Synthesized in a Planetary Ball Mill. *Journal of Molecular Structure*, 2007, *834–836*, 430–434. https://doi.org/10.1016/j.molstruc.2006.10.031.

43. Bolm, C.; Hernández, J. G. Mechanochemistry of Gaseous Reactants. *Angewandte Chemie International Edition*, 2019, *58*(11), 3285–3299. https://doi.org/10.1002/anie.201810902.

44. Pfennig, V. S.; Villella, R. C.; Nikodemus, J.; Bolm, C. Mechanochemical Grignard Reactions with Gaseous CO_2 and Sodium Methyl Carbonate. *Angewandte Chemie International Edition*, 2022, *61*(9), e202116514. https://doi.org/10.1002/anie.202116514.

45. Bilke, M.; Losch, P.; Vozniuk, O.; Bodach, A.; Schüth, F. Methane to Chloromethane by Mechanochemical Activation: A Selective Radical Pathway. *Journal of the American Chemical Society*, 2019, *141*(28), 11212–11218. https://doi.org/10.1021/jacs.9b04413.

46. Eckert, R.; Felderhoff, M.; Schüth, F. Preferential Carbon Monoxide Oxidation over Copper-Based Catalysts under In Situ Ball Milling. *Angewandte Chemie International Edition*, 2017, *56*(9), 2445–2448. https://doi.org/10.1002/anie.201610501.

47. Grätz, S.; Beyer, D.; Tkachova, V.; Hellmann, S.; Berger, R.; Feng, X.; Borchardt, L. The Mechanochemical Scholl Reaction – A Solvent-Free and Versatile Graphitization Tool. *Chemical Communications*, 2018, *54*(42), 5307–5310. https://doi.org/10.1039/C8CC01993B.

48. Brekalo, I.; Yuan, W.; Mottillo, C.; Lu, Y.; Zhang, Y.; Casaban, J.; Holman, K. T.; James, S. L.; Duarte, F.; Williams, P. A. et al. Manometric Real-Time Studies of the Mechanochemical Synthesis of Zeolitic Imidazolate Frameworks. *Chemical Science*, 2020, *11*(8), 2141–2141. https://doi.org/10.1039/C9SC05514B.

49. Bonn, P. van; Bolm, C.; Hernández, J. G. Mechanochemical Palladium-Catalyzed Carbonylative Reactions Using $Mo(CO)_6$. *Chemistry – A European Journal*, 2020, *26*(12), 2576–2580. https://doi.org/10.1002/chem.201904528.

50. Casco, M. E.; Badaczewski, F.; Grätz, S.; Tolosa, A.; Presser, V.; Smarsly, B. M.; Borchardt, L. Mechanochemical Synthesis of Porous Carbon at Room Temperature with a Highly Ordered Sp^2 Microstructure. *Carbon*, 2018, *139*, 325–333. https://doi.org/10.1016/j.carbon.2018.06.068.

51. Casco, M. E.; Kirchhoff, S.; Leistenschneider, D.; Rauche, M.; Brunner, E.; Borchardt, L. Mechanochemical Synthesis of N-Doped Porous Carbon at Room Temperature. *Nanoscale*, 2019, *11*(11), 4712–4718. https://doi.org/10.1039/C9NR01019J.

52. Troschke, E.; Grätz, S.; Lübken, T.; Borchardt, L. Mechanochemical Friedel–Crafts Alkylation—A Sustainable Pathway Towards Porous Organic Polymers. *Angewandte Chemie International Edition*, 2017, *56*(24), 6859–6863. https://doi.org/10.1002/anie.201702303.

53. Takacs, L. Self-Sustaining Reactions Induced by Ball Milling. *Progress in Materials Science*, 2002, *47*(4), 355–414. https://doi.org/10.1016/S0079-6425(01)00002-0.

54. Petersen, H.; Reichle, S.; Leiting, S.; Losch, P.; Kersten, W.; Rathmann, T.; Tseng, J.; Etter, M.; Schmidt, W.; Weidenthaler, C. In Situ Synchrotron X-ray Diffraction Studies of the Mechanochemical Synthesis of ZnS from Its Elements. *Chemistry – A European Journal*, 2021, *27*(49), 12558–12565. https://doi.org/10.1002/chem.202101260.

55. Germann, L. S.; Katsenis, A. D.; Huskić, I.; Julien, P. A.; Užarević, K.; Etter, M.; Farha, O. K.; Friščić, T.; Dinnebier, R. E. Real-Time In Situ Monitoring of Particle and Structure Evolution in the Mechanochemical Synthesis of UiO-66 Metal-Organic Frameworks. *Crystal Growth & Design*, 2020, *20*(1), 49–54. https://doi.org/10.1021/acs.cgd.9b01471.

56. Lampronti, G. I.; Michalchuk, A. A. L.; Mazzeo, P. P.; Belenguer, A. M.; Sanders, J. K. M.; Bacchi, A.; Emmerling, F. Changing the Game of Time Resolved X-ray Diffraction on the Mechanochemistry Playground by Downsizing. *Nature Communications*, 2021, *12*(1), 6134. https://doi.org/10.1038/s41467 -021-26264-1.

57. Hill, R. J.; Howard, C. J. Quantitative Phase Analysis from Neutron Powder Diffraction Data Using the Rietveld Method. *Journal of Applied Crystallography*, 1987, *20*(6), 467–474. https://doi.org/10.1107/ S0021889887086199.

58. Beldon, P. J.; Fábián, L.; Stein, R. S.; Thirumurugan, A.; Cheetham, A. K.; Friščić, T. Rapid Room-Temperature Synthesis of Zeolitic Imidazolate Frameworks by Using Mechanochemistry. *Angewandte Chemie International Edition*, 2010, *49*(50), 9640–9643. https://doi.org/10.1002/anie.201005541.

59. Huang, X.-C.; Lin, Y.-Y.; Zhang, J.-P.; Chen, X.-M. Ligand-Directed Strategy for Zeolite-Type Metal–Organic Frameworks: Zinc(II) Imidazolates with Unusual Zeolitic Topologies. *Angewandte Chemie International Edition*, 2006, *45*(10), 1557–1559. https://doi.org/10.1002/anie.200503778.

60. Tian, Y.-Q.; Cai, C.-X.; Ren, X.-M.; Duan, C.-Y.; Xu, Y.; Gao, S.; You, X.-Z. The Silica-Like Extended Polymorphism of Cobalt(II) Imidazolate Three-Dimensional Frameworks: X-ray Single-Crystal Structures and Magnetic Properties. *Chemistry – A European Journal*, 2003, *9*(22), 5673–5685. https:// doi.org/10.1002/chem.200304951.

61. Stolar, T.; Batzdorf, L.; Lukin, S.; Žilić, D.; Motillo, C.; Friščić, T.; Emmerling, F.; Halasz, I.; Užarević, K. In Situ Monitoring of the Mechanosynthesis of the Archetypal Metal-Organic Framework HKUST-1: Effect of Liquid Additives on the Milling Reactivity. *Inorganic Chemistry*, 2017, *56*(11), 6599–6608. https://doi.org/10.1021/acs.inorgchem.7b00701.

62. Katsenis, A. D.; Puškarić, A.; Štrukil, V.; Mottillo, C.; Julien, P. A.; Užarević, K.; Pham, M.-H.; Do, T.-O.; Kimber, S. A. J.; Lazić, P. et al. In Situ X-ray Diffraction Monitoring of a Mechanochemical Reaction Reveals a Unique Topology Metal-Organic Framework. *Nature Communications*, 2015, *6*(1), 6662. https://doi.org/10.1038/ncomms7662.

63. Vaughan, G. B. M.; Baker, R.; Barret, R.; Bonnefoy, J.; Buslaps, T.; Checchia, S.; Duran, D.; Fihman, F.; Got, P.; Kieffer, J. et al. ID15A at the ESRF – A Beamline for High Speed Operando X-Ray Diffraction, Diffraction Tomography and Total Scattering. *Journal of Synchrotron Radiation*, 2020, *27*(2), 515–528. https://doi.org/10.1107/S1600577519016813.

64. Ban, V.; Sadikin, Y.; Lange, M.; Tumanov, N.; Filinchuk, Y.; Černý, R.; Casati, N. Innovative In Situ Ball Mill for X-ray Diffraction. *Analytical Chemistry*, 2017, *89*(24), 13176–13181. https://doi.org/10 .1021/acs.analchem.7b02871.

65. Scarlett, N. V. Y.; Madsen, I. C. Effect of Microabsorption on the Determination of Amorphous Content via Powder X-ray Diffraction. *Powder Diffraction*, 2018, *33*(1), 26–31. https://doi.org/10.1017/ S0885715618000052.

66. Wilke, M.; Casati, N. Insight into the Mechanochemical Synthesis and Structural Evolution of Hybrid Organic–Inorganic Guanidinium Lead(II) Iodides. *Chemistry – A European Journal*, 2018, *24*(67), 17701–17711. https://doi.org/10.1002/chem.201804066.

67. Stinton, G. W.; Evans, J. S. O. Parametric Rietveld Refinement. *Journal of Applied Crystallography*, 2007, *40*(1), 87–95. https://doi.org/10.1107/S0021889806043275.

68. Halasz, I.; Dinnebier, R. E.; Angel, R. Parametric Rietveld Refinement for the Evaluation of Powder Diffraction Patterns Collected as a Function of Pressure. *Journal of Applied Crystallography*, 2010, *43*(3), 504–510. https://doi.org/10.1107/S0021889810005856.

69. Descamps, M.; Willart, J. F. Perspectives on the Amorphisation/Milling Relationship in Pharmaceutical Materials. *Advanced Drug Delivery Reviews*, 2016, *100*, 51–66. https://doi.org/10.1016/j.addr.2016.01.011.

70. Kosanović, C.; Bronić, J.; Čižmek, A.; Subotić, B.; Šmit, I.; Stubičar, M.; Tonejc, A. Mechanochemistry of Zeolites: Part 2. Change in Particulate Properties of Zeolites during Ball Milling. *Zeolites*, 1995, *15*(3), 247–252. https://doi.org/10.1016/0144-2449(94)00022-K.

71. Lukin, S.; Tireli, M.; Stolar, T.; Barišić, D.; Blanco, M. V.; di Michiel, M.; Užarević, K.; Halasz, I. Isotope Labeling Reveals Fast Atomic and Molecular Exchange in Mechanochemical Milling Reactions. *Journal of the American Chemical Society*, 2019, *141*(3), 1212–1216. https://doi.org/10.1021/jacs.8b12149.

72. Mørup, S.; Jiang, J. Z.; Bødker, F.; Horsewell, A. Crystal Growth and the Steady-State Grain Size during High-Energy Ball-Milling. *Europhysics Letters*, 2001, *56*(3), 441–446. https://doi.org/10.1209/epl/i2001-00538-1.

73. Hasa, D.; Schneider Rauber, G.; Voinovich, D.; Jones, W. Cocrystal Formation through Mechanochemistry: From Neat and Liquid-Assisted Grinding to Polymer-Assisted Grinding. *Angewandte Chemie International Edition*, 2015, *54*(25), 7371–7375. https://doi.org/10.1002/anie.201501638.

74. Lisac, K.; Topić, F.; Arhangelskis, M.; Cepić, S.; Julien, P. A.; Nickels, C. W.; Morris, A. J.; Friščić, T.; Cinčić, D. Halogen-Bonded Cocrystallization with Phosphorus, Arsenic and Antimony Acceptors. *Nature Communications*, 2019, *10*(1), 61. https://doi.org/10.1038/s41467-018-07957-6.

75. Esmonde-White, K. A.; Cuellar, M.; Uerpmann, C.; Lenain, B.; Lewis, I. R. Raman Spectroscopy as a Process Analytical Technology for Pharmaceutical Manufacturing and Bioprocessing. *Analytical & Bioanalytical Chemistry*, 2017, *409*(3), 637–649. https://doi.org/10.1007/s00216-016-9824-1.

76. Simon, L. L.; Pataki, H.; Marosi, G.; Meemken, F.; Hungerbühler, K.; Baiker, A.; Tummala, S.; Glennon, B.; Kuentz, M.; Steele, G. et al. Assessment of Recent Process Analytical Technology (PAT) Trends: A Multiauthor Review. *Organic Process Research & Development*, 2015, *19*(1), 3–62. https://doi.org/10.1021/op500261y.

77. Lukin, S.; Užarević, K.; Halasz, I. Raman Spectroscopy for Real-Time and In Situ Monitoring of Mechanochemical Milling Reactions. *Nature Protocols*, 2021, *16*(7), 3492–3521. https://doi.org/10.1038/s41596-021-00545-x.

78. Gebrekidan, M. T.; Knipfer, C.; Stelzle, F.; Popp, J.; Will, S.; Braeuer, A. A Shifted-Excitation Raman Difference Spectroscopy (SERDS) Evaluation Strategy for the Efficient Isolation of Raman Spectra from Extreme Fluorescence Interference. *Journal of Raman Spectroscopy*, 2016, *47*(2), 198–209. https://doi.org/10.1002/jrs.4775.

79. Long, D. A. *Raman Spectroscopy*. McGraw-Hill: New York, 1971.

80. Užarević, K.; Štrukil, V.; Mottillo, C.; Julien, P. A.; Puškarić, A.; Friščić, T.; Halasz, I. Exploring the Effect of Temperature on a Mechanochemical Reaction by In Situ Synchrotron Powder X-ray Diffraction. *Crystal Growth & Design*, 2016, *16*(4), 2342–2341. https://doi.org/10.1021/acs.cgd.6b00131.

81. Torre, F.; Carta, M.; Barra, P.; Cincotti, A.; Porcheddu, A.; Delogu, F. Mechanochemical Ignition of Self-Propagating Reactions in Zn-S Powder Mixtures. *Metallurgical & Materials Transactions. Part B*, 2021, *52*(2), 830–839. https://doi.org/10.1007/s11663-021-02056-2.

82. Ardila-Fierro, K. J.; Lukin, S.; Etter, M.; Užarević, K.; Halasz, I.; Bolm, C.; Hernández, J. G. Direct Visualization of a Mechanochemically Induced Molecular Rearrangement. *Angewandte Chemie - International Edition*, 2020, *59*(32), 13458–13462. https://doi.org/10.1002/anie.201914921.

83. Lieber, C. A.; Mahadevan-Jansen, A. Automated Method for Subtraction of Fluorescence from Biological Raman Spectra. *Applied Spectroscopy*, 2003, *57*(11), 1363–1361.

84. Cai, Y.; Yang, C.; Xu, D.; Gui, W. Baseline Correction for Raman Spectra Using Penalized Spline Smoothing Based on Vector Transformation. *Analytical Methods*, 2018, *10*(28), 3525–3533. https://doi.org/10.1039/C8AY00914G.

85. Wang, X.; Fan, X.; Xu, Y.; Wu, J.; Liang, J.; Zuo, Y. [Baseline Correction Method for Raman Spectroscopy Based on B-Spline Fitting]. *Guang Pu Xue Yu Guang Pu Fen Xi*, 2014, *34*(8), 2117–2121.

86. Wang, T.; Dai, L. Background Subtraction of Raman Spectra Based on Iterative Polynomial Smoothing. *Applied Spectroscopy*, 2017, *71*(6), 1169–1179. https://doi.org/10.1177/0003702816670915.

87. He, S.; Zhang, W.; Liu, L.; Huang, Y.; He, J.; Xie, W.; Wu, P.; Du, C. Baseline Correction for Raman Spectra Using an Improved Asymmetric Least Squares Method. *Analytical Methods*, 2014, *6*(12), 4402–4401. https://doi.org/10.1039/C4AY00068D.

88. Ye, J.; Tian, Z.; Wei, H.; Li, Y. Baseline Correction Method Based on Improved Asymmetrically Reweighted Penalized Least Squares for the Raman Spectrum. *Applied Optics*, 2020, *59*(34), 10933–10943. https://doi.org/10.1364/AO.404863.

89. Saveliev, A. A.; Galeeva, E. V.; Semanov, D. A.; Galeev, R. R.; Aryslanov, I. R.; Falaleeva, T. S.; Davletshin, R. R. Adaptive Noise Model Based Iteratively Reweighted Penalized Least Squares for Fluorescence Background Subtraction from Raman Spectra. *Journal of Raman Spectroscopy*, 2022, *53*(2), 247–255. https://doi.org/10.1002/jrs.6275.

90. Chen, H.; Chen, H.; Xu, W.; Xu, W.; Neil, G. R. B.; Neil, G. R. B. An Adaptive and Fully Automated Baseline Correction Method for Raman Spectroscopy Based on Morphological Operations and Mollification. *Applied Spectroscopy*, 2019, *73*(3), 284–293.

91. Mosier-Boss, P. A.; Lieberman, S. H.; Newbery, R. Fluorescence Rejection in Raman Spectroscopy by Shifted-Spectra, Edge Detection, and FFT Filtering Techniques. *Applied Spectroscopy*, 1995, *49*(5), 630–638.

92. Ramos, P. M.; Ruisánchez, I. Noise and Background Removal in Raman Spectra of Ancient Pigments Using Wavelet Transform. *Journal of Raman Spectroscopy*, 2005, *36*(9), 848–856. https://doi.org/10.1002/jrs.1370.

93. Savitzky, A.; Golay, M. J. E. Smoothing and Differentiation of Data by Simplified Least Squares Procedures. *Analytical Chemistry*, 1964, *36*(8), 1627–1639. https://doi.org/10.1021/ac60214a041.

94. Eilers, P. H. C. A Perfect Smoother. *Analytical Chemistry*, 2003, *75*(14), 3631–3636. https://doi.org/10.1021/ac034173t.

95. Whitaker, D. A.; Hayes, K. A Simple Algorithm for Despiking Raman Spectra. *Chemometrics & Intelligent Laboratory Systems*, 2018, *179*, 82–84. https://doi.org/10.1016/j.chemolab.2018.06.009.

96. Ehrentreich, F.; Sümmchen, L. Spike Removal and Denoising of Raman Spectra by Wavelet Transform Methods. *Analytical Chemistry*, 2001, *73*(17), 4364–4373. https://doi.org/10.1021/ac0013756.

97. Automatic Spike Removal Algorithm for Raman Spectra - Yao Tian, Kenneth S. Burch, 2016. https://journals.sagepub.com/doi/10.1177/0003702816671065 (accessed May 20, 2022).

98. Rinnan, Å.; Nørgaard, L.; Berg, F. van den; Thygesen, J.; Bro, R.; Engelsen, S. B. Chapter 2 Data Pre-processing. In: *Infrared Spectroscopy for Food Quality Analysis and Control*; Sun, D.-W., Ed.; San Diego: Academic Press, 2009; pp 29–50. https://doi.org/10.1016/B978-0-12-374136-3.00002-X.

99. Geladi, P.; MacDougall, D.; Martens, H. Linearization and Scatter-Correction for Near-Infrared Reflectance Spectra of Meat. *Applied Spectroscopy*, 1985, *39*(3), 491–500.

100. Lukin, S.; Tireli, M.; Lončarić, I.; Barišić, D.; Šket, P.; Vrsaljko, D.; di Michiel, M.; Plavec, J.; Užarević, K.; Halasz, I. Mechanochemical Carbon–Carbon Bond Formation That Proceeds *via* a Cocrystal Intermediate. *Chemical Communications*, 2018, *54*(94), 13216–13219. https://doi.org/10.1039/C8CC07853J.

101. Kulla, H.; Greiser, S.; Benemann, S.; Rademann, K.; Emmerling, F. Knowing When to Stop—Trapping Metastable Polymorphs in Mechanochemical Reactions. *Crystal Growth & Design*, 2017, *17*(3), 1190–1196. https://doi.org/10.1021/acs.cgd.6b01572.

102. Carta, M.; Colacino, E.; Delogu, F.; Porcheddu, A. Kinetics of Mechanochemical Transformations. *Physical Chemistry Chemical Physics*, 2020, *22*(26), 14489–14502. https://doi.org/10.1039/D0CP01658F.

103. Pelletier, M. J. Quantitative Analysis Using Raman Spectrometry. *Applied Spectroscopy*, 2003, *57*(1), 20A–42A. https://doi.org/10.1366/000370203321165133.

104. de Juan, A.; Tauler, R. Chapter 2 Multivariate Curve Resolution-Alternating Least Squares for Spectroscopic Data. In: *Data Handling in Science and Technology*; Ruckebusch, C., Ed.; Resolving Spectral Mixtures; Elsevier, 2016; Vol. 30, pp 5–51. https://doi.org/10.1016/B978-0-444-63638-6.00002-4.

105. Bjelopetrović, A.; Lukin, S.; Halasz, I.; Užarević, K.; Đilović, I.; Barišić, D.; Budimir, A.; Juribašić Kulcsár, M.; Ćurić, M. Mechanism of Mechanochemical C–H Bond Activation in an Azobenzene Substrate by PdII Catalysts. *Chemistry – A European Journal*, 2018, *24*(42), 10672–10682. https://doi.org/10.1002/chem.201802403.

106. Haferkamp, S.; Paul, A.; Michalchuk, A. A. L.; Emmerling, F. Unexpected Polymorphism during a Catalyzed Mechanochemical Knoevenagel Condensation. *Beilstein Journal of Organic Chemistry*, 2019, *15*(1), 1141–1148. https://doi.org/10.3762/bjoc.15.110.

107. Kulla, H.; Haferkamp, S.; Akhmetova, I.; Röllig, M.; Maierhofer, C.; Rademann, K.; Emmerling, F. In Situ Investigations of Mechanochemical One-Pot Syntheses. *Angewandte Chemie International Edition*, 2018, *57*(20), 5930–5933. https://doi.org/10.1002/anie.201800141.

108. Carta, M.; Delogu, F.; Porcheddu, A. A Phenomenological Kinetic Equation for Mechanochemical Reactions Involving Highly Deformable Molecular Solids. *Physical Chemistry Chemical Physics: PCCP*, 2021. https://doi.org/10.1039/d1cp01361k.

109. Ma, X.; Yuan, W.; Bell, S. E. J.; James, S. L. Better Understanding of Mechanochemical Reactions: Raman Monitoring Reveals Surprisingly Simple 'Pseudo-Fluid' Model for a Ball Milling Reaction. *Chemical Communications*, 2014, *50*(13), 1585–1581. https://doi.org/10.1039/C3CC47898J.

110. Prout, E. G.; Tompkins, F. C. The Thermal Decomposition of Potassium Permanganate. *Transactions of the Faraday Society*, 1944, *40*, 488–498. https://doi.org/10.1039/TF9444000488.

111. Brown, M. E.; Glass, B. D. Pharmaceutical Applications of the Prout–Tompkins Rate Equation. *International Journal of Pharmaceutics*, 1999, *190*(2), 129–131. https://doi.org/10.1016/S0378 -5173(99)00292-6.

112. Shiryayev, A. N. On the Statistical Theory of Metal Crystallization. In: *Selected Works of A. N. Kolmogorov: Volume II Probability Theory and Mathematical Statistics*; Shiryayev, A. N., Ed.; Mathematics and Its Applications (Soviet Series); Dordrecht: Springer Netherlands, 1992; pp 188–192. https://doi.org/10.1007/978-94-011-2260-3_22.

113. Avrami, M. Kinetics of Phase Change. I General Theory. *The Journal of Chemical Physics*, 1939, *7*(12), 1103–1112. https://doi.org/10.1063/1.1750380.

114. Delogu, F.; Takacs, L. Information on the Mechanism of Mechanochemical Reaction from Detailed Studies of the Reaction Kinetics. *Journal of Materials Science*, 2018, *53*(19), 13331–13342. https://doi .org/10.1007/s10853-018-2090-1.

115. Delogu, F.; Deidda, C.; Mulas, G.; Schiffini, L.; Cocco, G. A Quantitative Approach to Mechanochemical Processes. *Journal of Materials Science*, 2004, *39*(16), 5121–5124. https://doi.org/10.1023/B:JMSC .0000039194.07422.be.

116. Colacino, E.; Carta, M.; Pia, G.; Porcheddu, A.; Ricci, P. C.; Delogu, F. Processing and Investigation Methods in Mechanochemical Kinetics. *ACS Omega*, 2018, *3*(8), 9196–9209. https://doi.org/10.1021/ acsomega.8b01431.

117. Deidda, C.; Delogu, F.; Cocco, G. In Situ Characterisation of Mechanically-Induced Self-Propagating Reactions. *Journal of Materials Science*, 2004, *39*(16), 5315–5318. https://doi.org/10.1023/B:JMSC .0000039236.48464.8f.

118. Lukin, S.; Lončarić, I.; Tireli, M.; Stolar, T.; Blanco, M. V.; Lazić, P.; Užarević, K.; Halasz, I. Experimental and Theoretical Study of Selectivity in Mechanochemical Cocrystallization of Nicotinamide with Anthranilic and Salicylic Acid. *Crystal Growth & Design*, 2018, *18*(3), 1539–1541. https://doi.org/10 .1021/acs.cgd.7b01512.

119. Germann, L. S.; Arhangelskis, M.; Etter, M.; Dinnebier, R. E.; Friščić, T. Challenging the Ostwald Rule of Stages in Mechanochemical Cocrystallisation. *Chemical Science*, 2020, *11*(37), 10092–10100. https:// doi.org/10.1039/D0SC03629C.

120. Julien, P. A.; Germann, L. S.; Titi, H. M.; Etter, M.; Dinnebier, R. E.; Sharma, L.; Baltrusaitis, J.; Friščić, T. In Situ Monitoring of Mechanochemical Synthesis of Calcium Urea Phosphate Fertilizer Cocrystal Reveals Highly Effective Water-Based Autocatalysis. *Chemical Science*, 2020, *11*(9), 2350–2355. https://doi.org/10.1039/C9SC06224F.

121. Kulla, H.; Greiser, S.; Benemann, S.; Rademann, K.; Emmerling, F. In Situ Investigation of a Self-Accelerated Cocrystal Formation by Grinding Pyrazinamide with Oxalic Acid. *Molecules*, 2016, *21*(7), 911. https://doi.org/10.3390/molecules21070911.

122. Cinčić, D.; Friščić, T.; Jones, W. A Stepwise Mechanism for the Mechanochemical Synthesis of Halogen-Bonded Cocrystal Architectures. *Journal of the American Chemical Society*, 2008, *130*(24), 7524–7525. https://doi.org/10.1021/ja801164v.

123. Puccetti, F.; Lukin, S.; Užarević, K.; Colacino, E.; Halasz, I.; Bolm, C.; Hernández, J. G. Mechanistic Insights on the Mechanosynthesis of Phenytoin, a WHO Essential Medicine**. *Chemistry – A European Journal*, 2022, e202104409. https://doi.org/10.1002/chem.202104409.

124. Arhangelskis, M.; Topić, F.; Hindle, P.; Tran, R.; Morris, A. J.; Cinčić, D.; Friščić, T. Mechanochemical Reactions of Cocrystals: Comparing Theory with Experiment in the Making and Breaking of Halogen Bonds in the Solid State. *Chemical Communications*, 2020, *56*(59), 8293–8296. https://doi.org/10.1039/ D0CC02935A.

125. Caira, M. R.; Nassimbeni, L. R.; Wildervanck, A. F. Selective Formation of Hydrogen Bonded Cocrystals between a Sulfonamide and Aromatic Carboxylic Acids in the Solid State. *Journal of the Chemical Society. Perkin Transactions 2*, 1995, (12), 2213–2216. https://doi.org/10.1039/P29950002213.

126. Ojala, W. H.; Etter, M. C. Polymorphism in Anthranilic Acid: A Reexamination of the Phase Transitions. *Journal of the American Chemical Society*, 1992, *114*(26), 10288–10293. https://doi.org/10.1021/ ja00052a026.

127. Trask, A. V.; Shan, N.; Motherwell, W. D. S.; Jones, W.; Feng, S.; Tan, R. B. H.; Carpenter, K. J. Selective Polymorph Transformation via Solvent-Drop Grinding. *Chemical Communications*, 2005, (7), 880–882. https://doi.org/10.1039/B416980H.

128. Belenguer, A. M.; Lampronti, G. I.; Sanders, J. K. M. Implications of Thermodynamic Control: Dynamic Equilibrium Under Ball Mill Grinding Conditions. *Israel Journal of Chemistry*, 2021, *61*(11–12), 764–773.

129. Martinetto, P.; Bordet, P.; Descamps, M.; Dudognon, E.; Pagnoux, W.; Willart, J.-F. Structural Transformations of D-mannitol Induced by In Situ Milling Using Real Time Powder Synchrotron Radiation Diffraction. *Crystal Growth & Design*, 2017, *17*(11), 6111–6122. https://doi.org/10.1021/acs.cgd.7b01283.

130. Andersen, J. M.; Mack, J. Decoupling the Arrhenius Equation via Mechanochemistry. *Chemical Science*, 2017, *8*(8), 5447–5453. https://doi.org/10.1039/C7SC00538E.

131. Losev, E. A.; Boldyreva, E. V. The Role of a Liquid in "Dry" Co-grinding: A Case Study of the Effect of Water on Mechanochemical Synthesis in a "L-serine–Oxalic Acid" System. *CrystEngComm*, 2014, *16*(19), 3857–3866. https://doi.org/10.1039/C3CE42321B.

132. Shan, N.; Toda, F.; Jones, W. Mechanochemistry and Co-crystal Formation: Effect of Solvent on Reaction Kinetics. *Chemical Communications*, 2002, (20), 2372–2373. https://doi.org/10.1039/B207369M.

133. Friščić, T.; Reid, D. G.; Halasz, I.; Stein, R. S.; Dinnebier, R. E.; Duer, M. J. Ion- and Liquid-Assisted Grinding: Improved Mechanochemical Synthesis of Metal-Organic Frameworks Reveals Salt Inclusion and Anion Templating. *Angewandte Chemie International Edition*, 2010, *49*(4), 712–715. https://doi.org/10.1002/anie.200906583.

134. Hernández, J. G.; Bolm, C. Altering Product Selectivity by Mechanochemistry. *Journal of Organic Chemistry*, 2017, *82*(8), 4007–4019. https://doi.org/10.1021/acs.joc.6b02881.

135. Emmerling, S. T.; Germann, L. S.; Julien, P. A.; Moudrakovski, I.; Etter, M.; Friščić, T.; Dinnebier, R. E.; Lotsch, B. V. In Situ Monitoring of Mechanochemical Covalent Organic Framework Formation Reveals Templating Effect of Liquid Additive. *Chem*, 2021, *7*(6), 1639–1652. https://doi.org/10.1016/j.chempr.2021.04.012.

136. Chui, S. S.-Y.; Lo, S. M.-F.; Charmant, J. P. H.; Orpen, A. G.; Williams, I. D. A Chemically Functionalizable Nanoporous Material [Cu3(TMA)2(H2O)3]. *New Science*, 1999.

137. Germann, L. S.; Emmerling, S. T.; Wilke, M.; Dinnebier, R. E.; Moneghini, M.; Hasa, D. Monitoring Polymer-Assisted Mechanochemical Cocrystallisation through In Situ X-ray Powder Diffraction. *Chemical Communications*, 2020, *56*(62), 8743–8746. https://doi.org/10.1039/D0CC03460F.

138. Friščić, T.; Childs, S. L.; Rizvi, S. A. A.; Jones, W. The Role of Solvent in Mechanochemical and Sonochemical Cocrystal Formation: A Solubility-Based Approach for Predicting Cocrystallisation Outcome. *CrystEngComm*, 2009, *11*(3), 418–426. https://doi.org/10.1039/B815174A.

139. Tireli, M.; Juribašić Kulcsár, M.; Cindro, N.; Gracin, D.; Biliškov, N.; Borovina, M.; Ćurić, M.; Halasz, I.; Užarević, K. Mechanochemical Reactions Studied by In Situ Raman Spectroscopy: Base Catalysis in Liquid-Assisted Grinding. *Chemical Communications*, 2015, *51*(38), 8058–8061. https://doi.org/10.1039/C5CC01915J.

140. Arhangelskis, M.; Bučar, D.-K.; Bordignon, S.; Chierotti, M. R.; Stratford, S. A.; Voinovich, D.; Jones, W.; Hasa, D. Mechanochemical Reactivity Inhibited, Prohibited and Reversed by Liquid Additives: Examples from Crystal-Form Screens. *Chemical Science*, 2021, *12*(9), 3264–3269. https://doi.org/10.1039/D0SC05071G.

141. Hutchings, B. P.; Crawford, D. E.; Gao, L.; Hu, P.; James, S. L. Feedback Kinetics in Mechanochemistry: The Importance of Cohesive States. *Angewandte Chemie International Edition*, 2017, *56*(48), 15252–15256. https://doi.org/10.1002/anie.201706723.

142. Stolle, A.; Schmidt, R.; Jacob, K. Scale-Up of Organic Reactions in Ball Mills: Process Intensification with Regard to Energy Efficiency and Economy of Scale. *Faraday Discussions*, 2014, *170*(0), 267–286. https://doi.org/10.1039/C3FD00144J.

143. Ardila-Fierro, K. J.; Rubčić, M.; Hernández, J. G. Cocrystal Formation Precedes the Mechanochemically Acetate-Assisted C–H Activation with [Cp*RhCl2]2. *Chemistry – A European Journal*, 2022, *8*(27), e202200737. https://doi.org/10.1002/chem.202200737.

144. Štrukil, V.; Gracin, D.; Magdysyuk, O. V.; Dinnebier, R. E.; Friščić, T. Trapping Reactive Intermediates by Mechanochemistry: Elusive Aryl N-Thiocarbamoylbenzotriazoles as Bench-Stable Reagents. *Angewandte Chemie International Edition*, 2015, *54*(29), 8440–8443. https://doi.org/10.1002/anie.201502026.

145. Julien, P. A.; Užarević, K.; Katsenis, A. D.; Kimber, S. A. J.; Wang, T.; Farha, O. K.; Zhang, Y.; Casaban, J.; Germann, L. S.; Etter, M. et al. In Situ Monitoring and Mechanism of the Mechanochemical Formation of a Microporous MOF-74 Framework. *Journal of the American Chemical Society*, 2016, *138*(9), 2929–2932. https://doi.org/10.1021/jacs.5b13038.

146. Beamish-Cook, J.; Shankland, K.; Murray, C. A.; Vaqueiro, P. Insights into the Mechanochemical Synthesis of MOF-74. *Crystal Growth & Design*, 2021, *21*(5), 3047–3055. https://doi.org/10.1021/acs.cgd.1c00213.

147. Wilke, M.; Casati, N. A New Route to Polyoxometalates via Mechanochemistry. *Chemical Science*, 2022, *13*(4), 1146–1151. https://doi.org/10.1039/D1SC05111C.

148. Kulla, H.; Michalchuk, A. A. L.; Emmerling, F. Manipulating the Dynamics of Mechanochemical Ternary Cocrystal Formation. *Chemical Communications*, 2019, *55*(66), 9793–9796. https://doi.org/10.1039/C9CC03034D.

149. Kralj, M.; Lukin, S.; Miletić, G.; Halasz, I. Using Desmotropes, Cocrystals, and Salts to Manipulate Reactivity in Mechanochemical Organic Reactions. *Journal of Organic Chemistry*, 2021, *86*(20), 14160–14168. https://doi.org/10.1021/acs.joc.1c01811.

150. Bjelopetrović, A.; Barišić, D.; Duvnjak, Z.; Džajić, I.; Juribašić Kulcsár, M.; Halasz, I.; Martínez, M.; Budimir, A.; Babić, D.; Ćurić, M. A Detailed Kinetico-Mechanistic Investigation on the Palladium C–H Bond Activation in Azobenzenes and Their Monopalladated Derivatives. *Inorganic Chemistry*, 2020, *59*(23), 17123–17133. https://doi.org/10.1021/acs.inorgchem.0c02418.

151. Juribašić, M.; Užarević, K.; Gracin, D.; Ćurić, M. Mechanochemical C–H Bond Activation: Rapid and Regioselective Double Cyclopalladation Monitored by In Situ Raman Spectroscopy. *Chemical Communications*, 2014, *50*(71), 10287–10290. https://doi.org/10.1039/C4CC04423A.

152. Chatziadi, A.; Skorepova, E.; Kohout, M.; Ridvan, L.; Soos, M. Exploring the Polymorphism of Sofosbuvir via Mechanochemistry: Effect of Milling Jar Geometry and Material. *CrystEngComm*, 2022, *24*, 2107–2117.

153. Michalchuk, A. A. L.; Tumanov, I. A.; Boldyreva, E. V. Ball Size or Ball Mass – What Matters in Organic Mechanochemical Synthesis? *CrystEngComm*, 2019, *21*(13), 2174–2179. https://doi.org/10.1039/C8CE02109K.

154. Martins, I. C. B.; Carta, M.; Haferkamp, S.; Feiler, T.; Delogu, F.; Colacino, E.; Emmerling, F. Mechanochemical N-Chlorination Reaction of Hydantoin: In Situ Real-Time Kinetic Study by Powder X-ray Diffraction and Raman Spectroscopy. *ACS Sustainable Chemistry & Engineering*, 2021, *9*(37), 12591–12601.

155. Fischer, F.; Fendel, N.; Greiser, S.; Rademann, K.; Emmerling, F. Impact Is Important—Systematic Investigation of the Influence of Milling Balls in Mechanochemical Reactions. *Organic Process Research & Development*, 2017, *21*(4), 655–659. https://doi.org/10.1021/acs.oprd.6b00435.

156. Takahashi, R.; Hu, A.; Gao, P.; Gao, Y.; Pang, Y.; Seo, T.; Jiang, J.; Maeda, S.; Takaya, H.; Kubota, K. et al. Mechanochemical Synthesis of Magnesium-Based Carbon Nucleophiles in Air and Their Use in Organic Synthesis. *Nature Communications*, 2021, *12*(1), 6691. https://doi.org/10.1038/s41467-021-26962-w.

157. Soldatov, D. V.; Henegouwen, A. T.; Enright, G. D.; Ratcliffe, C. I.; Ripmeester, J. A. Nickel(II) and Zinc(II) Dibenzoylmethanates: Molecular and Crystal Structure, Polymorphism, and Guest- or Temperature-Induced Oligomerization. *Inorganic Chemistry*, 2001, *40*(7), 1626–1636. https://doi.org/10.1021/ic000981g.

158. Vugrin, L.; Carta, M.; Lukin, S.; Mestrovic, E.; Delogu, F.; Halasz, I. Mechanochemical Reaction Kinetics Scales Linearly with Impact Energy. *Faraday Discussions*, 2023, *241*, 217–229.

159. Ferguson, M.; Moyano, M. S.; Tribello, G. A.; Crawford, D. E.; Bringa, E. M.; James, S. L.; Kohanoff, J.; Pópolo, M. G. D. Insights into Mechanochemical Reactions at the Molecular Level: Simulated Indentations of Aspirin and Meloxicam Crystals. *Chemical Science*, 2019, *10*(10), 2924–2929. https://doi.org/10.1039/C8SC04971H.

160. Pladevall, B. S.; Aguirre, A.; Maseras, F. Understanding Ball Milling Mechanochemical Processes with DFT Calculations and Microkinetic Modeling. *ChemSusChem*, 2021, *14*(13), 2763–2768. https://doi.org/10.1002/cssc.202100491.

161. Gomollón-Bel, F. Ten Chemical Innovations That Will Change Our World: IUPAC Identifies Emerging Technologies in Chemistry with Potential to Make Our Planet More Sustainable. *Chemistry International*, 2019, *41*(2), 12–11. https://doi.org/10.1515/ci-2019-0203.

162. Do, J.-L.; Friščić, T. Chemistry 2.0: Developing a New, Solvent-Free System of Chemical Synthesis Based on Mechanochemistry. *Synlett*, 2017, *28*(16), 2066–2092. https://doi.org/10.1055/s-0036-1590854.

163. Ardila-Fierro, K. J.; Hernández, J. G. Sustainability Assessment of Mechanochemistry by Using the Twelve Principles of Green Chemistry. *ChemSusChem*, 2021, *14*(10), 2145–2162. https://doi.org/10.1002/cssc.202100478.

2

The Thermodynamics and Kinetics of Mechanochemical Reactions

A Computational Approach

Bruna S. Pladevall and Feliu Maseras

CONTENTS

2.1 Introduction

The sustained growth of data availability and computer power over the last decades has led to a steady increase in the fields of application for computational techniques. Nowadays, computational chemistry is a solidly established standard tool for the rationalization of chemical reactivity and is widely used in nearly all subdisciplines in chemistry, including homogeneous[1] and heterogeneous[2] processes. The application of computational methods has allowed a better mechanistic understanding[3–5] and rationalization of reactivity trends,[6,7] pushing the community towards more driven research. Still, there are many challenges that need to be addressed to achieve a more accurate representation of chemical reactivity.[8]

This is the case of mechanochemical reactions, where the extensive number of variables involved implies an extra layer of complexity for its computational modelling.[9] The complex nature of mechanochemistry is exemplified in Figure 2.1, where mechanochemical reactions are classified according to five crucial factors:

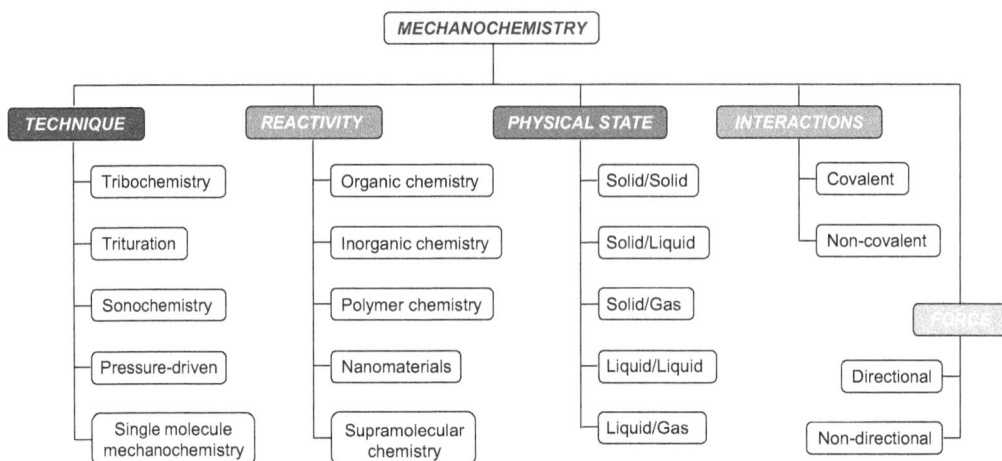

FIGURE 2.1   Classification of mechanochemistry according to the five most important features.

1. **Technique** employed to perform the transformation: (i) tribochemistry,[10] transformations activated by surface friction, (ii) trituration,[11, 12] which includes reactions induced by grinding, milling, or analogous techniques, (iii) sonochemistry,[13] those reactions activated through ultrasounds, (iv) pressure-driven transformations,[14] induced by the application of high-pressure treatments, and (v) single-molecule mechanochemistry,[15] which involves all mechanochemical techniques that apply a tensile force between different regions of a single molecule.

2. **Reactivity** that is taking place, since the effect of a force treatment might differ depending on the material or molecule employed.[16]

3. **Physical state** of the reagents involved (solid, liquid, or gas), which can strongly influence the reactivity, for instance, due to the modification of diffusion events.[17]

4. **Interaction** that is broken during the transformation can be either covalent (require forces higher than 1nN) or non-covalent (require forces lower than 1nN).[18]

5. **Force** applied can be either directional or isotropic.[19]

It is noteworthy that each mechanochemical reaction involves a specific combination of the aforementioned factors, which will directly influence the outcome of mechanochemical transformations. Consequently, the computational treatment required will also depend on the specific system that is being investigated. This chapter aims to shed light on the computational methodologies that are applied to different mechanochemical techniques. The discussion will provide a general perspective on the topic, without going deep into the technical details, since other specialized reviews have already been reported.[20–23] The chapter is organized into three different sections. After a brief introduction to the most important concepts of modern computational chemistry, two sections focus on computational mechanochemistry, which has been divided into directional and isotropic.

2.2 Modern Computational Chemistry

Before engaging directly on the computational modelling of mechanochemical reactions, we would like to provide a brief introduction to central concepts in computational chemistry in order to facilitate the discussion for readers less familiar with computational modelling. The explanation provided here is merely introductory, for a more thorough discussion on the basics of computational chemistry, the reader is referred to the book by Jensen.[24]

When a chemical process takes place, there is a reorganization of the system driven by the cleavage and formation of bonds. All these processes come accompanied by energy changes, which are central to describing a chemical transformation. The energy landscape of a particular system is based on the

Potential Energy Surface (PES), which is a hypersurface of 3N−6 dimensions (N being the number of atoms), where the energy changes according to the relative positions of the nuclei. Due to the high complexity of the PES, the chemical interpretation usually focuses on a single dimension, which corresponds to the reaction coordinate. This reaction coordinate describes the transformation from reactants to products, and the associated energy barrier is directly related to the rate constant of the process. This relationship can be done thanks to the Transition State Theory (TST),[25] which defines an equilibrium between the reactants and an activated complex, an intermediate structure between reactants and products (also known as transition state), which has higher energy (Figure 2.2, left).[26]

The activation energy that is required to achieve a given transformation can be translated into rate constants, and these rate constants, used in conjunction with concentrations, build the rate law of a particular system. The introduction of rate laws in microkinetic modelling software allows the obtention of the evolution of the concentrations of the different species through time. This leads to a theoretical kinetic model that includes the prediction of reaction times and yields, that at the same time can be easily compared with experimental data (Figure 2.2, right).[27]

The PES of a system can be explored through either static or dynamic calculations. An easy way to visualize the difference is to imagine static calculations as pictures, which define a system on a specific position of the PES, and dynamic calculations as movies, where the system evolves during a specific timelapse.[28] The method of choice will strictly depend on the system under study and the information that is needed. As a general rule, static calculations, more affordable, are employed when the behaviour of the system can be well characterized by a few selected structures, namely intermediates and transition states, while dynamic calculations, more demanding, are used when the consideration of the detailed displacement of atoms through time is necessary. In the current chapter we will mainly discuss static calculations for mechanochemical reactions, although some references to dynamic methods appear throughout the text.

The PES is based on energy calculations, where a given structure is associated with energy. There are a wide range of computational methods that can be used to perform these energy calculations, varying in price and quality of the results, among the most affordable there is molecular mechanics;[29] among the most demanding, post-Hartree–Fock (HF) methods.[30] The computational cost of such calculations scales with the size of the system and, therefore, one needs to find a good compromise between the accuracy level and the required computation time.

Figure 2.3 shows a summary of some of the most common computational methodologies which can be applied to static or dynamic calculations depending on the focus of each specific investigation. From the methods presented, the most employed one for systems of medium size like those discussed in this chapter is Density Functional Theory (DFT) calculations, which uses the electron density to describe the system rather than the wavefunction.[24] This modification with respect to HF and post-HF methods reduces the dimensionality of the problem. DFT is used in conjunction with basis functions, which can be centred in the atomic nuclei (gaussian, Slater type), for the study of molecular systems;[31] or periodic (plane waves), applied mostly for the study of solid-state reactions that involve repetitive structures.[32]

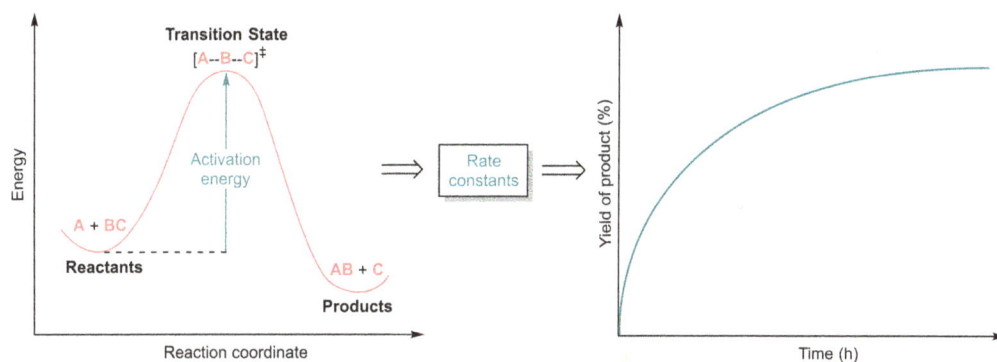

FIGURE 2.2 Schematic representation of a reaction coordinate (left) and its conversion to rate constants to obtain a theoretical kinetic model (right).

COMPUTATIONAL METHODS

Molecular mechanics
- Do not consider electrons
- Do not use Schrödinger equation
- NOT ab-initio

Quantum mechanics
- Consider electrons
- Uses Schrödinger equation

ab-initio
Without empirical parametrization

NOT ab-initio
With empirical parametrization

Hartree-Fock

Post Hartree-Fock

Density functional theory

Semiempirical

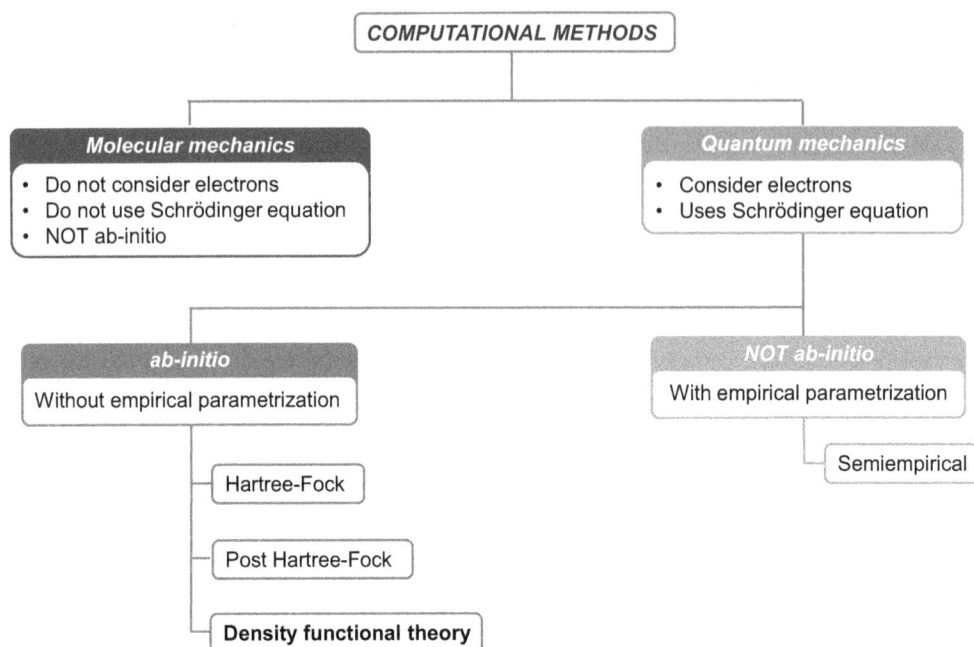

FIGURE 2.3   Summary of the most common computational methods.

2.3 Computational Studies on Directional Mechanochemistry

For the purpose of this chapter, we define directional mechanochemistry as the situation where a molecule is subject to a single directional tensile force at the microscopic level (Figure 2.4). It has been presented as single-molecule mechanochemistry or molecular mechanochemistry.[22] The simulation of these reactions is conducted through well-established techniques, which focus on the definition of a force-modified potential energy surface which determines the structural changes of the molecules or materials as well as the implications on the chemical reactivity.[21]

This scenario can be achieved through various single-molecule experimental techniques such as the use of Atomic Force Microscopy (AFM),[33] optical and magnetic tweezers,[34] or molecular force probes[35] and provides a completely different reactivity in comparison with thermochemical reactions. For example, it has been demonstrated the possibility to access electrocyclic reactions that are 'forbidden' by the Woodward–Hoffmann rules.[36]

The computational treatment for those reactions is based on the analysis of the effect of the applied force at the molecular level, which is strongly dependent on its direction.[15] If the force is projected into the direction and sense of the desired reaction coordinate, it mostly leads to an acceleration of the reaction rate. However, if we apply the force from another direction, the molecule might present a completely distinct reactivity.[37] In both situations, the focus of computational treatments is on the deformation of the molecule, which leads to modifications in the PES due to the application of the force.

The effect of an external force can be easily visualized on a diatomic molecule which presents a single nuclear coordinate, the interatomic distance (r), and therefore it can be modelled by employing the Morse potential of the dissociation of the chemical bond (Figure 2.5, left). Upon the application of an external force (pink line), the bond moves away from the equilibrium (purple line), leading to a lower dissociation energy (D'). This simple concept was employed by Kauzmann and Eyring to introduce an extension to the Transition State Theory (TST) that considered the effect of the application of an external force into the system.[38] The concept was further refined, and is usually referred to as the Bell model.[39]

With this simple concept, it is possible to understand the effect of a force on a specific bond. Still, the assumption that the force only affects the activation barrier and not the entire PES leads to the wrong

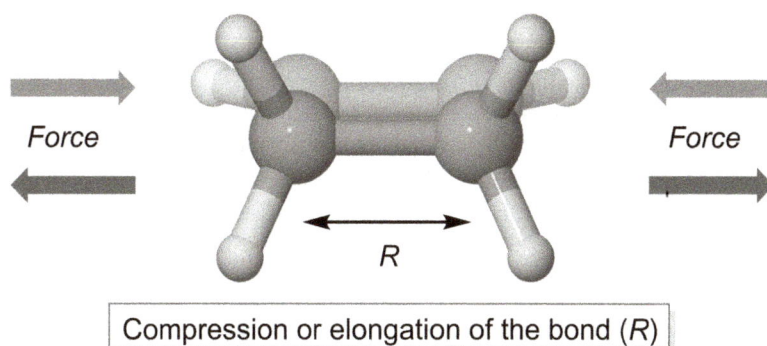

FIGURE 2.4 Schematic representation of the application of a directional tensile force to a cyclobutene molecule.

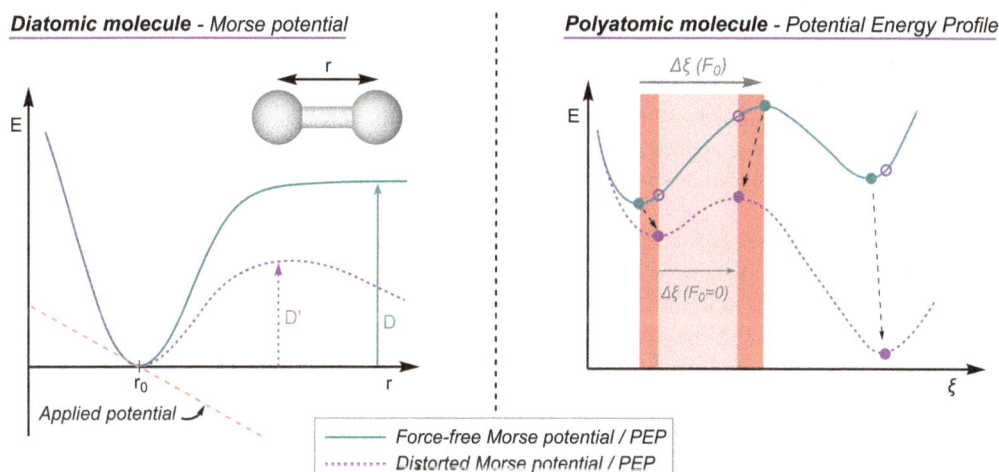

FIGURE 2.5 Force-free vs distorted energy profiles for a diatomic molecule and a polyatomic molecule.

prediction of almost all mechanochemical systems of interest. Kauzmann and Eyring addressed this problem by defining the tilted Potential Energy Profile (t-PEP) model, which accounts, in an approximate manner, for the deformation of molecule geometries (Figure 2.5, right).[40] The t-PEP model introduces the force in a more elaborate manner in which both intermediates and transition states are distorted due to the force. However, once again the force is introduced on a single reaction coordinate (ξ), and all the other molecular coordinates are ignored. This leads to an oversimplified scenario that brings the model to failure when a description of multidimensional deformations of the PES is required. This prompted the development of what is known as first-principle methods.[41]

First-principle methods are usually divided into isometric and isotensional approaches. To understand this division, we must introduce the two most important parameters when investigating directional mechanochemical reactions: (i) the breaking point distance R_{bp}, which is the inflection point where the system inevitably evolves towards the formation of the product, and (ii) the rupture force f_{max}, which corresponds to the minimum force that needs to be applied to the system to achieve the inflection point.[42] The investigation of mechanochemical systems through the modification of both variables at the same time would be extremely demanding. Thus, most treatments define an ideal system in which one of the variables is fixed. On the one hand, in isometric approaches, the bond distance is fixed, and the force is the variable measured. By contrast, in isotensional methods, the force is fixed, and the bond distance is the variable measured. Both methods have been widely employed, either separately or in combination to rationalize the outcome of directed mechanochemical transformations.[43, 44]

Although these methods have been largely applied for the investigation of mechanochemical reactions, they cannot provide information on the manner in which the force is stored in the molecule. Such information is crucial to understanding which region of the molecule is more susceptible to the external force, thereby permitting to predict mechanochemical reaction outcomes. With the aim to address this challenge, force analysis tools were developed, and are now among the most used methods to describe mechanochemical reactions, together with the aforementioned isometric and isotensional approaches.

2.3.1 Isometric Approaches

When a system is treated with an isometric approach, the force required to achieve a given bond distance is analyzed as a function of said distance, which is the control parameter. These approaches provide distorted geometries as a function of the specific coordinate under study. A representative example of isometric methodologies is the COnstrained Geometries simulate External Force (COGEF) method.

The COGEF method was developed by Beyer in 2000 and consists of a constrained optimization of the structure in which specific parameters are fixed to simulate the force.[45] This method allows a straightforward identification of the bond-breaking point, which will be the inflection point of the COGEF potential. Once the inflection point is localized, it is possible to assign the maximum force applied to the system, which as mentioned above will be the one needed to reach this inflection point. Thus, it is possible to identify which bond will be broken by comparing the forces needed in different scenarios. For example, the method was applied to $SiH_3CH_2CH_3$ and permitted the identification of the Si–C as the weakest bond within all the bonds stretched in the COGEF coordinate which are H–Si, Si–C, C–C, and C–H, as shown in Figure 2.6.

The COGEF method has provided extremely useful information on mechanochemical systems. For example, Robb and co-workers were able to predict the mechanochemical reactivity of various mechanophores previously reported in the literature, demonstrating a success rate of 88%.[46] However, it still presents some drawbacks: the rupture force determined by COGEF is highly dependent on the selection of the reaction coordinate. Moreover, the method overlooks the thermal effects, which leads to a generalized overestimation of the rupture forces. This issue has been further addressed by the development of dynamic methods, which permit the incorporation of thermal oscillations. Since dynamic methods are outside the scope of this chapter, the reader is referred to reference 23 for a more thorough discussion on the topic. Even with these accuracy limitations, the clear simplicity and reduced computational cost of COGEF have led to an extended use of the tool to investigate mechanochemical reactions, and it is still the current method of choice for many investigations.[47, 48]

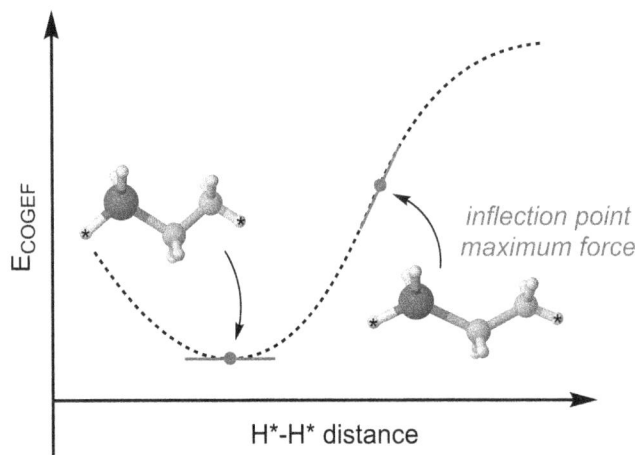

FIGURE 2.6 Qualitative representation of the COGEF potential of the $SiH_3CH_2CH_3$ molecule elongating between the two H* atoms.

2.3.2 Isotensional Approaches

The application of an isotensional approach entails the explicit consideration of the force. In these methods, the interest lies in the structural response of a molecule to the application of a mechanical input. In fact, these methods allow the determination of the exact distortion suffered by a system due to the application of an external force along a determinate coordinate, providing detailed information on both energetics and mechanisms of mechanochemical transformations. The most relevant and used isotensional approaches appeared during 2009, almost simultaneously, and were named Force-Modified Potential Energy Surface (FMPES), External Force is Explicitly Included (EFEI), and Enforced Geometry Optimization (EGO). All these techniques have been largely applied to the study of mechanochemical reactions and enabled the rationalization of counterintuitive experimental results providing insight into the chemical rearrangements induced by the absorption of mechanical energy.[49,][50] In the following sections we will briefly introduce the three methodologies and present the main differences between them.

2.3.2.1 Force-Modified Potential Energy Surface

The FMPES method was introduced by Martínez and co-workers when in the investigation of a cyclobutene ring opening they identified the poor description of ab initio Steered Molecular Dynamic methods to investigate low-force regime mechanochemical transformations.[51] The authors claimed that in these scenarios, a description based on the TST is optimal; therefore, they performed a multidirectional extension of the preliminary approach defined by Kauzmann–Eyring.[38]

The FMPES is the first method that incorporates an explicit force for the calculation of a deformed potential energy surface, required for the computational treatment of mechanochemical reactions. In this approach, the added external force is a constant force defined through various unit vectors between the atoms of interest, named the attachment point (AP) and a given position in space called the pulling point (PP).[22] Although in the initial method the force applied to the system was constant, the concept was further extended to consider non-constant pulling forces, which are critical for the representation of some mechanochemical systems.[52] A schematic representation of FMPES is shown in Figure 2.7, where the external forces are being applied to atoms 1 and 2, and the cross represents the pulling points that determine the direction of the vector.

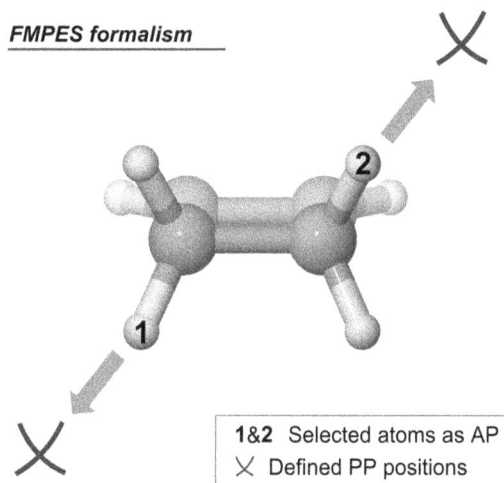

FMPES formalism

1&2 Selected atoms as AP
X Defined PP positions

FIGURE 2.7 Schematic representation of the method employed to apply the external force within the FMPES formalism.

EFEI formalism

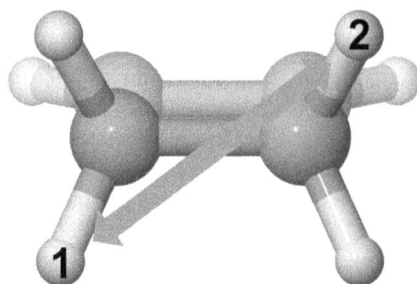

1&2 Selected atoms as AP and PP

FIGURE 2.8   Schematic representation of the method employed to apply the external force within the EFEI formalism.

2.3.2.2 External Force Is Explicitly Included

The EFEI method was introduced by Marx and co-workers,[53] and has a similar basis to the previously described FMPES approach. The most notorious difference is the manner in which the force is introduced. While in FMPES it is necessary to select the AP and the PP to determine the direction of the force (Figure 2.7), in EFEI the force is defined as a vector between two selected atoms which are at the same time the AP and PP. A schematic representation of the EFEI method on cyclobutene is shown in Figure 2.8. This use of the internal molecular coordinates to define the direction is probably the highest advantage of this method, since it allows what we call an adaptative pulling; the direction of the force is modified, while the molecule is stretched. This strategy has been successfully applied to numerous systems, such as the investigation of the mechanically induced Hückel–Möbius interconversions on hexaphyrin, which helped in the rationalization of the topology preferred depending on the pulling scenario.[54]

2.3.2.3 Enforced Geometry Optimization

The EGO method was developed by the group of Baker.[55, 56] The EGO approach resembles the EFEI in the manner in which the external force is introduced, by defining a vector between two selected atoms. However, EGO permits the incorporation of multiple forces. This can be seen in Figure 2.9, where EGO has been applied to a benzene ring where all the atoms are affected by the external force. This method has been also applied in several investigations such as the conformational changes in the structural units of glycosaminoglycans (GAGs).[57]

2.3.3 Force Analysis Tools

Both isometric and isotensional approaches described in Sections 2.3.1 and 2.3.2 have been shown as highly useful in the description of mechanically deformed molecules providing access to distorted geometries, as well as force-modified potential energy surfaces. This led to fundamental insights into mechanically activated transformations. However, these methods fail to determine which part of a molecule is more susceptible to being deformed by the application of external force. This information would be extremely useful to drive experimental mechanochemistry towards the achievement of a desired product, for the rational design of mechanochemical systems such as stress-responsive materials, and to understand diverse mechanochemical pathways. In this context, force analysis tools were developed with the aim of quantifying forces and determining which part of a molecule will store the largest part of mechanical stress. In this way this approach allows us to understand how these modifications on the molecule will affect the mechanochemical transformations.[23]

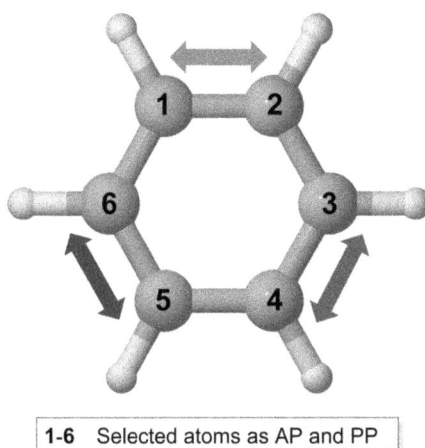

FIGURE 2.9 Schematic representation of the method employed to apply the external force within the EGO formalism.

The identification of mechanically relevant degrees of freedom requires the calculation of the restoring force for every internal coordinate of a molecule. This implies a major problem which lies on the selection of the coordinate system.[58] In order to describe accurately the mechanical stress, it is necessary a highly localized coordinate system. In this regard, the use of internal coordinates would be optimal; however, the redundancy of internal coordinates involves a huge concern. In mechanochemistry, the employment of redundant coordinates implies that a displacement of one coordinate might generate additional forces in other coordinates, leading to wrong and even sometimes unphysical results.[59] In the following section we will describe some of the methods developed aiming to circumvent this problem by permitting a cost-reduced treatment of mechanochemical degrees of freedom, and for a more throughout discussion, the reader is referred to the references.[60]

One of the first approximations was to consider very few or one single coordinate within the force analysis, thus ignoring the effect of the mechanical stress on the rest of the molecule.[61, 62] However, there is no standardized procedure for choosing the relevant coordinate, and it is necessary to evaluate each system independently to select the best option. Such a choice will have critical effects on the results.[63] Therefore, the analysis must be done with extreme care.

Another option to solve the aforementioned problem related to the coordinate system was reported by Avdoshenko and co-workers, and consists of the quantification of internal forces.[59] Their approach defines a set of interatomic distances that are particularly important for the mechanical process, and then, new coordinates are added following a random strategy. After each step, the redundant coordinates are located and removed to avoid interdependency problems. However, this implies a high dependency on the coordinates included since there are multiple possibilities in the definition of the system. This leads to failure on a high-force regime, where it can even provide unphysical results. Hence, the authors claimed that the strategy is limited to a weak force regime applied to large molecules, where the molecule distortion is small and the coordinate system is not highly affected by the distribution of the force.

The use of compliance forces is another option to define the coordinate system of mechanochemical reactions. With this methodology, instead of using internal coordinates, one calculates the force constants which can only be defined as normal modes. Hence, the problem we face here is different, since it lies in the difficulty of localizing the mechanical stress and thus, a highly difficult analysis of the results. An interesting point to take into account is that the compliance forces can be related to the bond strength, with stronger bonds presenting small forces.[64, 65] Therefore, this analysis can be used to quantify the strength of chemical bonds, which is crucial to rationalize mechanochemical transformations.

An alternative approach to those presented above is to focus on the stress energy instead of the forces. Since the energy is a scalar, the analysis of the results is facilitated. Stauch and Dreuw employed this methodology to define the Judgement of Energy Distribution (JEDI) analysis,[60, 66] which allows the quantification of the distribution of stress energy in a mechanically deformed molecule. This force analysis tool has been used to rationalize numerous systems, and the different applications were reviewed by the developers of the method.[67]

2.4 Computational Studies on Isotropic Mechanochemistry

Isotropic mechanochemistry, also called bulk mechanochemistry,[22] includes all the mechanochemical techniques not discussed in the previous section: tribochemistry, trituration, sonochemistry, and high-pressure transformations. In these reactions, the force is not transferred in a single direction (it has an isotropic nature); hence, its simulation is in principle not straightforward. Furthermore, each experimental technique has its own peculiarities and must be computed in different. It becomes more complex to match the microscopic with the macroscopic phenomena, a fact that complicates its computational analysis. Indeed, isotropic mechanochemistry is by far less understood from a computational view than the directional variety described in the previous section.[22]

The most remarkable techniques within isotropic mechanochemistry are tribochemistry, trituration, sonochemistry, and high-pressure transformations. Although they are classified under the same category, the mechanical action is transferred through extremely different tools in each of these treatments. Thus, the features that must be taken into account as well as the computational treatment required differ from one technique to another. For this reason, we will introduce them in different sections.

2.4.1 Tribochemistry

The term tribochemistry refers to those reactions that are activated by compression and shear which typically occur between a lubricant and a surface (Figure 2.10). While the force applied in tribochemical reactions is directional, the stress cannot be converted into a directional force at a molecular level. Indeed, every atom of the system will feel the stress from a specific direction depending on its orientation. Thus, it is not possible to treat it through the same methods used for single-molecule mechanochemistry.

The frictional process involved in tribochemical reactions leads to the formation of the so-called 'tribofilm', a protective surface formed upon the submission of a system to tribological stress. The understanding of tribofilms as well as tribochemical mechanisms is key to the control of frictional processes, and the improvement of lubricant design.[68] The computational modelling of tribochemical reactions focuses on the rationalization of interactions between different molecules in the lubricant, between the lubricant and the surface, or between the two contacting surfaces.[69, 70] This can be investigated through static calculations, which focus on structural, thermodynamic, and kinetic features, or dynamic calculations, which give insight into dynamical and temperature effects. The current section will only introduce the basis of static calculations for tribochemical processes. For a more comprehensive overview of computational tribochemistry, we refer the reader to the review by Ta and co-workers.[71]

The investigation of tribochemical reactions through static calculations typically involves the use of plane waves. There is a setup with specific boundary conditions in which the lubricant is delimited by two slabs that will apply the specific shear or compression force (Figure 2.10). The PES of the system can be obtained through a set of static calculations where the slabs are moved following the direction of the force. Through the analysis of the PES obtained, it is possible to quantify the slip or compression barrier

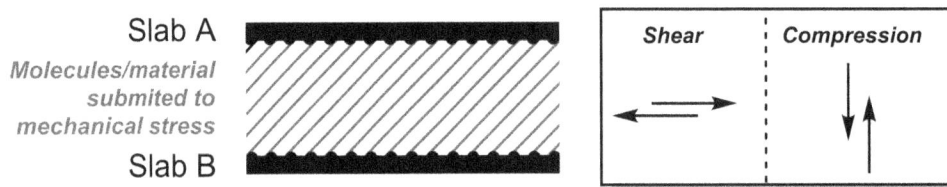

FIGURE 2.10 Schematic presentation of the setup of a tribochemical reaction.

and determine the preferred pathways along competing directions.[72] Such modelling through static calculations presents a major drawback: the large number of atoms involved in tribochemical transformations. This fact makes the use of highly accurate methods difficult due to the huge computational cost, and hence, it complicates the study of some tribochemical systems. This issue has been further addressed using Quantum Mechanics/Molecular Mechanics (QM/MM) calculations;[73, 74] however, a lot of challenges remain in this area.

2.4.2 Trituration

Trituration techniques, including grinding, milling and all the analogous methods, are the most applied in synthesis. However, the effect of the force in this type of transformation has been barely analyzed from a computational perspective. In fact, most of the scarce computational studies on trituration-activated systems utilize standard DFT calculations in the gas phase, without justifying the method of choice.[75] The development of a standardized method to compute these transformations is further hampered by the confusion on how the macroscopic phenomenon is transferred to a molecular level.

In trituration reactions the force is applied in a completely random manner. Hence, the mechanical stress is conferred to the molecule from different directions, which moreover changes along the reaction course (Figure 2.11). If we assume that the macroscopic force is directly translated into a microscopic level, the non-directional nature of the force combined with the anisotropic character of most molecules would imply that every single collision leads to a different mechanical effect. An oversimplistic interpretation could indicate that those collisions following the same direction as the reaction coordinate would increase the rate, and the others modify the PES, leading presumably to a complex mixture of products and a huge number of competing pathways. This scenario is not consistent with trituration processes where the reaction outcome is often similar to that obtained in solution.[76] Indeed, the commonly reported differences between trituration and solution chemistry are reaction times, yields and selectivity, and the reporting of scenarios with completely different reaction outcomes has been scarce.[77, 78]

Other hypotheses to understand this type of reactivity have been proposed over the last few years. Some of them, such as the presence of extreme local pressures or temperatures in the collision point, have already been discarded at least for some trituration systems.[79] Others, such as an effect on the pre-exponential factor of the Arrhenius equation driven by the collisions, are still on the table and represent a primary topic of discussion.[80] This term would be linked to the frequency applied to the system, a crucial parameter of trituration reactions whose role is still far from being fully understood. To the best of our knowledge, these treatments have not yet led to a satisfactory unified view of the computational modelling of these transformations.

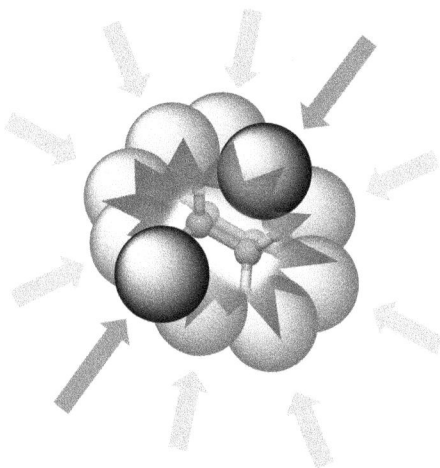

FIGURE 2.11 Schematic presentation of the mechanical stress conferred in a trituration setup.

Recently, our group proposed a methodology to compute ball milling reactions, presumably extensible to other trituration techniques, through the combination of standard DFT tools and microkinetic modelling.[81] Our approach is based on the following hypothesis: 'Ball-milling transformations are equivalent to those in solution, with differences found in the dielectric medium and the concentrations'. With this idea as a starting point, we defined an approach to tackle the two main differences between ball milling reactions and solution chemistry.

On the one hand, we defined the medium of a ball milling transformation as a dielectric environment described through the mixture of the dielectric constants of the reagents (and the Liquid Assisted Grinding [LAG] additive, when applicable). In this way, the medium would have an impact on the reaction which could lead to modifications in reaction times upon the modification of the LAG employed, a scenario that would agree with experimental observations.[82]

On the other side, the concentrations were approximated using the densities of the reagents. This approach defines an 'ideal' concentration in which the system would be fully homogenized. This is far from being exact, as the homogenization is likely only partial. It should depend on the frequency applied, which determines the collisions that promote the mixture of the reagents, as well as the breakage of the solid, when applicable. In a real system, both the physical state and nature of the reagent will directly affect the manner in which the reagents are homogenized. Although this could be approached through molecular dynamic calculations,[83] we focus here on the modelling of ball milling reactions in the most straightforward manner possible and the analysis of how accurate the method can be. Overall, the calculation of ball milling reactions would follow the workflow shown in Figure 2.12.

The method described above was found to be accurate for the reproduction of reaction times and yields of two different organic reactions that had been previously reported in the literature: Diels Alder[84] and N-sulfonylguanidine synthesis.[85] These positive results support the validity of the methodology and point to important similarities between ball milling and solution chemistry in terms of mechanism. Overall, this simple approach to ball milling reactions is a promising tool to be able to gain more understanding of this type of reactivity leading to a more rational design.

2.4.3 Sonochemistry

Sonochemical reactions are those transformations activated with ultrasounds. The treatment of a liquid phase with ultrasounds creates the so-called acoustic cavitation phenomenon, a process exemplified in Figure 2.13. That is to say, the generation of compression waves which lead to the formation and growth of gaseous microbubbles. At a certain size, these bubbles collapse creating extreme local conditions that are the driving force of sonochemical transformations. These extreme conditions created due to the bubble implosion are a central topic of discussion and are not fully understood yet, since it is not

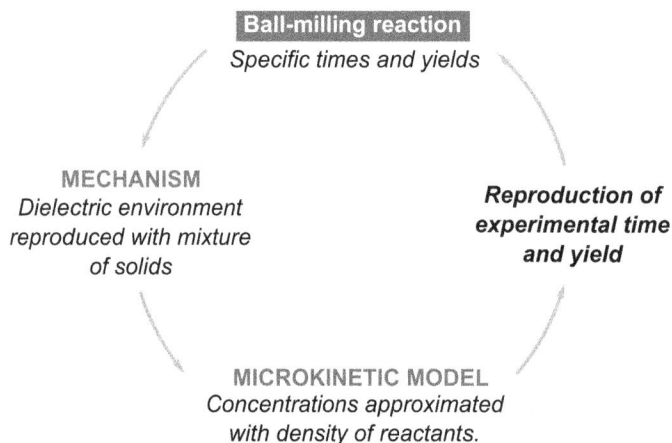

Ball-milling reaction
Specific times and yields

MECHANISM
*Dielectric environment
reproduced with mixture
of solids*

**Reproduction of
experimental time
and yield**

MICROKINETIC MODEL
*Concentrations approximated
with density of reactants.*

FIGURE 2.12 Workflow for the calculation of ball milling reactions.

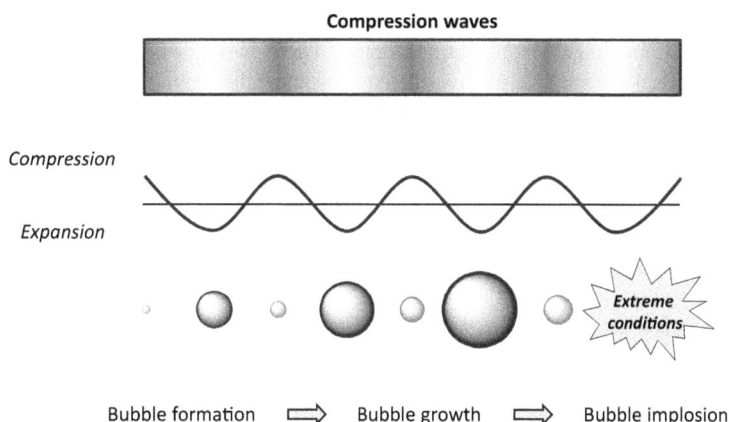

FIGURE 2.13 Schematic representation of the acoustic cavitation phenomenon, created due to the propagation of ultrasounds in a liquid phase.

straightforward to bridge the ultrasound with its molecular effect. However, in this case, in contrast to the ball milling reactions described in the previous section, the presence of these extreme local conditions is supported by the formation of radicals, sonoluminescence events, and in general, the obtention of completely different reactivity in comparison to thermochemical reactions.[86]

The acoustic cavitation process involves the creation of a non-directional force, specifically in ultrasound baths, and therefore has been classified in this section. In spite of that, several sonochemical systems have been computed through the directional methods explained in Section 2.2.[87] Overall, the computational investigation of sonochemical reactions is non-trivial. In fact, the strategy has been referred to as a *black art*, since it is difficult to predict what is going to be the outcome of a transformation in comparison to other activation methods.[88] Different numerical simulations for sonochemistry have been proposed, which have been recently reviewed by Yasui.[89] This theoretical vision of sonochemical reactions involves the use of dynamic studies to simulate the effect of the shock wave, and the bubble evolution.

2.4.4 High-Pressure Mechanochemistry

High-pressure calculations are another option for modelling mechanochemical reactions.[90] Although these techniques have also been included as a subsection, the treatment of these transformations is generally more straightforward. Still, we believe they help to exemplify the vast amount of reactions included under the term mechanochemistry. A brilliant example of this type of reactivity is that of Yan et al. in which the authors were able to convert the hydrostatic pressure, which involves a macroscopic isotropic force into a directional force at a molecular level. This was done through the design of engineered molecules composed of mechanically 'soft' and 'hard' parts.[91]

2.5 Conclusions

Computational chemistry has shown its potential as a tool for the understanding and prediction of mechanochemical processes, and it has a promising role to fulfil in the future. The diversity of phenomena included under the mechanochemistry label requires the application of significantly different techniques. Specifically, two different families of processes can be identified. In directional mechanochemistry, molecules are submitted to a single directional force, and the successful computational treatment requires the explicit introduction of the forces in the Hamiltonian, either in a direct or indirect form. In isotropic mechanochemistry, which includes the widely applied ball milling, forces act in all directions on the molecules. In the isotropic case, the main role of the external force seems to be the homogenization of the system, with the difference between mechanochemical and solvent conditions coming down to changes in concentrations and dielectric environment. There is still work ahead to further refine the computational methods necessary for the treatment of mechanochemical processes, but early returns are promising, and we should expect steady progress in the field in the next few years.

REFERENCES

1. Funes-Ardoiz, I.; Schoenebeck, F. (2020). Established and Emerging Computational Tools to Study Homogeneous Catalysis-From Quantum Mechanics to Machine Learning. *Chem* 6(8), 1904–1912. (DOI: 10.1016/k.chempr.2020.07.008)
2. Chen, B. W. J.; Xu, L.; Mavrikakis, M. (2021). Computational Methods in Hetetogeneous Catalysis. *Chem. Rev.* 121(2), 1007–1048. (DOI: 10.1021/acs.chemrev.0c01060)
3. Rull, S. G.; Funes-Ardoiz, I.; Maya, C.; Maseras, F.; Fructos, M. R.; Belderrain, T. R.; Nicasio, C. (2018). Elucidating the Mechanism of Aryl Aminations Mediated by NHC-Supported Nickel Complexes: Evidence for a Nonradical Ni(0)/Ni(II) Pathway. *ACS Catal.* 8(5), 3733–3742. (DOI: 10.1021/acscatal.8b00856)
4. Mato, M.; García-Morales, C.; Echavarren, A. M. (2020). Synthesis of Trienes by Rhodium-Catalyzed Assembly and Disassembly of Non-acceptor Cyclopropanes. *ACS Catal.* 10(6), 3564–3570. (DOI: 10.1021/acscatal.0c00006)
5. López-Resano, S.; Martínez de Salinas, S.; Garcés-Pineda, F. A.; Moneo-Corcuera, A.; Galán-Mascarós, J. R.; Maseras, F.; Pérez-Temprano, M. H. (2021). Redefining the Mechanistic Scenario of Carbon–Sulfur Nucleophilic Coupling via High-Valent Cp*Co^{IV} Species. *Angew. Chem. Int. Ed.* 60(20), 11217–11221. (DOI: 10.1002/anie.202101390)
6. Wodrich, M. D.; Sawatlon, B.; Busch, M.; Corminboeuf, C. (2021). The Genesis of Molecular Volcano Plots. *Chem. Sci.* 54(5), 1107–1117. (DOI: 10.1021/acs.accounts.0c00857)
7. Morán-González, L.; Besora, M.; Maseras, F. (2022). Seeking the Optimal Descriptor for SN2 Reactions through Statistical Analysis of Density Functional Theory Results. *J. Org. Chem.* 87(1), 363–372. (DOI: 10.1021/acs.joc.1c02387)
8. Harvey, J. N.; Himo, F.; Maseras, F.; Perrin, L. (2019). Scope and Challenge of Computational Methods for Studying Mechanism and Reactivity in Homogeneous Catalysis. *ACS Catal.* 9(8), 6803–6812. (DOI: 10.1021/acscatal.9b01537)
9. Suslick, K. S. (2014). Mechanochemistry and Sonochemistry: Concluding Remarks. *Faraday Discuss.* 170, 411–422. (DOI: 10.1039/C4FD00148F)
10. He, X.; Kim, S. H. (2017). Mechanochemistry of Physisorbed Molecules at Tribological Interfaces: Molecular Structure Dependence of Tribochemical Polymerization. *Langmuir* 33(11), 2717–2724. (DOI: 10.1021/acs.langmuir.6b04028)
11. Nicholson, W. I.; Howard, J. L.; Magri, G.; Seastram, A. C.; Khan, A.; Bolt, R. R. A.; Morrill, L. C.; Richards, E.; Browne, D. L. (2021). Ball-Milling-Enabled Reactivity of Manganese Metal. *Angew. Chem. Int. Ed.* 133, 23312–23317. (DOI: 10.1002/anie.202108752)
12. Yang, L.; Moores, A.; Friščić, T.; Provatas, N. (2021). Thermodynamics Model for Mechanochemical Synthesis of Gold Nanoparticles: Implications for Solvent-Free Nanoparticle Production. *ACS Appl. Nano Mater.* 4(2), 1886–1897. (DOI: 10.1021/acsanm.0c03255)
13. Boswell, B. R.; Mansson, C. M. F.; Cox, J. M.; Jin, Z.; Romaniuk, J. A. H.; Lindquist, K. P.; Cegelski, L.; Xia, Y.; Lopez, S. A.; Burns, N. Z. (2021). Mechanochemical Synthesis of an Elusive Fluorinated Polyacetylene. *Nat. Chem.* 13(1), 41–46. (DOI: 10.1038/s41557-020-00608-8)
14. Zeidler, A.; Salmon, P. S. (2016). Pressure-Driven Transformation of the Ordering in Amorphous Network-Forming Materials. *Phys. Rev. B* 93(21), 214204. (DOI: 10.1103/PhysRevB.92.214204)
15. Li, H.; Walker, G. C. (2017). Twist and Shout: Single-Molecule Mechanochemistry. *ACS Nano* 22(1), 28–30. (DOI: 10.1021/acsnano.6b08562)
16. Tan, D.; García, F. (2019). Main Group Mechanochemistry: From Curiosity to Established Protocols. *Chem. Soc. Rev.* 48(8), 2274–2292. (DOI: 10.1039/C7CS00813A)
17. Bolm, C.; Hernández, J. G. (2019). Mechanochemistry of Gaseous Reactants. *Angew. Chem. Int. Ed.* 58(11), 3285–3299. (DOI: 10.1002/anie.201810902)
18. Neill, R. T. O.; Boulatov, R. (2021). The Many Flavours of Mechanochemistry and Its Plausible Conceptual Underpinnings. *Nat. Rev. Chem.* 5(3), 148–167. (DOI: 10.1038/s41570-020-00249-y)

19. Makarov, D. E. (2016). Perspective: Mechanochemistry for Biological and Synthetic Molecules. *J. Chem. Phys.* 144(3), 030901. (DOI: 10.1063/1.4939791)

20. Beyer, M. K.; Clausen-Schaumann, H. (2005). Mechanochemistry: The Mechanical Activation of Covalent Bonds. *Chem. Rev.* 105(8), 2921–2948. (DOI: 10.1021/cr030697h)

21. Ribas-Arino, J.; Marx, D. (2012). Covalent Mechanochemistry: Theoretical Concepts and Computational Tools with Applications to Molecular Nanomechanics. *Chem. Rev.* 112(10), 5412–5487. (DOI: 10.1021/cr200399q)

22. Kochhar, G. S.; Heverly-Coulson, G. S.; Mosey, N. J. (2015). Theoretical Approaches for Understanding the Interplay Between Stress and Chemical Reactivity. *Top. Curr. Chem.* 369, 37–96. (DOI: 10.1007/128_2015_648)

23. Neudecker, T.; Dreuw, A. (2016). Advances in Quantum Mechanochemistry: Electronic Structure Methods and Force Analysis. *Chem. Rev.* 116(22), 14137–14180. (DOI: 10.1021/acs.chemrev.6b00458)

24. Jensen, F. (2017). *Introduction to Computational Chemistry*, 3rd edition. Chichester: John Wiley & Sons Ltd. (ISBN: 978-1-118-82599-0)

25. Eyring, H. (1935). The Activated Complex in Chemical Reactions. *J. Chem. Phys.* 3(2), 107–115. (DOI: 10.1063/1.1749604)

26. Dougherty, D. A.; Anslyn, E. (2006). *Modern Physical Organic Chemistry*. Mill Valley, CA: University Science Books. (ISBN: 978-1891389313)

27. Besora, M.; Maseras, F. (2018). Microkinetic Modeling in Homogeneous Catalysis. *WIREs Comput. Mol. Sci.* 8(6), e1372. (DOI: 10.1002/wcms.1372)

28. Paquet, E.; Viktor, H. L. (2014). Molecular Dynamics, Monte Carlo Simulations, and Langevin Dynamics: A Computational Review. *BioMed Res. Int.* 2015, 183918. (DOI: 10.1155/2015/183918)

29. Poltev, M. (2015). Molecular Mechanics: Principles, History, and Current Status. In: Leszczynski, J. (ed.) *Handbook of Computational Chemistry*. Dordrecht: Springer. (DOI: 10.1007/978-94-007-6169-8_9-2)

30. Blinder, S. M.; House, J. E. (eds.) (2019). Mathematical Physics in Theoretical Chemistry. Amsterdam: Elsevier. (ISBN 978-0-12-813651-5)

31. Lledós, A. (2021). Computational Organometallic Catalysis: Where We Are, Where We Are Going. *Eur. J. Inorg. Chem.* 26, 2547–2555. (DOI: 10.1002/ejic.202100330)

32. Monteiro, M. C. O.; Dattila, F.; López, N.; Koper, M. T. M. (2022). The Role of Cation Acidity on the Competition between Hydrogen Evolution and CO_2 Reduction on Gold Electrodes. *J. Am. Chem. Soc.* 144(4), 1589–1602. (DOI: 10.1021/jacs.1c10171)

33. Schulz, F.; Ritala, J.; Krejčí, O.; Seitsonen, A. P.; Foster, A. S.; Liljeroth, P. (2018). Elemental Identification by Combining Atomic Force Microscopy and Kelvin Probe Force Microscopy. *ACS Nano* 12(6), 5274. (DOI: 10.1021/acsnano.7b08997)

34. Conroy, R. (2008). Force Spectroscopy with Optical and Magnetic Tweezers. In: Noy, A. (ed.) *Handbook of Molecular Force Spectroscopy*. Springer. (ISBN: 978-1-4419-4323-1)

35. Qing-Zheng, Y.; Huang, Z.; Kucharski, T. J.; Khvostichenko, D.; Chen, J.; Boulatov, R. (2009). A Molecular Force Probe. *Nat. Nanotechnol.* 4(5), 302–306. (DOI: 10.1038/nnano.2009.55)

36. Hickenboth, C. R.; Moore, J. S.; White, S. R.; Sottos, N. R.; Baudry, J.; Wilson, S. R. (2007). Biasing Reaction Pathways with Mechanical Force. *Nature* 446(7134), 423–427. (DOI: 10.1038/nature05681)

37. Akbulatov, S.; Tian, Y.; Huang, Z.; Kucharski, T. J.; Yang, Q.-Z.; Boulatov, R. (2017). Experimentally Realized Mechanochemistry Distinct from Force-Accelerated Scission of Loaded Bonds. *Science* 357(6348), 299–302. (DOI: 10.1126/science.1193412)

38. Kauzmann, W.; Eyring, H. (1940). The Viscous Flow of Large Molecules. *J. Am. Chem. Soc.* 62(11), 3113–3125. (DOI: 10.1021/ja01868a059)

39. Bell, G. I. (1978). Models for the Specific Adhesion of Cells to Cells. *Science* 200(4342), 618–627. (DOI: 10.1126/science.347575)

40. Evans, E.; Ritchie, K. (1997). Dynamic Strength of Molecular Adhesion Bonds. *Biophys. J.* 72(4), 1541–1555. (DOI: 10.1016/S0006-3495(97)78802-7)

41. Kulik, H. J. (2018). Modeling Mechanochemistry from First Principles. In: Parrill, A. L.; Lipkowitz, K. B. (eds.) *Reviews in Computational Chemistry*, Volume 31. (DOI: 10.1002/9781119518068.ch6)

42. Quapp, W.; Bofill, J. M.; Ribas-Ariño, J. (2018). Towards a Theory of Mechanochemistry: Simple Models from the Very Beginnings. *Int. J. Quantum Chem.* 118(23), e25775. (DOI: 10.1002/qua.25775)

43. Dopieralski, P.; Ribas-Arino, J.; Anjukandi, P.; Krupicka, M.; Marx, D. (2017). Unexpected Mechanochemical Complexity in the Mechanistic Scenarios of Disulfide Bond Reduction in Alkaline Solution. *Nat. Chem.* 9(2), 164–170. (DOI: 10.1038/nchem.2632)

44. Nixon, R.; De Bo, G. (2020). Three Concomitant C–C Dissociation Pathways During the Mechanical Activation of an N-Heterocyclic Carbene Precursor. *Nat. Chem.* 12(9), 826–831. (DOI: 10.1038/s41557-020-0509-1)

45. Beyer, M. K. (2000). The Mechanical Strength of a Covalent Bond Calculated by Density Functional Theory. *J. Chem. Phys.* 112(17), 7307–7312. (DOI: 10.1063/1.481330)

46. Klein, I. M.; Husic, C. C.; Kovács, D. P.; Choquette, N. J.; Robb, M. J. (2020). Validation of the CoGEF Method as a Predictive Tool for Polymer Mechanochemistry. *J. Am. Chem. Soc.* 142(38), 16364–16381. (DOI: 10.1021/jacs.0c06868)

47. Sha, Y.; Zhang, Y.; Xu, E.; McAlister, C. W.; Zhu, T.; Craig, S. L.; Tang, C. (2019). Generalizing Metallocene Mechanochemistry to Ruthenocene Mechanophores. *Chem. Sci.* 10(19), 4959–4965. (DOI: 10.1039/C9SC01347D)

48. Bettens, T.; Alonso, M.; Geerlings, P.; De Proft, F. (2019). Implementing the Force into the Conceptual DFT Framework: Understanding and Predicting Molecular Mechanochemical Properties. *Phys. Chem. Chem. Phys.* 21(14), 7378–7388. (DOI: 10.1039/C8CP07349J)

49. Beedle, A. E. M.; Mora, M.; Davis, C. T.; Snijders, A. P.; Stirnemann, G.; Garcia-Manyes, S. (2018). Forcing the Reversibility of a Mechanochemical Reaction. *Nat. Commun.* 9(1), 3155–3164. (DOI: 10.1038/s41467-018-05115-6)

50. Chen, Y.; Mellot, G.; van Luijk, D.; Creton, C.; Sijbesma, R. P. (2021). Mechanochemical Tools for Polymer Materials. *Chem. Soc. Rev.* 50(6), 4100. (DOI: 10.1039/D0CS00940G)

51. Ong, M. T.; Leiding, J.; Tao, H.; Virshup, A. M.; Martínez, T. J. (2009). First Principles Dynamics and Minimum Energy Pathways for Mechanochemical Ring Opening of Cyclobutene. *J. Am. Chem. Soc.* 131(18), 6377–6379. (DOI: 10.1021/ja8095834)

52. Subramanian, G.; Mathew, N.; Leiding, J. (2015). A Generalized Force-Modified Potential Energy Surface for Mechanochemical Simulations. *J. Chem. Phys.* 143(13), 134109. (DOI: 10.1063/1.4932103)

53. Ribas-Arino, J.; Shiga, M.; Marx, D. (2009). Understanding Covalent Mechanochemistry. *Angew. Chem. Int. Ed.* 48(23), 4190–4192. (DOI: 10.1002/anie.200900673)

54. Bettens, T.; Hoffmann, M.; Alonso, M.; Geerlings, P.; Dreuw, A.; De Proft, F. (2021). Mechanochemically Triggered Topology Changes in Expanded Porphyrins. *Chem. Eur. J.* 27(10), 3397–3406. (DOI: 10.1002/chem.202003869)

55. Wolinski, K.; Baker, J. (2009). Theoretical Predictions of Enforced Structural Changes in Molecules. *Mol. Phys.* 107(22), 2403–2417. (DOI: 10.1080/00268970903321348)

56. Wolinski, K.; Baker, J. (2010). Geometry Optimization in the Presence of External Forces: A Theoretical Model for Enforced Structural Changes in Molecules. *Mol. Phys.* 108(14), 1845–1856. (DOI: 10.1080/00268976.2010.492795)

57. Brzyska, A.; Woliński, K. (2014). Enforced Conformational Changes in the Structural Units of Glycosaminoglycan (Non-sulfated Heparin-Based Oligosaccharides). *RSC Adv.* 4(69), 36640–36648. (DOI: 10.1039/C4RA05530F)

58. Avdoshenko, S. M.; Makarov, D. E. (2016). On the Reaction Coordinates and Pathways of Mechanochemical Transformations. *J. Phys. Chem. B* 120(8), 1537–1545. (DOI: 10.1021/acs.jpcb.5b07613)

59. Avdoshenko, S. M.; Konda, S. S. M.; Makarov, D. E. (2014). On the Calculation of Internal Forces in Mechanically Stressed Polyatomic Molecules. *J. Chem. Phys.* 141(13), 134115. (DOI: 10.1063/1.4896944)

60. Stauch, T.; Dreuw, A. (2015). On the Use of Different Coordinate Systems in Mechanochemical Force Analyses. *J. Chem. Phys.* 143(7), 074118. (DOI: 10.1063/1.4928973)

61. Koo, B.; Chattopadhyay, A.; Dai, L. (2016). Atomistic Modeling Framework for a Cyclobutane-Based Mechanophore-Embedded Nanocomposite for Damage Precursor Detection. *Comput. Mater. Sci.* 120, 135–141. (DOI: 10.1016/j.commatsci.2016.04.003)

62. Akbulatov, S.; Tian, Y.; Boulatov, R. (2012). Force-Reactivity Property of a Single Monomer Is Sufficient to Predict the Micromechanical Behavior of Its Polymer. *J. Am. Chem. Soc.* 134(18), 7620–7622. (DOI: 10.1021/ja301928d)

63. Makarov, D. E. (2016). Perspective: Mechanochemistry of Biological and Synthetic Molecules. *J. Chem. Phys.* 144(3), 030901. (DOI: 10.1063/1.4939791)

64. Brandhorst, K.; Grunenberg, J. (2007). Characterizing Chemical Bond Strengths Using Generalized Compliance Constants. *Chem. Phys. Chem.* 8(8), 1151–1156. (DOI: 10.1002/cphc.200700038)

65. Brandhorst, K.; Grunenberg, J. (2008). How Strong Is It? The Interpretation of Force and Compliance Constants as Bond Strength Descriptors. *Chem. Soc. Rev.* 37(8), 1558–1567. (DOI: 10.1039/B717781J)

66. Stauch, T.; Dreuw, A. A. (2014). A Quantitative Quantum-Chemical Analysis Tool for the Distribution of Mechanical Force in Molecules. *J. Chem. Phys.* 140(13), 134107. (DOI: 10.1063/1.4870334)

67. Stauch, T.; Dreuw, A. (2017). Quantum Chemical Strain Analysis for Mechanochemical Processes. *Acc. Chem. Res.* 50(4), 1041–1048. (DOI: 10.1021/acs.accounts.7b00038)

68. Wu, H.; Khan, A. M.; Johnson, B.; Sasikumar, K.; Chung, Y.-W.; Wang, Q. J. (2019). Formation and Nature of Carbon-Containing Tribofilms. *ACS Appl. Matter Interfaces* 11(17), 16139–16146. (DOI: 10.1021/acsami.8b22496)

69. Ta, H. T. T.; Tieu, A. K.; Zhu, H.; Yu, H.; Tran, N. V.; Tran, B. H.; Wan, S.; Ta, T. D. (2020). Ab Initio Study on Physical and Chemical Interactions at Borates and Iron Oxide Interface at High Temperature. *Chem. Phys.* 529, 110548. (DOI: 10.1016/j.chemphys.2019.110548)

70. Wang, L. F.; Ma, T. B.; Hu, Y. Z.; Wang, H. (2012). Atomic-Scale Friction in Graphene Oxide: An Interfacial Interaction Perspective from First-Principles Calculations. *Phys. Rev. B: Condens. Matter Mater. Phys.* 86(12), 125436. (DOI: 10.1103/PhysRevB.86.125436)

71. Ta, H. T. T.; Tran, N. V.; Tieu, A. K.; Zhu, H.; Yu, H.; Ta, T. D. (2021). Computational Tribochemistry: A Review from Classical and Quantum Mechanics Studies. *J. Phys. Chem. C* 125(31), 16875–16891. (DOI: 10.1021/acs.jpcc.1c03725)

72. Ta, H. T. T.; Tieu, A. K.; Zhu, H.; Yu, H.; Ta, T. D.; Wan, S.; Tran, N. V.; Le, H. M. (2018). Chemical Origin of Sodium Phosphate Interactions on Iron and Iron Oxide Surfaces by First Principle Calculations. *J. Phys. Chem. C* 122(1), 635–647. (DOI: 10.1021/acs.jpcc.7b10731)

73. Restuccia, P.; Ferrario, M.; Righi, M. C. (2020). Quantum Mechanics/Molecular Mechanics (QM/MM) Applied to Tribology: Real-Time Monitoring of Tribochemical Reactions of Water at Graphene Edges. *Comput. Mater. Sci.* 173, 109400. (DOI: 10.1016/j.commatsci.2019.109400)

74. Peeters, S.; Restuccia, P.; Loehlé, S.; Thiebaut, B.; Righi, M. C. (2020). Tribochemical Reactions of MoDTC Lubricant Additives with Iron by Quantum Mechanics/Molecular Mechanics Simulations. *J. Phys. Chem. C* 124(25), 13688–13694. (DOI: 10.1021/acs.jpcc.0c02211)

75. Krištofíková, D.; Mečiarová, M.; Rakovský, E.; Šebesta, R. (2020). Mechanochemically Activated Asymmetric Organocatalytic Domino Mannich Reaction-Fluorination. *ACS Sustain. Chem. Eng.* 8(38), 14417–14424. (DOI: 10.1021/acssuschemeng.0c04260)

76. Williams, M. T. J.; Morrill, L. C.; Browne, D. L. (2021). Mechanochemical Organocatalysis: Do High Enantioselectivities Contradict What We Might Expect? *ChemSusChem* 15(2), e202102157. (DOI: 10.1002/cssc.202102157)

77. Howard, J. L.; Cao, Q.; Browne, D. L. (2018). Mechanochemistry as an Emerging Tool for Molecular Synthesis: What Can It Offer? *Chem. Sci.* 9(12), 3080–3094. (DOI: 10.1039/c7sc05371a)

78. Hernández, J. G.; Bolm, C. (2017). Altering Product Selectivity by Mechanochemistry. *J. Org. Chem.* 82(8), 4007–4019. (DOI: 10.1021/acs.joc.6b02887)

79. Fischer, F.; Wenzel, K.-J.; Rademann, K.; Emmerling, F. (2016). Quantitative Determination of Activation Energies in Mechanochemical Reactions. *Phys. Chem. Chem. Phys.* 18(33), 23320–23325. (DOI: 10.1039/C6CP04280E)

80. Latorre, C. A.; Remias, J. E.; Moore, J. D.; Spikes, H. A.; Dini, D.; Ewen, J. P. (2021). Mechanochemistry of Phosphate Esters Confined Between Sliding Iron Surfaces. *Commun. Chem.* 4(1), 178. (DOI: 10.1038/s42004-021-00615-x)

81. Pladevall, B. S.; de Aguirre, A.; Maseras, F. (2021). Understanding Ball Milling Mechanochemical Processes with DFT Calculations and Microkinetic Modeling. *ChemSusChem* 14(13), 2763–2768. (DOI: 10.1002/cssc.202100497)

82. Howard, J. L.; Brand, M. C.; Browne, D. L. (2018). Switching Chemoselectivity: Using Mechanochemistry o Alter Reaction Kinetics. *Angew. Chem. Int. Ed.* 57(49), 16104–16108. (DOI: 10.1002/anie.201810141)

83. Ferguson, M.; Moyano, M. S.; Tribello, G. A.; Crawford, D. E.; Bringa, E. M.; James, S. L.; Kohanoff, J.; Del Pópolo, M. G. (2019). Insights into Mechanochemical Reactions at the Molecular Level: Simulated Indentations of Aspirin and Meloxicam Crystals. *Chem. Sci.* 10(10), 2924–2929. (DOI: 10.1039/c8sc04971h)

84. Andersen, J. M.; Mack, J. (2017). Decoupling the Arrhenius Equation via Mechanochemistry. *Chem. Sci.* 8(8), 5447–5452. (DOI: 10.1039/C7SC00538E)

85. Tan, D.; Mottillo, C.; Katsenis, A. D.; Štrukil, V.; Friščić, T. (2014). Development of C-N Coupling Using Mechanochemistry: Catalytic Coupling of Arylsulfonamides and Carbodiimides. *Angew. Chem. Ind.*, 53rd edition, 9321–9324. (DOI: 10.1002/anie.201404120)

86. Li, J.; Nagamani, C.; Moore, J. S. (2015). Polymer Mechanochemistry: From Destructive to Productive. *Acc. Chem. Res.* 48(8), 2181–2190. (DOI: 10.1021/acs.accounts.5b00184)

87. Stauch, T.; Dreuw, A. (2017). Force-Induced Retro-clock Reaction of Triazoles Competes with Adjacent Single-Bond Rupture. *Chem. Sci.* 8(8), 5567–5575. (DOI: 10.1039/C7SC01562C)

88. Martínez, R. F.; Cravotto, G.; Cintas, P. (2021). Organic Sonochemistry: A Chemist's Timely Perspective on Mechanisms and Reactivity. *J. Org. Chem.* 86(20), 13833–13856. (DOI: 10.1021/acs.joc.1c00805)

89. Kyuichi, Y. (2021). Numerical Simulations for Sonochemistry. *Ultrason. Sonochem.* 78, 105728. (DOI: 10.1016/j.ultsonch.2021.105728)

90. Miao, M.; Sun, Y.; Zurek, E.; Lin, H. (2020). Chemistry Under High Pressure. *Nat. Rev. Chem.* 4(10), 508–527. (DOI: 10.1038/s41570-020-0213-0)

91. Yan, H.; Yang, F.; Pan, D.; Lin, Y.; Hohman, J.; Solis-Ibarra, D.; Li, F.; Dahl, J.; Carlson, R.; Tkachenko, B.; Fokin, A.; Schreiner, P.; Galli, G.; Mao, W.; Shen, Z.; Melosh, N. (2018). Sterically Controlled Mechanochemistry Under Hydrostatic Pressure. *Nature* 554(7693), 505–510. (DOI: 10.1038/nature25765)

3

The Thermodynamics and Kinetics of Mechanochemical Reactions

An Experimental Approach

Adam A. L. Michalchuk

CONTENTS

Since pre-history, humans have made use of mechanically driven chemical and physical transformations.[1, 2] From creating fire by friction and the invention of chemical explosives (e.g. black powder in 3000 BCE),[3] through to modern-day methods of extractive metallurgy,[4] our ability to manipulate materials with mechanical force has formed a cornerstone of human civilisation. Yet, only within the last century have such mechanically induced transformations caught the eye of scientists,[5] leading to the establishment of the scientific discipline now known as *mechanochemistry*. To date, mechanochemical approaches have been devised to tackle many challenges across academia and industry. Tools of mechanochemistry such as ball milling[6] and twin-screw extrusion[7] are now routinely applied to synthesise and process new molecules and materials (ranging from inorganic complexes to large biomacromolecules). These tools have demonstrated immense potential as rapid, high-yielding, and environmentally benign technologies placing them amongst the 2019 International Union of Pure and Applied Chemistry (IUPAC) list of the "ten chemical innovations that will change our world".[8]

As with any chemical process, our ability to use mechanochemistry for a real-world application depends on our ability to understand and control it. However, despite many decades of research on mechanochemical technologies,[9] many questions remain open, and controlling mechanochemical reactions is still a real and significant problem. For example, we cannot yet know *a priori* what mechanochemical conditions will be needed to drive forward a given reaction, nor can we easily transfer a reaction from one mechanochemical reactor to another. From the perspective of industrial applications, it is even not easy to scale up a mechanochemical process from milligram to gram (or kilogram) scale.[10, 11] These difficulties are indicative of the fact that we do not yet understand how or why processing parameters affect mechanochemical transformation. In fact, in many cases we do not even yet know which processing parameters *can* affect a given mechanochemical transformation in the first place. That is to say that we still do not know what we do not know about mechanochemical reactions.

Aristotle famously stated that "corpora non agunt nisi fluida seu solute", that is that compounds that are not fluid or dissolved do not react.[12] This statement has greatly shaped the way the physical and chemical sciences have evolved over the centuries to focus on fluid phase reactivity. With growing interest in mechanochemical technologies, there is a pressing need to better understand the driving forces that underpin these reactions. Only in this way can we develop mechanochemistry as a robust and reliable chemical technology. In analogy to conventional (solution and gas phase) chemistry, such knowledge can be obtained through dedicated investigations into the thermodynamic and kinetic aspects of mechanochemical reactions. Comparatively little is known about the kinetics and thermodynamics (or energetics) of solid-state mechanochemical transformations. In fact, even the experimental and theoretical tools needed to study such phenomena are still being actively developed.[13] Although we cannot hope to provide a complete view of the kinetics and thermodynamic investigation of mechanochemical reactions, this chapter aims to highlight the major approaches being currently applied to study these fundamental aspects of mechanochemical transformations.

A philosophical discussion on the thermodynamics and kinetics of mechanochemical transformations is out with the scope of this contribution, and some discussion can be found elsewhere.[14–17] We will instead introduce some of the general concepts in the thermodynamic (energetic) and kinetic aspects of mechanochemical reactions and use them to demonstrate how various aspects of mechanochemical reactions are being studied and understood in practice, focusing primarily on recent examples. The field of mechanochemistry is very broad, ranging from stretching of single molecules by AFM,[18] to high-pressure phenomena[19] and bulk ball milling reactions. It is not therefore possible to cover the entire scope of transformations here. The focus will remain on what is currently the most quickly growing area of mechanochemistry, the processing of powders by dynamic mechanical strain (e.g. ball milling). For broader discussions on the general aspects of mechanochemistry, excellent texts are available elsewhere.[1, 6, 20, 21]

3.1 Considerations on the Types of Mechanochemical Reactions

Before delving into a discussion on the study of mechanochemical transformations, it is prudent to paint a picture of *what* a mechanochemical transformation in fact describes. A detailed discussion on this topic has been recently summarised in Ref. [1], and only a brief introduction is given here. In the current paradigm, it is understood that a mechanochemical reaction is any that is in some way driven or facilitated by mechanical force. On the one hand this may include thermal transformations in powders where mechanical force facilitates mixing or provides heating to the system, e.g. through friction, and includes also transformations in which the mechanical energy itself drives directly the chemical reaction. Very broadly, mechanochemical reactions can be thought of as belonging to one of three categories: Type 1, mechanically facilitated thermal reactions; Type 2, reactions made possible (or mediated) by mechanical action; and Type 3, reactions driven directly by mechanical energy.[1] It is worth noting that, in principle, both Type 1 and Type 2 share similar underlying elementary mechanisms, with the transformations driven primarily by thermal processes. Their separation into distinct categories aids largely to differentiate the role of the mechanical action itself.

In the first category, the mechanical action works primarily to create new reactive sites, *via* mixing and the creation of new, unreacted surfaces. However, the transformations are themselves driven by thermal energy, rather than by the mechanical energy itself (although mechanical action can cause heating by friction).[22] Such reactions are common across the mechanochemistry of organic solids, for example in the case of the reaction between solid powders of 4-aminotoluene and 2-hydroxy-3-methoxy-benzaldehyde.[23] In this example, simply bringing the powders of the starting reagents into direct contact at room temperature led to the formation of a melt, which recrystallised as the product phase, Figure 3.1a. Similarly, the cocrystallisation of thymol and hexamethylenetetramine was shown to pass through

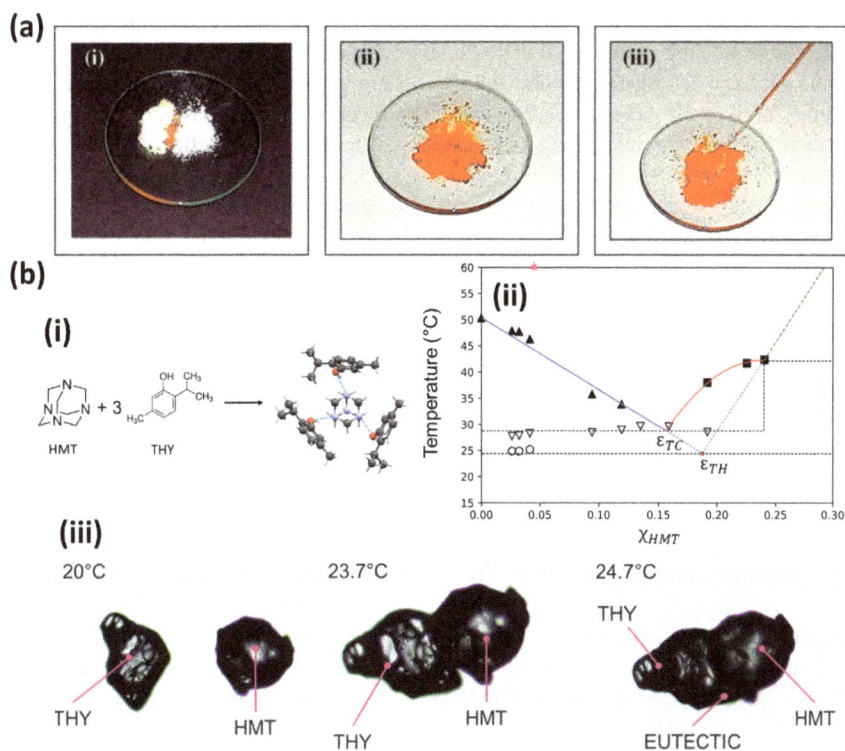

FIGURE 3.1 Thermally driven solid + solid mechanochemical reactions. (a) The cocrystallisation of 4-aminotoluene and 2-hydroxy-3-methoxybenzaldehyde, driven through contact melting at room temperature. Figure reprinted with permission from Ref. [23]. Copyright 2001 American Chemical Society. (b) The mechanochemical cocrystallisation of thymol (THY) and hexamethylenetetramine (HMT) driven by the formation of a eutectic phase. (i) Reaction scheme for cocrystallisation of THY and HMT. (ii) Binary phase diagram for the melting temperature of mixtures of the starting reagents. (iii) Photographs for eutectic formation at particle contact upon mild heating, easily achievable through friction in a ball mill (see e.g. Ref. [22]). Figures reprinted with permission from Ref. [24]. Copyright 2002 Royal Society of Chemistry.

a eutectic melt at temperatures slightly above room temperature and hence driven by mild heating during ball milling, Figure 3.1b.[24] In such cases it is essential to understand the thermal behaviour of the reacting system, for example by exploring the binary phase diagram of the reagents, Figure 3.1b. These thermally driven mechanisms are presumably also responsible for many liquid-assisted grinding (LAG) processes, particularly where materials are partially soluble in the liquid additive. In the mechanochemical cocrystallisation of glycine and malonic acid,[25] liquid-assisted grinding using polar solvents such as water and ethanol (in which the reagents are soluble) greatly increased the rate of cocrystallisation. In contrast, using apolar solvents such as benzene (in which the reagents are insoluble) impeded the transformation. Importantly, this liquid additive may not always be explicitly added to the mixture and may appear inadvertently through atmospheric moisture, as was shown for the mechanochemical cocrystallisation of glycine and malonic acid.[26] It is likely that mechanically facilitated transformations are also responsible for the growing number of reports of "ball free" mechanochemical transformations being conducted by Resonant Acoustic Mixing (most of which can so far be only achieved by LAG),[11, 27–29] or other media-free mixing devices.[30]

In the second class of transformation, the mechanical energy works to modify the underlying potential energy surface, although the reaction itself still proceeds by a "thermal" (or equilibrium) route. This is to say that the mechanical action opens new reaction pathways, by lowering (or imposing) new energetic barriers.[31] These reactions can be generally considered as "mechanical activation",[32] wherein mechanical energy is stored in the material to enhance/alter its reactivity.[32] For example reactions may be driven or enhanced through the mechanical generation of defects (bulk or surface) that decrease kinetic barriers and hence favour the thermal reaction in the solid state.[33, 34] Reactions and transformations driven by slow compression (e.g. using diamond anvil cell technologies) also fall within this category. Under these conditions, the nuclei re-arrange atop a new potential energy surface, imposed by the external mechanical force. Importantly, this class of reaction can depend critically on the rate of the applied load and hence the structure of the potential energy surface on which thermal relaxation occurs.[35]

The third category of mechanochemical reactions is somewhat unique as compared with the first two. In these reactions, the mechanical force is directly responsible for the transformation, which typically passes via pathways that are "forbidden" by thermal routes.[36] One can discuss here the direct mechanical fracture of chemical bonds (for example by AFM),[37] or the excitation of chemical bonds through adiabatic[38] or ultrafast non-adiabatic processes.[39]

In practice it is not usually obvious (or even possible) to distinguish between different types of mechanochemical transformations. In fact, most transformations likely do not fit cleanly into one "classification", making the interpretation of kinetic and energetic aspects of these reactions a significant challenge. It is however instructive to keep these classifications in mind, allowing us a view of the *types* of processes that should be considered when investigating mechanochemical transformations and establishing models to describe them, Figure 3.2.

Generally, all mechanochemical transformations are based on an interplay between the generation of a stress field and its subsequent relaxation, Figure 3.2.[32] The nature of the stress depends on many factors, including the size and shape of solid particles, the parameters of the mechanochemical reactor (e.g. temperature, regime, stress rate), and the physical properties of the material such as its compressibility or its melting point. In Type 1 (mechanically facilitated) reactions, the stress field is applied, and its relaxation leads to the grinding of solid material. This grinding ultimately facilitates the mixing between unreacted materials and hence allows the transformation to continue. The relaxation of the stress can also lead to

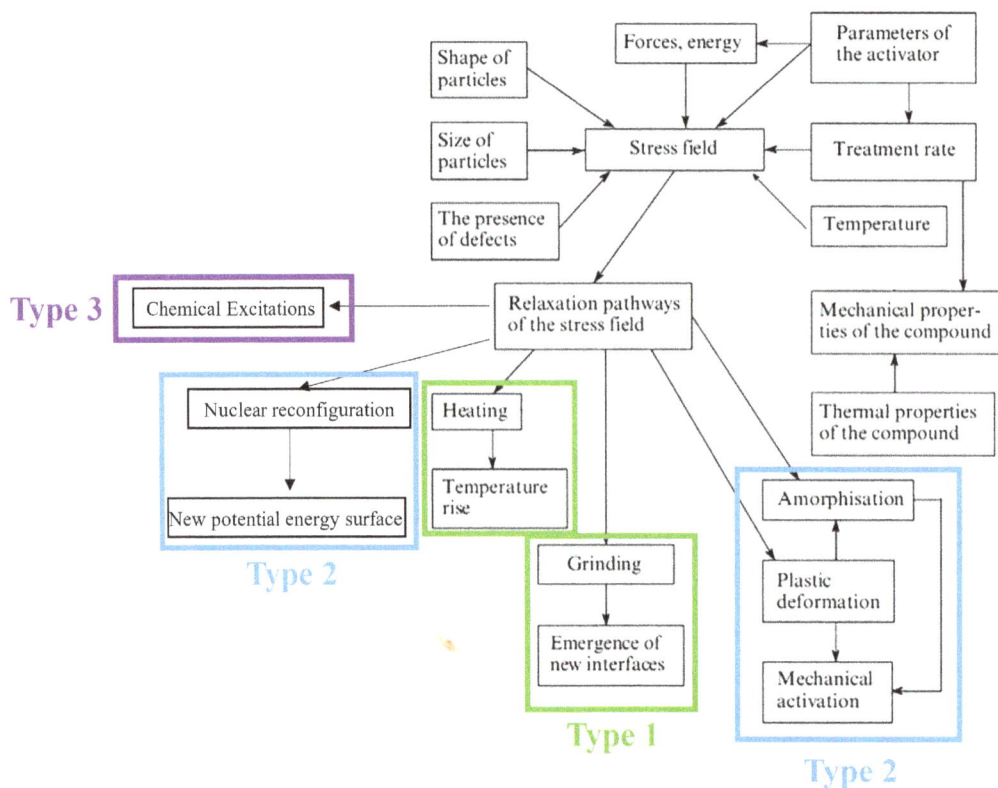

FIGURE 3.2   Some routes for the generation and relaxation of stress, leading to mechanochemical reactions in their various forms. Figure modified from Ref. [32].

heating, further facilitating the transformation. For Type 2 (mechanically mediated) transformations, the applied stress can be either static or dynamic. Where the stress is static (e.g. using a diamond anvil cell or pulling with AFM), the stress relaxes *via* reconfiguration of the atoms in the material.[19] This leads to a new potential energy surface upon which subsequent transformations can occur. When the stress is dynamic (e.g. during ball milling), relaxation of stress can lead to the formation of defects.[40] In turn this increases the internal energy of the system, enhancing its reactivity (*i.e.* activates the material).[41] Finally, in Type 3 (mechanically driven) reactions, relaxation of the stress leads to direct excitation of bonds or electronic states, directly driving the transformation.[38]

In all cases, the relaxation of the perturbed system occurs very quickly and under non-equilibrium conditions. It is therefore often difficult to separate the effects of kinetics and thermodynamics (or energetics) in the study of mechanochemical reactions. Moreover, the relative importance of kinetic and thermodynamic features will differ for different types of mechanochemical transformations, and the processes that dictate the kinetics and thermodynamics will also be different for different reaction types, Table 3.1. In fact, the exact mode of action of mechanical energy is also largely dependent on the system being investigated.[20] For example, it is well known that the mechanical action of low melting point materials tends not to pass *via* direct mechanochemical mechanisms.[43] Instead, when stimulated by mechanical action (e.g. friction), these materials tend to melt and are therefore available for subsequent thermal reactions. In contrast, high melting point materials have been frequently shown to undergo various forms of mechanical activation. Similarly, the formation and participation of true amorphous phases can only be considered where the glass transition temperature is higher than the temperature of the mill.[44] Importantly, it is usually non-trivial to determine unambiguously the role of mechanical energy in driving mechanochemical reactions. In fact, it is likely that many reactions are driven at different stages through different types of mechanochemical mechanisms. For example, it may be that solids will spontaneously melt when mixed (i.e. a mechanically facilitated process) but that the liquid phase becomes directly activated by mechanical energy (i.e. a mechanically driven process). Such cases must be considered possible, particularly for reactions that combine multiple phases, where "many-body collisions" are highly unlikely to occur (i.e. reactions involving more than two solid phases).

In this spirit, through this chapter we will not attempt to classify reactions by type. Instead, we will outline trends in the experimental and theoretical tools currently being employed to study the various kinetic and thermodynamic (or energetic) aspects of mechanochemical reactions. Although extensive work in this area has been done over the years, we cannot hope to cover it all here.[16, 17] We will instead focus on recent trends in the study of mechanochemistry, where possible. An initial introduction will be given to the recent advances in experimental methods currently being used to analyse mechanochemical transformations. The chapter will then outline some major aspects of the mechanistic features currently believed to underpin mechanochemical reactions, highlighting their intimate interrelation in terms of their influence on the kinetics and thermodynamics of mechanochemical transformations. In this way we hope to provide a comprehensive view of the complexity of mechanochemical reactions and the importance of studying them within a holistic mindset.

TABLE 3.1

Dominant Features Responsible for the Kinetic and Energetic Aspects of Mechanochemical Transformations

	Dominant Kinetic Features	**Dominant Energetic Features**
Mechanically facilitated (Type 1)	• Mixing (*inc.* comminution) • Chemical kinetics	• Phase stability, including surface contribution
Mechanically mediated (Type 2)	• Comminution • Activation/Deformation • Chemical kinetics	• Formation and energetics of defective phases • Force-modified PES
Mechanically driven (Type 3)	• Elastic-chemical energy transition	• Bond strengths[a]

Note this list is purely illustrative and by no means exhaustive

[a] The mechanisms of direct mechanochemical reactions are still debated, with discussion open regarding direct athermal[36, 42] or indirect thermal mechanisms.

3.2 Mechanochemical Reaction Profiles and Experimental Methods for Their Study

To appreciate the experimental approaches currently being used for studying the kinetic and energy aspects of mechanochemical transformations, it is first worth considering the overall reaction profile of such a transformation. In a typical mechanochemical transformation – say in a vibratory ball mill – one generally finds a reaction profile as shown in Figure 3.3. The initial stage of the profile is marked by an induction period, wherein the reaction appears not to progress.[45, 46] After some induction period, the transformation begins to occur until all available reagent has been consumed and the reaction stops, leading to a final plateau. Importantly, because not all reagent is necessarily available to the reaction volume and is only exposed to the reaction volume through comminution, the final plateau does not necessarily reach the quantitative conversion. For this same reason, the final plateau can also depend on the regime of mechanical action being applied (see further discussion in Section 3.4).[41, 47]

The induction period is generally accompanied by physical changes in the bulk powder and the solid-state reagents, including mixing,[48] reduction of particle size (comminution), and the generation of defects in the solid particles. While mixing (and analogously comminution) presumably facilitates the formation of interfaces between reagent particles, the generation of smaller (often nano-) particles[49, 50] and internal defects can make the solid state more reactive (*i.e.* change the relative stability of the reagents).[41, 51] Hence the induction period represents a complex series of events wherein the mechanical stress alters the energy landscape of the reagents (through mixing, comminution, or defect formation) and drives the reaction forward. This is to say that the induction period itself represents a complex interplay between kinetic and thermodynamic (energetic) aspects and has a significant influence on the overall kinetic profile of the transformation. Details of the induction period will be discussed further in Section 3.3. The reaction period is equally complex, and its nature depends on the type of mechanochemical reaction being investigated (see Figure 3.2 and Section 3.2). For example, Type 1 reactions will typically involve the dissolution or melting of the powder, and the subsequent reaction in the liquid phase and their kinetics, therefore, depend on the rate of dissolution/melting (and hence mixing kinetics[48, 52]), as well as the kinetics of the liquid phase reaction, Table 3.1. In contrast, Type 2 reactions will depend on the relative kinetics with which mechanical strain is applied to the solid, and the kinetics of thermal processes on the modified potential energy surface. Details of the reaction period will be discussed further in Section 3.4. Finally, the product period – and the nature of the product that results – depends on a complex interplay between kinetic and thermodynamic (energetic) aspects of the system. For example, following a

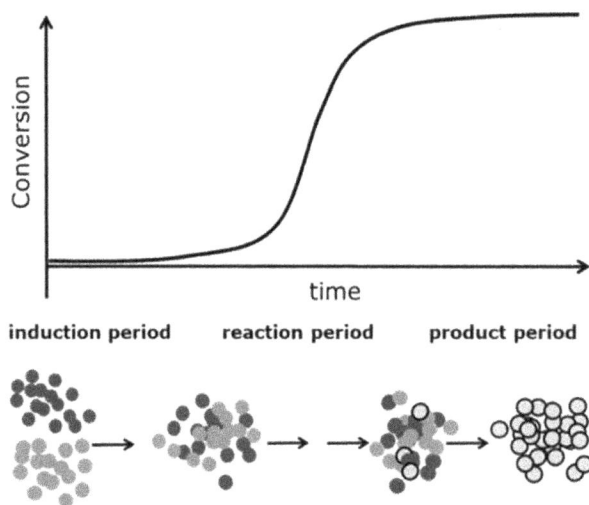

FIGURE 3.3   Schematic reaction profile for a typical mechanochemical transformation. Figure adapted with permission from Ref. [13].

chemical reaction, nucleation and growth phenomena drive the formation of the final solid product. The structure of this final solid product depends both on the rates of these nucleation/growth processes and the relative stability of the final product including the existence of surfaces and internal defects.[49] These aspects will be discussed in greater detail in Section 3.5.

The growing need to understand the mechanisms that underpin mechanochemistry has been met with a surge in the development of experimental techniques to probe them.[13, 53] Historically, mechanochemical transformations have been studied using *ex situ* analysis. In this paradigm, the sample is exposed to mechanical action and then removed from the reactor for analysis. By removing the sample from the reactor, any analytical technique can be in principle used to investigate the sample, from positron annihilation spectroscopy,[54] through to optical[24, 26] and electron microscopy,[55] X-ray[56, 57] and neutron diffraction,[58] calorimetry,[41, 59] Mössbauer spectroscopy,[60] and many fluid phase analyses including high-performance liquid chromatography,[61] gas chromatography, solution NMR,[62] and mass spectrometry.[63] With a plethora of analytical techniques available, the *ex situ* paradigm of mechanochemistry has offered significant advancement in the field. The analytical techniques used for studying the kinetic and energetic aspects of mechanochemical transformations can be grouped into those which probe directly the solid state, and those which probe materials in the solution state. The former has been undoubtedly most common, owing to the need to characterise the structural and thermodynamic features of the solid reagent and products that are typical of most mechanochemical transformations. However, interest in using solution-phase characterisation techniques is growing as mechanochemistry becomes increasingly applied to organic molecular synthesis. We here briefly highlight only some of the major experimental techniques currently being used to study mechanochemical transformations, and further details are available in dedicated texts including Ref. [53].

3.2.1 Ex Situ Solid-State Characterisation

The most common choice for solid-state ex situ analysis of mechanochemical transformations is certainly powder X-ray diffraction (PXRD). By providing a direct measure for the crystal and atomic structure of a material, PXRD has found widespread use for characterising new solid phases generated by mechanical action, including entirely new materials (e.g. alloys, cocrystals) and new crystal forms (polymorphs) which cannot be analysed by conventional fluid phase methods.[64] Identifying the presence and structure of new crystalline phases has been essential for mapping energy landscapes of mechanochemical transformations by *ab initio* methods, and hence for elucidating energetic aspects of these reactions.[22, 65, 66] Moreover, PXRD allows for the quantification of microstructure (i.e. defects and distortions) within crystal structures, which are essential features for understanding the reactivity of solids exposed to mechanical stress.[45, 56] PXRD has been also irreplaceable for following the kinetics of mechanochemical transformations,[45] offering a direct route to quantify the amount of each crystallographic phase in a given powder. For example, by using PXRD the kinetic profile for mechanochemical cocrystallisation was followed,[67] including revealing the role of intermediate crystallographic phases in the overall kinetic profile of the transformation.

Additional to PXRD, a lot can be learned about the structure of solids prepared by mechanochemistry from other structural probes. For example, Raman and infrared spectroscopies can be used to provide indications for the formation of new solid forms through the (dis)appearance or shifting of new vibrational bands. While this is most visible when new covalent bonds are formed, or where strong intermolecular interactions form during material formation (e.g. hydrogen bonds),[46] advanced data analysis has been shown to be able to identify even minor structural changes in solid form (e.g. polymorphs) through small changes in their Raman spectra.[63] These conventional vibrational spectroscopies can be also used to investigate the migration of heavy isotopes and hence are useful for looking at dynamic mass transport effects during, e.g. milling.[68] Instead, low-frequency vibrational spectroscopy such as THz spectroscopy is particularly powerful for analysing changes in solid structure,[69] as these low-frequency vibrations correspond to vibrations of the crystal lattice itself. Vibrational spectroscopy has been also essential for characterising structural distortions in materials activated by mechanical action.[70, 71]

Other spectroscopic techniques such as X-ray absorption spectroscopy (XAS) or X-ray photoelectron spectroscopy (XPS), electron spin resonance spectroscopy (ESR), and Mössbauer spectroscopy have proved to be essential methods for characterising the electronic states of ions within mechanochemically prepared solids.[55, 70, 72, 73] For example, XAS has been a central tool for following the kinetics of redox processes involved in the mechanochemical synthesis of metal nanoparticles.[74] Mössbauer spectroscopy has been often applied for exploring the structure of mechanochemically prepared products and thus for contributing to studying the thermodynamic aspects of mechanochemical reactions.[75, 76] Additionally, methods like XAS provide also access to local structural information,[77, 78] both in terms of immediate coordination sphere and extended coordination sphere. In this way XAS has contributed also to studying the effects of mechanical action on the structure (and hence thermodynamic stability) of mechanochemical transformations.

Studying the structural evolution of solids during mechanochemical transformations has been essential to probe the overall reaction profile of the transformation. However, such studies have offered little in terms of experimental probes into the thermodynamic (or energetic) driving forces in mechanochemical transformations. To this end, thermal analysis has proved to be an essential tool for understanding mechanochemical transformations. Differential scanning calorimetry (DSC) has been widely used to investigate the effects of heat on the stability and interconversion of solid forms, in attempts to identify and discriminate between mechanical and thermal mechanisms. DSC has been also essential for mapping the thermodynamic stability of solid forms, providing an experimental probe into the thermodynamic landscape of transformations.[44, 79] Moreover, DSC (and other calorimetric approaches) has been used to quantify the enthalpy and entropy of solid forms, for example after mechanical activation.[41] These techniques have therefore proved essential in the study of induction periods as well as of the final product phase obtained by mechanochemistry. It is worth highlighting that, while DSC is primarily used as a probe for phase stability, it has been also used to follow macroscopic kinetics of mechanochemical transformations, with particular application for following mechanochemical polymorphic transitions and amorphisation.[44]

3.2.2 Ex Situ Solution-Phase Characterisation

With the growing application of mechanochemical routes for molecular synthesis, researchers are increasingly turning towards solution-phase characterisation techniques for *ex situ* analysis. By using solution methods, one inherently loses a significant amount of information regarding the driving forces that underpin the reaction, which is dictated largely by the solid material itself and can differ significantly from conventional solution stabilities.[80] However, in many cases solution methods have proved complementary to solid-state analysis[50, 81] or provide important chemical information where relatively insensitive solid-state characterisation techniques are insufficient. For example, in addition to PXRD, studies on the mechanochemical disulfide exchange reactions[80, 82] have relied heavily on the use of HPLC for high-sensitivity quantification of the molecular composition following a mechanochemical synthesis, offering them unprecedented insight into the sensitivity of mechanochemical reactions to various processing conditions such as liquid additive. Similarly, both solution NMR and HPLC have been used for the sensitive quantification of molecular synthesis and for mapping out energy landscapes of organic mechanochemical reactions, Figure 3.4.[83] Hence, although solution methods do omit analysis of solid-state aspects of mechanochemical reactions, they have an important role to play for extracting intricate and sensitive details where solid-state characterisation techniques fail.

3.2.3 Time-Resolved In Situ (TRIS) Characterisation

As discussed above (see Figure 3.2), mechanochemical transformations involve a balance between the generation and relaxation of mechanical stress. Correspondingly, a system that is exposed to repetitive mechanical stressing (e.g. in a ball mill) will eventually reach a steady state wherein the physical and chemical state of the material remain constant so long as the mechanical treatment is not altered or stopped. In this light it is not surprising that growing evidence has suggested that stop-start *ex situ* analyses may not provide the "correct" picture of the transformation.[25] For example, the apparent mechanism for the mechanochemical synthesis of the Cu-based metal-organic framework, HKUST-1, differed when

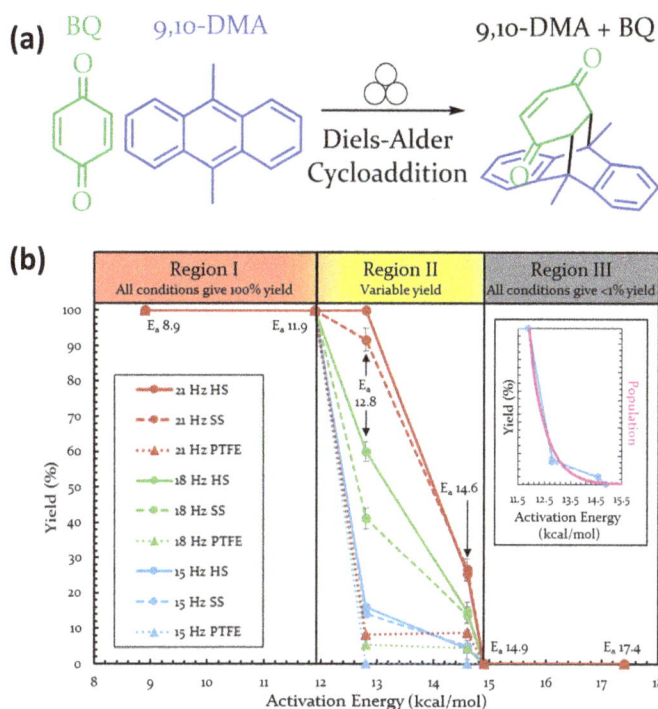

FIGURE 3.4 Investigation of reaction yield vs activation energy for model Diels–Alder reactions. (a) Example Diels–Alder reaction between benzoquinone (BQ) and 9,10-dimethylanthracene (9,10-DMA). (b) The reaction yields (obtained by solution NMR spectroscopy) of cycloaddition for Diels–Alder reaction with different activation energies at various ball milling conditions, including in hardened steel (HS), stainless steel (SS), and Teflon (PTFE) milling jars. Figure reproduced from Ref. [83] with permission from the Royal Society of Chemistry.

probed by in situ and *ex situ* analysis.[84] Such studies have inspired the development of new methods for time-resolved in situ (TRIS) monitoring of mechanochemical reactions.

Interest in following mechanochemical transformations by TRIS methods dates to the late 20th century when the focus was placed on measuring the evolution of the bulk temperature and pressure over the course of the transformation. Attempts at directly probing the structure of the material inside the mechanochemical reactor appeared more recently, dominated by the development of TRIS-XRPD[57] and TRIS Raman spectroscopy.[85] A detailed discussion on the developments of TRIS analysis in mechanochemistry can be found elsewhere.[13] Since these early studies, significant advances have been made in the methodology of these techniques.[56, 86] For example, recent advances in TRIS-XRPD have now allowed for the reliable extraction of phase composition, crystal size, and microstrain,[56] opening entirely new dimensions for studying the kinetics and energetic aspects of mechanochemical transformations. Similarly, methodological developments[86] have now made TRIS Raman spectroscopy an accurate probe to extract kinetic curves from ball milling reactions,[68, 87] making it an irreplaceable laboratory tool for TRIS mechanochemical investigation. To date, methods available for TRIS monitoring of mechanochemical reactions have grown to include XRPD,[56, 57] neutron diffraction,[58] Raman spectroscopy,[71, 86] X-ray absorption spectroscopy,[88] thermometry,[89] manometry,[90] solid-state NMR,[91] and fluorescence spectroscopy.[92] Each of these tools provides a unique view into the evolution of a mechanochemically reacting system, whether through direct analysis of the crystal structure and its distortions (XRPD, neutron diffraction, Raman spectroscopy, and solid-state NMR), changes in the electronic behaviour of the material (X-ray absorption and fluorescence spectroscopies), or the bulk evolution of the system by means of macroscopic parameters (manometry and thermometry). Hence, it is now possible to obtain a broad and multi-dimensional view of mechanochemical reactions using TRIS technologies. Although many different mechanochemical reactors are used, most TRIS methods have so far been applied to studies using vibratory ball mills, with limited examples reported to study the kinetics of transformations in other reactors such as the Resonant Acoustic Mixer (RAM).[27, 58] Further developments in TRIS methodology are needed before the technologies can be applied to follow transformations in more complex mechanochemical reactors such as planetary ball mills.

Despite the incredible advances made by TRIS methods, it is worth highlighting a number of caveats. TRIS analyses work by placing a probe at a fixed position, and moving the jar through the probing beam. As such, TRIS methods only measure a small fraction of the material that passes through the beam at any given time. Changes in material rheology that can lead to caking[93] can therefore lead to significant distortion of apparent reaction profiles measured by TRIS.[94] Such issues can be in principle overcome by thorough testing of materials to avoid caking (e.g. by reducing sample loading or reducing milling intensity) over the course of the transformation.[56] In addition to potential issues with sampling, TRIS methods also impose limitations on the nature of the sample jar that can be used. For example, TRIS PXRD requires the use of a jar made of a material that is transparent to X-rays (typically polymethylmethacrylate),[57] whereas TRIS Raman requires the use of optically transparent vessels.[86] Growing evidence is demonstrating that changes in sample vessel material can have significant consequences for the rates,[95] or even the outcomes[65] and phase stabilities[22] of mechanochemical transformations. It is not yet clear why jar material can play such a significant role, although numerical simulations suggest impact energy may play a role.[96] However, it is clear that the TRIS kinetic data in polymer jars must be interpreted carefully when compared with *ex situ* data obtained using conventional stainless steel jars.

3.2.4 A Need for Multi-Method Approaches

There are clear advantages and disadvantages to following mechanochemical transformations with different analytical techniques, Figure 3.5. While some methods provide direct information on the structure of the crystallographic phases involved in the reaction (e.g. PXRD), they do not provide any direct probe for the *chemical* processes that occur outside the crystal lattice. In contrast, other techniques are more sensitive to changes in the molecular (e.g. Raman spectroscopy) or electronic (e.g. XAS) structure but provide little insight into the bulk crystallographic information. It follows that to obtain a complete analysis of the kinetics and thermodynamic driving forces for a given mechanochemical transformation, one must make use of a combination of complementary analytical techniques. In this regard, the use of multiple analytical techniques has been a common trend in the study of mechanochemical reactions.

While XRD is commonly used to follow the evolution of crystallographic phases, and hence to map overall reaction kinetics, complementary thermal analysis has been essential for simultaneously mapping out the thermodynamic evolution of the system.[41, 44, 59] Such investigations are often complemented by *ab initio* computational simulations for a more detailed understanding of the thermodynamic landscape of the system being investigated.[22, 50, 66, 97] In a recent example exploring the ball milling-induced polymorphic conversion between two cocrystal polymorphs, the kinetics of the polymorphic transformation was investigated by PXRD, Figure 3.6. At the same time, the thermodynamic relationship between the different polymorphic cocrystals was studied through a combination of *ab initio* DFT simulations, variable temperature PXRD, and DSC.[41] Together these techniques provided a complete picture of both the kinetic and thermodynamic aspects of the ball milling transformation, mapping how ball milling energy alters the rate of polymorphic conversion from a metastable to thermodynamically bulk stable polymorph.

Owing to the intricate and complex landscape of crystal structure and its evolution during mechanochemical reactions, even multiple structural techniques can be required to obtain realistic insights into the system's overall kinetic behaviour. For example, while the kinetics at the scale of elementary redox chemistry can be followed by XAS, the subsequent formation and growth of the product solid phase can be only followed by X-ray diffraction. Correspondingly, following the kinetic profile of mechanochemical nanoparticle synthesis requires simultaneous analyses by both techniques.[88] Similarly, while XRD is highly effective to follow the kinetic evolution of crystalline phases, it is incapable of following the behaviour and participation of non-crystalline materials. In systems where both crystalline and non-crystalline materials play essential roles in the overall reaction profile, a combination of techniques such as XRD and Raman spectroscopy is therefore needed.[98]

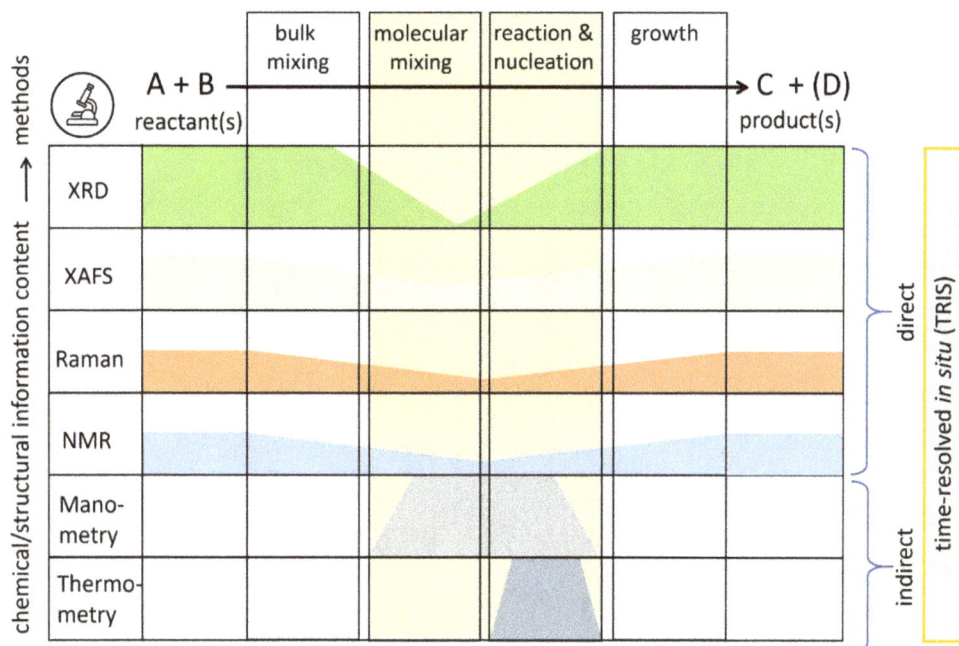

FIGURE 3.5 Schematic representation of the information content provided about a mechanochemical transformation for different TRIS methods. Both direct (i.e. providing structural information) and indirect (i.e. providing only secondary evidence for reaction progress) methods are shown. Figure reproduced from Ref. [13].

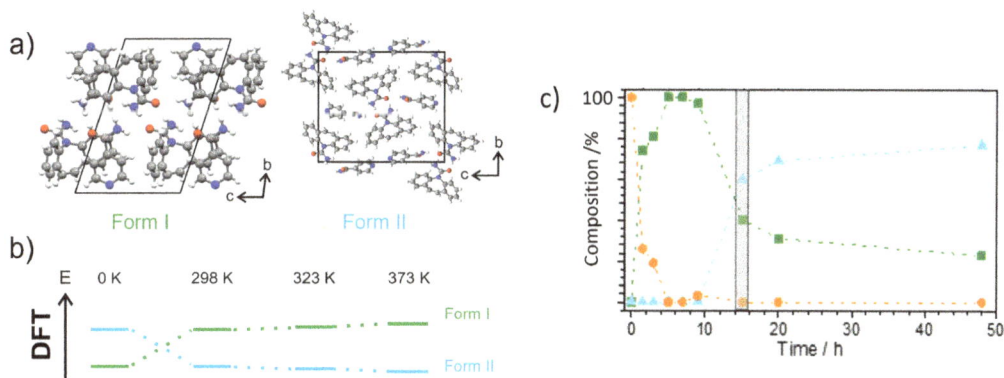

FIGURE 3.6 Mapping the kinetic and thermodynamic landscape of mechanochemical polymorphism using a combination of DFT and XRPD. (a) The polymorphic forms for the stoichiometric cocrystal of carbamazepine: nicotinamide. (b) The relative stability of the two forms derived from *ab initio* (DFT) simulations. (c) The kinetic profile for the formation of cocrystal Form I (green) from the starting materials (orange), and the subsequent conversion to Form II (blue), as obtained from quantitative analysis of PXRD profiles. Figure adapted with permission from Ref. [41].

In many cases complementary techniques are also used to reinforce data interpretation.[87] This was exemplified for the TRIS monitoring of the cocrystallisation of theobromine and oxalic acid, Figure 3.7.[89] TRIS PXRD data suggested that, upon ball milling, one of the starting reagents began to lose crystallinity during the induction period. This was confirmed by TRIS thermometry, which revealed a discontinuity in the temperature profile at the same point during the reaction. After prolonged milling, the cocrystal product formed from the reagents was clearly visible by PXRD and a second marked discontinuity in the thermometric curves. Finally, as PXRD was unable to follow the evolution of the amorphous phases during the induction period, TRIS Raman spectroscopy verified that no important

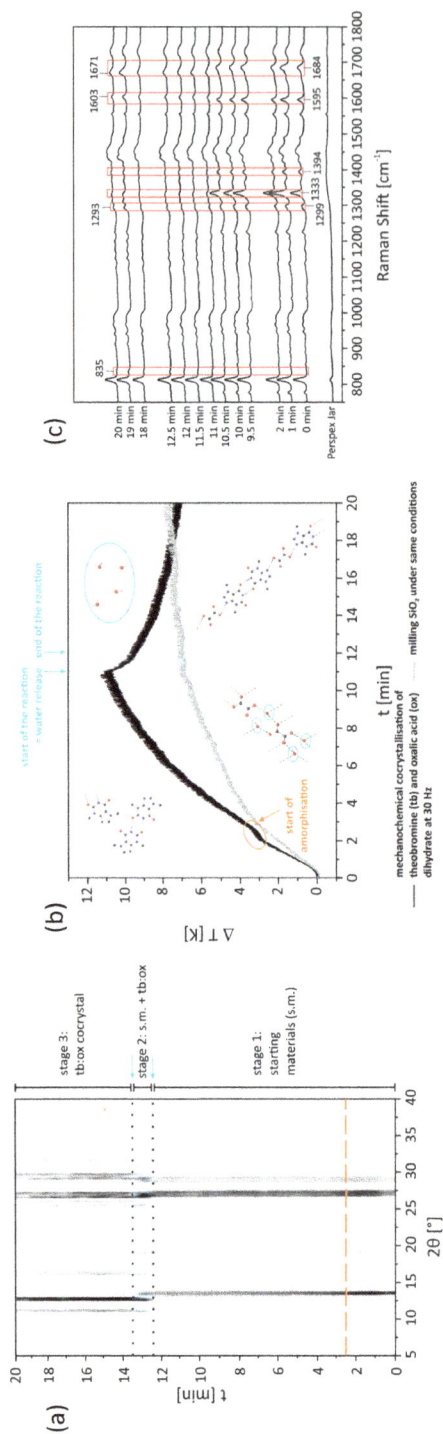

FIGURE 3.7 The use of multiple analytical techniques to follow the kinetics of cocrystallisation of theobromine (tb) and oxalic acid (ox). (a) Following cocrystallisation by TRIS PXRD, (b) following cocrystallisation by TRIS thermometry, and (c) following cocrystallisation by TRIS Raman spectroscopy. Figure adapted from Ref. [89] with permission from the Royal Society of Chemistry.

chemical transformations occurred during this time. Various other examples have been reported which demonstrate the importance of combining TRIS Raman spectroscopy and TRIS PXRD to follow changes in molecular structure,[71] for example, within non-crystalline phases,[98] and has been even essential for locating material that seemed to be "missing" from the PXRD data alone.[99]

3.3 Kinetics and Energy of Induction Period in Mechanochemical Reactions

With regard to a general mechanochemical reaction profile, Figure 3.3, the first kinetic stage of the reaction is the induction period. The induction period – where no apparent transformation occurs – is marked by physical processes, including mixing and material activation. Understanding both these processes is particularly important for understanding the kinetic and energetic (or reactive) behaviour of both Type 1 and Type 2 reactions, while Type 3 reactions typically do not exhibit this kinetic feature. This is because Type 1 reactions typically depend on the formation of mobile fluid intermediate phases at heterogeneous contacts, for example by melting either on contact[23] or by mechanical heating, by dissolving in liquid additive or atmospheric moisture[26], by forming non-crystalline intermediate phases, or by volatilisation.[100] Correspondingly, ensuring efficient mixing is the key to maximising the kinetics of these transformations. Similarly, Type 2 reactions depend both on the formation of heterogeneous contacts and the activation of the material.

To better understand this kinetic dependence, we can take an analogy to reactions between molecules in a fluid phase, which are conceptually well understood. In a fluid, molecules diffuse randomly through the medium and undergo collisions. The rate of fluid phase reaction is defined by the Arrhenius equation:

$$k = A.e^{-E_a/RT} \tag{3.1}$$

The pre-exponential factor, A, describes the rate of collision and probability of favourable collision orientation, and the exponential describes the probability of the collision having enough energy to react at the constant equilibrium temperature, T. This mathematical treatment of fluid phase kinetics assumes a homogeneous distribution of molecules throughout the bulk. Under these conditions, the probability of reaction is identical at all points, thereby allowing the bulk transformation rate to be described as a function of the mean concentration. For example, the rate of reaction between two fluid components, A and B, can be described by:

$$\text{rate} \propto k\left[A\right]^n\left[B\right]^m \tag{3.2}$$

The kinetic description of a reaction between solids is, however, significantly different from fluid phase reactions. First, from the molecular perspective a solid phase system comprising at least two components is never homogeneous,[101] Figure 3.8. Even where, at the particle level, mixing appears to be homogeneous, there exist localised regions of unmixed molecules within each particle: it is not possible to achieve 100% mixing in a solid system.[48] All particles have an intrinsic minimum size that can be achieved through mechanical comminution, which depends on the energy required to fracture the particle: the comminution limit.[40, 102] In any multi-phase solid system, it is only the material at the surface of the particles that can react, Figure 3.8b. Moreover, in a realistic (non-ideally mixed) sample, many contacts are "non-reactive" as they involve the direct interaction between particles of the same phase. Hence, at the elementary level, it is not realistic to directly link bulk "concentration" to the rate of reaction. In fact, recent TRIS PXRD kinetic data have demonstrated that it is not the bulk "concentration" that dictates the reaction profile, but rather the effective *local* composition that is important.[94]

Unlike molecules in a fluid, particles (and hence molecules) in a dry powder sample do not mix as a result of thermal diffusion. Correspondingly, the pre-exponential factor of Equation (3.1) is no longer a statistical thermal average but rather requires the explicit input of energy from the mechanochemical reactor. Moreover, from the perspective of the *molecule*, mixing comprises at least two stages: (1) reducing particle size to maximise surface area and hence the number of molecules available to react; and (2)

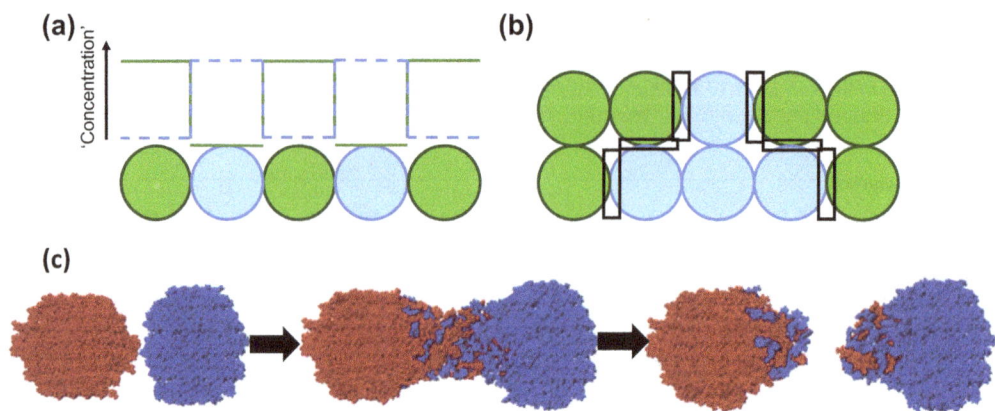

FIGURE 3.8   Schematic representation of particle mixing and solid–solid reaction zones in a binary powder mixture. (a) The local "concentration" profile for a chain of "perfectly mixed" powder particles showing the inherent inhomogeneity of the material and the molecular level; (b) the presence of reactive contacts within a binary powder, indicated by black boxes. (c) The intimate molecular-level mixing that occurs at heterogeneous contacts, simulated by molecular dynamics simulations. Part (c) adapted from Ref. [103] with permission from the Royal Society of Chemistry.

homogenisation of particles. A recent breakthrough in computational molecular dynamics simulations[103] showed that once heterogeneous contacts are formed between particles, their collision can lead to a ready exchange of material from one particle to the other, creating a non-crystalline "molecular alloy" at the particle surfaces in which a reaction can subsequently occur, Figure 3.8c.

A reaction between solids, therefore, comprises at least two superimposed phenomena, mixing (at the bulk and molecular levels) and the chemical or physical transformation, leading to a kinetic rate constant:

$$k_{total} \propto f\left(k_{mix}, k_{rxn}\right) \tag{3.3}$$

This dependence of solid-state reaction kinetics on mixing and surface area has been known for over two centuries,[104] and interest in elucidating the mechanisms of mixing dates to at least the 1940s,[105] with significant work since then to develop a detailed understanding of the complexities of particle mixing.[52, 101, 106–108] TRIS PXRD has been also used to consider the effects of mixing on the kinetics of mechanochemical transformations, using for example the polymorphic conversion of caffeine.[109] In this case the rate of polymorph conversion was studied as a function of milling ball diameter. Semi-quantitative kinetic models revealed that, for larger milling balls, mixing played a less significant role in the macroscopic reaction rate, as the probability of collision on unreacted powder was greater for a greater surface area of the milling ball.

While mixing is certainly central to control the kinetics of a mechanochemical transformation, mixing alone cannot easily explain the sometimes extensive (10 s of minutes, to hours) induction period.[41, 45] In this light, it is worth noting that the kinetics of mechanochemical reactions become even more complex when noting that the energy term (kT) in Equation (3.1) is also not constant,[16] as it is in thermo-chemistry. Instead, mechanochemical reactions are subject to dynamic and localised "bursts" of energy, thereby placing the system in a highly non-equilibrium state that formally requires non-equilibrium thermodynamical concepts to model (although great success has been made through the use of modified concepts of equilibrium thermodynamics as will be discussed briefly below[14]). In this case a reaction can proceed only at a heterogeneous contact if: (i) the global temperature of the vessel is sufficiently high, or (ii) at the moment where mechanical impact is applied to the contact. This concept has led to the development of specialist energy balance and kinetic equations which account for mixing (bulk and comminution), the probability for ball-powder collisions, and the probability that the collision has sufficient energy to drive the desired chemical or physical transformation.[17, 110] Recent models[111–113] have been shown to provide excellent agreement with experimental kinetic data and to provide exceptional new insight into the bulk aspects of mechanochemical transformations.

The process of comminution is itself an excellent example of the intimate connection between the *kinetics* of a mechanochemical reaction and its thermodynamic (energetic) aspects. Comminution is itself bound by energy conservation laws.[17] During comminution, the excess energy driven by the mechanoreactor must lead to the formation of new surfaces, with associated surface free energies. Originally these excess energies were determined from continuum mechanics models developed by Griffith (where $\Delta G_{surf} \propto 2 \times E_{surf}$). A recent breakthrough in TRIS PXRD has for the first time allowed for the accurate measurement of comminution rates during ball milling,[56] demonstrating decisively that comminution is an essential aspect that determines the length of the induction period, Figure 3.9. Using as a model system the organic cocrystallisation of theophylline and benzamide, TRIS PXRD showed that comminution of the starting reagents occurred during the marked induction period, before any signs of the product phases were observed. This strongly suggests that the ability to comminute the material is essential for effective mixing that is necessary to drive forward the subsequent reaction.

It is in the discussion of comminution kinetics that we quickly see parallels between Type 1 and Type 2 mechanochemical reactions. Comminution only occurs according to the simple rules of energy conservation at the macro- and meso-scales. Once the fine grinding regime is reached,[40] inelastic (plastic) processes begin to dominate, wherein microscopic distortion of structural and electronic states of the material result after stressing: mechanical activation.[32] Mechanically activated states were first proposed in the 1940s as thermodynamically and structurally metastable arrangements of lattice elements.[14] In a "classical" thermodynamic view, one can define an activated state in terms of its excess Gibbs free energy, G, taken as the difference in free energies of the equilibrium and activated (asterisked) states:

$$\Delta G = G^* - G = \left(H^* - H\right) - T\left(S^* - S\right) \tag{3.4}$$

This is to say that the activated state will have some increase in the enthalpy of the system, for example defined by the types, concentration, and degree of defects being generated in the structure, and a change in the entropy of the system, also depending on the nature of the defects being developed. The metastability of activated states can be measured by thermal analysis,[114] and both the enthalpy and entropy excess can be measured by thermal analysis (DSC, for example). This was demonstrated in a recent work exploring the origins of extensive induction periods associated with the polymorphic conversion of a stoichiometric cocrystal of carbamazepine and nicotinamide.[41] It was shown that the crystal size of the initial polymorph quickly reaches a constant size (on the nanometre scale) after which continued grinding has no detectable change. Instead, the continued milling leads to a stepwise increase in the lattice enthalpy and the entropy of the starting polymorph, systematically destabilising the phase and driving it towards polymorphic transformation, Figure 3.10.

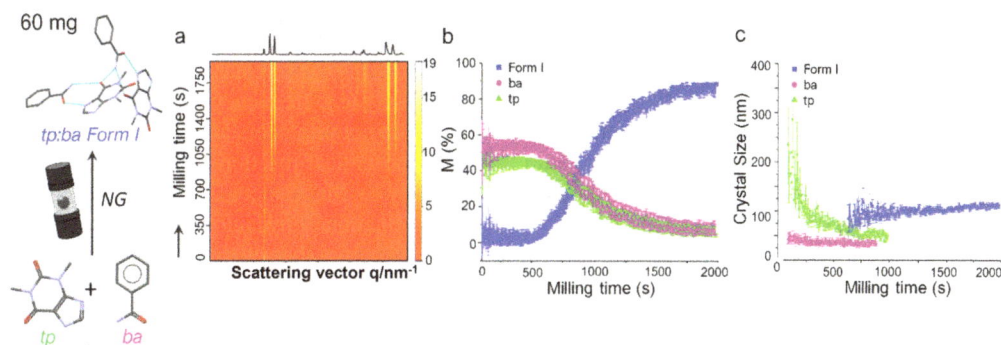

FIGURE 3.9 TRIS PXRD monitoring of the mechanochemical cocrystallisation of theophylline (tp) and benzamide (ba). (a) Reaction scheme for the neat grinding reaction. (b) TRIS PXRD heat map showing the time evolution of the system during ball milling. (c) The quantitative analysis of phase composition obtained by Rietveld refinement of TRIS PXRD profiles. (d) The evolution of crystal size as a function of ball milling, clearly showing that the comminution of the reagents precedes the reaction. Figure reproduced from Ref. [56] with permission.

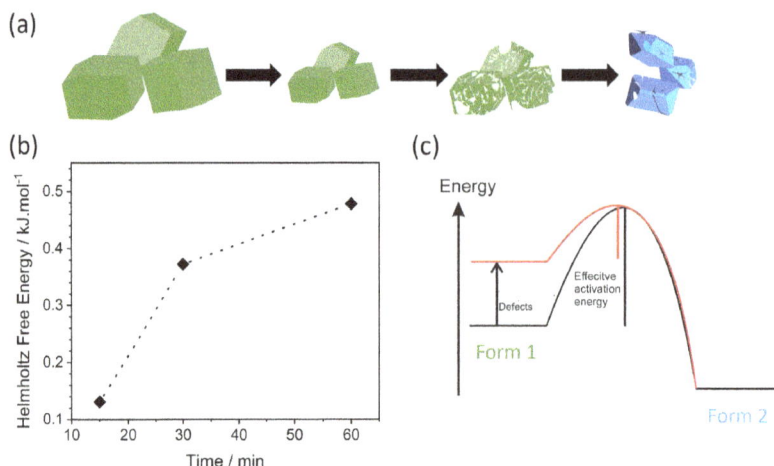

FIGURE 3.10 Mechanical activation of powders leads is responsible for significant induction periods in the polymorphic transformation of a stoichiometric cocrystal of carbamazepine: nicotinamide from Form I to Form II. (a) Schematic representation for the mechanical activation of crystals through the inclusion of defects. (b) The effects of milling time on the free energy of the starting polymorph (Form I) obtained from DSC measurements. (c) Schematic representation for the energy profile of mechanically activated polymorphism. Figure adapted from Ref. [41] with permission.

The need to activate material in this way has been also proposed based on numerical analysis of reaction kinetics from *ex situ* PXRD and HPLC analysis.[45] Similarly, the activation of materials has been studied indirectly, for example through measuring the temperatures required for subsequent thermochemical reactions in mechanically activated powders. As an example, we can highlight the effects of mechanical activation on the thermal decomposition of ammonium perchlorate.[34] As a result of ball milling, the decomposition of mechanically activated powders of ammonium perchlorate decreased by over 100 K, believed to be related to the increased ease of ion migration through mechanically generated defects.[115] Works on self-sustaining reactions (where defect generation leads to the onset of a self-propagating reaction) have been also used to investigate the kinetics of mechanical activation and its role in dictating the reactivity of solids.[116] It is important to highlight that there is strong evidence[117] to suggest that the structure of defective (mechanically activated) solids is close, but distinct, from defective solids prepared from fluids such as a melt, thereby providing yet further evidence that solid-state mechanochemical transformations are unique from their fluid phase counterparts.

Despite decades of research, it is still not always clear how best to design a mechanochemical reaction. It is clear however that a significant aspect of controlling mechanochemical reactions will come once we learn to better control their induction periods. However, relatively simple questions regarding how to control this kinetic regime still do not have answers. For example, it is not yet clear how different reactors affect the onset of mechanochemical reactions, or at what intensity (e.g. frequency, ball mass, number of balls) mechanical treatment should be applied. In addition, we do not yet understand the role of additives (solids, liquids, polymers) in altering the kinetics of mechanochemical reactions. TRIS methods have so far proved useful in studying such questions. A recent study by Martins et al.[46] used TRIS-XRD to follow the macroscopic kinetics of the mechanochemical chlorination of hydantoin as a function of the size (and mass) of the milling balls used. Of particular note, the authors showed that the induction period (as well as the time to reach a final steady state) became longer as the milling ball size decreased, albeit with a non-linear trend, Figure 3.11. Analytical kinetic analysis of the data showed that this non-linearity related to the volume of the ball, and thus to the amount of material hit with each impact. This is similar to earlier studies using TRIS PXRD that explored the kinetics of mixing and its role in determining the rate of polymorphic conversion in caffeine, where it was found that mixing kinetics were more prevalent when a smaller milling ball was used.[109] These in situ findings are consistent with additional *ex situ* studies which showed that both the volume and mass of the milling bodies are influential over the rate of mechanochemical transformations.[118] An alternative analysis was performed by Kulla et al.[119] who used

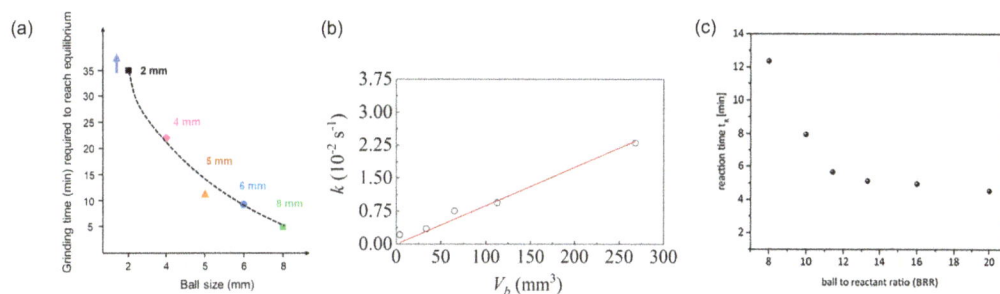

FIGURE 3.11 Influencing the kinetics of mechanochemical transformations by ball size. (a and b) The milling time required for the chlorination of hydantoin as a function of milling time, and the apparent rate constants as a function of milling ball volume. Figures adapted with permission from Ref. [46]. Copyright 2021 American Chemical Society. (c) Effect of ball-to-reactant ratio on the rate of mechanochemical synthesis of a 1:1 cocrystal of theophylline + benzamide. Figure reproduced from Ref. [119] with permission from the Royal Society of Chemistry.

TRIS-XRD to study the effects of ball-to-powder mass ratio on the overall kinetics of the transformation. To a similar effect, it was found that a higher ball-to-powder mass ratio also led to faster reaction kinetics, Figure 3.11c, presumably due to a reduced need for mixing (i.e. more probable collisions on unreacted powders). Most recently using TRIS Raman spectroscopy, attempts to decouple mixing from mechanical impact energy on a mechanochemical reaction were reported, showing a linear relationship between kinetics and impact energy. Such studies suggest that if mixing (comminution and bulk mixing) can be controlled, there is real hope for being able to selectively design mechanochemical reactions.[120] Further dedicated studies by TRIS methods are essential for resolving outstanding issues in understanding the role of mixing, comminution, and mechanical activation. Already using TRIS, we have obtained a deeper understanding of how to control mechanochemical reactions, and further studies promise exciting developments in this direction.

3.4 Kinetics and Energy of Reaction Period in Mechanochemistry

When a solid reaches an appropriate energy condition for reaction (see, e.g. Section 3.3), the reaction period of the mechanochemical transformation can begin, Figure 3.3. It is worth noting that in many systems – notably Type 3 reactions like the mechanical initiation of explosive materials – there is no need for an activation or mixing step, and the material enters the reaction period immediately. The mechanisms and conditions of the solid in the reaction period are highly dependent on the nature of the reaction (see Table 3.1). As Type 1 reactions are purely thermal, and proceed through mixing (see Section 3.3), we here outline only some of the prevailing theories that can dictate the kinetics and energetics of Type 2 and Type 3 reactions.

3.4.1 Force-Modified Potential Energy Surface

When mechanical stress is applied to a system, it has the effect of modifying the underlying potential energy surface (PES). Various theoretical tools have been developed to investigate the effects of such distortion on the energy landscapes of molecular systems. The COGEF (Constrained Geometry to Simulate Forces) approach – or its variations based on modelling force-transformed potential energy surfaces – has been of particular note in this regard.[121] Within the COGEF scheme, the distance between a pair of nuclear coordinates is perturbed and fixed (to replicate a mechanical strass), and the remainder of the system is allowed to relax under the strained potential energy surface using either *ab initio* or empirical methods. This approach has proved to be exceptionally powerful for modelling single-molecule mechanochemistry, for example the stretching disulfide using atomic force microscopy,[37] and has been also applied, for example to the study of the reactivity of covalent bonds under mechanical strain,[122] the effects of mechanical strain on chemical reactions,[123, 124] and the effects of strain on polymers.[125] Similar approaches, considering how molecular distortions affect chemical reactivity, have been also explored using *ab initio* models in other chemical reactions.[126]

For crystalline systems, it has been instead more common to explore mechanochemical reactions under strains applied to the unit cell, either through hydrostatic or uniaxial compression. In this case the unit cell dimensions are distorted to mimic the mechanical stress and the rest of the system is allowed to "relax" in response to the distortion. Such methods are routinely applied to study the high-pressure behaviour of materials[127–130] and have been also used to explore the effects of strain on the chemical reactivity of solids. For example, this approach was used to demonstrate the shear-driven chemical reactivity of a model explosive material, FOX-7, Figure 3.12.[131] We note that such deformations are akin to mechanical activation (see discussion in Section 3.4), highlighting the intimate interplay between the kinetics of the induction period and the onset of mechanochemical reactivity, Figure 3.3. Similar mechanisms were responsible for the enhanced thermal decomposition of, e.g. ammonium perchlorate after mechanical activation, where energy barriers for ion diffusion were reduced, thereby facilitating chemical transformations in the solid state, [115] including when considering tribochemical reactions at interfaces.[132]

The ability to distort molecular and crystal structures selectively and controllably *in silico* and probe the effects of this distortion on the structure, dynamics, and electronic properties has been a critical advancement in studying the energetic aspects of mechanochemical transformations. This is exemplified by the experimental synthesis of dibenzophenazine by ball milling.[133] Conventional frontier orbital theory failed to rationalise the reaction mechanism for the formation of dibenzophenazine. Instead, by applying selective mechanical force to the reagents, Figure 3.13, the authors could show that a new – previously inaccessible – reaction path becomes possible *via* the dissolution of an imaginary frequency. In this way the authors provided a non-classical reaction scheme based on a force-modified potential energy surface that is unique to mechanochemistry to rationalise the experimental synthesis.

While theoretical tools have become essential for the fundamental study of mechanical force on the stability of molecules and solids, the effects of force-modified potential energy surfaces on the kinetics and thermodynamics of solid-state reactions have been also probed experimentally. Such investigations have been routinely investigated using controllable high-pressure devices, such as diamond anvil cells. In these devices, pressure can be applied hydrostatically or non-hydrostatically, and the material inside the cell can be measured using, e.g. X-ray diffraction or vibrational spectroscopy. In this light high-pressure science has been a cornerstone technique for understanding how mechanical strain can alter the potential energy surface, and hence reactivity, of materials. Excellent dedicated reviews on high-pressure science are available elsewhere.[19, 134] An excellent example of driving mechanochemical

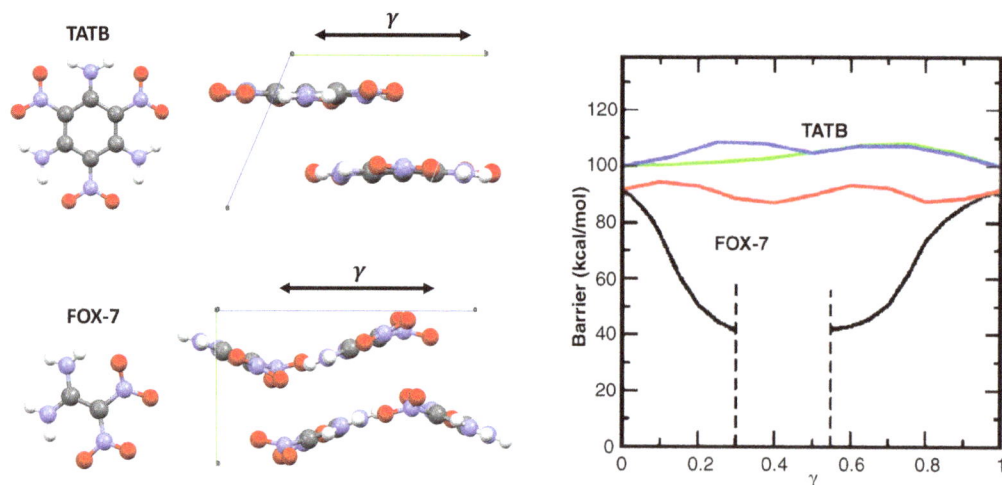

FIGURE 3.12 Effects of mechanical strain on the energy barrier to covalent bond rupture in model organic materials FOX-7 and TATB. As interlayer planes are sheared, the energy barrier to thermal NO_2 bond scission is reduced. Between shear magnitudes (γ) of 0.3 and 0.55, there is no energy barrier, leading to a spontaneous bone rupture in FOX-7. Figure reprinted from Ref. [131] with permission of AIP Publishing.

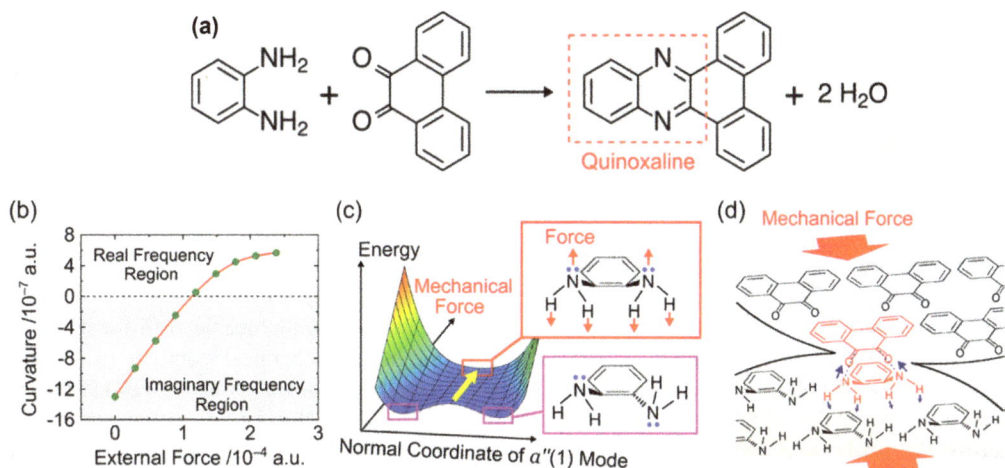

FIGURE 3.13 The effects of mechanical force on the potential energy surface of o-phenylenediamine, a reagent in the synthesis of dibenzophenazine. (a) Reaction scheme for the synthesis of dibenzophenazine. (b) The effects of mechanical force on the curvature of the potential energy surface of o-phenylenediamine. (c) The structure of o-phenylenediamine at different positions on the force-modified potential energy surface, and (d) schematic representation for the putative mechanism of reaction under mechanochemical conditions. Figure adapted with permission from Ref. [133]. Copyright 2019 the American Chemical Society.

reactions through force-modification of the potential energy surface is the redox reaction in a metal-organic chalcogenide, wherein bending bond angles through mechanical force activated the metal chalcogen bonds in Cu(I) m-carborane-9-thiolate crystals.[135] Although less common, experimental devices have been also established to apply simultaneously pressure and shear, attempting to better identify the role of anisotropic strains in driving physico-chemical transformations in solids.[136] Reactions driven by shear diamond anvil cells have been also modelled using *ab initio* modelling, where it was demonstrated that the simple amino acid glycine could be dimerised simply by allowing the molecules to relax under an applied mechanical force.[137] These studies suggest that prebiotic chemistry may in fact have a mechanochemical origin.

Using high-pressure devices, it has been also possible to carefully control the rate of applying stress to the sample. Such studies have clearly shown that the rate of applying stress can play a critical role in determining the outcome of a mechanochemical transformation, for example the polymorphism of L-serine.[35] In these studies, where the pressure is applied slowly, the rate of stress affects the structure of the PES on which the nuclei relax, thereby dictating the outcome of the transformation. Ultrafast compression, however, drives a different mode of action (see Section 3.4.2).

3.4.2 Elasto-Chemical Transitions

Some mechanochemical reactions are initiated directly by the dynamic application of mechanical strain, without the need for any activation stage or equilibration of nuclear coordinates on a modified potential energy surface. This class (Type 3) of reaction poses an exciting fundamental question for mechanochemistry, as their mechanism of action relies heavily on the microscopic kinetics of energy transfer from a mechanically excited solid into chemically active bonds. Moreover, mechanisms exist wherein no thermalisation (or equilibration) of this excitation occurs, thereby bypassing any conventional consideration for the rupture of the "weakest" or "thermodynamically most favourable" bonds.[36] Hence, many Type 3 reactions appear to be purely kinetic (out of equilibrium) in nature. Such mechanisms fall largely into the field of shock-driven chemistry, where readers are referred to elsewhere for a thorough picture of the field.[39] Here we elect to discuss only one growing theory in the field of mechanochemistry.

In this theory for Type 3 reactions, mechanical excitation of the lattice occurs by super-heating the lattice vibrations in the material.[38] This excess energy scatters and leads to the subsequent activation of molecular vibrations, driving them into highly excited vibrational states. Where this energy excites a molecular vibration, it can drive transient, dynamical metallization[138] and hence cause chemistry to occur on ultrafast timescales. Early works on this mechanistic framework[139, 140] were accompanied by exciting experimental evidence based on ultrafast spectroscopies. For example, the ultrafast excitation of a molecular dye was used to selectively excite low-frequency vibrations in a liquid sample of nitromethane, Figure 3.14a. This low-frequency excitation is analogous to the early stages of a mechanical shock.[139, 141] Through the use of ultrafast incoherent anti-Stokes Raman scattering spectroscopy, the evolution of this excess energy was followed through the material. Within 10 s of picoseconds, direct evidence for vibrational "up-pumping" was observed, wherein the excitation is internally transferred into higher frequency vibrational modes, capable of inducing a chemical response.[138]

Based on recent advances in *ab initio* electronic structure methods, this mechanism for mechanochemistry has been recently explored as a tool to predict the mechanochemical reactivity of explosive compounds.[141–143] By studying the relative rate of energy transfer from the lattice vibrations into the manifold of molecular vibrations, the mechanochemistry of a diverse class of energetic materials was successfully predicted, Figure 3.14b.[142] Importantly, it was shown that not only could the kinetics of vibrational energy transfer predict the reactivity of different molecules, but it could capture and explain how different crystal structures exhibit different mechanochemical reactivity.[143] This success provides convincing evidence for the foundational basis of ultrafast vibrational energy transfer mechanisms as being responsible for Type 3 mechanochemical reactions.

Growing evidence from in situ thermal analysis using thermometry has suggested that bulk temperatures do not rise significantly during ball milling.[22, 89] It is however important to note that such findings do not discount the formation of highly localised, short-lived, extreme quasi-temperatures at the side of mechanical stress. Following from concepts of the kinetics of vibrational energy transfer, these extremely "hot" localised states were recently measured in polymer-coated single organic crystals exposed to ultrasonic shock, which revealed quasi-temperatures in excess of 600 K over the period of 10 s of ms, resulting from friction between the polymer coating and the single crystal, Figure 3.15. Similar local temperatures were measured by using grits of defined melting temperatures added to explosive powders and monitored by photographic paper to locate successful explosive initiation.[145] It was found that only grits with sufficiently high melting temperatures could initiate the explosives, thereby indicating the local temperatures (>500 °C) were required.[43] It is such studies that continue to cause debate as to the true differences between mechanochemical reactions and thermochemical reactions. Although there is clear evidence that demonstrates that thermo- and mechano-chemical reactions are different,

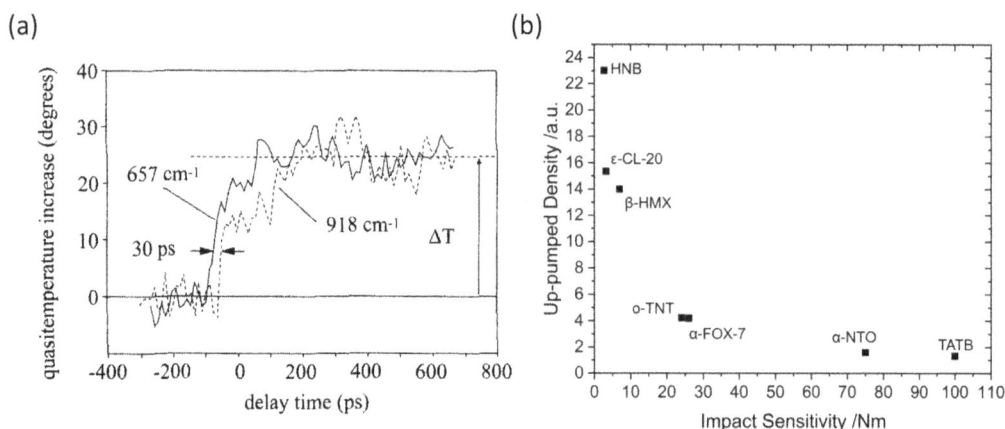

FIGURE 3.14 Vibrational up-pumping as a mechanism for mechanochemical reactions. (a) Up-pumping quasi-temperatures in liquid nitromethane measured through incoherent anti-Stokes Raman scattering. Figure reproduced with permission from Ref. [144]. Copyright 1994 the American Chemical Society. (b) Prediction of the mechanochemical reactivity of energetic materials through *ab initio* vibrational up-pumping models. Figure adapted from Ref. [142] with permission from the Royal Society of Chemistry.

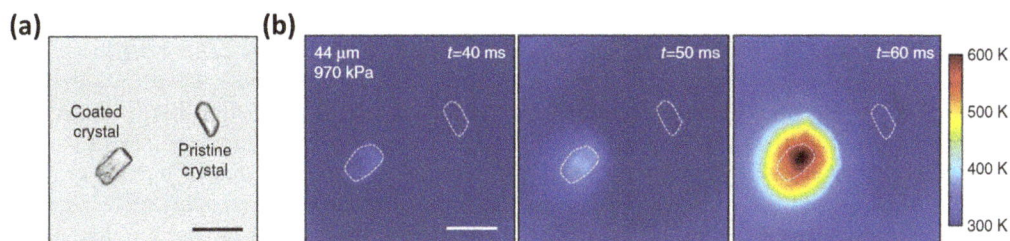

FIGURE 3.15 The evolution of local dynamic quasi-temperature in response to dynamic mechanical stress. (a) Microphotographs of organic crystals, and (b) their temperature in response to a shock impulse after some time. Figure reprinted from Springer Nature, Ref. [146], Copyright 2015.

identifying elementary mechanisms to unambiguously differentiate them promises to remain an exciting and dynamic area of research for years to come.

3.5 Thermodynamic and Energetic Aspects of Mechanochemical Product Phase

The final stage of the reaction profile for a mechanochemical reaction is the formation of the final product, which should – in principle – remain stable under continued mechanical treatment (although subsequent reactions are not uncommon[147]). The results of the chemical/molecular reactions are determined by the reaction mechanism itself and the structure of the potential energy surface upon which the reaction occurs (see Sections 4 and 5). What is unique to mechanochemical reactions, however, is that the product phases typically crystallise, leading to the formation of a solid product. This leads to an exciting range of potential final product phases that can result, with a given product molecule/atomic composition being able to crystallise as very different crystal forms (polymorphs,[148] solvates, etc.) with very different physico-chemical properties. We will focus here on outlining some of the implications of this solid product on the final solid product that is obtained from common mechanochemical transformations.

Unlike in solution where a reaction occurs through the bulk, mechanochemical reactions in a powder occur in very localised regions,[149] and only at the point of heterogeneous contacts, Figure 3.8. This likely has important implications for the composition of the final powder, as the reaction zones are not in a state of mutual equilibrium. The effects of this mean that the outcomes of a solid-state mechanochemical transformation and a solution-phase reaction are not necessarily the same. A striking example was demonstrated for a reversible disulfide exchange reaction.[80] In solution, a thermodynamic equilibrium is reached wherein a statistical mixture of the starting reagents and product form is obtained. In contrast, when the same reaction was performed by ball milling, the reactants were converted entirely to the product phase. While the lack of dynamic equilibrium in the solid state is presumably in part responsible for this effect, the authors also demonstrated that the crystallisation of the product form also contributes to excess stabilisation of the product phase over the reagents.

A key challenge in the discussion of the energetics of mechanically facilitated reactions is to identify the relative stabilities of the species involved. A common approach to identifying the thermodynamic landscape of mechanochemical reactions has been through the use of *ab initio* calculations. Once the crystal structure is known for each of the starting materials and product phases, it becomes possible to compute the "ideal" thermodynamic stability of each phase and hence to obtain the energy landscape for the system under investigation.[22, 66, 150] Such analysis has been used, for example to identify which polymorphic forms of a given mechanochemical product are thermodynamically stable and which are metastable. In this way it has helped to elucidate some features of mechanochemical reactions (such as the jar material) that can be used to selectively stability unconventional products.[65] Experimentally, thermal analysis and slurry experiments have often been the standard methods for assessing the relative stability of solid forms.

All the above methods assume bulk thermodynamic stability of the reacting system. However it is well known for inorganic materials,[60, 151, 152] and becoming clearer for organic solids,[50] that mechanochemical reactions are dominated by nano-scale materials, and hence that such studies of bulk thermodynamic stabilities might not be appropriate for rationalising mechanochemical reactions. This is because, unlike in solution where thermodynamic equilibrium is generally achieved, mechanochemical reactions are predominantly kinetically driven (*i.e.* metastable products can be easily trapped[114]). To this end, a number of reports have used so-called "competitive milling" experiments as a means to test the relative stability of different crystallographic phases under milling conditions.[153] In a similar mindset, extensive work has demonstrated that different solid forms can be very reliably and reversibly produced if the appropriate milling conditions are known.[97] Such studies have the important implication that, despite the seemingly stochastic nature of mechanochemical transformations, the thermodynamic landscape of mechanochemical reactions can be controlled to a high degree.

However, developing control over the final products of a mechanochemical transformation remains largely trial and error, with the critical need to better understand what factors influence the outcome of the transformation. To this end, one can perhaps think of mechanochemical products as being similar to crystallisation in confined conditions,[154] where the reaction is confined to a small amount of material within a small volume (*i.e.* only the material available at the heterogeneous particle contacts, Figure 3.8). Nano-crystalline phases are rarely the "thermodynamically stable" products (*i.e.* they would prefer to grow larger if possible), but the enormous kinetic barriers to mixing and diffusion in a powdered system offer a phenomenal opportunity to trap nanoparticles. It is well understood that the stability of nanocrystalline phases can differ significantly from bulk phases.[155] This results from the dominating surface free energy in nanocrystals, as compared with the dominating bulk free energy of larger crystals.[49] It follows that when considering bulk mechanochemical transformations in solids, the thermodynamic stability of bulk phases unlikely dominates the result. Instead, the kinetic trapping of nano-crystallites under ball milling conditions demands an additional term be added to traditional "thermodynamic" analysis. Lin et al.[49] have derived such a function, noting that for small (e.g. nanocrystalline) phases, the free energy is given by:

$$G^* = G_b + A_\gamma + E \tag{3.5}$$

where G_b is the free energy of the bulk, γ is the surface energy, A denotes the surface area, and E describes any additional strain energy contained within the crystal as a result of mechanical deformation.

Belenguer and colleagues realised that conventional mechanochemical processing (e.g. ball milling) in fact leads to the formation of nanocrystalline materials in organic solids.[50] In this light, they identified that a complete understanding of the outcome of such transformations requires consideration of these surface-dominating effects, with emphasis on their role in the polymorphism in organic molecular solids.[50, 81, 81, 97, 157] Studying a series of polymorphic transformations demonstrated a highly reproducible correlation between the obtained polymorphic form and its crystal size (on the order of <100 nm). *Ab initio* simulations suggested instead that it is the excess surface energy at a given polymorph size that correlates with the obtained polymorph, Figure 3.16.[50] This concept has been further extended towards considering non-ideal surfaces through the addition of surface rugosity.[158] Importantly, it was also shown that the stability of these surfaces can be manipulated by very small changes in the quantity of additives like solvents, Figure 3.16b,[157] thereby offering an exciting and diverse opportunity to control the products of mechanochemical transformations. The importance of crystal size/exposed surface area on the solid form stability of inorganic compounds is also known, for example as in the polymorphs of Al_2O_3 and AlOOH.[156] Moreover, the surfaces of these inorganic polymorphs can be also affected by the ball milling process, which can lead to changes in the relative stability of the solid forms. Recent thermodynamic arguments[159] have suggested that crystal size might be even more important than just controlling the relative stability of crystalline forms, indicating that, by milling crystals to sizes below a critical threshold, one can even cause crystals to spontaneously amorphise. Thus, controlling and characterising the size of mechanically prepared crystals, understanding the structure of their surfaces, and how mechanical treatment affects these surfaces are essential for controlling the outcome of a mechanochemical reaction.

Recognising the dominating effects of understanding surface free energies and their influence on the resulting solid form, the inclusion of surface energies has been recently extended into crystal structure

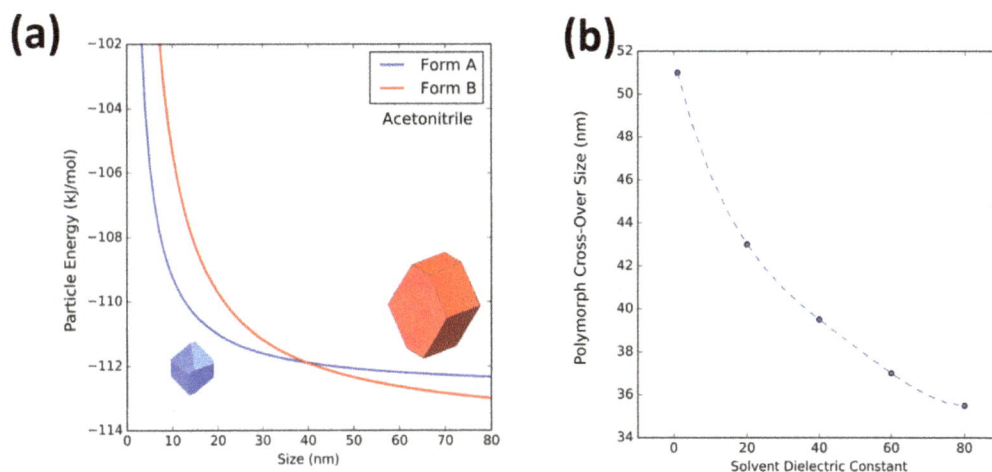

FIGURE 3.16 The effects of particle size on the stability of two polymorphs of an organic solid, with energies dominated by excess surface energy. (a) The effects of particle size on the relative stability of two polymorphic forms. (b) The effects of different solvents on stabilising the relative stability of the polymorphs. Figure adapted from Ref. [80] with permission from the Royal Society of Chemistry.

prediction methods.[160] This transformative view of crystal structure-directing effects is challenging our current understanding of how to pre it is now possible to extract reliable crystal sizes and microstructure from TRIS data;[56] further studies are needed to understand the morphology of mechanochemically prepared solids, and hence to get a better understanding of the influence of their surfaces on the stability of the resulting product.

In addition to the effects of surface energy on the relative phase stabilities, local, transient conditions of pressure and temperature that form, e.g. on impact in a ball mill[161] are also likely to influence the nature of the solid that forms. From model studies using high-pressure devices, crystallisation under these extreme thermodynamic conditions has been shown to yield unconventional crystal forms,[162, 163] many of which are recoverable to ambient conditions. These local extreme thermodynamic conditions may also play an important role in facilitating the chemical reactions in the first place.[164] While these extreme conditions have been thoroughly investigated using model devices, the exact conditions achieved in conventional mechanochemical reactors are not well characterised. As a result, the thermodynamic conditions under which mechanochemically prepared materials form are largely unknown. As mechanochemical reactors are better characterised, for example by using piezoelectric sensing[165, 166] and numerical simulations,[96] we will undoubtedly learn more about the thermodynamic and kinetic features that dictate the outcome of mechanochemical transformations.

3.6 Summary and Outlook

Mechanochemical transformations promise to change the way academic and industrial chemical and materials manufacturing is being done. Before this transformation can be fully realised, however, significant advancement is needed in our understanding of how mechanochemical transformations occur and how we can control them. At the core of this is a need to better elucidate the driving forces that underpin mechanochemical reactions, and how these driving forces affect the reaction kinetics and the resulting products. The current paradigm of mechanochemistry encompasses an enormous range of reaction types that are somehow driven or facilitated by mechanical action. On one side there are reactions that occur thermally at the interface between solid particles, where mechanical action mainly provides a mechanism for mixing and creating new particle–particle contacts. On the other side are reactions that are driven directly by the absorption and transformation of the dynamically applied mechanical energy. Somewhere in between are thermal reactions that are possible (or enhanced) because of how mechanical stress changes the underlying potential energy surface. While these are all considered to be mechano-chemical reactions, their mechanisms clearly differ. Hence the approaches needed to study them and control their kinetic and thermodynamic features differ as well. As we continue to push the boundaries of mechanochemical technologies, it becomes increasingly important to appreciate the immense complexity of this class of reaction, from which we can begin to identify the appropriate and targeted strategies to selectively manipulate and control their course and result.

In the spirit of pushing the boundaries of understanding and controlling mechanochemical reactions, enormous efforts have been devoted to the development of new experimental and computational mechanisms to probe reactions. Of particular note are the developments in time-resolved in situ (TRIS) monitoring of mechanochemical reactions. TRIS has opened the door to entirely new dimensions in our ability to follow mechanochemical reactions in real time and hence to extract unprecedented detail of their reaction course. This has allowed us to identify conserved kinetic features in reaction profiles without ambiguity, including features such as the induction period, reaction (growth) period, and product period. The induction period – consisting largely of mixing, comminution, and mechanical activation phenomena – can be monitored by various TRIS methods, including TRIS X-ray diffraction. Instead, the reaction period – wherein the chemical or physical transformation occurs – often requires the use of other experimental probes such as X-ray absorption spectroscopy, Raman spectroscopy, or fluorescence spectroscopy, which follow more closely the local chemical or electronic structure. Finally, the product period is being also regularly investigated by TRIS X-ray diffraction, coupled closely with *ab initio* simulations. However, there remain many open questions as to the nature of the thermodynamic conditions in which products appear, and how we can begin to selectively control the outcomes.

It is very clear that a complete study of the kinetics and thermodynamic aspects of a mechanochemical reaction requires a complementary set of analytical probes. This is to say that different probes have very different sensitivities and provide often different viewpoints of the reaction course. Although many TRIS probes have been already developed, there is still plenty of space for the development of new techniques, and room for improvement on existing methods to provide faster, more localised, and more detailed analysis of mechanochemical reactions in real time. These developments will continue to go hand-in-hand with well-established and powerful ex situ analytical methods to provide a complete picture of mechanochemistry. Mechanochemical transformations are inherently non-equilibrium processes, associated with highly transient and likely extreme local conditions. Correspondingly, to elucidate the complete picture of mechanochemical transformations will require the development both of faster, more sensitive, and more sophisticated experimental techniques and of new theories of the kinetics and thermodynamics of mechanochemical transformations. This promises to be an exciting challenge for years to come. As we continue to push the boundaries in our understanding of how and why mechanochemical reactions progress, we will undoubtedly acquire new capabilities to control them. In doing so, mechanochemistry will continue to grow as a dynamic and exciting field with the real potential to change the world.

Acknowledgements

I am grateful for the many years of fruitful discussions with Prof. E. V. Boldyreva and Acad. Prof. V. V. Boldyrev whose insights into mechanochemical reactions have shaped much of our understanding of the field today. I would also like to express my thanks to Dr P. F. M. de Oliveira and Dr F. Emmerling for their critical reading of the manuscript and for their helpful comments and discussions.

Note

1. This categorization represents an oversimplification and aims only to facilitate discussion of the types of processes to be considered when studying mechanochemical reactions. Moreover, we stress that there are no hard boundaries between these classes.

REFERENCES

1. Michalchuk, A. A. L.; Boldyreva, E. V.; Belenguer, A. M.; Emmerling, F.; Boldyrev, V. V. Tribochemistry, Mechanical Alloying, Mechanochemistry: What Is in a Name? *Front. Chem.* 2021, *9*, 29. https://doi.org /10.3389/fchem.2021.685789.

2. Takacs, L. The Historical Development of Mechanochemistry. *Chem. Soc. Rev.* 2013, *42*(18), 7649. https://doi.org/10.1039/c2cs35442j.

3. Akhavan, J. *The Chemistry of Explosives*, 3. Cambridge: RSC Publishing, 2011.

4. Alex, T. C.; Kumar, R.; Roy, S. K.; Mehrotra, S. P. Mechanical Activation of Al-Oxyhydroxide Minerals–A Review. *Miner. Process. Extr. Metall. Rev.* 2016, *37*(1), 1–26. https://doi.org/10.1080 /08827508.2015.1055626.

5. Ostwald, W. *Die Chemische Literatur Und Die Organisation Der Wissenschaft.* Leipzig, Germany: Akademie-Verlag, 1919.

6. Andersen, J.; Mack, J. Mechanochemistry and Organic Synthesis: From Mystical to Practical. *Green Chem.* 2018, *20*(7), 1435–1443. https://doi.org/10.1039/c7gc03797j.

7. Crawford, D. E.; Casaban, J. Recent Developments in Mechanochemical Materials Synthesis by Extrusion. *Adv. Mater.* 2016, *28*(27), 5747–5754. https://doi.org/10.1002/adma.201505352.

8. Gomollón-Bel, F. Ten Chemical Innovations That Will Change Our World: IUPAC Identifies Emerging Technologies in Chemistry with Potential to Make Our Planet More Sustainable. *Chem. Int.* 2019, *41*(2), 12–17. https://doi.org/10.1515/ci-2019-0203.

9. Boldyrev, V. V.; Tkáčová, K. Mechanochemistry of Solids: Past, Present, and Prospects. *J. Mater. Synth. Process.* 2000, *8*(3/4), 121–132. https://doi.org/10.1023/A:1011347706721.

10. Iwasaki, T.; Yabuuchi, T.; Nakagawa, H.; Watano, S. Scale-Up Methodology for Tumbling Ball Mill Based on Impact Energy of Grinding Balls Using Discrete Element Analysis. *Adv. Powder Technol.* 2010, *21*(6), 623–629. https://doi.org/10.1016/j.apt.2010.04.008.

11. am Ende, D. J.; Anderson, S. R.; Salan, J. S. Development and Scale-Up of Cocrystals Using Resonant Acoustic Mixing. *Org. Process Res. Dev.* 2014, *18*(2), 331–341. https://doi.org/10.1021/op4003399.

12. Reichardt, C. Solvents and Solvent Effects: An Introduction. *Org. Process Res. Dev.* 2007, *11*(1), 105–113. https://doi.org/10.1021/op0680082.

13. Michalchuk, A. A. L.; Emmerling, F. Time-Resolved In Situ Monitoring of Mechanochemical Reactions. *Angew. Chem. Int. Ed.* 2022, *61*(21), anic.202117270 https://doi.org/10.1002/anie.202117270.

14. Heinicke, G. *Tribochemistry.* Berlin: Akademie-Verlag, 1984.

15. Urakaev, F. K.; Boldyrev, V. V. Mechanism and Kinetics of Mechanochemical Processes in Comminuting Devices 1. Theory. *Powder Technol.* 2000, *107*(1–2), 93–107. https://doi.org/10.1016/ S0032-5910(99)00175-8.

16. Butyagin, P. Y. Kinetics and Nature of Mechanochemical Reactions. *Russ. Chem. Rev.* 1971, *40*(11), 901–915. https://doi.org/10.1070/RC1971v040n11ABEH001982.

17. Butyagin, P. Y. The Kinetics and Energy Balance of Mechanochemical Transformations. *Phys. Solid State* 2005, *47*(5), 856–862. https://doi.org/10.1134/1.1924845.

18. Weymouth, A. J.; Riegel, E.; Gretz, O.; Giessibl, F. J. Strumming a Single Chemical Bond. *Phys. Rev. Lett.* 2020, *124*(19), 196101. https://doi.org/10.1103/PhysRevLett.124.196101.

19. Zakharov, B. A.; Boldyreva, E. V. High Pressure: A Complementary Tool for Probing Solid-State Processes. *CrystEngComm* 2019, *21*(1), 10–22. https://doi.org/10.1039/C8CE01391H.

20. Boldyreva, E. Mechanochemistry of Inorganic and Organic Systems: What Is Similar, What Is Different? *Chem. Soc. Rev.* 2013, *42*(18), 7719. https://doi.org/10.1039/c3cs60052a.

21. Boldyrev, V. V.; Avvakumov, E. G. Mechanochemistry of Inorganic Solids. *Russ. Chem. Rev.* 1967, *40*(10), 847.

22. Kulla, H.; Becker, C.; Michalchuk, A. A. L.; Linberg, K.; Paulus, B.; Emmerling, F. Tuning the Apparent Stability of Polymorphic Cocrystals through Mechanochemistry. *Cryst. Growth Des.* 2019, *19*(12), 7271–7279. https://doi.org/10.1021/acs.cgd.9b01158.

23. Rothenberg, G.; Downie, A. P.; Raston, C. L.; Scott, J. L. Understanding Solid/Solid Organic Reactions. *J. Am. Chem. Soc.* 2001, *123*(36), 8701–8708. https://doi.org/10.1021/ja0034388.

24. Mazzeo, P. P.; Prencipe, M.; Feiler, T.; Emmerling, F.; Bacchi, A. On the Mechanism of Cocrystal Mechanochemical Reaction via Low Melting Eutectic: A Time-Resolved In Situ Monitoring Investigation. *Cryst. Growth Des.* 2022, *22*, 4260–4267. https://doi.org/10.1021/acs.cgd.2c00262.

25. Michalchuk, A. A. L.; Tumanov, I. A.; Drebushchak, V. A.; Boldyreva, E. V. Advances in Elucidating Mechanochemical Complexities via Implementation of a Simple Organic System. *Faraday Discuss.* 2014, *170*, 311–335. https://doi.org/10.1039/C3FD00150D.

26. Tumanov, I. A.; Michalchuk, A. A. L.; Politov, A. A.; Boldyreva, E. V.; Boldyrev, V. V. Inadvertent Liquid Assisted Grinding: A Key to "Dry" Organic Mechano-Co-Crystallisation? *CrystEngComm* 2017, *19*(21), 2830–2835. https://doi.org/10.1039/C7CE00517B.

27. Michalchuk, A. A. L.; Hope, K. S.; Kennedy, S. R.; Blanco, M. V.; Boldyreva, E. V.; Pulham, C. R. Ball-Free Mechanochemistry: *In Situ* Real-Time Monitoring of Pharmaceutical Co-crystal Formation by Resonant Acoustic Mixing. *Chem. Commun.* 2018, *54*(32), 4033–4036. https://doi.org/10.1039/C8CC02187B.

28. Anderson, S. R.; am Ende, D. J.; Salan, J. S.; Samuels, P. Preparation of an Energetic-Energetic Cocrystal Using Resonant Acoustic Mixing. *Propellants Explos. Pyrotech.* 2014, *39*(5), 637–640. https://doi.org/10.1002/prep.201400092.

29. Titi, H. M.; Do, J.-L.; Howarth, A. J.; Nagapudi, K.; Friščić, T. Simple, Scalable Mechanosynthesis of Metal–Organic Frameworks Using Liquid-Assisted Resonant Acoustic Mixing (LA-RAM). *Chem. Sci.* 2020, *11*(29), 7578–7584. https://doi.org/10.1039/D0SC00333F.

30. Teoh, Y.; Ayoub, G.; Huskic, I.; Titi, H.; Nickels, C.; Herrmann, B.; Friscic, T. SpeedMixing: Rapid Synthesis and Discovery of Model Pharmaceutical Cocrystals without Milling or Grinding Media. *ChemRxiv* 2022. https://doi.org/10.26434/chemrxiv-2022-dgdfr.

31. Beyer, M. K.; Clausen-Schaumann, H. Mechanochemistry: The Mechanical Activation of Covalent Bonds. *Chem. Rev.* 2005, *105*(8), 2921–2948. https://doi.org/10.1021/cr030697h.

32. Boldyrev, V. V. Mechanochemistry and Mechanical Activation of Solids. *Russ. Chem. Rev.* 2006, *75*(3), 177–189. https://doi.org/10.1070/rc2006v075n03abeh001205.

33. Rossberg, M.; Khairetdinov, E. F.; Linke, E.; Boldyrev, V. V. Effect of Mechanical Pretreatment on Thermal Decomposition of Silver Oxalate under Nonisothermal Conditions. *J. Solid State Chem.* 1982, *41*(3), 266–271. https://doi.org/10.1016/0022-4596(82)90145-1.

34. Dolgoborodov, A. Y.; Streletskii, A. N.; Shevchenko, A. A.; Vorobieva, G. A.; Val'yano, G. E. Thermal Decomposition of Mechanoactivated Ammonium Perchlorate. *Thermochim. Acta* 2018, *669*, 60–65. https://doi.org/10.1016/j.tca.2018.09.007.

35. Fisch, M.; Lanza, A.; Boldyreva, E.; Macchi, P.; Casati, N. Kinetic Control of High-Pressure Solid-State Phase Transitions: A Case Study on L-Serine. *J. Phys. Chem. C* 2015, *119*(32), 18611–18617. https://doi.org/10.1021/acs.jpcc.5b05838.

36. Coffey, C. S.; Toton, E. T. A Microscopic Theory of Compressive Wave-Induced Reactions in Solid Explosives. *J. Chem. Phys.* 1982, *76*(2), 949–954. https://doi.org/10.1063/1.443065.

37. Dopieralski, P.; Ribas-Arino, J.; Anjukandi, P.; Krupicka, M.; Kiss, J.; Marx, D. The Janus-Faced Role of External Forces in Mechanochemical Disulfide Bond Cleavage. *Nat. Chem.* 2013, *5*(8), 685–691. https://doi.org/10.1038/nchem.1676.

38. Michalchuk, A. A. L.; Morrison, C. A. From Lattice Vibrations to Molecular Dissociation. In: *Theoretical and Computational Chemistry*; Elsevier, 2022; Vol. 22, pp 215–232. https://doi.org/10.1016/B978-0-12-822971-2.00010-3.

39. Hamilton, B. W.; Sakano, M. N.; Li, C.; Strachan, A. Chemistry Under Shock Conditions. *Annu. Rev. Mater. Res.* 2021, *51*(1), 101–130. https://doi.org/10.1146/annurev-matsci-080819-120123.

40. Boldyrev, V. V.; Pavlov, S. V.; Goldberg, E. L. Interrelation between Fine Grinding and Mechanical Activation. *Int. J. Miner. Process.* 1996, *44*, 181–185. https://doi.org/10.1016/0301-7516(95)00028-3.

41. Linberg, K.; Szymoniak, P.; Schönhals, A.; Emmerling, F.; Michalchuk, A. A. L. The Origin of Delayed Polymorphism in Molecular Crystals under Mechanochemical Conditions. *ChemRxiv*, 10. https://doi.org/10.26434/chemrxiv-2022-04jdf.

42. Michalchuk, A. A. L.; Fincham, P. T.; Portius, P.; Pulham, C. R.; Morrison, C. A. A Pathway to the Athermal Impact Initiation of Energetic Azides. *J. Phys. Chem. C* 2018, *122*(34), 19395–19408. https://doi.org/10.1021/acs.jpcc.8b05285.

43. Bowden, F. P.; Gurton, O. A. Initiation of Solid Explosives by Impact and Friction: The Influence of Grit. *Proc. R. Soc. Lond. A* 1949, *198*(1054), 337–349. https://doi.org/10.1098/rspa.1949.0105.

44. Descamps, M.; Willart, J. F. Perspectives on the Amorphisation/Milling Relationship in Pharmaceutical Materials. *Adv. Drug Deliv. Rev.* 2016, *100*, 51–66. https://doi.org/10.1016/j.addr.2016.01.011.

45. Belenguer, A. M.; Michalchuk, A. A. L.; Lampronti, G. I.; Sanders, J. K. M. Understanding the Unexpected Effect of Frequency on the Kinetics of a Covalent Reaction under Ball-Milling Conditions. *Beilstein J. Org. Chem.* 2019, *15*, 1226–1235. https://doi.org/10.3762/bjoc.15.120.

46. Martins, I. C. B.; Carta, M.; Haferkamp, S.; Feiler, T.; Delogu, F.; Colacino, E.; Emmerling, F. Mechanochemical *N*-Chlorination Reaction of Hydantoin: *In Situ* Real-Time Kinetic Study by Powder X-Ray Diffraction and Raman Spectroscopy. *ACS Sustain. Chem. Eng.* 2021, *9*(37), 12591–12601. https://doi.org/10.1021/acssuschemeng.1c03812.

47. Rojac, T.; Kosec, M.; Malič, B.; Holc, J. The Application of a Milling Map in the Mechanochemical Synthesis of Ceramic Oxides. *J. Eur. Ceram. Soc.* 2006, *26*(16), 3711–3716. https://doi.org/10.1016/j.jeurceramsoc.2005.11.013.

48. Lapshin, O. V.; Boldyreva, E. V.; Boldyrev, V. V. Role of Mixing and Milling in Mechanochemical Synthesis (Review). *Russ. J. Inorg. Chem.* 2021, *66*(3), 433–453. https://doi.org/10.1134/S0036023621030116.

49. Lin, I. J.; Nadiv, S. Review of the Phase Transformation and Synthesis of Inorganic Solids Obtained by Mechanical Treatment (Mechanochemical Reactions). *Mater. Sci. Eng.* 1979, *39*(2), 193–209. https://doi.org/10.1016/0025-5416(79)90059-4.

50. Belenguer, A. M.; Lampronti, G. I.; Cruz-Cabeza, A. J.; Hunter, C. A.; Sanders, J. K. M. Solvation and Surface Effects on Polymorph Stabilities at the Nanoscale. *Chem. Sci.* 2016, *7*(11), 6617–6627. https://doi.org/10.1039/C6SC03457H.

51. Ahmed, E.; Karothu, D. P.; Pejov, L.; Commins, P.; Hu, Q.; Naumov, P. From Mechanical Effects to Mechanochemistry: Softening and Depression of the Melting Point of Deformed Plastic Crystals. *J. Am. Chem. Soc.* 2020, *142*(25), 11219–11231. https://doi.org/10.1021/jacs.0c03990.

52. Lapshin, O. V.; Boldyrev, V. V.; Boldyreva, E. V. Mathematical Model of the Grinding and Mixing of Powder Binary Solids in a High-Energy Mill. *Russ. J. Phys. Chem.* 2019, *93*(8), 1592–1597. https://doi.org/10.1134/S0036024419080181.

53. Michalchuk, A. A. L.; Kabelitz, A.; Emmerling, F. 4.8 In Situ Methods for Monitoring Solid-State Processes in Molecular Materials. In: *Methods and Principles in Medicinal Chemistry*; Gruss, M., Ed.; Wiley, 2021; Vol. 79, pp 215–248. https://doi.org/10.1002/9783527823048.ch4-8.

54. Kajcsos, Z.; Marczis, L.; Tshakarov, C.; Gospodinov, G.; Horvath, D.; Vertes, A. Positron Annihilation Study of the Mechanochemical Reaction between Zn + Se and Zn + S. *Nucl. Instrum. Methods Phys. Res.* 1982, *199*(1–2), 273–275. https://doi.org/10.1016/0167-5087(82)90214-9.

55. De Oliveira, P. F. M.; Michalchuk, A. A. L.; Marquardt, J.; Feiler, T.; Prinz, C.; Torresi, R. M.; Camargo, P. H. C.; Emmerling, F. Investigating the Role of Reducing Agents on Mechanosynthesis of Au Nanoparticles. *CrystEngComm* 2020, *22*(38), 6261–6267. https://doi.org/10.1039/d0ce00826e.

56. Lampronti, G. I.; Michalchuk, A. A. L.; Mazzeo, P. P.; Belenguer, A. M.; Sanders, J. K. M.; Bacchi, A.; Emmerling, F. Changing the Game of Time Resolved X-ray Diffraction on the Mechanochemistry Playground by Downsizing. *Nat. Commun.* 2021, *12*(1), 6134. https://doi.org/10.1038/s41467-021-26264-1.

57. Friščić, T.; Halasz, I.; Beldon, P. J.; Belenguer, A. M.; Adams, F.; Kimber, S. A. J.; Honkimäki, V.; Dinnebier, R. E. Real-Time and In Situ Monitoring of Mechanochemical Milling Reactions. *Nat. Chem.* 2013, *5*(1), 66–73. https://doi.org/10.1038/nchem.1505.

58. Hope, K. S. High-Pressure Studies of Energetic Co-Crystals, 2020. https://doi.org/10.7488/ERA/247.

59. Oliveira, P. F. M.; Willart, J.-F.; Siepmann, J.; Siepmann, F.; Descamps, M. Using Milling to Explore Physical States: The Amorphous and Polymorphic Forms of Dexamethasone. *Cryst. Growth Des.* 2018, *18*(3), 1748–1757. https://doi.org/10.1021/acs.cgd.7b01664.

60. Sepelak, V.; Tkacova, K.; Boldyrev, V. V.; Becker, K.-D. Mechanically Induced Cation Redistribution in $ZnFe_2O_4$ and Its Thermal Stability. *Physica B: Cond. Mat* 1997, *234–236*, 617–619. https://doi.org/10.1016/S0921-4526(96)01061-7.

61. Belenguer, A. M.; Michalchuk, A. A. L.; Lampronti, G. I.; Sanders, J. K. M. Using Solid Catalysts in Disulfide-Based Dynamic Combinatorial Solution- and Mechanochemistry. *ChemSusChem* 2021. https://doi.org/10.1002/cssc.202102416.

62. Burton, T. F.; Pinaud, J.; Pétry, N.; Lamaty, F.; Giani, O. Simple and Rapid Mechanochemical Synthesis of Lactide and 3S-(Isobutyl)Morpholine-2,5-Dione-Based Random Copolymers Using DBU and Thiourea. *ACS Macro Lett.* 2021, *10*(12), 1454–1459. https://doi.org/10.1021/acsmacrolett.1c00617.

63. Haferkamp, S.; Paul, A.; Michalchuk, A. A. L.; Emmerling, F. Unexpected Polymorphism during a Catalyzed Mechanochemical Knoevenagel Condensation. *Beilstein J. Org. Chem.* 2019, *15*, 1141–1148. https://doi.org/10.3762/bjoc.15.110.

64. Fischer, F.; Scholz, G.; Batzdorf, L.; Wilke, M.; Emmerling, F. Synthesis, Structure Determination, and Formation of a Theobromine: Oxalic Acid 2: 1 Cocrystal. *CrystEngComm* 2015, *17*(4), 824–829. https://doi.org/10.1039/C4CE02066A.

65. Germann, L. S.; Arhangelskis, M.; Etter, M.; Dinnebier, R. E.; Friščić, T. Challenging the Ostwald Rule of Stages in Mechanochemical Cocrystallisation. *Chem. Sci.* 2020, *11*(37), 10092–10100. https://doi.org/10.1039/D0SC03629C.

66. Arhangelskis, M.; Topić, F.; Hindle, P.; Tran, R.; Morris, A. J.; Cinčić, D.; Friščić, T. Mechanochemical Reactions of Cocrystals: Comparing Theory with Experiment in the Making and Breaking of Halogen Bonds in the Solid State. *Chem. Commun.* 2020, *56*(59), 8293–8296. https://doi.org/10.1039/D0CC02935A.

67. Tumanov, I. A.; Achkasov, A. F.; Boldyreva, E. V.; Boldyrev, V. V. Following the Products of Mechanochemical Synthesis Step by Step. *CrystEngComm* 2011, *13*(7), 2213. https://doi.org/10.1039/c0ce00869a.

68. Lukin, S.; Tireli, M.; Stolar, T.; Barišić, D.; Blanco, M. V.; di Michiel, M.; Užarević, K.; Halasz, I. Isotope Labeling Reveals Fast Atomic and Molecular Exchange in Mechanochemical Milling Reactions. *J. Am. Chem. Soc.* 2019, *141*(3), 1212–1216. https://doi.org/10.1021/jacs.8b12149.

69. Lien Nguyen, K.; Friščić, T.; Day, G. M.; Gladden, L. F.; Jones, W. Terahertz Time-Domain Spectroscopy and the Quantitative Monitoring of Mechanochemical Cocrystal Formation. *Nat. Mater.* 2007, *6*(3), 206–209. https://doi.org/10.1038/nmat1848.

70. Tsuchiya, N.; Isobe, T.; Senna, M.; Yoshioka, N.; Inoue, H. Mechanochemical Effects on the Structures and Chemical States of $[Fe(Phen)_3](NCS)_2 \cdot H_2O$. *Solid State Commun.* 1996, *99*(8), 525–529. https://doi.org/10.1016/0038-1098(96)00359-6.

71. Batzdorf, L.; Fischer, F.; Wilke, M.; Wenzel, K.-J.; Emmerling, F. Direct In Situ Investigation of Milling Reactions Using Combined X-ray Diffraction and Raman Spectroscopy. *Angew. Chem. Int. Ed.* 2015, *54*(6), 1799–1802. https://doi.org/10.1002/anie.201409834.

72. Tsuchiya, N.; Tsukamoto, A.; Ohshita, T.; Isobe, T.; Senna, M.; Yoshioka, N.; Inoue, H. Anomalous Spin Crossover of Mechanically Strained Iron(II) Complexes with 1,10-Phenanthroline with Their Counterions, NCS− and PF−6. *J. Solid State Chem.* 2000, *153*(1), 82–91. https://doi.org/10.1006/jssc.2000.8750.

73. Fiss, B. G.; Richard, A. J.; Douglas, G.; Kojic, M.; Friščić, T.; Moores, A. Mechanochemical Methods for the Transfer of Electrons and Exchange of Ions: Inorganic Reactivity from Nanoparticles to Organometallics. *Chem. Soc. Rev.* 2021, *50*(14), 8279–8318. https://doi.org/10.1039/D0CS00918K.

74. de Oliveira, P. F. M.; Quiroz, J.; de Oliveira, D. C.; Camargo, P. H. C. A Mechano-Colloidal Approach for the Controlled Synthesis of Metal Nanoparticles. *Chem. Commun. (Camb)* 2019, *55*(95), 14267–14270. https://doi.org/10.1039/C9CC06199A.

75. Šepelák, V.; Becker, K. D.; Bergmann, I.; Suzuki, S.; Indris, S.; Feldhoff, A.; Heitjans, P.; Grey, C. P. A One-Step Mechanochemical Route to Core−Shell $Ca_2 SnO_4$ Nanoparticles Followed by 119 Sn MAS NMR and 119 Sn Mössbauer Spectroscopy. *Chem. Mater.* 2009, *21*(12), 2518–2524. https://doi.org/10.1021/cm900590d.

76. Indris, S.; Scheuermann, M.; Becker, S. M.; Šepelák, V.; Kruk, R.; Suffner, J.; Gyger, F.; Feldmann, C.; Ulrich, A. S.; Hahn, H. Local Structural Disorder and Relaxation in SnO_2 Nanostructures Studied by 119 Sn MAS NMR and 119 Sn Mössbauer Spectroscopy. *J. Phys. Chem. C* 2011, *115*(14), 6433–6437. https://doi.org/10.1021/jp200651m.

77. Wilke, M.; Kabelitz, A.; Gorelik, T. E.; Buzanich, A. G.; Reinholz, U.; Kolb, U.; Rademann, K.; Emmerling, F. The Crystallisation of Copper(Ii) Phenylphosphonates. *Dalton Trans.* 2016, *45*(43), 17453–17463. https://doi.org/10.1039/C6DT02904C.

78. Al-Terkawi, A.-A.; Scholz, G.; Prinz, C.; Emmerling, F.; Kemnitz, E. Ca-, Sr-, and Ba-Coordination Polymers Based on Anthranilic Acid *via* Mechanochemistry. *Dalton Trans.* 2019, *48*(19), 6513–6521. https://doi.org/10.1039/C9DT00991D.

79. Descamps, M.; Willart, J. F.; Dudognon, E.; Lefort, R.; Desprez, S.; Caron, V. Phase Transformations Induced by Grinding: What Is Revealed by Molecular Materials. *MRS Proc.* 2006, *979*, 606. https://doi.org/10.1557/PROC-979-0979-HH06-06.

80. Belenguer, A. M.; Friščić, T.; Day, G. M.; Sanders, J. K. M. Solid-State Dynamic Combinatorial Chemistry: Reversibility and Thermodynamic Product Selection in Covalent Mechanosynthesis. *Chem. Sci.* 2011, *2*(4), 696. https://doi.org/10.1039/c0sc00533a.

81. Belenguer, A. M.; Lampronti, G. I.; Sanders, J. K. M. Implications of Thermodynamic Control: Dynamic Equilibrium Under Ball Mill Grinding Conditions. *Isr. J. Chem.* 2021. https://doi.org/10.1002/ijch.202100090.

82. Belenguer, A. M.; Lampronti, G. I.; Wales, D. J.; Sanders, J. K. M. Direct Observation of Intermediates in a Thermodynamically Controlled Solid-State Dynamic Covalent Reaction. *J. Am. Chem. Soc.* 2014, *136*(46), 16156–16166. https://doi.org/10.1021/ja500707z.

83. Andersen, J. M.; Mack, J. Decoupling the Arrhenius Equation via Mechanochemistry. *Chem. Sci.* 2017, *8*(8), 5447–5453. https://doi.org/10.1039/C7SC00538E.

84. Stolar, T.; Batzdorf, L.; Lukin, S.; Žilić, D.; Motillo, C.; Friščić, T.; Emmerling, F.; Halasz, I.; Užarević, K. In Situ Monitoring of the Mechanosynthesis of the Archetypal Metal–Organic Framework HKUST-1: Effect of Liquid Additives on the Milling Reactivity. *Inorg. Chem.* 2017, *56*(11), 6599–6608. https://doi.org/10.1021/acs.inorgchem.7b00707.

85. Gracin, D.; Štrukil, V.; Friščić, T.; Halasz, I.; Užarević, K. Laboratory Real-Time and In Situ Monitoring of Mechanochemical Milling Reactions by Raman Spectroscopy. *Angew. Chem. Int. Ed.* 2014, *53*(24), 6193–6197. https://doi.org/10.1002/anie.201402334.

86. Lukin, S.; Užarević, K.; Halasz, I. Raman Spectroscopy for Real-Time and In Situ Monitoring of Mechanochemical Milling Reactions. *Nat. Protoc.* 2021, *16*(7), 3492–3521. https://doi.org/10.1038/s41596-021-00545-x.

87. Ardila-Fierro, K. J.; Lukin, S.; Etter, M.; Užarević, K.; Halasz, I.; Bolm, C.; Hernández, J. G. Direct Visualization of a Mechanochemically Induced Molecular Rearrangement. *Angew. Chem. Int. Ed.* 2020, *59*(32), 13458–13462. https://doi.org/10.1002/anie.201914921.

88. de Oliveira, P. F. M.; Michalchuk, A. A. L.; Buzanich, A. G.; Bienert, R.; Torresi, R. M.; Camargo, P. H. C.; Emmerling, F. Tandem X-ray Absorption Spectroscopy and Scattering for *In Situ* Time-Resolved Monitoring of Gold Nanoparticle Mechanosynthesis. *Chem. Commun. (Camb)* 2020, *56*(71), 10329. https://doi.org/10.1039/D0CC03862H.

89. Kulla, H.; Wilke, M.; Fischer, F.; Röllig, M.; Maierhofer, C.; Emmerling, F. Warming up for Mechanosynthesis – Temperature Development in Ball Mills during Synthesis. *Chem. Commun. (Camb)* 2017, *53*(10), 1664–1667. https://doi.org/10.1039/C6CC08950J.

90. Brekalo, I.; Yuan, W.; Mottillo, C.; Lu, Y.; Zhang, Y.; Casaban, J.; Holman, K. T.; James, S. L.; Duarte, F.; Williams, P. A.; Harris, K. D. M.; Friščić, T. Manometric Real-Time Studies of the Mechanochemical Synthesis of Zeolitic Imidazolate Frameworks. *Chem. Sci.* 2020, *11*(8), 2141–2147. https://doi.org/10.1039/C9SC05514B.

91. Schiffmann, J. G.; Emmerling, F.; Martins, I. C. B.; Van Wüllen, L. In-Situ Reaction Monitoring of a Mechanochemical Ball Mill Reaction with Solid State NMR. *Solid State Nucl. Magn. Reson.* 2020, *109*, 101687. https://doi.org/10.1016/j.ssnmr.2020.101687.

92. Julien, P. A.; Arhangelskis, M.; Germann, L. S.; Etter, M.; Dinnebier, R. E.; Morris, A. J.; Friščić, T. Illuminating Mechanochemical Reactions by Combining Real-Time Fluorescence Emission Monitoring and Periodic Time-Dependent Density- Functional Calculations. *ChemRxiv.* https://doi.org/10.26434/chemrxiv-2021-lw8sm.

93. Hutchings, B. P.; Crawford, D. E.; Gao, L.; Hu, P.; James, S. L. Feedback Kinetics in Mechanochemistry: The Importance of Cohesive States. *Angew. Chem. Int. Ed.* 2017, *56*(48), 5. https://doi.org/10.1002/anie.201706723.

94. Michalchuk, A. A. L.; Tumanov, I. A.; Konar, S.; Kimber, S. A. J.; Pulham, C. R.; Boldyreva, E. V. Challenges of Mechanochemistry: Is In Situ Real-Time Quantitative Phase Analysis Always Reliable? A Case Study of Organic Salt Formation. *Adv. Sci.* 2017, *4*(9), 1700132. https://doi.org/10.1002/advs.201700132.

95. Losev, E.; Arkhipov, S.; Kolybalov, D.; Mineev, A.; Ogienko, A.; Boldyreva, E.; Boldyrev, V. Substituting Steel for a Polymer in a Jar for Ball Milling Does Matter. *CrystEngComm* 2022, *24*(9), 1700–1703. https://doi.org/10.1039/D1CE01703A.

96. Chatziadi, A.; Skořepová, E.; Kohout, M.; Ridvan, L.; Šoóš, M. Exploring the Polymorphism of Sofosbuvir *via* Mechanochemistry: Effect of Milling Jar Geometry and Material. *CrystEngComm* 2022, *24*(11), 2107–2117. https://doi.org/10.1039/D1CE01561C.

97. Belenguer, A. M.; Lampronti, G. I.; Michalchuk, A. A. L.; Emmerling, F.; Sanders, J. K. M. Quantitative Reversible One Pot Interconversion of Three Crystalline Polymorphs by Ball Mill Grinding. *CrystEngComm* 2022. https://doi.org/10.1039/D2CE00393G.

98. Boldyreva, E. V. Combined X-ray Diffraction and Raman Spectroscopy Studies of Phase Transitions in Crystalline Amino Acids at Low Temperatures and High Pressures: Selected Examples. *Phase Transit.* 2009, *82*(4), 303–321. https://doi.org/10.1080/01411590902838656.

99. Wilke, M.; Kabelitz, A.; Zimathies, A.; Rademann, K.; Emmerling, F. Crystal Structure and In Situ Investigation of a Mechanochemical Synthesized 3D Zinc N-(Phosphonomethyl)Glycinate. *J. Mater. Sci.* 2017, *52*(20), 12013–12020. https://doi.org/10.1007/s10853-017-1121-7.

100. Mikhailenko, M. A.; Shakhtshneider, T. P.; Boldyrev, V. V. On the Mechanism of Mechanochemical Synthesis of Phthalylsulphathiazole. *J. Mater. Sci.* 2004, *39*(16/17), 5435–5439. https://doi.org/10.1023/B:JMSC.0000039261.66084.a3.

101. Lacey, P. M. C. Developments in the Theory of Particle Mixing. *J. Appl. Chem.* 1954, *4*(5), 257–268. https://doi.org/10.1002/jctb.5010040504.

102. Tkáčová, K.; Heegn, H.; Števulová, N. Energy Transfer and Conversion during Comminution and Mechanical Activation. *Int. J. Miner. Process.* 1993, *40*(1–2), 17–31. https://doi.org/10.1016/0301-7516(93)90037-B.

103. Ferguson, M.; Moyano, M. S.; Tribello, G. A.; Crawford, D. E.; Bringa, E. M.; James, S. L.; Kohanoff, J.; Del Pópolo, M. G. Insights into Mechanochemical Reactions at the Molecular Level: Simulated Indentations of Aspirin and Meloxicam Crystals. *Chem. Sci.* 2019, *10*(10), 2924–2929. https://doi.org/10.1039/C8SC04971H.

104. Wenzel, C. F. *Lehre von der Verwandtshaft der Koerper.* Hansebooks, 1777.

105. Brothman, A.; Wollan, F.; New, S. M. Analysis Provides Formula to Solve Mixing Problems. *Chem. Met. Eng.* 1945, *52*, 102.

106. Blumberg, R.; Maritz, J. S. Mixing of Solid Particles. *Chem. Eng. Sci.* 1953, *2*(6), 240–246.

107. Lai, F.; Hersey, J. A.; Staniforth, J. N. Segregation and Mixing of Fine Particles in an Ordered Mixture. *Powder Technol.* 1981, *28*(1), 17–23. https://doi.org/10.1016/0032-5910(81)87004-0.

108. Tang, P.; Puri, V. M. Methods for Minimizing Segregation: A Review. *Part. Sci. Technol.* 2004, *22*(4), 321–337. https://doi.org/10.1080/02726350490501420.

109. Michalchuk, A. A. L.; Tumanov, I. A.; Boldyreva, E. V. The Effect of Ball Mass on the Mechanochemical Transformation of a Single-Component Organic System: Anhydrous Caffeine. *J. Mater. Sci.* 2018, *53*(19), 13380–13389. https://doi.org/10.1007/s10853-018-2324-2.

110. Delogu, F.; Orrù, R.; Cao, G. A Novel Macrokinetic Approach for Mechanochemical Reactions. *Chem. Eng. Sci.* 2003, *58*(3–6), 815–821. https://doi.org/10.1016/S0009-2509(02)00612-7.

111. Sanna, A. L.; Carta, M.; Pia, G.; Garroni, S.; Porcheddu, A.; Delogu, F. Chemical Effects Induced by the Mechanical Processing of Granite Powder. *Sci. Rep.* 2022, *12*(1), 9445.

112. Colacino, E.; Carta, M.; Pia, G.; Porcheddu, A.; Ricci, P. C.; Delogu, F. Processing and Investigation Methods in Mechanochemical Kinetics. *ACS Omega* 2018, *3*(8), 9196–9209. https://doi.org/10.1021/acsomega.8b01431.

113. Carta, M.; Delogu, F.; Porcheddu, A. A Phenomenological Kinetic Equation for Mechanochemical Reactions Involving Highly Deformable Molecular Solids. *Phys. Chem. Chem. Phys.* 2021, *23*(26), 14178–14194. https://doi.org/10.1039/D1CP01361K.

114. Šepelák, V.; Steinike, U.; Uecker, D. C.; Wißmann, S.; Becker, K. D. Structural Disorder in Mechanosynthesized Zinc Ferrite. *J. Solid State Chem.* 1998, *135*(1), 52–58. https://doi.org/10.1006/jssc.1997.7589.

115. Boldyrev, V. V. Thermal Decomposition of Ammonium Perchlorate. *Thermochim. Acta* 2006, *443*(1), 1–36. https://doi.org/10.1016/j.tca.2005.11.038.

116. Takacs, L. Self-Sustaining Reactions Induced by Ball Milling. *Prog. Mater. Sci.* 2002, *47*(4), 355–414. https://doi.org/10.1016/S0079-6425(01)00002-0.

117. Bordet, P.; Bytchkov, A.; Descamps, M.; Dudognon, E.; Elkaïm, E.; Martinetto, P.; Pagnoux, W.; Poulain, A.; Willart, J.-F. Solid State Amorphization of β-Trehalose: A Structural Investigation Using Synchrotron Powder Diffraction and PDF Analysis. *Cryst. Growth Des.* 2016, *16*(8), 4547–4558. https://doi.org/10.1021/acs.cgd.6b00660.

118. Michalchuk, A. A. L.; Tumanov, I. A.; Boldyreva, E. V. Ball Size or Ball Mass – What Matters in Organic Mechanochemical Synthesis? *CrystEngComm* 2019, *21*(13), 2174–2179. https://doi.org/10.1039/C8CE02109K.

119. Kulla, H.; Fischer, F.; Benemann, S.; Rademann, K.; Emmerling, F. The Effect of the Ball to Reactant Ratio on Mechanochemical Reaction Times Studied by In Situ PXRD. *CrystEngComm* 2017, *19*(28), 3902–3907. https://doi.org/10.1039/C7CE00502D.

120. Vugrin, L.; Carta, M.; Lukin, S.; Mestrovic, E.; Delogu, F.; Halasz, I. Mechanochemical Reaction Kinetics Scales Linearly with Impact Energy. *Faraday Discuss.* 2022. https://doi.org/10.1039/D2FD00083K.

121. Beyer, M. K. The Mechanical Strength of a Covalent Bond Calculated by Density Functional Theory. *J. Chem. Phys.* 2000, *112*(17), 7307–7312. https://doi.org/10.1063/1.481330.

122. Iozzi, M. F.; Helgaker, T.; Uggerud, E. Influence of External Force on Properties and Reactivity of Disulfide Bonds. *J. Phys. Chem. A* 2011, *115*(11), 2308–2315. https://doi.org/10.1021/jp109428g.

123. Kryger, M. J.; Munaretto, A. M.; Moore, J. S. Structure–Mechanochemical Activity Relationships for Cyclobutane Mechanophores. *J. Am. Chem. Soc.* 2011, *133*(46), 18992–18998. https://doi.org/10.1021/ja2086728.

124. Ribas-Arino, J.; Shiga, M.; Marx, D. Understanding Covalent Mechanochemistry. *Angew. Chem. Int. Ed.* 2009, *48*(23), 4190–4193. https://doi.org/10.1002/anie.200900673.

125. Klein, I. M.; Husic, C. C.; Kovács, D. P.; Choquette, N. J.; Robb, M. J. Validation of the CoGEF Method as a Predictive Tool for Polymer Mechanochemistry. *J. Am. Chem. Soc.* 2020, *142*(38), 16364–16381. https://doi.org/10.1021/jacs.0c06868.

126. Pradipta, M. F.; Watanabe, H.; Senna, M. Semiempirical Computation of the Solid Phase Diels–Alder Reaction between Anthracene Derivatives and p-Benzoquinone via Molecular Distortion. *Solid State Ionics* 2004, *172*(1–4), 169–172. https://doi.org/10.1016/j.ssi.2004.02.070.

127. Konar, S.; Michalchuk, A. A. L.; Sen, N.; Bull, C. L.; Morrison, C. A.; Pulham, C. R. High-Pressure Study of Two Polymorphs of 2,4,6-Trinitrotoluene Using Neutron Powder Diffraction and Density Functional Theory Methods. *J. Phys. Chem. C* 2019, *123*(43), 26095–26105. https://doi.org/10.1021/acs.jpcc.9b07658.

128. Warren, L. R.; McGowan, E.; Renton, M.; Morrison, C. A.; Funnell, N. P. Direct Evidence for Distinct Colour Origins in ROY Polymorphs. *Chem. Sci.* 2021, *12*(38), 12711–12718. https://doi.org/10.1039/D1SC04051K.

129. Hunter, S.; Coster, P. L.; Davidson, A. J.; Millar, D. I. A.; Parker, S. F.; Marshall, W. G.; Smith, R. I.; Morrison, C. A.; Pulham, C. R. High-Pressure Experimental and DFT-D Structural Studies of the Energetic Material FOX-7. *J. Phys. Chem. C* 2015, *119*(5), 2322–2334. https://doi.org/10.1021/jp5110888.

130. Liu, X.; Michalchuk, A. A. L.; Bhattacharya, B.; Emmerling, F.; Pulham, C. R. High-Pressure Reversibility in a Plastically Flexible Coordination Polymer Crystal. *Nat. Commun.* 2021, *12*(1), 3871.

131. Kuklja, M. M.; Rashkeev, S. N. Shear-Strain-Induced Chemical Reactivity of Layered Molecular Crystals. *Appl. Phys. Lett.* 2007, *90*(15), 151913. https://doi.org/10.1063/1.2719031.

132. Adams, H.; Miller, B. P.; Furlong, O. J.; Fantauzzi, M.; Navarra, G.; Rossi, A.; Xu, Y.; Kotvis, P. V.; Tysoe, W. T. Modeling Mechanochemical Reaction Mechanisms. *ACS Appl. Mater. Interfaces* 2017, *9*(31), 26531–26538. https://doi.org/10.1021/acsami.7b05440.

133. Haruta, N.; de Oliveira, P. F. M.; Sato, T.; Tanaka, K.; Baron, M. Force-Induced Dissolution of Imaginary Mode in Mechanochemical Reaction: Dibenzophenazine Synthesis. *J. Phys. Chem. C* 2019, *123*(35), 21581–21587. https://doi.org/10.1021/acs.jpcc.9b05582.

134. Katrusiak, A. Lab in a DAC – High-Pressure Crystal Chemistry in a Diamond-Anvil Cell. *Acta Crystallogr. B Struct. Sci. Cryst. Eng. Mater.* 2019, *75*(6), 918–926. https://doi.org/10.1107/S2052520619013246.

135. Yan, H.; Yang, F.; Pan, D.; Lin, Y.; Hohman, J. N.; Solis-Ibarra, D.; Li, F. H.; Dahl, J. E. P.; Carlson, R. M. K.; Tkachenko, B. A.; Fokin, A. A.; Schreiner, P. R.; Galli, G.; Mao, W. L.; Shen, Z.-X.; Melosh, N. A. Sterically Controlled Mechanochemistry under Hydrostatic Pressure. *Nature* 2018, *554*(7693), 505–510. https://doi.org/10.1038/nature25765.

136. Ciezak-Jenkins, J. A.; Jenkins, T. A. Mechanochemical Induced Structural Changes in Sucrose Using the Rotational Diamond Anvil Cell. *J. Appl. Phys.* 2018, *123*(8), 085901. https://doi.org/10.1063/1.5020231.

137. Steele, B. A.; Goldman, N.; Kuo, I.-F. W.; Kroonblawd, M. P. Mechanochemical Synthesis of Glycine Oligomers in a Virtual Rotational Diamond Anvil Cell. *Chem. Sci.* 2020, *11*(30), 7760–7771. https://doi .org/10.1039/D0SC00755B.

138. Michalchuk, A. A. L.; Rudić, S.; Pulham, C. R.; Morrison, C. A. Vibrationally Induced Metallisation of the Energetic Azide α-NaN₃. *Phys. Chem. Chem. Phys.* 2018, *20*(46), 29061–29069. https://doi.org/10 .1039/C8CP06161K.

139. Tokmakoff, A.; Fayer, M. D.; Dlott, D. D. Chemical Reaction Initiation and Hot-Spot Formation in Shocked Energetic Molecular Materials. *J. Phys. Chem.* 1993, *97*(9), 1901–1913. https://doi.org/10.1021 /j100111a031.

140. Dlott, D. D.; Fayer, M. D. Shocked Molecular Solids: Vibrational Up Pumping, Defect Hot Spot Formation, and the Onset of Chemistry. *J. Chem. Phys.* 1990, *92*(6), 3798–3812. https://doi.org/10.1063 /1.457838.

141. Michalchuk, A. A. L.; Trestman, M.; Rudić, S.; Portius, P.; Fincham, P. T.; Pulham, C. R.; Morrison, C. A. Predicting the Reactivity of Energetic Materials: An *Ab Initio* Multi-phonon Approach. *J. Mater. Chem. A* 2019, *7*(33), 19539–19553. https://doi.org/10.1039/C9TA06209B.

142. Michalchuk, A. A. L.; Hemingway, J.; Morrison, C. A. Predicting the Impact Sensitivities of Energetic Materials through Zone-Center Phonon Up-Pumping. *J. Chem. Phys.* 2021, *154*(6), 064105. https://doi .org/10.1063/5.0036927.

143. Michalchuk, A. A. L.; Rudić, S.; Pulham, C. R.; Morrison, C. A. Predicting the Impact Sensitivity of a Polymorphic High Explosive: The Curious Case of FOX-7. *Chem. Commun. (Camb)* 2021, *57*(85), 11213–11216. https://doi.org/10.1039/D1CC03906G.

144. Dlott, D. D. Ultrafast Spectroscopy of Shockwaves in Molecular Materials. *Annu. Rev. Phys. Chem.* 1999, *50*(1), 251–278. https://doi.org/10.1146/annurev.physchem.50.1.251.

145. Bowden, F. P.; Stone, M. A.; Tudor, G. K. Hot Spots on Rubbing Surfaces and the Detonation of Explosives by Friction. *Proc. R. Soc. Lond. A* 1947, *188*(1014), 329–349. https://doi.org/10.1098/rspa .1947.0012.

146. You, S.; Chen, M.-W.; Dlott, D. D.; Suslick, K. S. Ultrasonic Hammer Produces Hot Spots in Solids. *Nat. Commun.* 2015, *6*, 6581. https://doi.org/10.1038/ncomms7581.

147. Katsenis, A. D.; Puškarić, A.; Štrukil, V.; Mottillo, C.; Julien, P. A.; Užarević, K.; Pham, M.-H.; Do, T.-O.; Kimber, S. A. J.; Lazić, P.; Magdysyuk, O.; Dinnebier, R. E.; Halasz, I.; Friščić, T. In Situ X-ray Diffraction Monitoring of a Mechanochemical Reaction Reveals a Unique Topology Metal-Organic Framework. *Nat. Commun.* 2015, *6*(1), 6662. https://doi.org/10.1038/ncomms7662.

148. Cruz-Cabeza, A. J.; Reutzel-Edens, S. M.; Bernstein, J. Facts and Fictions about Polymorphism. *Chem. Soc. Rev.* 2015, *44*(23), 8619–8635. https://doi.org/10.1039/C5CS00227C.

149. Traversari, G.; Porcheddu, A.; Pia, G.; Delogu, F.; Cincotti, A. Coupling of Mechanical Deformation and Reaction in Mechanochemical Transformations. *Phys. Chem. Chem. Phys.* 2021, *23*(1), 229–245. https://doi.org/10.1039/D0CP05647B.

150. Bygrave, P. J.; Case, D. H.; Day, G. M. Is the Equilibrium Composition of Mechanochemical Reactions Predictable Using Computational Chemistry? *Faraday Discuss.* 2014, *170*, 41–57. https://doi.org/10 .1039/C3FD00162H.

151. Delogu, F.; Gorrasi, G.; Sorrentino, A. Fabrication of Polymer Nanocomposites via Ball Milling: Present Status and Future Perspectives. *Prog. Mater. Sci.* 2017, *86*, 75–126. https://doi.org/10.1016/j.pmatsci .2017.01.003.

152. Baláž, P.; Achimovičová, M.; Baláž, M.; Billik, P.; Cherkezova-Zheleva, Z.; Criado, J. M.; Delogu, F.; Dutková, E.; Gaffet, E.; Gotor, F. J.; Kumar, R.; Mitov, I.; Rojac, T.; Senna, M.; Streletskii, A.; Wieczorek-Ciurowa, K. Hallmarks of Mechanochemistry: From Nanoparticles to Technology. *Chem. Soc. Rev.* 2013, *42*(18), 7571. https://doi.org/10.1039/c3cs35468g.

153. Fischer, F.; Lubjuhn, D.; Greiser, S.; Rademann, K.; Emmerling, F. Supply and Demand in the Ball Mill: Competitive Cocrystal Reactions. *Cryst. Growth & Des.* 2016, *16*(10), 5843–5851. https://doi.org/10 .1021/acs.cgd.6b00928.

154. Guinet, Y.; Paccou, L.; Danède, F.; Hédoux, A. Confinement of Molecular Materials Using a Solid-State Loading Method: A Route for Exploring New Physical States and Their Subsequent Transformation Highlighted by Caffeine Confined to SBA-15 Pores. *RSC Adv.* 2021, *11*(55), 34564–34571. https://doi .org/10.1039/D1RA05757J.

155. van Steen, E.; Claeys, M.; Dry, M. E.; van de Loosdrecht, J.; Viljoen, E. L.; Visagie, J. L. Stability of Nanocrystals: Thermodynamic Analysis of Oxidation and Re-reduction of Cobalt in Water/Hydrogen Mixtures. *J. Phys. Chem. B* 2005, *109*(8), 3575–3577. https://doi.org/10.1021/jp045136o.

156. Amrute, A. P.; Łodziana, Z.; Schreyer, H.; Weidenthaler, C.; Schüth, F. High-Surface-Area Corundum by Mechanochemically Induced Phase Transformation of Boehmite. *Science* 2019, *366*(6464), 485–489. https://doi.org/10.1126/science.aaw9377.

157. Belenguer, A. M.; Lampronti, G. I.; De Mitri, N.; Driver, M.; Hunter, C. A.; Sanders, J. K. M. Understanding the Influence of Surface Solvation and Structure on Polymorph Stability: A Combined Mechanochemical and Theoretical Approach. *J. Am. Chem. Soc.* 2018, *140*(49), 17051–17059. https://doi.org/10.1021/jacs.8b08549.

158. Belenguer, A. M.; Cruz-Cabeza, A. J.; Lampronti, G. I.; Sanders, J. K. M. On the Prevalence of Smooth Polymorphs at the Nanoscale: Implications for Pharmaceuticals. *CrystEngComm* 2019, *21*(13), 2203–2211. https://doi.org/10.1039/C8CE02098A.

159. Dujardin, N.; Willart, J. F.; Dudognon, E.; Danède, F.; Descamps, M. Mechanism of Solid State Amorphization of Glucose upon Milling. *J. Phys. Chem. B* 2013, *117*(5), 1437–1443. https://doi.org/10.1021/jp3069267.

160. Montis, R.; Davey, R. J.; Wright, S. E.; Woollam, G. R.; Cruz-Cabeza, A. J. Transforming Computed Energy Landscapes into Experimental Realities: The Role of Structural Rugosity. *Angew. Chem. Int. Ed.* 2020, *59*(46), 20357–20360. https://doi.org/10.1002/anie.202006939.

161. Boldyrev, V. V. Hydrothermal Reactions under Mechanochemical Action. *Powder Technol.* 2002, *122*(2–3), 247–254. http://doi.org/10.1016/S0032-5910(01)00421-1.

162. Fabbiani, F. P. A.; Allan, D. R.; David, W. I. F.; Moggach, S. A.; Parsons, S.; Pulham, C. R. High-Pressure Recrystallisation—A Route to New Polymorphs and Solvates. *CrystEngComm* 2004, *6*(82), 504–511. https://doi.org/10.1039/B406631F.

163. Fabbiani, F. P. A.; Pulham, C. R. High-Pressure Studies of Pharmaceutical Compounds and Energetic Materials. *Chem. Soc. Rev.* 2006, *35*(10), 932. https://doi.org/10.1039/b517780b.

164. Sobczak, S.; Drożdż, W.; Lampronti, G. I.; Belenguer, A. M.; Katrusiak, A.; Stefankiewicz, A. R. Dynamic Covalent Chemistry under High-Pressure: A New Route to Disulfide Metathesis. *Chem. Eur. J.* 2018, *24*(35), 8769–8773. https://doi.org/10.1002/chem.201801740.

165. Cocco, G.; Delogu, F.; Schiffini, L. Towards a Quantitative Understanding of the Mechanical Alloying Process. *Journal of Materials Synthesis and Processing*, 2000, *8*, 167–180.

166. Caravati, C.; Delogu, F.; Cocco, G.; Rustici, M. Hyperchaotic Qualities of the Ball Motion in a Ball Milling Device. *Chaos* 1999, *9*(1), 219–226. https://doi.org/10.1063/1.166393.

4

Life Cycle Assessment

A Tool for Sustainability Evaluation of Emerging Mechanochemical Process Engineering

Sabrina Spatari, Or Galant and Charles E. Diesendruck

CONTENTS

4.1 Introduction

For more than three decades, life cycle assessment (LCA) has stood as an analytical method to support sustainability evaluation and design of products, processes and activities. Originally conceived of and used by industry in the 1960s[1] as a method to understand the environmental impacts of product manufacture and consumption of mineral resources, LCA was later formalized as a method with the development of the framework by the Society for Environmental Toxicology and Chemistry (SETAC)[2] and the International Organization for Standardization (ISO).[3, 4] Since the earliest case studies at the beginning of the 1990s,[5] LCA has been used to guide and support industrial manufacturing of multiple products ranging from automobiles,[6] their parts[7–9] and propulsion and energy systems;[10, 11] buildings and building materials,[12–15] and multiple forms of infrastructure including electricity,[16] transportation[17] and water.[18]

For commercial products, LCAs[19] aim is to quantify their environmental impacts using a "cradle-to-grave" framework, where life cycle environmental impacts are traced from resource extraction to material production, fabrication and product manufacture, and finally to use and retirement, known as end-of-life (Figure 4.1). Within the cradle-to-grave framework, products, materials or processes are often assessed through cradle-to-gate or gate-to-gate boundaries depending on the goal of the work.

To assess complex and multi-component/multi-ingredient products on the market like computers, electronics and pharmaceutical and personal care products, LCA requires data through audits of manufacturing processes that are often proprietary. Moreover, commercially manufactured products depend on supply chains of materials and energy from multiple sectors of the economy. For this reason, the manufacture of almost any given product on the market will involve the supply of inputs from the chemical sector. For example, most plastic materials are ultimately produced from petroleum or fossil energy resources. In special cases they are made from unconventional resources, such as platform chemicals[20, 21] obtained from the fermentation of sugars in/from crops or lignocellulose (biomass).[22] Therefore, to

DOI: 10.1201/9781003178187-5

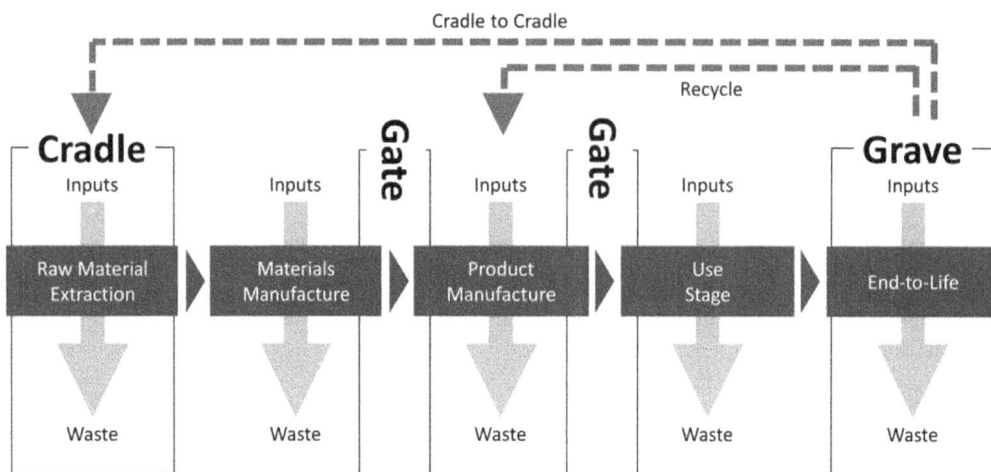

FIGURE 4.1 Cradle-to-grave, cradle-to-gate, and gate-to-gate framework of life cycle assessment.

quantify and understand the environmental "footprint" of most products, LCA must trace the environmental impacts of chemicals. Those environmental "footprints" or impacts calculated using LCA need to trace sequential material and energy balances related to the production of each chemical. Because this can involve acquiring proprietary datasets, increasingly chemical process models[23, 24] and in some cases machine learning[25–27] approaches are used to quantify the environmental impact of chemicals and complex materials. Furthermore, when evaluating the environmental impact of not-yet manufactured products or early-stage technologies, commercial scale data do not exist and assessments can be highly uncertain[28] as is the case for biofuels[29–32] and bio-derived products like polymers[33, 34] and platform chemicals (e.g. bioethanol).[35]

This chapter describes the LCA method in the context of early-stage technology, and specifically, mechanochemical process engineering methods. We focus on three manufacturing routes where mechanochemical methods are described as a promising methodology to achieve the goals of a low-carbon economy: (1) biomass and the bioeconomy; (2) energy materials; and (3) the role of mechanochemistry in recycling to achieve a circular economy. Each area could realize sustainability gains through developing and integrating mechanochemical processes industrially.

4.2 Life Cycle Assessment of Early-Stage Technologies and Product Systems

Over the last two decades, LCA has been expanded towards use as a tool to evaluate early-stage technology and products.[36] Described as prospective or ex ante LCA,[37] early-stage LCA is used to assess technologies and product systems that are evolving from points of low technology readiness level (TRL). Mechanochemical processes are a prime example of promising early-stage technologies for synthetic chemistry and manufacturing in the chemical, pharmaceutical and material industries.

LCA research aims to quantify the environmental impacts of full-scale chemical production by creating detailed predictive models based on material and energy balances. Following the ISO framework for process-based LCA,[4] evaluating an early-stage product or technology follows a four-stage framework defined by: (1) goal and scope definition, (2) life cycle inventory (LCI) analysis, (3) life cycle impact assessment (LCIA), and (4) interpretation (Table 4.1).

Recent work by Galant et al.[46] applied the four-stage LCA framework to build an LCI model for the active pharmaceutical ingredient (API) nitrofurantoin produced via a continuous mechanochemical synthesis method and a conventional solvent-batch process. Figure 4.2 illustrates and summarizes those four stages of LCA with the example of nitrofurantoin synthesis via continuous twin-screw extrusion (TSE), where each step in the four-stage framework is described from goal and scope definition to LCI analysis,

TABLE 4.1

The Four Stages of Life Cycle Assessment (LCA)

LCA Framework	Steps
Step 1: Goal and Scope Definition	• Define the purpose of the study and temporal and spatial boundaries that describe cradle-to-grave; cradle-to-gate; gate-to-gate or cradle-to-cradle boundary conditions • Define functional unit, the basis of the analysis • Define the system boundary, through a life cycle process flow diagram, often referred to as a system diagram
Step 2: Life Cycle Inventory Analysis	• Collect data to build sequential material and energy balances of all materials and energy consumed at all stages of the life cycle • The LCI audits all material flows within the product life cycle as defined through the system boundary in Goal and Scope definition
Step 3: Life Cycle Impact Assessment	• Calculate midpoint or endpoint life cycle impact assessment (LCIA) metrics to describe the resource and environmental impacts using methods such as ReCiPe, USEtox and TRACI[38–40] • LCA software such as SimaPro[41] and GaBi[42, 43] allow direct computation of midpoint and endpoint LCIA metrics through matrix algebra • LCIA metrics describe midpoint environmental impacts such as climate change impact measured through quantified greenhouse gas emissions, and end-points describe environmental damages such as the long-term damage effects resulting from anthropogenic greenhouse gas emissions on climate change
Step 4: Interpretation	• The results of steps 1–3 are critically evaluated and interpreted through qualitative and quantitative means, including statistical analysis of uncertainty (e.g., analysis of parametric uncertainty through Monte Carlo simulation[44, 45]) and scenario development to test boundary conditions

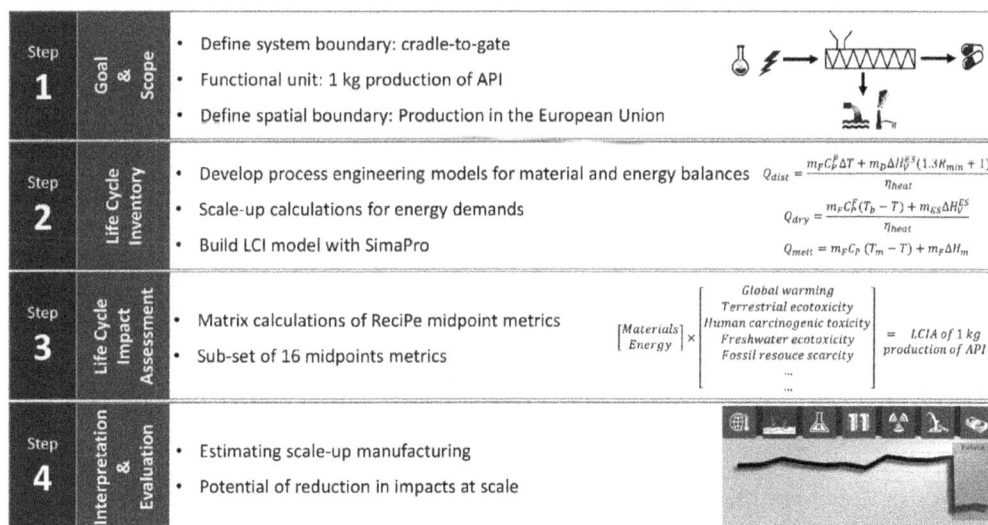

FIGURE 4.2 The four-stage LCA framework with the example of active pharmaceutical ingredient (API) manufacture via twin-screw extrusion study by Galant et al.[46]

LCIA and finally interpretation. Figure 4.2 summarizes the process of conducting LCA for one chemical product via one mechanochemical process based on the results from Galant et al.;[46] results from the LCA can be used to benchmark production of the API using one mechanochemical process, in this case TSE. This same framework can be used to evaluate other early-stage mechanochemical processes for producing organic chemicals, pharmaceuticals and inorganic chemicals or materials at laboratory scale or higher based on ball milling, acoustic mixing and other techniques evolving and poised to enter industrial scale-up.[47]

In the next section we describe three promising research areas related to renewable feedstocks (biomass) for organic chemical and pharmaceutical synthesis, inorganic energy materials and recycling technology that can involve mechanochemical processing methods.

4.2.1 Case 1: Biomass and the Bioeconomy

Transitioning to a low-carbon economy will demand pursuing multiple strategies to limit GHG emissions from multiple sectors. Liquid fuels, chemicals, pharmaceuticals and polymeric materials currently derived from petroleum resources could be sourced from sustainably harvested biomass resources. This presents an opportunity for combining no- or low-chemical/solvent input manufacturing via mechanochemical methods with renewable feedstocks from biomass to manufacture biofuels, biochemicals and bioproducts.[48]

Over the last 20 years, there has been major progress in research on synthesizing biofuels via synthetic biology and bioproducts from biomass. Likewise, recent literature on mechanochemistry has identified methods for converting biomass into value-added bioproducts through ball-milling and twin-screw extrusion.[49] Deconstructing biomass into 5- and 6-carbon fermentable sugars and lignin for valorization requires pretreatment to dissolve the hemicellulose sugars and fractionate the lignin, followed by enzymatic hydrolysis to access the cellulose sugars. Two recent studies employed milling to greatly reduce or eliminate aqueous and organic solvents to enzymatically recover sugars from cellulose[50] and hemicellulose,[51] leading to high sugar yields from agricultural residues like corn (*Zea mays* L.) stover. If scaled, these techniques would reduce the enormous quantities of aqueous solvents needed to access lignocellulosic sugars and reduce the energetic steps involved in heating and drying.

Hajiali et al. reviewed mechanochemical methods for converting biomass into functional materials through ball mill, grinding, extrusion and ultrasonic techniques.[52] Multiple value-added materials can be produced through mechanochemical processing that preserves the polymeric structure of the cellulose (e.g., nanocellulose) and deconstructs hemicellulose via ball-milling or twin-screw extrusion as a pretreatment step and secondary treatment with enzymes to recover xylose without the use of any solvent to produce platform chemicals. The authors also describe opportunities for using ball milling as a post-treatment of pyrolyzed biochar with metal oxides to produce biochar–metal composites.

The promise of scaling a bioeconomy is the opportunity to replace the fossil fuel-sourced hydrocarbon economy with biogenic carbon that can achieve neutrality and in some cases, like if combining bioenergy with carbon capture and storage (BECCS), negative carbon emissions.[53] The biomass and bioenergy literature has many examples where LCA has demonstrated low carbon emission balance opportunities for fuels and chemicals. There are great prospects to further investigate biomass as a feedstock in pharmaceutical manufacturing to produce platform chemicals as raw materials and reactants for APIs.[54] If combined with industrial-scale mechanochemical methods like twin-screw extrusion (TSE), this could greatly reduce supply chain-related environmental impacts for the industry. As a guiding example, Minowa et al. summarized the use of mechanochemical methods for the preparation of bioethanol, reducing or eliminating the need for sulfuric acid.[55]

4.2.2 Case 2: Energy Materials

Mechanochemical techniques have been demonstrated at the laboratory scale to also eliminate solvents while producing high-purity chemicals and materials needed for a renewable energy transition. For example, metal–organic frameworks[56, 57] are identified as promising materials for energy storage and conversion and for carbon capture and sequestration, and materials for solid-state batteries[58] can be synthesized via mechanochemistry. Moreover, lead halide perovskite materials are promising next-generation solar photovoltaic materials with power conversion efficiencies surpassing 25%.[59] Laboratory studies using ball milling have demonstrated kilogram-scale production of solvent-free phase-pure metal–halide perovskite materials at room temperatures via solid-phase reaction.[60] While laboratory-scale proofs-of-concept like this clearly demonstrate the direct gate-to-gate benefits of synthesizing perovskites mechanochemically, perovskite film deposition onto multi-layered PV panel will likely involve either solution- or vapor-based thin film coating[61] and the perovskite material still requires industrial synthesis of the raw materials, which are precursor salts that would be mined and purified in solvent-based processes as described by Khalifa et al.[62]

While LCA studies have been published on MOF,[63] energy storage materials[64, 65] and perovskite PV cells,[66, 67] a critical step is needed for understanding whether industrial-scale mechanochemical methods could reduce environmental footprints during the manufacturing of materials and final products.

4.2.3 Case 3: Recycling Technology

Mechanical processes are ubiquitous in recycling.[68–70] Naturally, while some modern recycling practices involve chemical processes,[71, 72] the recycling industry tries to avoid them for multiple reasons – safety, cost, specialized knowledge and equipment, and all the regulatory requirements that come with becoming a chemical industry. Alternatively, physical separation, shredding and grinding are much simpler in all the relative aspects. Therefore, moving into typical mechanochemical equipment[73] – mills, extruders – can still be done without (formally) entering the "chemical industry" regulatory realm. Here lay many opportunities in using mechanochemistry during recycling[74–77] – milling/extruding a reagent together with the waste to produce an improved recycled material, and all that without chemical reactors, organic solvents, etc. Therefore, in recent years, numerous research programs have focused on using mechanochemistry in recycling. While all aspects of recycling are being explored using mechanochemistry, two, in particular, are more prominent: metal and semi-metal recovery from spent electrochemical devices,[78–80] and plastic recycling.[81, 82]

Metal extraction is one of the most energy-intensive and environmentally damaging processes in the chemical industry.[83, 84] In the 2020s, we are only a few decades since lithium-ion batteries entered and dominated the energy storage market for portable electronic devices,[85] and they are currently entering and dominating the electrical vehicle field as well,[86] which is expected to grow enormously in coming decades. The combined result is a large requirement for lithium extraction and production, but also the creation of increasing amounts of battery waste from which lithium and other metals in the cathode can be recycled. While dealing with metal waste already puts this type of recycling site into the chemical industry domain, mechanochemistry can offer numerous advantages both in terms of production per volume and safety, as the combination of metals and organic solvents can easily lead to combustion and explosions. A recent representative example of such a process was presented by Li et al.[87] In their work, the authors showed that lithium iron phosphate cathode materials, after separation from the rest of the battery, can undergo a mechanochemical solid-state oxidation with sodium persulfate in 5 min using a zirconia ball mill. No acid or any liquid is required for the oxidation, but then acid extraction is done to separate the iron phosphate, which is followed by lithium phosphate precipitation via the addition of sodium phosphate. Unfortunately, this study included an economic analysis, but not an LCA. In more recent examples, different groups showed that combining different waste sources could produce a synergy. Zhou et al. combined two cathode materials to obtain iron phosphate, cobalt oxide and lithium oxide which could be easily recycled,[88] while Zhang et al. showed that milling the cathode materials with polyvinyl chloride (PVC) could lead to the formation of lithium chloride while dechlorinating the PVC.[89] Yet, again, none of these studies include LCAs. In terms of lithium recovery, very recently a first evaluation of different technologies (including mechanochemical) comparing different works from different groups was shown by Bontempi et al.[90] While the authors did not perform a complete LCA, they used a partial environmental evaluation described as ESCAPE analysis (Evaluation of Sustainability of material substitution using CArbon footPrint by a simplified approach), which focuses on embodied energy (an issue in mechanochemical processes) and carbon footprint. Interestingly, 35 technologies were compared, including two mechanochemical processes, one of which was detailed here.[89] As described above, the mechanochemical processes are typically followed by a solution process, be it acid extraction or water wash, in the end producing some amount of liquid waste in addition to the desired products. In the end, in most of the processes, the chemicals used had the largest impact in terms of carbon footprint, and therefore the advantages of mechanochemical processes did not bring such studies to the top 10, as the chemicals used in such processes had severe environmental costs.

In terms of plastic recycling, mechanical processes can be used for different "types of mechanochemistries": bond forming, or bond scission.[91] In polymer mechanochemistry, instead of combining different compounds to make a chemical reaction, chemical bonds are cleaved due to accumulated mechanical stress in the long polymer molecules.[92] Historically, polymer mechanochemistry is considered a "degradative" process, reducing the length of polymer chains, and, as a consequence, the plastic properties.[93] Polymer mechanochemistry is one of the reasons why paper recycling is limited to a number of times, as cellulose chains decay in size and crystallinity.[94] However, in recent years interesting new concepts have been proposed and demonstrated, looking at polymer mechanochemistry as an opportunity. As mechanochemistry focuses on weak bonds and short cross-links,[95, 96] it has been proposed as an approach to recycle thermosets, including epoxy composites.[97] Moore et al. have shown that polymers with a low ceiling temperature can be completely depolymerized to monomers using polymer mechanochemistry;[98] however, this could also be achieved with a small amount of acid,[99] and therefore it is not clear if mechanical processes present an environmental advantage.

Unlike metal recovery from batteries, numerous LCAs have been carried out to analyze mechanical processes in thermoplastic recycling,[100] but not necessarily mechanochemical processes. Mechanical processes perform badly in such LCAs, as the plastic obtained has poor properties due to plastic incompatibility and reduced properties, and also due to polymer mechanochemistry, although details (changes in molecular weight or crystallinity) are typically missing. Induced mechanochemical processes were shown to potentially improve the compatibility between different types of thermoplastics through the creation of block copolymers, for example, using solid-state shear pulverization; yet, no LCAs have been done comparing such processes to chemical processes.[101] Still, LCAs have shown that even simple mechanical processes are always preferable compared to incineration or landfills. Therefore, given there are no good chemical processes for the chemical recycling of common thermosets (epoxy, cross-linked polyethylene, vulcanized rubber, etc.), mechanochemical processes are expected to perform well in thermoset recycling. An LCA comparing such concepts is required to confirm this.

4.3 Conclusions: LCA as a Tool for Benchmarking the Sustainability of Emerging Technologies

Understanding the potential impact of mechanochemistry demands benchmarking this new technology against current manufacturing methods. Here, life cycle assessment (LCA) is described as a critical tool to measure and monitor progress towards more sustainable manufacturing. Overall, mechanochemical processes can yield dramatic improvements in product manufacture due to solvent elimination and as one published case study showed,[46] by greatly reducing direct energy consumption along with changing from thermal to electrical energy inputs; both can have an immediate reduction in energy and GHG emissions in the product supply chain. However, reducing environmental impacts in product supply chains can be challenging where often solvent elimination or high-temperature processing is unavoidable due to accessing raw materials for production. A future goal of scaling mechanochemical processes is applying LCA methods to understand the environmental implications in feedstock supply for product design, including for promising renewable resources like biomass as well as non-renewables that are critical to scaling renewable electricity supply. Finally, mechanochemistry has a critical role in the recycling of high-value materials that are needed for efficient energy carriers such as Li batteries. For all three cases, environmental benchmarking through LCA is critical to inform decision-making.

Further Reading on Life Cycle Assessment

Curran, M. A., Life Cycle Assessment: A Systems Approach to Environmental Management and Sustainability. *Chem. Eng. Prog.* 2015, pp 26–36.

Curran, M. A., *Life-cycle assessment: inventory guidelines and principles.* CRC Press: 2020.

Galant, O.; Cerfeda, G.; McCalmont, A. S.; James, S. L.; Porcheddu, A.; Delogu, F.; Crawford, D. E.; Colacino, E.; Spatari, S., Mechanochemistry Can Reduce Life Cycle Environmental Impacts of Manufacturing Active Pharmaceutical Ingredients. *ACS Sustain. Chem. Eng.* 2022, *10*, (4), 1430–1439.

REFERENCES

1. Guinée, J. B.; Heijungs, R.; Huppes, G.; Zamagni, A.; Masoni, P.; Buonamici, R.; Ekvall, T.; Rydberg, T., Life cycle assessment: Past, present, and future. *Environ. Sci. Technol.* 2011, *45*(1), 90–96.

2. Society of Environmental Toxicology and Chemistry (SETAC), *A Technical Framework for Life-Cycle Assessment*. Pensacola, FL: SETAC Foundation, p. 124, 1991.

3. ISO. *ISO 14044: Environmental Management — Life Cycle Assessment — Requirements and Guidelines*; ISO 14044:2006(E). Geneva: International Organization for Standardization, 2006a.

4. ISO. *ISO 14040: Environmental Management — Life Cycle Assessment — Principles and Framework*; ISO 14040:2006(E). Geneva: International Organization for Standardization, 2006b.

5. Hocking, M. B., Paper versus polystyrene: A complex choice. *Science* 1991, *251*(4993), 504–505.

6. Sullivan, J. L.; Williams, R. L.; Yester, S.; Cobas-Flores, E.; Chubbs, S. T.; Hentges, S. G.; Pomper, S. D., Life cycle inventory of a generic U. S. Family Sedan overview of results USCAR AMP project. *SAE Trans.* 1998, *107*, 1909–1923.

7. Keoleian, G.; Spatari, S.; Beal, R.; Stephens, R.; Williams, R., Application of life cycle inventory analysis to fuel tank system design. *Int. J. Life Cycle Assess.* 1998, *3*(1), 18–28.

8. Saur, K.; Fava, J. A.; Spatari, S., Life cycle engineering case study: Automobile fender designs. *Environ. Prog.* 2000, *19*(2), 72–82.

9. Keoleian, G. A.; Kar, K., Elucidating complex design and management tradeoffs through life cycle design: Air intake manifold demonstration project. *J. Clean. Prod.* 2003, *11*(1), 61–77.

10. Lave, L.; Maclean, H.; Hendrickson, C.; Lankey, R., Life-cycle analysis of alternative automobile fuel/propulsion technologies. *Environ. Sci. Technol.* 2000, *34*(17), 3598–3605.

11. MacLean, H. L.; Lave, L. B., Life cycle assessment of automobile/fuel options. *Environ. Sci. Technol.* 2003, *37*(23), 5445–5452.

12. Bilec, M. M.; Ries, R. J.; Matthews, H. S., Life-cycle assessment modeling of construction processes for buildings. *J. Infrastruct. Syst.* 2010, *16*(3), 199–205.

13. Junnila, S.; Horvath, A.; Guggemos, A. A., Life-cycle assessment of office buildings in Europe and the United States. *J. Infrastruct. Syst.* 2006, *12*(1), 10–17.

14. Napolano, L.; Menna, C.; Graziano, S. F.; Asprone, D.; D'Amore, M.; de Gennaro, R.; Dondi, M., Environmental life cycle assessment of lightweight concrete to support recycled materials selection for sustainable design. *Constr. Build. Mater.* 2016, *119*, 370–384.

15. Tian, Y.; Spatari, S., Environmental life cycle evaluation of prefabricated residential construction in China. *J. Build. Eng.* 2022, *57*, 104776.

16. Siler-Evans, K.; Azevedo, I. L.; Morgan, M. G., Marginal emissions factors for the U.S. Electricity system. *Environ. Sci. Technol.* 2012, *46*(9), 4742–4748.

17. Chester, M. V.; Horvath, A., Environmental assessment of passenger transportation should include infrastructure and supply chains. *Environ. Res. Lett.* 2009, *4*(2), 024008.

18. Stokes, J.; Horvath, A., Life cycle energy assessment of alternative water supply systems (9 pp). *Int. J. Life Cycle Assess.* 2006, *11*(5), 335–343.

19. Curran, M. A., Life cycle assessment: A systems approach to environmental management and sustainability. *Chem. Eng. Prog.* 2015, 26–36.

20. Kumar, S.; Posmanik, R.; Spatari, S.; Ujor, V. C., Repurposing anaerobic digestate for economical biomanufacturing and water recovery. *Appl. Microbiol. Biotechnol.* 2022.

21. Dusselier, M.; Van Wouwe, P.; Dewaele, A.; Makshina, E.; Sels, B. F., Lactic acid as a platform chemical in the biobased economy: The role of chemocatalysis. *Energy Environ. Sci.* 2013, *6*(5), 1415–1442.

22. Batten, R.; Karanjikar, M.; Spatari, S., Bio-based polyisoprene can mitigate climate change and deforestation in expanding rubber production. *Fermentation* 2021, *7*(4), 204.

23. Riazi, B.; Zhang, J.; Yee, W.; Ngo, H.; Spatari, S., Life cycle environmental and cost implications of isostearic acid production for pharmaceutical and personal care products. *ACS Sustain. Chem. Eng.* 2019, *7*(18), 15247–15258.

24. Parvatker, A. G.; Eckelman, M. J., Simulation-based estimates of life cycle inventory gate-to-gate process energy use for 151 organic chemical syntheses. *ACS Sustain. Chem. Eng.* 2020, *8*(23), 8519–8536.

25. Cashman, S. A.; Meyer, D. E.; Edelen, A. N.; Ingwersen, W. W.; Abraham, J. P.; Barrett, W. M.; Gonzalez, M. A.; Randall, P. M.; Ruiz-Mercado, G.; Smith, R. L., Mining available data from the United States environmental protection agency to support rapid life cycle inventory modeling of chemical manufacturing. *Environ. Sci. Technol.* 2016, *50*(17), 9013–9025.

26. Smith, R. L.; Ruiz-Mercado, G. J.; Meyer, D. E.; Gonzalez, M. A.; Abraham, J. P.; Barrett, W. M.; Randall, P. M., Coupling computer-aided process simulation and estimations of emissions and land use for rapid life cycle inventory modeling. *ACS Sustain. Chem. Eng.* 2017, *5*(5), 3786–3794.

27. Song, R.; Keller, A. A.; Suh, S., Rapid life-cycle impact screening using artificial neural networks. *Environ. Sci. Technol.* 2017, *51*(18), 10777–10785.

28. Bergerson, J. A.; Brandt, A.; Cresko, J.; Carbajales-Dale, M.; MacLean, H. L.; Matthews, H. S.; McCoy, S.; McManus, M.; Miller, S. A.; Morrow, W. R.; Posen, I. D.; Seager, T.; Skone, T.; Sleep, S., Life cycle assessment of emerging technologies: Evaluation techniques at different stages of market and technical maturity. *J. Ind. Ecol.* 2020, *24*(1), 11–25.

29. Spatari, S.; Bagley, D. M.; MacLean, H. L., Life cycle evaluation of emerging lignocellulosic ethanol conversion technologies. *Bioresour. Technol.* 2010, *101*(2), 654–667.

30. Spatari, S.; MacLean, H. L., Characterizing model uncertainties in the life cycle of lignocellulose-based ethanol fuels. *Environ. Sci. Technol.* 2010, *44*(22), 8773–8780.

31. Sorunmu, Y.; Billen, P.; Spatari, S., A review of thermochemical upgrading of pyrolysis bio-oil: Techno-economic analysis, life cycle assessment, and technology readiness. *GCB Bioenergy* 2020, *12*(1), 4–18.

32. Nguyen, L.; Cafferty, K.; Searcy, E.; Spatari, S., Uncertainties in life cycle greenhouse gas emissions from advanced biomass feedstock logistics supply chains in Kansas. *Energies* 2014, *7*(11), 7125–7146.

33. Posen, I. D.; Jaramillo, P.; Griffin, W. M., Uncertainty in the life cycle greenhouse gas emissions from U.S. Production of three biobased polymer families. *Environ. Sci. Technol.* 2016, *50*(6), 2846–2858.

34. Nguyen, L. K.; Na, S.; Hsuan, Y. G.; Spatari, S., Uncertainty in the life cycle greenhouse gas emissions and costs of HDPE pipe alternatives. *Resour. Conserv. Recy.* 2020, *154*, 104602.

35. Ögmundarson, Ó.; Sukumara, S.; Laurent, A.; Fantke, P., Environmental hotspots of lactic acid production systems. *GCB Bioenergy* 2020, *12*(1), 19–38.

36. Arvidsson, R.; Tillman, A.-M.; Sandén, B. A.; Janssen, M.; Nordelöf, A.; Kushnir, D.; Molander, S., Environmental assessment of emerging technologies: Recommendations for prospective LCA. *J. Ind. Ecol.* 2018, *22*(6), 1286–1294.

37. Buyle, M.; Audenaert, A.; Billen, P.; Boonen, K.; Van Passel, S., The future of ex-ante LCA? Lessons learned and practical recommendations. *Sustainability* 2019, *11*(19), 5456.

38. Huijbregts, M. A. J.; Steinmann, Z. J. N.; Elshout, P. M. F.; Stam, G.; Verones, F.; Vieira, M.; Zijp, M.; Hollander, A.; van Zelm, R., ReCiPe2016: A harmonised life cycle impact assessment method at midpoint and endpoint level. *Int. J. Life Cycle Assess.* 2017, *22*(2), 138–147.

39. Rosenbaum, R. K.; Huijbregts, M. A. J.; Henderson, A. D.; Margni, M.; McKone, T. E.; van de Meent, D.; Hauschild, M. Z.; Shaked, S.; Li, D. S.; Gold, L. S.; Jolliet, O., USEtox human exposure and toxicity factors for comparative assessment of toxic emissions in life cycle analysis: Sensitivity to key chemical properties. *Int. J. Life Cycle Assess.* 2011, *16*(8), 710.

40. Henderson, A. D.; Hauschild, M. Z.; van de Meent, D.; Huijbregts, M. A.; Larsen, H. F.; Margni, M.; McKone, T. E.; Payet, J.; Rosenbaum, R. K.; Jolliet, O., USEtox fate and ecotoxicity factors for comparative assessment of toxic emissions in life cycle analysis: Sensitivity to key chemical properties. *Int. J. Life Cycle Assess.* 2011, *16*(8), 701.

41. Pre Consultants. *SimaPro 8.4.* The Netherlands, 2018.

42. Spatari, S.; Betz, M.; Florin, H.; Baitz, M.; Faltenbacher, M., Using GaBi 3 to perform life cycle assessment and life cycle engineering. *Int. J. Life Cycle Assess.* 2001, *6*(2), 81–84.

43. *Sphera GaBi 9: The Software System for Life Cycle Assessment.* Stuttgart, 2020.

44. Huijbregts, M., Uncertainty and variability in environmental life-cycle assessment. *Int. J. Life Cycle Assess.* 2002, *7*(3), 173–173.

45. Heijungs, R.; Huijbregts, M. A. J., A review of approaches to treat uncertainty in LCA. In: *iEMSs 2004 - Complexity and Integrated Resources Management*, International Environmental Modelling and Software Society. Germany: University of Osnabrück, 2004.

46. Galant, O.; Cerfeda, G.; McCalmont, A. S.; James, S. L.; Porcheddu, A.; Delogu, F.; Crawford, D. E.; Colacino, E.; Spatari, S., Mechanochemistry can reduce life cycle environmental impacts of manufacturing active pharmaceutical ingredients. *ACS Sustain. Chem. Eng.* 2022, *10*(4), 1430–1439.

47. Sopicka-Lizer, M.: *High-Energy Ball Milling: Mechanochemical Processing of Nanopowders*; Elsevier, 2010.

48. Calcio Gaudino, E.; Cravotto, G.; Manzoli, M.; Tabasso, S., Sono- and mechanochemical technologies in the catalytic conversion of biomass. *Chem. Soc. Rev.* 2021, *50*(3), 1785–1812.

49. Razumovskii, S. D.; Podmaster'ev, V. V.; Zelenetskii, A. N., Mechanochemical methods of activating processes of biomass pretreatment. *Cat. Ind.* 2011, *3*(1), 23–27.

50. Hammerer, F.; Ostadjoo, S.; Dietrich, K.; Dumont, M.-J.; Del Rio, L. F.; Friščić, T.; Auclair, K., Rapid mechanoenzymatic saccharification of lignocellulosic biomass without bulk water or chemical pretreatment. *Green Chem.* 2020, *22*(12), 3877–3884.

51. Ostadjoo, S.; Hammerer, F.; Dietrich, K.; Dumont, M.-J.; Friscic, T.; Auclair, K., Efficient enzymatic hydrolysis of biomass hemicellulose in the absence of bulk water. *Molecules* 2019, *24*(23), 4206.

52. Hajiali, F.; Jin, T.; Yang, G.; Santos, M.; Lam, E.; Moores, A., Mechanochemical transformations of biomass into functional materials. *ChemSusChem* 2022, *15*(7), e202102535.

53. Fuss, S.; Lamb, W. F.; Callaghan, M. W.; Hilaire, J.; Creutzig, F.; Amann, T.; Beringer, T.; de Oliveira Garcia, W.; Hartmann, J.; Khanna, T.; Luderer, G.; Nemet, G. F.; Rogelj, J.; Smith, P.; Vicente, J. L. V.; Wilcox, J.; del Mar Zamora Dominguez, M.; Minx, J. C., Negative emissions—Part 2: Costs, potentials and side effects. *Environ. Res. Lett.* 2018, *13*(6), 063002.

54. Keasling, J. D., Manufacturing molecules through metabolic engineering. *Science* 2010, *330*(6009), 1355–1358.

55. Fujimoto, S.; Inoue, H.; Yano, S.; Sakaki, T.; Minowa, T.; Endo, T.; Sawayama, S.; Sakanishi, K., Bioethanol production from lignocellulosic biomass requiring no sulfuric acid: Mechanochemical pretreatment and enzymic saccharification. *J. Jpn Petrol. Inst.* 2008, *51*(5), 264–273.

56. Stock, N.; Biswas, S., Synthesis of metal-organic frameworks (MOFs): Routes to various MOF topologies, morphologies, and composites. *Chem. Rev.* 2012, *112*(2), 933–969.

57. Klimakow, M.; Klobes, P.; Thünemann, A. F.; Rademann, K.; Emmerling, F., Mechanochemical synthesis of metal−organic frameworks: A fast and facile approach toward quantitative yields and high specific surface areas. *Chem. Mater.* 2010, *22*(18), 5216–5221.

58. Schlem, R.; Burmeister, C. F.; Michalowski, P.; Ohno, S.; Dewald, G. F.; Kwade, A.; Zeier, W. G., Energy storage materials for solid-state batteries: Design by mechanochemistry. *Adv. Energy Mater.* 2021, *11*(30), 2101022.

59. Tian, X.; Stranks, S. D.; You, F., Life cycle energy use and environmental implications of high-performance perovskite tandem solar cells. *Sci. Adv.* 2020, *6*(31), eabb0055.

60. Hong, Z.; Tan, D.; John, R. A.; Tay, Y. K. E.; Ho, Y. K. T.; Zhao, X.; Sum, T. C.; Mathews, N.; García, F.; Soo, H. S., Completely solvent-free protocols to access phase-pure, metastable metal halide perovskites and functional photodetectors from the precursor salts. *iScience* 2019, *16*, 312–325.

61. Celik, I.; Song, Z.; Cimaroli, A. J.; Yan, Y.; Heben, M. J.; Apul, D., Life Cycle Assessment (LCA) of perovskite PV cells projected from lab to fab. *Sol. Energy Mater. Sol. Cells* 2016, *156*, 157–169.

62. Khalifa, S. A.; Spatari, S.; Fafarman, A. T.; Baxter, J. B. *In Life Cycle Environmental Impacts of Precursors Used in the Supply Chain of Emerging Perovskite Solar Cells*, 2021 IEEE 48th Photovoltaic Specialists Conference (PVSC), 20–25 June 2021, 2021; pp 0569–0572.

63. Hu, J.; Gu, X.; Lin, L.-C.; Bakshi, B. R., Toward sustainable metal−organic frameworks for post-combustion carbon capture by life cycle assessment and molecular simulation. *ACS Sustain. Chem. Eng.* 2021, *9*(36), 12132–12141.

64. Baumann, M.; Peters, J. F.; Weil, M.; Grunwald, A., CO_2 footprint and life-cycle costs of electrochemical energy storage for stationary grid applications. *Energy Technol.* 2017, *5*(7), 1071–1083.

65. Bautista, S. P.; Weil, M.; Baumann, M.; Montenegro, C. T., Prospective life cycle assessment of a model magnesium battery. *Energy Technol.* 2021, *9*(4), 2000964.

66. Gong, J.; Darling, S. B.; You, F., Perovskite photovoltaics: Life-cycle assessment of energy and environmental impacts. *Energy Environ. Sci.* 2015, *8*(7), 1953–1968.

67. Billen, P.; Leccisi, E.; Dastidar, S.; Li, S.; Lobaton, L.; Spatari, S.; Fafarman, A. T.; Fthenakis, V. M.; Baxter, J. B., Comparative evaluation of lead emissions and toxicity potential in the life cycle of lead halide perovskite photovoltaics. *Energy* 2019, *166*, 1089–1096.

68. Yun, L.; Linh, D.; Shui, L.; Peng, X.; Garg, A.; Le, M. L. P.; Asghari, S.; Sandoval, J., Metallurgical and mechanical methods for recycling of lithium-ion battery pack for electric vehicles. *Resour. Conserv. Recy.* 2018, *136*, 198–208.

69. Vollmer, I.; Jenks, M. J. F.; Roelands, M. C. P.; White, R. J.; van Harmelen, T.; de Wild, P.; van der Laan, G. P.; Meirer, F.; Keurentjes, J. T. F.; Weckhuysen, B. M., Beyond mechanical recycling: Giving new life to plastic waste. *Angew. Chem. Int. Ed.* 2020, *59*(36), 15402–15423.

70. de Andrade Salgado, F.; de Andrade Silva, F., Recycled aggregates from construction and demolition waste towards an application on structural concrete: A review. *J. Build. Eng.* 2022, *52*, 104452.
71. Ellis, L. D.; Orski, S. V.; Kenlaw, G. A.; Norman, A. G.; Beers, K. L.; Román-Leshkov, Y.; Beckham, G. T., Tandem heterogeneous catalysis for polyethylene depolymerization via an olefin-intermediate process. *ACS Sustain. Chem. Eng.* 2021, *9*(2), 623–628.
72. Zhang, G.; Yuan, X.; He, Y.; Wang, H.; Zhang, T.; Xie, W., Recent advances in pretreating technology for recycling valuable metals from spent lithium-ion batteries. *J. Hazard. Mater.* 2021, *406*, 124332.
73. Colacino, E.; Isoni, V.; Crawford, D.; García, F., Upscaling mechanochemistry: Challenges and opportunities for sustainable industry. *J. Trends Chem.* 2021.
74. Guo, X.; Xiang, D.; Duan, G.; Mou, P., A review of mechanochemistry applications in waste management. *Waste Manag. (Oxf.)* 2010, *30*(1), 4–10.
75. Plescia, P.; Gizzi, D.; Benedetti, S.; Camilucci, L.; Fanizza, C.; De Simone, P.; Paglietti, F., Mechanochemical treatment to recycling asbestos-containing waste. *Waste Manag.* 2003, *23*(3), 209–218.
76. Baláž, M., *Environmental Mechanochemistry: Recycling Waste into Materials using High-Energy Ball Milling.* Springer Nature, 2021.
77. Tan, Q.; Li, J., Recycling metals from wastes: A novel application of mechanochemistry. *Environ. Sci. Technol.* 2015, *49*(10), 5849–5861.
78. Yang, Y.; Okonkwo, E. G.; Huang, G.; Xu, S.; Sun, W.; He, Y., On the sustainability of lithium ion battery industry – A review and perspective. *Energy Storage Mater.* 2021, *36*, 186–212.
79. Makuza, B.; Tian, Q.; Guo, X.; Chattopadhyay, K.; Yu, D., Pyrometallurgical options for recycling spent lithium-ion batteries: A comprehensive review. *J. Power Sources* 2021, *491*, 229622.
80. Peplow, M., Solar panels face recycling challenge. *ACS Cent. Sci.* 2022, *8*(3), 299–302.
81. Davidson, M. G.; Furlong, R. A.; McManus, M. C., Developments in the life cycle assessment of chemical recycling of plastic waste – A review. *J. Clean. Prod.* 2021, *293*, 126163.
82. Lange, J.-P., Managing plastic waste—sorting, recycling, disposal, and product redesign. *ACS Sustain. Chem. Eng.* 2021, *9*(47), 15722–15738.
83. Wang, J.; Yue, X.; Wang, P.; Yu, T.; Du, X.; Hao, X.; Abudula, A.; Guan, G., Electrochemical technologies for lithium recovery from liquid resources: A review. *Renew. Sustain. Energy Rev.* 2022, *154*, 111813.
84. Watari, T.; Nansai, K.; Nakajima, K., Major metals demand, supply, and environmental impacts to 2100: A critical review. *Resour. Conserv. Recy.* 2021, *164*, 105107.
85. Zubi, G.; Dufo-López, R.; Carvalho, M.; Pasaoglu, G., The lithium-ion battery: State of the art and future perspectives. *Renew. Sustain. Energy Rev.* 2018, *89*, 292–308.
86. Horowitz, Y.; Schmidt, C.; Yoon, D.-H.; Riegger, L. M.; Katzenmeier, L.; Bosch, G. M.; Noked, M.; Ein-Eli, Y.; Janek, J.; Zeier, W. G.; Diesendruck, C. E.; Golodnitsky, D., Between liquid and all solid: A prospect on electrolyte future in lithium-ion batteries for electric vehicles. *Energy Technol.* 2020, *8*(11), 2000580.
87. Liu, K.; Liu, L.; Tan, Q.; Li, J., Selective extraction of lithium from a spent lithium iron phosphate battery by mechanochemical solid-phase oxidation. *Green Chem.* 2021, *23*(3), 1344–1352.
88. Jiang, Y.; Chen, X.; Yan, S.; Ou, Y.; Zhou, T., Mechanochemistry-induced recycling of spent lithium-ion batteries for synergistic treatment of mixed cathode powders. *Green Chem.* 2022, *24*(15), 5987–5997.
89. Wang, M.-M.; Zhang, C.-C.; Zhang, F.-S., Recycling of spent lithium-ion battery with polyvinyl chloride by mechanochemical process. *Waste Manag. (Oxf.)* 2017, *67*, 232–239.
90. Fahimi, A.; Ducoli, S.; Federici, S.; Ye, G.; Mousa, E.; Frontera, P.; Bontempi, E., Evaluation of the sustainability of technologies to recycle spent lithium-ion batteries, based on embodied energy and carbon footprint. *J. Clean. Prod.* 2022, *338*, 130493.
91. Takacs, L., The historical development of mechanochemistry. *Chem. Soc. Rev.* 2013, *42*(18), 7649–7659.
92. Caruso, M. M.; Davis, D. A.; Shen, Q.; Odom, S. A.; Sottos, N. R.; White, S. R.; Moore, J. S., Mechanically-induced chemical changes in polymeric materials. *Chem. Rev.* 2009, *109*(11), 5755–5798.
93. Li, J.; Nagamani, C.; Moore, J. S., Polymer mechanochemistry: From destructive to productive. *Acc. Chem. Res.* 2015, *48*(8), 2181–2190.
94. Kuga, S.; Wu, M., Mechanochemistry of cellulose. *Cellulose* 2019, *26*(1), 215–225.
95. Kingsbury, C. M.; May, P. A.; Davis, D. A.; White, S. R.; Moore, J. S.; Sottos, N. R., Shear activation of mechanophore-crosslinked polymers. *J. Mater. Chem.* 2011, *21*(23), 8381–8388.

96. Wang, F.; Diesendruck, C. E., Effect of disulphide loop length on mechanochemical structural stability of macromolecules. *Chem. Commun.* 2020, *56*(14), 2143–2146.

97. Kang, P.; Yang, S.; Bai, S.; Wang, Q., Novel application of mechanochemistry in waste epoxy recycling via solid-state shear milling. *ACS Sustain. Chem. Eng.* 2021, *9*(35), 11778–11789.

98. Diesendruck, C. E.; Peterson, G. I.; Kulik, H. J.; Kaitz, J. A.; Mar, B. D.; May, P. A.; White, S. R.; Martínez, T. J.; Boydston, A. J.; Moore, J. S., Mechanically triggered heterolytic unzipping of a low-ceiling-temperature polymer. *Nat. Chem.* 2014, *6*(7), 623–628.

99. Kaitz, J. A.; Possanza, C. M.; Song, Y.; Diesendruck, C. E.; Spiering, A. J. H.; Meijer, E. W.; Moore, J. S.; Depolymerizable, Depolymerizable, adaptive supramolecular polymer nanoparticles and networks. *Polym. Chem.* 2014, *5*(12), 3788–3794.

100. Schwarz, A. E.; Ligthart, T. N.; Godoi Bizarro, D.; De Wild, P.; Vreugdenhil, B.; van Harmelen, T., Plastic recycling in a circular economy; determining environmental performance through an LCA matrix model approach. *Waste Manag.* 2021, *121*, 331–342.

101. Khait, K.; Torkelson, J. M., Solid-state shear pulverization of plastics: A green recycling process. *Polym. Plast. Technol. Eng.* 1999, *38*(3), 445–457.

5

Intellectual Property (IP) Strategy in Mechanochemistry

Tanja Bendele

CONTENTS

5.1 Definition of Mechanochemistry

According to IUPAC, mechanochemistry relates to "Chemical reaction that is induced by the direct absorption of mechanical energy" (shearing, stretching, and grinding are typical methods for the mechanochemical generation of reactive sites, usually macroradicals, in polymer chains that undergo mechanochemical reactions).[1]

5.1.1 Definition Mechanochemistry for Sustainable Industry

The following part is an attempt to define sustainable manufacturing, sustainable chemistry, or green chemistry. Both terms "Green Chemistry" and "Sustainable Chemistry" are used to describe chemistry avoiding unnecessary compounds, hazard initial substances, hazardous products, and solvents, saving energy, reducing the complexity of synthesis, and may consider many more environmental principles for sustainable conduct.

Most people believe that to achieve these goals, the development and use of completely novel techniques will be necessary. I believe we have to have a holistic view considering novel techniques, techniques used by the nature, and sometimes to consider very old techniques in a new set-up with novel amendments and supplements. We need an objective view, an objective analysation of the whole set-up, and a consequent holistic solution with a vast view into the future.

One key method to achieve this goal is to continually question: "Is the sustainable behaviour of the past and today also a sustainable behaviour in the future or can we perform more sustainably?" Therefore, a current sustainable innovation may be replaced by a more sustainable innovation of tomorrow or sometimes a look at the past may be also useful for finding an old but environmentally friendly synthesis or product that may be amended and/or supplemented.

A good source for classification of known industrial processes and products is the patent literature. To enable the evaluation and classification of known industrial processes and products with respect to the criteria for sustainable manufacturing, the basic principles will be repeated.

The 12 basic principles of Green Chemistry were developed by Paul Anastas and John Warner in 1998,[2] and the following list is a reduced extract of the framework for making a greener chemical, process, or product published by the American Chemical Society.[3]

> **Prevention:** It is better to prevent waste than to treat or clean up waste after it has been created.
>   Atom Economy: Synthetic methods should be designed to maximize incorporation of all materials used in the process into the final product.
>   **Less Hazardous Chemical Syntheses**: Wherever practicable, synthetic methods should be designed to use and generate substances that possess little or no toxicity to human health and the environment.
>   **Designing Safer Chemicals**: Chemical products should be designed to preserve efficacy of function while reducing toxicity.
>   **Safer Solvents and Auxiliaries**: The use of auxiliary substances (e.g., solvents, separation agents, etc.) should be made unnecessary wherever possible and, innocuous when used.
>   **Design for Energy Efficiency**: Energy requirements should be recognized for their environmental and economic impacts and should be minimized. Synthetic methods should be conducted at ambient temperature and pressure.
>   **Use of Renewable Feedstocks**: A raw material or feedstock should be renewable rather than depletable whenever technically and economically practicable.
>   **Reduce Derivates**: Unnecessary derivatization (use of blocking groups, protection/deprotection, temporary modification of physical/chemical processes) should be minimized or avoided if possible, because such steps require additional reagents and can generate waste.
>   **Catalysis**: Catalytic reagents (as selective as possible) are superior to stoichiometric reagents.
>   **Design for Degradation**: Chemical products should be designed so that at the end of their function they break down into innocuous degradation products and do not persist in the environment.
>   **Real-time analysis for pollution Prevention**: Analytical methodologies need to be further developed to allow for real-time, in-process monitoring and control prior to the formation of hazardous substances.
>   **Inherently Safer Chemistry for Accident Prevention**: Substances and the form of a substance used in a chemical process should be chosen to minimize the potential for chemical accidents, including releases, explosions, and fires.

Therefore, the objective must be to implement all or as many as possible of these 12 basic principles for greener chemistry with sustainable manufacturing. Mechanochemistry as well as gaseous reaction conditions are therefore per se best suited to achieve in combination with the use of renewable energy, and most of these principles are applicable to various reactions. A balanced and mixed IP strategy will also support the development, use, and dissemination of greener chemistry with sustainable manufacturing, as patents protect the inventions to the innovators over a long period but also makes the innovations public to the market companion, who may try to invent more improved novel developments.

5.2 Basics of Intellectual Property

5.2.1 Relevance of the IP Strategy for Implementation of Sustainable Manufacturing

As mentioned above, sustainable manufacturing is one of the most common demands the industry, such as the pharmaceutical, chemical, and biotechnological industry, is faced with today.

Before my brief introduction to the relevant intellectual property rights, a general attempt will be made to examine the factors that may hinder or restrict innovations.

Innovations, such as product innovations, may be hindered by their own subjective imagination or the limited imagination of other participants. A frequently used phrase may be: "We've never done it that way before!"

Furthermore, the general framework conditions, which also include laws, regulations, directives and other guidelines, established infrastructure, rules of the market, cost etc., can also hinder inventions or an unbiased view of the inventors or involved parties.

Also, approved development costs and the available equipment frame inventors' possibilities. Moreover, technical innovations are often very complex, so at least one or more cooperations with other market operators are necessary.

Additionally, costs must be taken into account for securing the developments via intellectual property rights well before disclosing the invention to a third party or a market launch. Necessary research collaborations with existing cooperation agreements and non-disclosure agreements will depend on a practised, open, trusting, and cooperative collaboration of the participants. The assignment of rights to inventions must be proactively regulated in cooperation agreements, whereby national special legal regulations must be taken into account in international inventor teams.

A further impediment to innovations may be a product policy aimed at product differentiation rather than product variation that may dominate a company's innovation strategy. Also, succumbing to the temptation of a "me-too-strategy", where a comparable product or process is developed that is already successfully launched on the market may hinder distinct innovations. Moreover, a corporate policy based on persistence, which, for example, waits until a third-party innovation has established itself on the market or relies on tradition out of conviction.

The aforementioned obstacles, which are not to be understood exhaustively, are only intended to indicate possible difficulties, to be solved, resolved, or overcome after a kick-off meeting for technical developments by enthusiastic inventors (a term from patent law) with a great deal of optimism, persuasion, perseverance as well as technical, legal, and rhetorical knowledge.

For the above reasons, the development of technical innovations or of disruptive technologies is a very special challenge.

5.2.2 Good Reasons for Industrial Property Rights

Valid IP rights secure an innovator a monopoly position in the markets and thus enable him to recoup his investments and make profits. Likewise, IP rights give the innovator or pioneer protection against imitators, such as companies that are considered early followers and late followers, or companies that want to market detailed improvements as a modification. Valid IP rights are also marketing tools to prove the innovative spirit of a company and they are an enormously important part of the value of companies.

Research-based companies, and in particular the classical research-based chemical, pharmaceutical as well as research-based companies in the pharmaceutical biotechnology field, have an enormous interest in the reliable protection of the progressive and cost-intensive developments of their products towards an imitation by third parties.

Moreover, after a patent has lapsed, the market encompasses a great market volume and creates an enormous potential for innovative solutions. For example, in the classical pharmaceutical field (small molecules), drugs with the same active ingredient that are marketed after the expiry of the active ingredient patent are called generic drugs. Whereas, biopharmaceuticals (biologicals) are biotechnologically produced protein-based medicinal products whose imitation products are called biosimilars after the patent of the biotechnologically produced original active substance has expired. Biosilimars often have different glycosylation of the active ingredient. In both fields, the generic and biosimilar market often uses new production processes, novel polymorphs, salts, hydrates, solvates, and innovative formulations which are developed by all market players.

5.2.3 Definitions of Intellectual Property

Hereinafter, different definitions of "Intellectual Property" are given that are disclosed by official organisations.

According to the DPMA[4] (Deutsches Patent und Markenamt, German Patent and Trademark Office), the term "intellectual property" – also referred to as IP – covers property rights in creations of the human intellect, e.g. inventions, know-how, software. The term "industrial property rights" refers to all legal rights that protect these individual intellectual achievements that provide creators, such as patent and utility model rights in respect of inventions or the copyrighting of works of science, literature, and art (including software).

WIPO[5] defines intellectual property (IP) as

> creations of the mind, such as inventions; literary and artistic works; designs; and symbols, names, and images used in commerce. Wherein, the IP is protected in law by, e.g. by patents, copyright, and trademarks, which enable people to earn recognition or financial benefit from what they invent or create. By striking the right balance between the interests of innovators and the wider public interest, the IP system aims to foster an environment in which creativity and innovation can flourish.

EPO[6] defines "Intellectual property (IP)" as

> Creations of the mind or intellect. Intellectual property is divided into two categories: industrial property, which includes patents, trademarks, industrial designs, and copyrights, which include literary and artistic works, such as novels, poems and plays, films, musical works, artistic works as well as drawings, paintings, photographs, sculptures, and architectural designs.

USPTO[7] defines "intellectual property" as

> Creations of the mind – creative works or ideas embodied in a form that can be shared or can enable others to recreate, emulate, or manufacture them. There are four ways to protect **intellectual property** – patents, trademarks, copyrights or trade secrets.

All definitions of DPMA, WIPO and EPO (European Patent Organisation) refer to the mind or intellect and therefore to a natural person, who creates intellectual property. Also, according to the USPTO, an inventor is a natural person and not an artificial intelligence. Therefore, intellectual property that is accessible to registered technical intellectual property and is reserved for inventors, who are per definition natural persons. The same may apply to designers of designs and copyrights etc. An exception seems to apply to trademarks that are registered for an owner without mentioning the creator.

5.2.3.1 *Historical International Developments of Intellectual Property*

For a better understanding of Intellectual Property rights, the general concept will be explained briefly in a historical and international context.

As patent rights are predominantly known and the corresponding national patent law has been established for decades, the historical overview will be made on patent law, wherein the trademark and design law typically follow within a certain period.

The first idea of establishing a monopoly privilege was conceived in Sybaris a Greek city (720 a.Ch.n.–510 a.Ch.n.)[8] known for the luxurious lifestyle of its citizens for which has been handed down:

> And if any confectioner or cook invented any peculiar and excellent dish, no other artist was allowed to make this for a year; but he alone who invented it was entitled; to all the profit to be derived from the manufacture of it for that time; in order that others might be induced to labour at excelling in such pursuits.[7]

Thus, the monopoly was given as a reward to the inventor and to catalyse new innovations, which is a continuing principle of the patent laws of all major states today.

A precursor of modern patent law goes back to 1475 and was established in Venice as the "Statute of Venice".[9] The Statute ruled that the devices must be novel, ingenious, and reduced to perfection. The term of protection was 10 years. Also, licensing, an infringement procedure with damages and a claim for destruction, was laid down.[10]

In England, when Elizabeth I ascended the throne in 1558, England was technologically backward compared with the other countries in Europe,[10] Elizabeth I devised a concept for introducing new technologies by foreign workers by giving them monopolies for introducing new technologies and teaching them to the indigenous population. The later Statute of Monopolies of 1624 defines in Section 5.6, that for new manners of manufacturing, a patent can be gained for a limited term.

In France the first Patent Law was enacted in 1791 based on the principle that no examination of any kind was required.[11] A modern patent system with a patent office and an examination procedure proving requirements for patentability and stipulating claims was established in the United States by the Patent Act of 1836. In 1817 the Netherlands adopted its first patent law that was abolished in 1869.[10]

Japan proclaimed in year 4 of the Meiji Era (1871)[10] its own law, called Provisional Regulations for Monopoly, which was the first patent law. In the following year, the enforcement was suspended due to problems of the government with the operation of the law and other problems. But it became apparent that a patent system was necessary, so the "Patent Monopoly Act" was officially announced, thanks to the efforts of the first Patent Office Director Korekiyo Takahashi, in year 18 of the Meiji Era.

The first comprehensive Patent Law in the world with a mandatory examination was enacted in Germany in 1877. Also, Turkey established a special Patent Law in 1879 during the Ottoman Emperorship granting patents without examination.[12]

From the end of the 19th century, international trade developed to an extent that the sole national view on patents and other industrial property rights had to be thought of by the states on an international level.

Due to these circumstances in 1873, a congress was held in Vienna on which debates on international aspects were held connected with the protection of inventions and paved the way for the Paris Convention adopted in 1883.[13]

The Paris Convention[14] was the first convention providing the right of priority for applications filed in another contracting state for an application in any of the other contracting states and will be regarded as if they had been filed on the same day as the first application. The priority term according to Paris Convention is 12 months for patents and utility models and 6 months for industrial designs and trademarks. The Paris Convention applies in general to industrial property in the broadest sense, comprising patents, utility models, trademarks, designs, service marks, trade names, geographical indications, and the avoidance of unfair competition

Further harmonisation was induced in the 20th century by the pressure of intensified internationalisation of international trade, triggering cumbersome processes for patent applications in different countries.

In 1970, in Washington, the Patent Cooperation Treaty (PCT)[15] was established, which initiated one procedural mechanism for a certain time, the international phase, for patentees wishing to obtain international protection in several contracting states. The international phase of such a patent application comprises an international search and optionally a preliminary examination. At the end of the international phase, the patentee has to decide to nationalise the application into a bundle of national and regional applications.

The European Patent Convention (EPC),[16] which was established in 1977, created the European Patent Organisation (EPO) and is a supranational treaty of the contracting states and was amended in 2000.

The EPC, with currently 38 member states (Jan. 2022), 2 extension states and 4 validation states, provides a harmonised substantive law on patentability and a single granting procedure for European patents. After the granting of the European patents, a national state has to be entered for each member state, where the patentee wishes an enforceable national patent. The EPC, therefore, provides a substantive examination procedure for granting a European patent that becomes a bundle of national patents and/or from the second half of 2022 a unified patent (see below the paragraph on IP rights patents) for the participating European Union (EU) member states. Moreover, the enforcement of an infringement can be performed in each member state for the bundle patent for a transitional period of 7 years, with an "opt-out" declaration after the Unified Patent System has been set up in 2022.

The multilateral TRIPS Agreement came into force in 1995 and the WTO (World Trade Organisation)[17] was established. The TRIPS Agreement frames substantive and comprehensive disciplines on intellectual property rights (IPRs) into the multilateral trading system and had a significant impact on national intellectual property laws all over the world. It sets out certain minimum standards for many intellectual property rights. According to Article 27, TRIPS patents shall be available for any inventions, whether products or processes, in all fields of technology, provided that they are new, involve an inventive step, and are capable of industrial application.[18] Furthermore, patents shall be available and patent rights enjoyable without discrimination as to the place of invention, the field of technology, and whether products are imported or locally produced. Article 27, TRIPS also allows to exclude *according to ordre public* or morality patentability of inventions as well as to exclude (a) diagnostic, therapeutic, and surgical methods for the treatment of humans and animals; (b) plants and animals other than micro-organisms, and essentially biological processes for the production of plants or animals other than non-biological and microbiological processes. The protection of plant varieties shall be possible either by patents or by an effective *sui generis* system or by any combination thereof.[19]

5.2.4 IP Rights

In the following passage, the IP rights will be briefly explained by European IP conventions or German IP law, as the law in the IP field is mainly harmonised with other national laws.

5.2.4.1 Patents, Unitary Patents, and SPCs (Supplementary Protection Certificates) and Utility Models

Patents relate to inventions in all fields of technology. They shall be granted provided that the claimed invention is new, involves an inventive step, and is susceptible to an industrial application (Art. 52 EPC). Therefore, a patent can be applied and shall be granted for technical inventions that comprise at least one technical feature that is new and inventive. The following exceptions will, according to Article 52 EPC, not be regarded as an invention for subject matter or activity as such:

(a) Discoveries, scientific theories, and mathematical methods
(b) Aesthetic creations
(c) Schemes, rules, and methods for performing mental acts, playing games or doing business, and programs for computers
(d) Presentations of information

As mentioned above, a priority right of 12 months is granted for patents and **utility models**. The EPC encompasses its own provisions on priority which correspond to the provisions of the Paris Convention. The patent term can be extended to 20 years from the filing date and a utility model of 10 years in most countries. A utility model is only available through national law and can be deposited or registered with a formal examination but does not need to pass a substantive examination procedure. A German utility model cannot be registered for processes or use. A search for state of the art can be requested at the national offices.

The EPC itself does not mention utility models and utility certificates, but they can be requested on the basis of the national patent law for a European patent application according to the specific law. A granted patent confers the patent proprietor to prohibit any third party from producing, offering, putting on the market, or using a product which is the subject matter of the patent, from importing or possessing for the purpose referred to, using a process which is the subject matter of the patent.[20] Moreover, further constellations may infringe the patent without the consent of the proprietor.

Inventions in the mechanochemical field can comprise, for example: (i) a product, substance, apparatus, instrument, first or further medical use, intermediate, computer program product, (ii) process, working method, process of manufacturing, or analogy process with further requirements and use. Typical examples in the life science field for (i) products are chemical or biological substances, their derivatives, salts, hydrates, solvates, polymorphs, cocrystals, molecular complexes, and their intermediates, formulations comprising them, alloys, kits, systems, reactor, milling beaker, extruder, barrel extruder, or parts thereof, computer program products or computer program products of AI (artificial intelligence) both with technical applications or inducing a technical effect and for (ii) processes are processes for the manufacture of substances, technical applications, or an effect by a software- or AI (artificial intelligence)-related process, (iii) kit-of-parts, (iii) and endless further possibilities.[21]

For inventions for specific pharmaceutical and plant protection products that have to pass an approval procedure by a regulatory authority, such as EMA or a national Federal Institute for Drugs and Medical Devices, e.g. BfarM, a Supplementary protection certificate (SPC)[22] can be applied. The SPC serves as a patent term extension for the loss of patent term due to compulsory testing and clinical trials before the application for the approval procedure can be requested with the regulatory authority. The patent term extension is limited to a maximum of 5 years, and limited to a maximum of 15 years of protection from the filing date of the patent, with an exception for paediatric drugs (drugs for children) for which a 6-month additional extension is available. A patent term extension by an SPC is possible in nearly every country in Europe, and according to specific national law in the United States, Japan, and many other countries worldwide.

The SPC can be requested for a pharmaceutical or plant protection product, i.e. a substance or a combination of substances and comprising of all derivates of the scope of the patent, e.g. salts, esters, hydrates, solvates. In cases, where the derivative, e.g. a cocrystal, is covered by its own patent in accordance with certain conditions, an additional SPC may be granted.

In many jurisdictions Bolar exemptions for necessary studies and trials exist to enable experiments and tests for medical approval procedures during the patent term. Furthermore, in many countries research exemptions exist with different interpretations, but the minimum requirements are ruled in Art. 30 TRIPS (e.g. German Supreme Court; Germany BGH X ZR 99/92; UK Monsanto/Stauffer; court of Appeal 11 June 1985).

In 2019, in the European Union, the so-called "SPC Manufacturing Waiver" entered into force, since then an SPC no longer confers protection against manufacturing in the European Union (EU) of pharmaceutical products that are intended for export to countries outside the EU. Additionally, no earlier than 6 months before the expiry of an SPC, a pharmaceutical product can be manufactured and stored, in order to launch the product on the market, after the expiry of the SPC.

More protection for innovations through a more uniform and cost-effective patent system can be gained from the second half of 2022 in participating member states of the EU through the Unified Patent System.

After almost 50 years of development, the Unitary Patent and the Unified Patent Court can be launched in 2022 as part of the enhanced cooperation of the participating EU member states.

5.2.4.2 *Unitary Patent and Unified Patent Court*

The current system at the European Patent Office (EPO) is based on a centralised examination system. If a European patent (bundle patent) is granted after its conclusion, it must be validated in the contracting states, in order to be able to develop its protection nationally as a bundle patent. For this purpose, parts or the entire patent must be translated into the national language or into English in various countries at a considerable cost. In addition, there are national representative fees and, in some cases, official fees.

The unitary patent, on the other hand, is a European patent with unitary effect and must be translated completely into English for at least the first 6 years, if the text version is available in French or German; if it is available in English, it must be translated into another official language of the European Union.

The new system now allows, with effect for the participating EU member states, to file a request for "unitary effect" with the EPO, within a time limit following the grant of the patent in the established central examination procedure.

The "unitary effect" of the unitary patent avoids the previously fragmented protection in European Union countries, as all EU member states except Spain and Croatia want to implement the system. In addition, the administrative and cost burden for all parties involved will be considerably reduced. The entire administration of the unitary patent will be bundled at the EPO and the annual fees will be equal to the sum of the four countries where the most patents were validated in 2015. Remember, the UK was one of these four countries and left the EU. However, to date, about 50% of the patents granted have been validated in Germany, France, and the UK, for which no translation had to be filed. When considering the renewal fees in isolation, the cost advantage of the unitary patent for the above example has thus melted away, as a result of Brexit.

However, the great advantage of the unitary patent is the territorial, uniform extension of protection of the granted patent in currently up to 24 participating EU member states. Until ratification in all member states, there will be different generations of unitary patents with extensions of protection in the member states, with deposited instruments of ratification. Currently, about 17 member states, such as France, Italy, the Netherlands, Germany, Denmark, etc., have initiated or completed the ratification process. In all other contracting states of the European Patent Convention, the granted European patent can still be validated nationally. Alternatively, a patent applicant can take the purely national route by applying for a national patent and thus also wanting to ensure a purely national jurisdiction.

The Unified Patent Court (UPC) will have jurisdiction over actions of infringement of unitary patents, granted European patents, and future granted European patents and unitary patents as well as SPCs. This ensures uniform jurisdiction even for coexisting nationally validated European patents (bundle patents) and unitary patents.

However, the parallel jurisdiction of national courts for nationally validated European patents will continue for a transitional period of 7 years. This must be communicated to the registry of the Unified Patent Court by means of an "opt-out" declaration before the end of the transitional period and before an action is brought before the Unified Patent Court.

Future innovation protection of inventions will thus be possible in a more uniform and cost-effective manner and can be enforced at the Unified Patent Court in a single procedure.

In the future, however, an action for revocation can centrally destroy the unitary patent and nationally validated European patents, provided that no opt-out has been declared for the nationally validated patents. Invalidity actions can also be brought without first having to file an opposition with the EPO. Particularly critical is the possibility of an action for revocation before the Unified Patent Court despite parallel opposition proceedings.[23]

For each patent portfolio, therefore, a strategy must be very carefully worked out for which patents the "opt-out" possibility before the Unified Patent Court will be used. From a cost point of view, it can be expected that German infringement proceedings will be significantly cheaper to conduct than proceedings before a Unified Patent Court.

With the new patent system, the participating EU member states have developed a modern patent system for SMEs and industries that is fit for the challenges of the advancing innovation decade.

5.2.4.3 Designs

A design is the appearance of a product resulting from the features of the product, in particular its shape, patterns, contour, texture, materials, and colours or ornamentation. The protection requires that the design is new and has individual character.

Sole technical dictated features of a product cannot be protected by a design in many design laws, in particular, features which must necessarily be reproduced (must-fit) in their exact form and dimensions in order to permit mechanical connection are excluded from protection by many design laws. But multiple assemblies or connections of mutually interchangeable products within a modular system (must-match) can be protected.

A product also comprises parts of the product assembled into a complex product, but once it has been incorporated into the complex product, the parts must remain visible during normal use of the latter to be regarded as new and to have an individual character to the extent that those visible features of the part fulfil in themselves the requirements as to novelty and individual character. Examples of designs are furniture, decorative objects, vehicles, wheel rims, tools, instruments, apparatus, shoes, fabrics, fonts, graphical symbols, pictograms, and many other creations. With respect to mechanochemistry, aspects of a design may be applied, for example, to the appearance of an apparatus in process engineering, such as an extruder, milling machine, milling beaker, special milling equipment, special milling balls, and the design of software applications. In each case the special presentation of a design has to be evaluated individually to confer the best scope of protection.

A priority right of 6 months is granted for designs. The so-called "grace period" (in other words, it is considered novel for 1 year after disclosure) gives a proprietor the possibility to launch a product and to check whether it is worth to seek protection for a design or not.

Furthermore, a registered design can be extended up to 25 years in the European Union, Germany, and many other jurisdictions.

Examples of design applications for mechanochemical applications may comprise a design of a beaker, surfaces of balls, e.g. in analogy to tyre treads, design of mills, housing of apparatus, design of software applications etc.

5.2.4.4 *Trademarks*

Trademarks, colloquially often named logos, distinguish and/or identify goods and services of a proprietor from goods and services of other enterprises, in particular from competitors. All signs can be registered as trademarks, if they are suitable for distinguishing goods and/or services of an application from those of another enterprise.[24] Typical signs of registered trademarks are words including personal names, designs, letters, numerals, sound marks, three-dimensional designs, the shape of goods or of their packaging as well as other wrapping, including colours and colour combinations. Trademark protection is not available for signs consisting exclusively of a shape which results from the nature of the goods themselves, is necessary to obtain a technical result, or gives substantial value to the goods. Goods and services are classified according to Nice Classification.

A priority right of 6 months is granted for trademarks. Furthermore, a registered trademark is registered for 10 years from the filing date and can be extended.

Examples of trademarks for mechanochemistry applications comprise words, designs, 3D-designs, positional marks, motives, shape of process engineering products or parts, appearance of desktop applications, sound of mills or extruders etc.

5.3 Innovations in Mechanochemistry – In the Past, Today, and in the Future

5.3.1 A Historic Approach of Sustainable Chemistry and/ or Mechanochemistry in Patent Literature

The view back into historical innovations of sustainable manufacturing methods and/or apparatus in patent databases.

5.3.1.1 *The First German Patent and an Outlook*

In 1877 the German Patent Act of 25 May 1877 and the first patent office as the Imperial Patent Office was established in Berlin on 1 July 1877. As early as 2 July 1877, the first German (DE 1) patent was granted for a "Process for the production of a red ultramarine dye" invented by Johann Zeltner of the Nuremberg Ultramarine Factory. The synthesis comprised the following steps: ultramarine violet is subjected to the action of vapours of a more or less concentrated nitric acid at 130–150°C. Highly concentrated nitric acid produces a colour rising to light pink; diluted nitric acid, on the other hand, produces a deep to darker red ultramarine.

From the digitalised *Polytechnisches Journal*, 1876, the following information is disclosed:

> Die concentrirte Salpetersäure entfärbt die drei Ultramarine unter Entwicklung rother Dämpfe, indem sie nicht nur als Säure, sondern auch als Oxydationsmittel einwirkt.

> **(German orthography 1876)**[25]

In English: "Concentrated nitric acid decolourises the three ultramarines by producing red vapours, acting not only as an acid but also as an oxidising agent".

Considering the above 12 principles for greener chemistry, many of these principles are already met by the disclosed synthesis as only two initial substances are used, a solid is converted with a gaseous reactant, and no solvent is used. Therefore, the authoress' view back in the history of former innovations aims to give the suggestion to consult the whole state of the art comprising patents, scientific publications, and freely available knowledge from the internet as a springboard for innovation. At least patent databases and freely available knowledge from the internet are free of charge and therefore accessible for all interested innovators.

For example, as the used nitric acid also acts as an oxidising agent toxic nitrogen dioxide (NO_2) will be a by-product. The combination of this old technology with brand-new technology may be a good solution to transfer the old well-known synthesis into a sustainable manufacturing process for today. For example, the new research, led by the University of Manchester, that developed a metal–organic framework (MOF) material named MFM-520 selectively offers a fully reversible and repeatable ability to capture nitrogen dioxide (NO_2) and can theoretically be combined with the old technology, by which the nitrogen dioxide can then easily be converted into nitric acid again.[26]

The authoress' theoretically designed process would enable a closed sustainable manufacturing process in which the by-products can be converted to initial reactants.

5.3.1.2 A Historic Approach in Mechanochemistry in Patent Literature

The following examples will give a rough view into the developments of grinding, milling apparatus, and apparatus for dispersions as well as to synthesis performed by them.

DE337429 dated 30 August 1918 from Plauson's Forschungsinstitut GmbH, Hamburg, discloses a process for the production of dispersions with improved stability produced with special fast-running mills. The patent addresses the disadvantage of previously known processes of converting solid colloids, starch, dextrin, or types of protein, such as casein, yeast etc. into finely divided soluble or insoluble powders. According to the patent, the disadvantage of all these processes is that a solution of the colloid is first prepared, and only then, by atomising the solution into a liquid which does not dissolve the colloid in question, the finely divided product is obtained. Therefore, two liquids are always necessary for the production of such powders: one as a solvent and the other as a dispersion agent. Furthermore, this results in the need to separate these two solvents from each other and from the powder. A particular disadvantage is the difficulty of removing the solvent completely which avoids the powder from sticking together again.

According to the invention, it has been found that solid colloidal substances, when mixed with a sufficient quantity of a liquid which does not dissolve the colloid, are treated for a longer or shorter period of time at room temperature or higher temperatures in so-called impact, cross, pin, distemper, or similar mills, can be converted into fine dispersions with improved stability. This effect is only achieved by treatment in very fast-running mills of the type mentioned with a speed of more than 300 m per minute, but most suitably in mills with a rotational speed of 2000–3000 m per minute.

The fact that the solid colloidal substances do indeed take on a high degree of dispersity and not only a grinding effect was deduced from the following effect, that the dispersions have with the correct choice of dispersants and adequate processing parameters in beater mills the typical properties of a sol. The dispersed particles float in the dispersing medium, wherein their separation can be affected by the addition of any electrolyte.

Already in 1918, the inventors sought to avoid waste and considered recycling the products. The patent discloses that: in the first place, the wastes which are produced in the manufacture of articles from the above-mentioned, solid colloidal substances can be processed and reused, bypassing the costly dissolving process, and from the resulting dispersed powders by pressing in a dry way new articles of any kind can be produced.

The disclosure of this patent clearly demonstrates that in the first part of the last century, the inventors already took the current aspects of sustainable manufacturing into account. At that time, the reason for this behaviour was not to act sustainable for the environment but to manufacture sustainable in view of the industry with limited resources.

Hereinafter, some patented mills invented in the last or the second last century for grinding and optionally other process steps are mentioned.

Published patent DE 2047244B (26 March 1981, Gebrüder Netzsch, Maschinenfabrik GmbH & Co) was also filed in the United Kingdom and discloses an agitator mill, for comminuting and/or dispersing a material, of the type comprising of a container of circular sections and, mounted co-axially within it, a rotor formed and driven to agitate a mixture of the material with a grinding medium, such as metal balls, the container having an outlet in the form of an annular separating gap, through which the processed material can flow and which is formed between the co-axial opposed surfaces of the container and the rotor. In further aspects, the agitator mill could comprise a hollow member or have a double-walled container for passing a coolant through it.

The patent application DE000R0006257 (21 May 1952, Rosenthal-Porzellan AG, **Selb** (Bay.) discloses a kind of extruder for the mixing of materials, e.g. ceramic material in two different directions.

GB 605,433 (filed 20 December 1937, accepted 30 May 1939, John Matthey & Company, Limited) discloses an invention which relates to the manufacture and production of hard brazing solders in a ball mill. The solder, in particular silver solder, is obtained in the form of fine metallic powders. Example 1 discloses the following process: a brittle alloy is prepared by melting copper and zinc. The molten alloy is poured into water. The granules are stamped to get a coarser powder which is ground with water in a ball mill until it passes through a 200-mesh screen. Thereafter, a malleable alloy comprising silver and copper is prepared in powder form as described in GB 506,432, 1937. The resulting powder is milled together with the powdered brittle alloy in a ball mill. The purification comprises treatment with water and sulphuric acid.

Patent CH 17502 18.07.1898, 6 ¼ Uhr p. (Carl Hoffmann, Breslau), discloses a mill with circling and oscillating grinding media, with oval grinding media that oscillate inwards and outwards during rotation.

The United Kingdom patent 332530 accepted 22 July 1930 (Henkel & Cie. GmbH) discloses a cleaning medium which is produced in an edge runner mill with 10 kg. Furthermore, a polishing medium is obtained which does not attack metals, 90 kg of dry sand, 5 kg of crystallised tri-sodium phosphate, 2 kg of soda, and 1 kg of lump water-glass (soluble in water) by mixing in a roller mill until it's a homogeneous mixture.

Patent DE 482598 dated 29 August 1929 (Fried. Krupp Grusonwerk AG, Magdeburg-Buckau) discloses a "Tube mill for grinding material containing magnetisable substance" for continuous processing instead of batch-wise processing of, in particular, Thomas slag, blast furnace slag, and the like. Object of the invention was to avoid the batch-wise process which was caused by metallic iron that entered the previous mills with the slag, accumulated there as it was ground with the slag and induced a considerable increase in the power requirement of the mill. Therefore, the mill had to be shut down in intervals to be cleaned. The patent furnished an innovative continuously operable tube mill for the grinding of material containing magnetisable, in particular ferrous, substances, wherein a magnetic separating device is provided in the mill itself. The mill comprised a special chamber for the magnetic separating device between a pre-grinding and fine-grinding chamber. The disclosed present state-of-the-art process may enable the recycling of magnetisable material in various processes tube mills adapted to the special purpose.

Patent DE 480090 A dated 9 April 1926 (Pfenning Schumacher Werke GmbH) with the title "Process for the direct production of commodities from casein, blood, horn meal and other protein and keratin substances" discloses the treatment of casein with water in a ball mill, e.g. a liquid-assisted ball milling.

Casein as such is a long-lasting and extensively used binding agent, which was already used by carpenters in ancient times in Egypt and China and up to the 1930s casein was used for the production of knobs, cutlery, and jewellery.

The patent of 1926 states that it is well known that the direct production of articles of daily use from casein powder, bypassing the forming of cubes, rods, tubes, plates, etc., has so far posed insuperable difficulties. Furthermore, at that time a published process which purports to overcome this difficulty by finely grinding casein, for the purpose of uniform impregnation with water, brought towards the water dust in apparatus specially constructed for this purpose and finally pressing the material hot into the moulded piece, shall comprise difficulties in large-scale production furnishing fine particles with even particle size in a batch-wise production. Furthermore, the main difficulty shall be that the fogging with water dust in such a way that each casein particle is soaked with water is hardly feasible from a production point of view and that this may be the reason that the process has not been adopted as a technical application. Moreover, the production of articles of daily use, such as buttons, combs, buckles, by direct pressing of the casein powder, hampered by the fact that the preforms easily develop cracks a few minutes after the removal from the mould on the surface, due to tension, and thus took on the appearance of being covered with a fine network of cracks. According to the invention, it was surprisingly discovered that these defects do not occur when the moulding (pressed part) made from powder is subjected to a higher pressure during cooling than during hot pressing and the cooling is continued until the temperature has been reduced to at least half of that required for hot pressing. The patent discloses additional information, regarding the treating of the casein powder, namely that the casein powder does not need to be dust-free and that the water content can be achieved in the usual way during mixing in a ball mill, but the water content should be limited to as little as possible, on average not more than 15%.

Therefore, in addition, a special liquid, here water, assisted casein powder treatment in a ball mill was disclosed to produce a material that is suitable to be subsequently used in a special moulding process. The process developed in 1926 and belonging to the state of the art complies with the 12 basic principles for greener chemistry with sustainable manufacturing established by Paul Anastas and John Warner in 1998.

5.3.1.3 Historic Approach of Alternative Solvent-Free Methods in Patent Literature

The following examples give a rough overview of historical process patents utilising the melt, water-based, plasma-based, or gaseous synthesis or natural products to obtain the desired products.

BASF previously known as Badische Anilin- & Sodafabrik (BASF) was founded in 1865. The patent DE 54626 (6.05.1890, Badische Anilin- & Sodafabrik) discloses a process for the manufacture of artificial indigo in a melt without the use of a solvent in the conversion process.

The synthesis of artificial indigo dye starts with the conversion of phenylamidoacetic acid by melting potash or natron, in particular with about twice the weight of an etching alkali at 270–300 °C. The melt was afterwards dissolved in water and the initially formed leuco compound was oxidised, e.g. by the oxygen in the air. The artificial indigo dye possesses very similar properties to ordinary indigo but has greater solubility, e.g. in alcohol.

US 2,779,062 A of BASF AG claiming a German priority of 1951 (DE973134B) discloses the production of porous-shaped articles true to shape and size from synthetic thermoplastic materials using a solvent-free process utilising ball mills and a gas-tight cylindrical mould.

> Example 1: 850 parts of polystyrene are finely ground with 150 parts of ammonium bicarbonate in a ball mill and the mixture introduced into a gas-tight cylindrical mould which is closed at both ends by a screwed cover and heated to 180 °C. After heating for an hour, the whole **mixture** is cooled to 25 °C and the gas-containing solid body removed from the mould and comminuted by grinding in a cross-beater mill to grain size. [...] [amended: **mixture**]
>
> Example 3: 850 parts of polymethacrylic acid methyl ester and 150 parts of azodiisobutyric acid dinitrile are finely ground in a ball mill. The mixture is then heated for 2 hours at 180 °C in a gas-tight pressure mould without the employment of external pressure. It is allowed to cool and a homogeneous solid article is removed and comminuted by grinding. 100 grams of this

comminuted material are introduced into a closable ball mould having a volume of 1000 cubic centimetres and exposed to the action of steam at 110' C. for 60 minutes. After cooling and removal from the mould there is obtained a spherical porous shaped article having a specific gravity of 0.1.

DE 18726 (12.10.1950, Chemische Werke Hüls GmbH) discloses the conversion of toluene to benzoic acid without producing undesired by-products such as benzaldehyde or benzyl alcohol. Oxidation mixtures consisting only of starting material and the desired carboxylic acids are produced. The patent describes an example in which through a solution of 10 parts by weight of cobalt naphthenate and 1000 parts by weight of pure toluene, an air stream of 6 atm is blown at 165 °C through a porous quartz sinter plate. The water produced during the oxidation distils off continuously and is separated from the over-distilled toluene, which is returned to the reaction vessel. After 12 hours, about 70% of the toluene used has been reacted. The unreacted toluene is distilled off, and 850 parts by weight of benzoic acid of high purity are obtained.

DE 22832 discloses a process for the removal of diacetylenes (23 November 1948, Chemische Werke Hüls GmbH) in a plasma-based process. The patent discloses that it is known to produce acetylene from hydrocarbons in an electric arc as so-called arc acetylene, which contains homologues of acetylene, such as methyl acetylene, and higher acetylene, such as diacetylene. Although the diacetylene is separated before further processing, residual amounts remain in the acetylene. When acetylene is reacted, acetaldehyde is produced, and acetone is produced from methyl acetylene, whereby the diacetylene is not reacted and thus ends up in the products. Although diacetylene is only present in less than 1%, it can render the aforementioned compounds ineffective during further processing, e.g. in the presence of catalysts. According to the invention, the diacetylene can be removed quantitatively by treating it with hydrogen in the presence of palladium-containing catalysts under mild conditions, so that hydrogenation of the carbonyl group does not occur. The reaction temperature is chosen as low as possible, while the pressure can be variable from unpressurised to high pressure. The palladium can be precipitated on silica gel, diatomaceous earth, etc. A palladium catalyst reduced in the hydrogen stream is hydrogenated on this catalyst with acetone prepared from the methyl acetylene obtained in the arc furnace to butane or butylene. The disclosed processes are solvent-free utilising electricity in an arc furnace process, with hydrogen and palladium as a catalyst.

US 4,379,871 A ot Degussa (DEUTSCHE GOLD- UND SILBER-SCHEIDEANSTALT) German priority 1975 discloses a process for the production of carbon black, containing pigment-synthetic resin concentrates, utilising a continuous system of two consecutive ball mills.

Example 1: The mixture of carbon black powder and synthetic resin powder together with the acetone that contained dissolved therein the plasticizer, stabilizer and wetting were charged to a twin-screw pump and removed as a homogenized solution. This solution was charged to two successive ball mills provided with different sizes of balls for preliminary and fine dispersion and the finished dispersion withdrawn at the top of the mills.

(a) The dispersion was freed of solvent on a vacuum drum drier at a drum temperature of 70 °C and a pressure of 60 Torr. The resultant carbon black-synthetic resin concentrate has a very low bulk density (apparent weight) was comminuted on a grater and subsequently ground to a powder.

(b) The dispersion was diluted to a solids content of 20% by the addition of acetone and atomized by passing through a binary nozzle operated by nitrogen at 120 °C in a nitrogen heating medium. The carbon black-synthetic resin concentrate accumulated as small, porous balls.

Patent DE 679 587 (20.07.1939, IG FARBENINDUSTRIE AKT GES, Frankfurt) discloses that the polymerisation of butadiene homopolymers or co-polymers with styrene or acrylic was performed in the presence of oxygen or substances that produce oxygen and can be enhanced by organic complex substances comprising of metals having several oxidation states. In Example 2 of the patent, a natural occurring complex, which could also be synthesis, was used for the polymerisation of butadiene.

Example 2: 25 parts by weight of styrene and 75 parts by weight of butadiene are treated in 200 parts, by weight, of 10% sodium oleate solution, in the presence of air at 21 °C for 5 ½ days. No polymerisation has occurred after this time. If, on the other hand, 0.01 part by weight of haemin is added, a yield of about 79 % is obtained under the same conditions. Haemin can be obtained from blood. In 1930 Hans Fischer obtained the Nobel Price in chemistry for the synthesis of haemin.[27]

CH140049 (filing date 6 October 1928, 16 ¼ hrs, granted 31 May 1930, Algemeene Kunstzijde Unie later applicant Akzo Nobel) discloses a process for the production of finished acetate silk by spinning acetate solution and evaporating the solvent. The process makes it possible to produce finished artificial silk as cellulose acetate without post-treatment and without an excess of the finishing agent. For this purpose, fat or oil is added to the spinning solution before spinning, in particular a vegetable or animal fat whose melting point is 20–40 °C. Suitable oils are olive oil, olein, coconut fat, butter fat, beef fat, and peanut oil. Particularly suitable are olein and coconut fat or olein and stearic acid or olive oil and coconut fat in equal parts by weight. These additives make it possible to immediately twist and hasp or twist and wind the spun yarn.

US 000002820252 A of Polymer Corp. filed in 1954 discloses a glass balls milled assisted blending of polytetrafluoroehtylen, water, and silica forming:

140 g of 50% polytetrafluoroethylene aqueous suspensoid (containing some wetting or dispersing agent and ammonia) are used, and to this 420 g of water are added, with stirring. 30 g of silica of particle size averaging from 3 to 7 microns are also added with stirring, after which glass mill balls are introduced and the mixture is tumbled until a dough-like mass precipitates (about 6 to 24 hours).

GB 19,670 (2 November 1900, Friedrich-Bayer & Co of Elberfeld in the Empire of Germany) discloses a procedure for eliminating sulphuretted compounds from dye stuffs. According to the invention, the commercial dye, e.g. katigen-black S.W., is boiled with zinc-dust and soda lye to reduce the dye and the sulphuretted compounds are transformed into zinc sulphide which precipitates and can be removed from the obtained paste by pressing the paste through a cotton cloth.

DE 69740 (8 September 1982, Society for Chemical Industry Basel [CIBA]) Azo dyes describes a process for the preparation of a beta-amidooxynaphthalene sulphonic acid by melting beta-amidonaphtalenedisulphonic acid with caustic soda or caustic potash in water at 200–240 °C in an autoclave under pressure until the foaming of the melt has subsided and a sample precipitates considerable amounts of the new acid during acidification. The melt obtained is divided into pieces and introduced into dilute hydrochloric acid, whereby beta-amidooxynaphtalinemonosulfonic acid is deposited as crystals.

The present patent specification is not cited on the basis of the product produced, but to show that, in principle, syntheses in an alkaline melt, in the presence of water in an autoclave and subsequent crystallisation of the product from water for the production of chemical substances, are also conceivable and a more environmentally friendly production without organic solvents. An attempt to transfer this process or an analogous process from the melt into a mill may be assisted by a certain amount of water under elevated temperatures and is imaginable.

5.3.1.4 History of Cocrystals or Molecular Complexes in Patent Databases

This section summarises some cocrystals from patent literature as cocrystals as long as they are mentioned in the cited patent as cocrystals even if they are salts. The intention of this chapter is not to discuss cocrystals in a correct scientific context but to give a compressed overview of methods for the production of cocrystals, salts, or molecular complexes in the food industry, life science, pharma, and biotechnology industry. Typically, the term "co-crystals or cocrystals" describes a crystalline phase wherein at least two components of the crystal interact by hydrogen bonding and possibly by other non-covalent interactions rather than by ion pairing.

In 1937 Hoffmann-La Roche invented and claimed in patent CH187826 potentially the first cocrystals, disclosed as two components in a molecular ratio.[28] The patent discloses a method for the production of

2,4-dioxo-3,3-diethyl-tetrahydropyridine and further discloses that this compound forms crystals in a molecular ratio with 1-phenyl-2,3-dimethyl-4-dimethylaminopyrazolone or 1-phenyl-2,3-diemthyl-4-isopropylpyrazolone.

Original:

Das so gewonnene 2,4-Dioxo-3,3-diäthyltetrahydropyridin bildet farblose Kristalle vom Schmelzpunkt 98 bis 99°, die sich in warmem Wasser leicht, in kaltem wenig lösen. [...] Mit Pyrazolonderivaten vereinigt es sich in molekularem Verhältnis zu gut kristallisierenden, beständigen Verbindungen, zum Beispiel mit 1-Phenyl-2,3-dimethyl-4-dimethylaminopyrazolon (F = 69 bis 70°) oder mit 1-Phenyl-2,3-dimethyl-4-isopropylpyrazolon (F = 93°). Die neue Verbindung soll als Arzneimittel Verwendung finden, da ihr eine starke Schlafwirkung zukommt.

In English:

The 2,4-dioxo-3,3-diethyltetrahydropyridine thus obtained, forms colourless crystals of melting point 98 to 99°, which dissolve readily in warm water and little in cold. [...] With pyrazolone derivatives it forms in a molecular ratio well-crystallising, stable compounds, for example with 1-phenyl-2,3-dimethyl-4-dimethylaminopyrazolone (F = 69 to 70°) or with 1-phenyl-2,3-dimethyl-4-isopropylpyrazolone (F = 93°). The new compound is to be used as a medicinal product because it has a strong sleep effect.

As the second compound in the cocrystal, namely, 1,2-dihydro-1,5-dimethyl-4-(1-methylethyl)-2-phenyl-3H-pyrazol-3-one (1-Phenyl-2,3-dimethyl-4-isopropylpyrazolon (F = 93°)) is also an API (active pharmaceutical ingredient) used as a non-steroidal anti-inflammatory drug the authoress reading the patent with a mind-set of a patent attorney, it seems to her that the formation of the crystals in a molecular ration was mentioned "by the way". Therefore, the authoress would not be surprised if the formation of cocrystals, in particular of APIs – disclosed as crystals being formed from components in a molecular ratio – maybe was more established as we know these days. It would be interesting to evaluate if the approval of APIs or their "combined" pharmacological effect impeded the usual use of cocrystals of two APIs.

EP2056798A2 of AMGEN INC with the earliest priority of 2006 describes cocrystals of sorbic acid and analogue cocrystals. In the instructions it is disclosed that:

under names such as organic molecular compounds or complexes, have been described in the literature as far back as the 1890's, where Ling investigated halogen derivatives of quinhydrone.[29] [...] Co-crystals have been prepared by a variety of techniques such as melt crystallization, grinding[30] and re-crystallization from solvents.[31] [...] Co-crystal formation of Carbamazepine has been investigated. Eight polymorphs and pseudo polymorphs for the epilepsy drug have been reported, thereby making the drug an excellent candidate for co-crystal formation[32, 33]

One interesting example describes cocrystal formation with Fluoxetine Hydrochloride, a salt, with organic acids such as benzoic acid, fumaric acid, and succinic acids. The approach is based on halide ions as hydrogen-bonding acceptors. The authors also performed powder dissolution experiments and showed that two of the three cocrystals (fumaric acid and succinic acids co- crystals) had a higher dissolution rate as compared to Fluoxetine Hydrochloride.[34] In another study the formation of fumaric acid, succinic acid, and L-malic acid cocrystals of an extremely water-insoluble anti-fungal drug, itraconazole, is described. The cocrystals were reported to have similar dissolution profiles to the amorphous drug and superior to the crystalline compound, thereby indicating the potential for enhanced bioavailability.[35]

API and co-crystal former were ball milled with or without approximately 20 [μmL] of isopropyl alcohol, acetone, methanol, ethyl acetate or 2-butanol in a mixer mill MM301 (Retsch Inc., Newton, PA) at a 1 : 1.2 ratio of API to co-crystal former in a 1.5 mL stainless steel grinding jar containing a 5 mm stainless steel grinding ball for 2 min.

Crystallizations were accomplished by slow cooling a saturated solution. API and co-crystal former were dissolved in a 1 :1.2 ratio in isopropyl alcohol, isopropyl acetate, acetone, methanol, ethyl acetate, dichloromethane, 1.2- dichloroethane or 2-butanol at 50 °C (or less depending on boiling point) then cooled at 2 °C/min in an Imperial V oven (Lab-Line Instruments Inc., Melrose Park, IL). If crystallization did not occur within 48–72 hrs, slow evaporation was also utilized.

An early patent disclosing a cocrystal is DE2000210A1 (priority GB193569A, 1969) of Squibb & Sons Inc. disclosing a complex of phenazone and paracetamol. The complex can be obtained from an equimolar melt, from water, acetone, and ethanol.

GB1568417A (priority HURI059376A, 1976) of Gedeon Richter discloses a molecular complex of Vincamine derivatised and D-tartaric acid that was obtained from ethanol.

A controlled-release composition was disclosed by EP0032004B1 of Euro-Celtique S.A. (priority IE248079A, 1979). This old invention relates to internal molecular chelate complexes or matrixes which are porous semi-permeable compositions that can be used to form pharmaceutical controlled-release formulations. The process for the production of the internal molecular chelate complex or matrix involves a direct reaction of the solvated cellulose with the aliphatic alcohol, without any intermediate granulation step.

The first patent that could be searched by the authoress in the field of foodstuff was from Procter & Gable for proteinaceous foodstuff comprising of an edible amino acid-deficient protein. US4,379,177 of Procter&Gamble with the earliest priority dated 1979 relates to cocrystals through spray-drying and discloses stable dehydrated cocrystalline amino acid food additives.

The object of the invention was to reduce or eliminate the undesirable off-flavours of free amino acid food additives and to provide a method for dehydrating an aqueous solution of amino acids and their salts and derivatives using conventional drying techniques for the delivery of an amino acid containing powder having a lower hygroscopicity.

The invention discovered that neutralisation improves the taste of some key amino acid materials because their salts are blander in many cases. But due to their strong hydrophilic and hygroscopic nature, the dehydration of those salts is difficult. Normal drying techniques on aqueous solutions of many amino acid materials and particularly neutralised salts of amino acid materials shall result in incomplete drying or decomposition. According to the invention, a certain amount of a cocrystalliser material mixed with amino acid solutions could be successfully dehydrated using conventional drying techniques. The preferred method comprised of dissolving an amino acid material in water along with an effective cocrystalliser material.

The following example represents the first example of the patent:

EXAMPLE 1

The following materials were combined:

- Solvent
400 grams of water
Amino acid materials
59 grams of L-lysine hydrochloride
31 grams of N-acetyl L-methionine
20 grams of L-threonine
Cocrystallizer material
120 grams sodium chloride

The amino acid materials and cocrystallizer material were dissolved in 400 grams of 150 DEG F water. All solids dissolved. The solution was frozen for 24 hours at 0 DEG F and placed in a vacuum freeze dryer for 24 hours. At the end of this period all of the water had been evaporated and a dry stable, uniform, dehydrated cocrystalline solid matrix remained. This was ground and added to a formula consisting of: peanut paste, sugar, molasses, stabilizer and emulsifier at a level of 23 grams of the improved cocrystalline amino acid food additive to 1000 grams of peanut butter mix. This level should give a protein efficiency ratio of 2.5. A control peanut butter consisting of the same ingredients and substituting sodium chloride for the cocrystallized material to obtain an equivalent sodium chloride level was prepared. These two materials were panelled by expert panellists and were determined to be equal in flavor.

Furthermore, according to Example 12, nitrogen was used to suspend the cocrystallate in the melted fat:

EXAMPLE 12

Sodium NALM, NaCl, Melted Fat, Spread Over the Surface of Extrudate

Two grams of NaCl and 0.5 g of NALM were dissolved in 30 ml of water and the pH of the solution brought to 6.75 by the addition of 0.1 N NaOH. The resulting solution was freeze dried. Eight tenths of a gram of the resulting freeze-dried cocrystallate was added to 4 g of melted fat, saturated triglyceride, mix contained in a test tube immersed in a silicone bath maintained at 80 DEG C. A fine stream of nitrogen was bubbled through the melted fat in order to suspend the sodium NALM-NaCl cocrystallate. One and two tenths' grams of the suspension were dripped on the surface of 3.75 g of freeze-dried soy meal based meat analog extrudate using a transfer pipet. The resulting sample was used in the rapid aging test.

Moreover, the next cocrystals from EP528604B1 (priority 1991) of ICI America Inc. are related to the food industry and disclose melt cocrystallised sorbitol/xylitol compositions. These cocrystallised compositions exhibit superior processing capabilities relative to blends produced by mixing equivalent amounts of melt-crystallised sorbitol with melt-crystallised xylitol. The following examples disclose the process of production of the cocrystals:

Examples 1–5 and Comparative Experiments A and B

Employing a laboratory size Readco mixer having counter-rotating mixing blades 2 inches in diameter and a barrel length of 18 inches, several batches of melt cocrystallized sorbitol/xylitol (Examples 1–5) having varying weight ratios of sorbitol to xylitol were prepared. For comparative purposes, as detailed below, batches of melt crystallized sorbitol (Comparative Experiment A) and of melt crystallized xylitol (Comparative Experiment B) were similarly prepared.

The molten sorbitol starting material, was produced by melting aqueous crystallized sorbitol having a water content of about 0.2% by weight and was maintained at 210°F prior to introduction into the mixer. The molten xylitol starting material was produced by melting commercially available aqueous crystallized xylitol and maintaining such material at 225°F in order to keep the moisture content of such molten xylitol to less than about 0.5 weight percent.

The heated starting materials were added to the mixer 9:1 to 1:9 and agitated at 20 revolutions per minute for 2 minutes while the jacketed mixer was cooled with water at the temperature indicated. At this point, the mixing was stopped to ensure that the melt temperature had cooled to a temperature below the melting point of the seed crystals to be added (i.e., for sorbitol less than 195°F; for xylitol less than 190°F). At this point, agitation was continued and the seed crystals were added.

Once crystallization was initiated and the mass began thickening, the mass was dumped into foil covered pans and placed in an oven at 100°F until fully crystallized.

EP0885961B1 of Hoechst Marion Roussell GmbH (earliest priority: 1997) discloses the biochemical expression of insulin derivatives with an accelerated onset of action compared to human insulin. Furthermore, a depot form for the purpose of miscibility is disclosed, wherein the depot excipient protamine sulphate and insulin are present in a cocrystallisate (see also EP1084248B1).

The above-mentioned examples from different fields of innovations from all over the world and over more than a century clearly prove that cocrystals have always been extensively used in the industry.

5.3.1.5 Current Mechanochemical Patents in Patent Databases

The following chapter summarises some cocrystals from patent literature as cocrystals as long as they are mentioned in the cited patent as cocrystals even if they are salts starting with priorities from the year

2000. As the previous chapter, intention of this chapter is not to discuss cocrystals in a correct scientific context but to give a compressed overview of methods for the production of cocrystals, salts, or molecular complexes in the food industry, life science, pharma, and biotechnology industry. Furthermore, an example of the mechanochemical treatment of polymorphs is mentioned. The search revealed that much more patents are applied in India and China than in Europe or the United States.

EP2860801B1 of Advanced Lithium Electrochemistry Co Ltd and earliest priority 2007 discloses a process for the production of an electrochemical composition and the composition, wherein a particle mixture is ball milled with an oxide of at least one element selected from zinc, magnesium, aluminium, silicon, titanium, vanadium, manganese, nickel, copper, cobalt, and chromium, in a ball mill jar to produce a semicrystalline particle mixture.

EP1426416B1 of EASTMAN KODAK CO and earliest priority 2002 discloses a two-step process for preparing cocrystals of titanyl fluorophthalocyanine and titanyl phthalocyanine, and electrophotographic element containing the same comprising milling of both crystalline components, to form an amorphous mixture and contacting the mixture with water.

EP1426415B1 of Nexpress Solutions LLC with a priority in 2002 discloses a process for forming low-concentration cocrystals of non-chlorinated titanyl phthalocyanine and titanyl fluorophthalocyanine using an organic grinding aid.

WO2005089375A2 discloses a screening method utilising the isolation of cocrystals of drugs.

WO2005072699A of Studiengesellschaft Kohle, which was transferred to Avantium Int BV filed in 2005 of inventors Tanja Bendele (authoress), Michael Felderhoff, and Claudia Weidenthaler, discloses a method for induction of a polymorphic phase transfer in organic solids by ball milling. The phase transformation was conducted on sulfamerazine form 1 into sulfamerazine form 2 and of acetazolamide form 2 into form 1 via high-energy ball milling (Fritsch Pulverisette P7, beaker: hardened steel, balls: hardened steel). Furthermore, chloropropamide was converted from form C into form A within 60 minutes. A partial phase transformation could also be observed starting with form A into form C when the milling was under different conditions. The patent discloses a process for inducing and/or accelerating at least one phase transformation in solid organic molecules, wherein the solid organic molecules are subjected to a tribochemical treatment to result in a phase transformation of the solid organic molecules, and wherein the phase transformation is achieved essentially by means of transmission of high kinetic energies of 20 G or higher.

DE102007030695A1 filed in 2007 by inventors Michael Felderhoff and Tanja Bendele (authoress) discloses the preparation of pharmaceutical cocrystals by ball milling of an amino compound with urea. The examples disclose the synthesis of a cocrystal of carbamazepine with urea in a ball mill.

EP2197278B1 of BASF AG with a priority date in 2007 discloses cocrystals two fungicides pyrimethanil and dithianon. Inter alia a process for the production of the cocrystals encompasses the mixing of the components before heating the mixture.

> The premix containing both, pyrimethanil and dithianon, is passed continuously over a bead-mill at elevated temperature. After half an hour, the complete suspension has turned dark green. The mill is cooled to 20° C.
> Dry dithianon and pyrimethanil are thoroughly mixed and kept at 50° C. under agitation. After some hours, the powdery product has turned its colour to olive green.

WO2012010316 of Grünenthal GmbH with a priority in 2010 relates to salt or cocrystal of 3-(3-dimethylamino-1-ethyl-2-methyl-propyl)-phenol such as (1R,2R)-3-(3-dimethylamino-1-ethyl-2-methyl-propyl)-phenol, which is also known as tapentadol (CAS no. 175591-23-8), are synthetic, centrally acting analgesics which are effective in the treatment of pain and at least one acid component or at least one acid component, wherein the salt or cocrystal is produced by classical solvent-based crystallisation.

EP2707366B1 of Janssen Pharmaceuticals with priority in 2012 claims L-proline and citric acid cocrystals of (2s, 3r, 4r, 5s, 6r)-2-(3-((5-(4-fluorophenyl)thiophen-2-yl) methyl)-4-methylphenyl)-6-(hydroxymethyl)tetrahydro-2h-pyran-3,4,5-triol for use in the treatment of glucose-related disorders such as Type 2 diabetes mellitus and Syndrome X are obtained by ball milling in the presence of a liquid.

EP3057953B1 of Vertrex Pharma Inc. with earliest priority in 2013 discloses cocrystals of DNA-dependent protein kinase (DNA-PK) inhibitors and adipic acid, as well as many other organic acids. The patent refers to known methods for the production of cocrystals with an active pharmaceutical ingredient and a CCF (cocrystal former) including hot-melt extrusion, ball milling, melting in a reaction block, evaporating the solvent, slurry conversion, blending, sublimation, or modelling. Moreover, the hot-meld extrusion is described as:

> In the hot-melt extrusion (HME) method a new material (the extrudate) is formed by forcing it through an orifice or die (extruder) under controlled conditions, such as temperature, mixing, feed-rate and pressure.

In particular, a cocrystal was synthesised in a holt melt extrusion of DNA-dependent protein kinase (DNA-PK) inhibitor (S)-N-methyl-8-(1-((2′-methyl-[4,5′-bipyrimidin]-6-yl)amino)propan-2-yl)quinoline-4-carboxamide) and adipinic acid.

EP2943454B1 of Kkrka with a priority in 2014 provides a process for the preparation of tapentadol maleate Form II from tapentadol maleate Form I by applying mechanical stress to tapentadol maleate, for example in a mill. Furthermore, it is disclosed that:

> Form II or amorphous form of tapentadol maleate can optionally be produced by grinding of the salt sample of Form I or mixture of Form I and II or Form II in a ball mill. Preferably, the grinding is performed in the presence of water soluble diluents such as sucrose, lactose in hydrated or anhydrous form, sugar alcohols such as mannitol, water soluble polymers such as povidone with K value 2 to 50, preferably 7 to 35, cellulose ether such as hypromelose, methyl cellulose, polyvinyl alcohol, graft copolymer of polyvinyl alcohol and polyethyleneglycol, copovidone, polyethyleneglycol or the like, inorganic materials such as colloidal silica (sold under trade name Aerosil), natural silicates such as bentonite or zeolite. During milling, the mechanical force exerted on the particle surface leads to particle size reduction, and in the case of tapentadol maleate, also to the conversion to polymorphic Form II of tapentadol maleate.

EP3424920B1 and further patents as family members of Vertrex Pharma, United States, with priority in 2014, disclose and claim cocrystals of (s)-N-methyl-8-(1-((2′-methyl-[4,5′-bipyrimidin]-6-yl)amino) propan-2-yl)quinoline-4-carboxamide and deuterated derivatives thereof as DNA-PK inhibitors. The cocrystal former is selected from adipic acid, citric acid, fumaric acid, maleic acid, succinic acid, or benzoic acid. Cocrystals were obtained by hot-melt extrusion (HME), ball milling and melting in a reaction block, evaporating solvent, slurry conversion, blending, sublimation, or modelling. For example, the pharmaceutic active compound and neat adipic acid were extruded with high shear mixing and elevated temperatures, e.g. 144°C or 155°C, to generate cocrystal.

EP3294074B1 of Nestec SA, CH, with a priority in 2015 claims a fast-dissolving cocrystalline lactose nutritional composition comprising lactose•calcium salt cocrystals. In an alternative, if the calcium salt is selected from a water-insoluble calcium salt, the cocrystals may be prepared by mechanical processes such as grinding, ball milling of a mixture etc.

EP3455229B1 of Grünenthal GmbH with priority of 2016 discloses different cocrystals of zoledronic acid. Furthermore, the patent discloses that zoledronic acid is administered as an intravenous (IV) dose of 4 mg over 15 minutes per month for hypercalcaemia of malignancy, multiple myeloma, and bone metastases from solid tumours, while an IV dose of 5 mg over 15 minutes is used for osteoporosis and Paget's disease. As zoledronic acid is sparingly soluble in water and 0.1 N HCl solution but is freely soluble in 0.1 N NaOH and is practically insoluble in various organic solvents. There was a need to improve the aqueous solubility to generate a novel oral formulation. The patent discloses the following complexes of zoledronic acid: zoledronic acid in the form of a zoledronic acid L-carnitine complex, or fatty acid L-carnitine, zoledronic acid in the form of zoledronic acid, o-palmitoyl-L-carnitine and water complex (1:1:1), zoledronic acid in the form of zoledronic acid, o-myristoyl-L-carnitine and water complex (1:1:1), zoledronic acid in the form of zoledronic acid, o-lauroyl-L-carnitine, and water complex (1:1:1), zoledronic acid in the form of zoledronic acid, o-decanoyl-L-carnitine, and water complex (1:1:1), zoledronic acid in the form of zoledronic acid, o-octanoyl-L-carnitine, and water complex (1:1:1), and zoledronic acid in the form of zoledronic, L-carnitine, and water complex (1:1:1).

Example 1: Preparation of zoledronic acid L-carnitine and water complex by solvent drop grinding

A solid mixture of 105 mg of zoledronic acid and 59 mg of L-carnitine was ground via mechanical milling with 50 µl of water. The solids gathered after grinding were stored in a screw cap vial for subsequent analysis.

CN107721900A of University China Pharma with priority in 2017 discloses an indomethacin cocrystal and a nanocrystallisation method realised by adopting a wet ball milling method.

CN110452156A of Institute Material Medica Cams with priority in 2018 discloses a eutectic substance of donepezil and irbesartan. The cocrystals of donepezil and irbesartan in a 1:1 molar ratio of donepezil and irbesartan can be prepared by a manual grinding or ball milling method, preferably a liquid-addition ball milling method, in particular, a liquid-added ball milling method.

EP3600255A1 of Tiefenbacher Alfred GmbH with priority in 2018 discloses a stable extrudate containing valsartan and sacubitril.

WO2020165807A1 of Dr Reddy's Laboratories Ltd with priority in 2019 discloses cocrystals of deutetrabenazine and vanillin, the process for the preparation of cocrystals and pharmaceutical composition for use in the treatment of tardive dyskinesia and chorea associated with Huntington's disease. The preparation of said cocrystal of deutetrabenazine and vanillin was performed by ball milling deutetrabenazine and vanillin in the presence of acetonitrile.

EP3718602A1 with priority in 2019 of University Limerick Ireland discloses an ionic multicomponent crystal or cocrystal of lamotrigine:valproic acid (1:2) that was obtained from grinding in a mortar with pestle and manually grounding.

WO2022067043A1 of University Southern Methodist, United States, with priority in 2020 discloses a cocrystal of creatine and citric acid that is obtained by milling. Previously, several attempts to cocrystalise creatine with various coformers by solvothermal methods were unsuccessful. Numerous crystallisations at various conditions resulted in poly-crystalline mixtures. Creatine is poorly soluble in various solvents, limiting the design of a broad screening procedure. Therefore, the inventors turned to mechanochemistry as an alternative strategy for cocrystal screening. The ball milling of equimolar amounts of creatine anhydrate with citric acid anhydrate led to a supramolecular reaction and the formation of a co-amorphous formulation. Figures 5.4 and 5.5 show a crystal-packing diagram of the cocrystal of creatine citric acid.

CN112209887A with priority in 2020 discloses a 1:1 cocrystal of 5-fluorouracil and kaempferol. The cocrystal was obtained by a crystallisation process and ball milling. The cocrystal reduces effectively the hygroscopicity of kaempferol and optimises the dissolution properties of 5-fluorouracil and kaempferol. Further, the solubility difference between 5-fluorouracil and kaempferol is reduced and the preparation compatibility of 5-fluorouracil and kaempferol was improved. It is mentioned that this is a research basis for playing a better anti-tumour synergistic effect of 5-fluorouracil and kaempferol.

Example 3

Weigh 13.0 mg of 5-fluorouracil and 28.6 mg of kaempferol, add them to a ball mill tank, then add 20 microliters of ethyl acetate, and grind at 20 Hz for 60 minutes. The white solid obtained is dried at 40°C to obtain 5-fluorouracil and kaempferol Eutectic solid sample.

CN112480287A with priority in 2020 discloses the oxidation of chitosan, which comprises the following steps: mixing solid chitosan with a solid oxidant, and carrying out ball milling on the obtained solid mixture under solvent-free conditions.

WO2022000265A1 of University Tianjin Technology, China, with priority in 2020, discloses 1:1 cocrystals of axitinib and glutaric acid. Compared with axitinib, the cocrystals of axitinib and glutaric acid possess better solubility and a faster dissolution rate and facilitate improving the oral bioavailability of axitinib. Preparation can be performed by liquid-assisted ball milling in the presence of acetone.

CN113476453A of Shenzhen Taili Bio Pharmacy Ltd with priority in 2021 discloses compositions of panatinib, ibrutinib, and vandetanib salt. The compositions of panatinib, ibrutinib, or vandetanib salt are prepared by mixing, in particular, by ball milling, panatinib, ibrutinib, or vandetanib with an acidic polymer, in particular, with hydroxypropyl cellulose acetate phthalate (HPCAP). The composition of panatinib, ibrutinib, or vandetanib salt is in a stable amorphous form for at least 3 months under accelerated stability conditions (40 °C, relative humidity of 75%) and possesses a higher instantaneous solubility and dissolution rate. Further, the composition has better thermal stability and can bear higher processing temperature in a preparation process according to the glass-transition temperature of the composition.

Example 1: Ibrutinib: HPMCAS salt composition preparation:
Weigh 0.2g ibrutinib and 0.3g HPMCAS into a ball milling jar, and place the ball milling jar at -20°C for 30 minutes and then start ball milling. The vibration frequency of the ball mill is set to 20 Hz, and the total milling time is 30 minutes. At the same time, after each ball milling for 10 minutes, freeze for 30 minutes. Through the high-energy mixing of ball milling, ionic bonds are formed between the free base of ibrutinib and the acidic polymer HPMCAS. An ibrutinib: HPMCAS salt composition with an ibrutinib content of 40% can be obtained.

CN113105367A of University Guangzouh with priority in 2021 discloses a metformin salt obtained by ball milling, wherein the pharmaceutical adjuvant acid is one of 1-naphthylacetic acid, 1-hydroxy-2-naphthoic acid, 2-hydroxy-3-naphthoic acid, 2-naphthoic acid, 1, 5-dihydroxy naphthalene, 3,7-dihydroxy-2-naphthoic acid, and 2,3-dihydroxy naphthalene. This metformin salt possesses a weaker hygroscopicity, good solubility in water, capability of slowing down the inherent dissolution rate of metformin, stability, and easiness in storage.

CN111777551A with priority in 2022 discloses a eutectic cocrystal of regorafenib and suberic acid that was obtained in liquid-assisted grinding.

CN111822106A filed in 2020 discloses a wet pulverisation method for preparing ultrafine pear stone cell powder in a planetary ball mill for use in toothpastes as an organic abrasive instead of inorganic abrasives with the following steps: (1) putting the pear stone cell extract particles into a plant pulveriser to pre-pulverise to prepare pear stone cell coarse powder and sieving; (2) adding pear cell coarse powder, water, dispersant (sodium methylene dinaphthalene sulfonate) and defoamer (1 ⁰/₀₀ of water tributyl phosphate) in a ratio of solid to liquid of 1:4 to a planetary ball mill for ball milling with ceramic beads (diameter 10 mm, 200 rpm, grinding temp. 70 °C). Ball mass ratio 1:1, filling factor 30%, grinding time 60 minutes; and (3) centrifugation of the obtained suspension (centrifuge speed: 6000 rpm, centrifugal time of 60 minutes). The obtained solid is washed with deionised water and ethanol one time, and freeze-dried overnight at −30°C. Obtained is an ultrafine pear stone cell powder.

CN114478673A with priority in 2022 discloses a pharmaceutical cocrystal of estriol. For the preparation of estriol, urea and ethyl acetate were ball milled to obtain a white solid powder of estriol–urea cocrystal. The cocrystal is defined through the formation of a supramolecular network of the drug molecules and a single melting point.

The above-mentioned examples from different fields of innovations over several years clearly prove that mechanochemical methods have always been extensively used in the industry for the development of cocrystals. The question arises, what are the reasons for them becoming less comprehensively used in the field of organic chemistry and in particular in the pharmaceutical industry? One reason may be the industrialisation and demands for continuous production processes that were much easier to be processable in liquid phases.

But miniaturisation in pharmaceutical and/or chemical reactors in combination with an application of digital twins and AI may allow implementation and rising resilience through innovative sustainable manufacturing processes relying on modular production sites using digital twins and/or AI.

For over 20 years, many pharmaceuticals are produced outside of Europe due to cost and eventually due to existing intellectual property rights. Furthermore, many pharmaceuticals are up to know produced in a batch-wise manner. This led to a situation in countries where some pharmaceuticals were not available for the patients. With innovative sustainable manufacturing processes relying on modular production sites using digital twins and/or AI new production sides may be cost effective in the healthcare industry, chemical industry, and other industrial areas in Europe.

Digital Twins a large research initiative in Europe and beyond aims at revolutionising healthcare and biomedical research for the benefit of citizens and society through the creation of digital twins. The digital twins are known to be computer models enabling a prediction of optimised chemical processes in view of all issues to be considered in a chemical production process. Artificial intelligence (AI) and digital twins are well developed and allow processes to be designed in a fully virtual environment. Moreover, costs and by-products can be estimated and optimised before any invest in a new production line is necessary.

Further, as an example testing modular standardised connectable continuously operated small-sized extruders, gaseous reactors, or mechanochemical reactors may allow a reduction of the plant size and may allow a switch from batch-wise production in the pharmaceutical industry to continuous production as it is typical for the chemical industry. Further, using digital twins for the estimation of a new process set-up switching from solvent-assisted batch-wise production to a continuous, e.g. extruder supported, production enables the development of nearly fully automated production processes in small-sized reaction plants in movable and scalable infrastructures. Also, the calculations of the digital twin can lead to heavily reduced production costs.

A side effect especially with sustainable manufacturing methods optionally designed by digital twins will strengthen the resilience of the healthcare system.

5.4 Mechanochemical Synthesis of Cocrystals – Ball Milling as Cocrystal Screening Method

During the authoress own Ph.D. thesis in 1999 on "Structure and Properties of Supramolecules form 1,4-Diaza-Compounds and bi-functional H-Bridgedonors"[36] as well as during the previously performed master thesis, many solid-state analytics on cocrystals with emerged photochromic properties and other new properties were synthesised by crystallisation from solvents.

Published cocrystals are inter alia "Phenazine and meso-1,2-Diphenyl-1,2-ethanediol – Partners in Photochromic Cocrystals",[37] "Design of a three Component Crystal based on the Cocrystal of Phenazine and 2,2′-Dihydroxybiphenyl" with Acridine,[38] "Light-Induced Cooperative Electron-Proton Transfer in Hydrogen-Bonded Networks of Diaryl Substituted 1,4-Bisimines and meso-1,2-Diaryl-1,2-ethanediols",[39] and "Structure and Properties of Cocrystals of Phenazine and Fumaric, 2,3-Dihydroxyfumaric- and Oxalic Acid"[40] as well as the further mentioned publications on cocrystals.

The work was strongly focused on the evaluation of new solid-state analytics to be applied on the cocrystals and being useful for the clarification of all emerging properties such as photochromism and to clarify the structure of the cocrystals and their emerging properties in relation to the initial substances without these properties. Therefore, it was crucial to crystallise crystals of a certain quality of the cocrystals from an appropriate solvent for single crystal studies and clarification of the structure of the cocrystals.

It was a very astonishing finding to obtain in the first crystallisation from solvent purple cocrystals of the yellow-coloured phenazine and colourless *meso*-1,2-diphenyl-1,2-ethandiole. ESR (electron spin resonance) showed paramagnetism of the cocrystals. The then-followed crystallisation with the exclusion of light resulted in diamagnetic, yellow-coloured, transparent cocrystals. The melting point of 121 °C was below the melting points of the educts. Figure 5.1 discloses the X-ray structure analysis and confirms the formation of the cocrystal.

The stepped chains from phenazine and *meso*-1,2-diphenyl-1,2-ethandiole are linked by H-bridges with an O···N-distance of 285.0 pm ($\mu = 167.2°$). Attractive C_{sp2}-H··· π-contacts are enabled by phenyl rings of the diol arranged at an angle of 63.4° in relation to the phenazine molecules. Further cocrystals of phenazine with other dihydroxy-compounds were obtained which are disclosed in the Ph.D. thesis and publications. Below is disclosed a single crystal structure of cocrystal of 5,10-Dihydrophenyzine and Phenazine of the Ph.D. thesis and is shown in Figure 5.2.

An additional cocrystal (Figure 5.3) could be obtained from 1,4-cis-(4-methoxyphenyl)-1,4-diaza-1,3-butadiene and *meso*-1,2-diphenyl-1,2-ethandiole. The diaza-compound possesses a conjugated π-system. If the cocrystal is illuminated through a WG360-edge filter, the diamagnetic cocrystals became paramagnetic with an ESR signal of a width of 90 G (g-value: 2.0038). Cocrystals that were illuminated for a

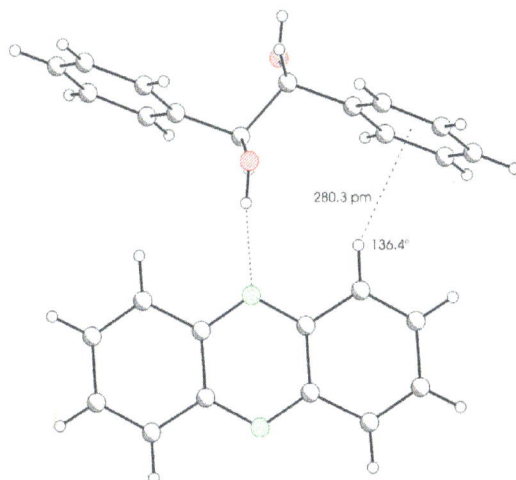

FIGURE 5.1 Extract of the crystal structure of phenazine and *meso*-1,2-diphenyl-1,2-ethandiols.

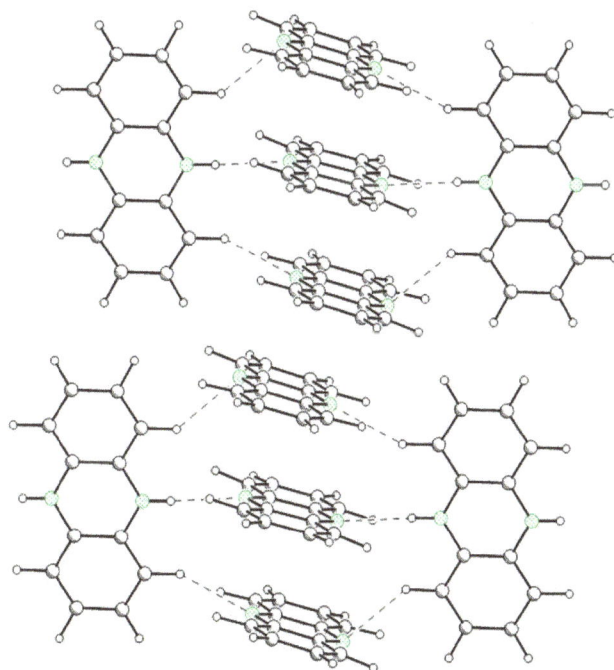

FIGURE 5.2 Extract of the crystal structure of 5,10-dihydrogen-phenazine and phenazine.

short period of 4 seconds could be reverted from the paramagnetic state back into the diamagnetic state by heating them up to 60 °C.

The next two cocrystals comprise a two-component cocrystal of phenazine and 2,2-dihydroxybinphenyl as shown in Figure 5.4, wherein the phenazine stacks in the right and left columns are only connected with one OH⋯N-bridge and therefore could be substituted by acridine as shown in Figure 5.5.

Furthermore, the obtained new cocrystals were analysed with solid-state analytics, which were not too common at that time. For example, solid-state infrared and UV/Vis-spectrograms were measured in potassium bromide matrix. To avoid regular reflections from potassium bromide crystals during solid-state UV/Vis-spectra measurement, first, the potassium bromide crystals were milled in a planetary ball

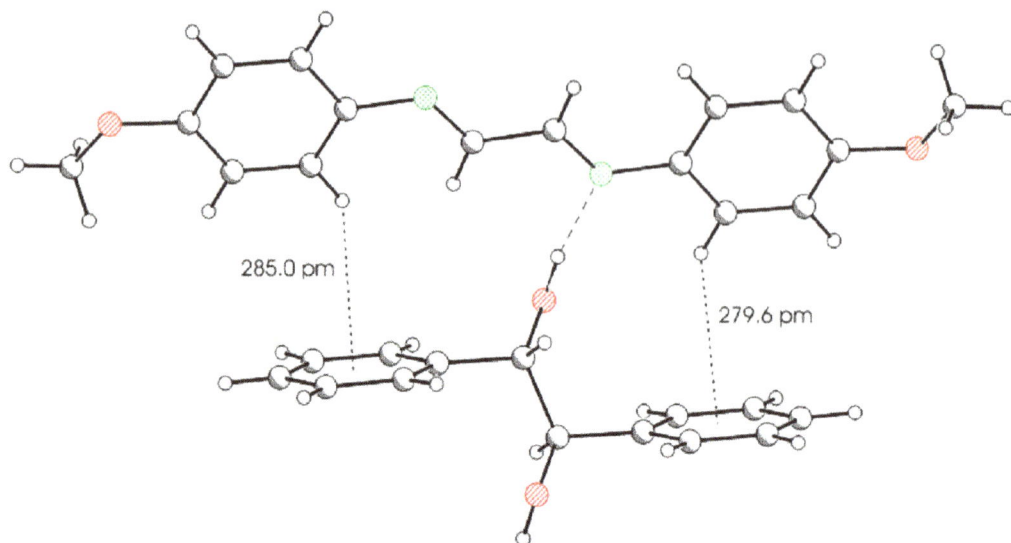

FIGURE 5.3 Extract of the crystal structure of the cocrystals of 1,4-bis-(4-methoxyphenyl)-1,4-diaza-1,3-butadiene and *meso*-1,2-diphenyl-1,2-ethandiols.

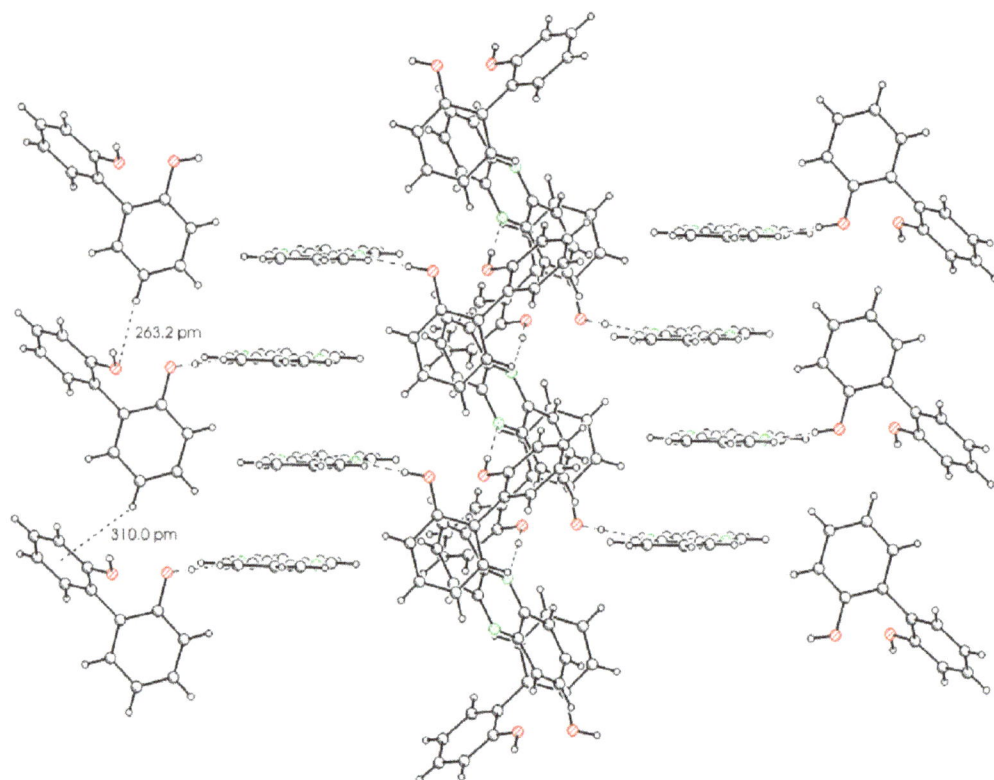

FIGURE 5.4 Extract of the structure of the cocrystal of phenazine and 2,2′-dihydroxy-biphenyl. View on the zipper-like stapled phenazine molecules with H-bridges.

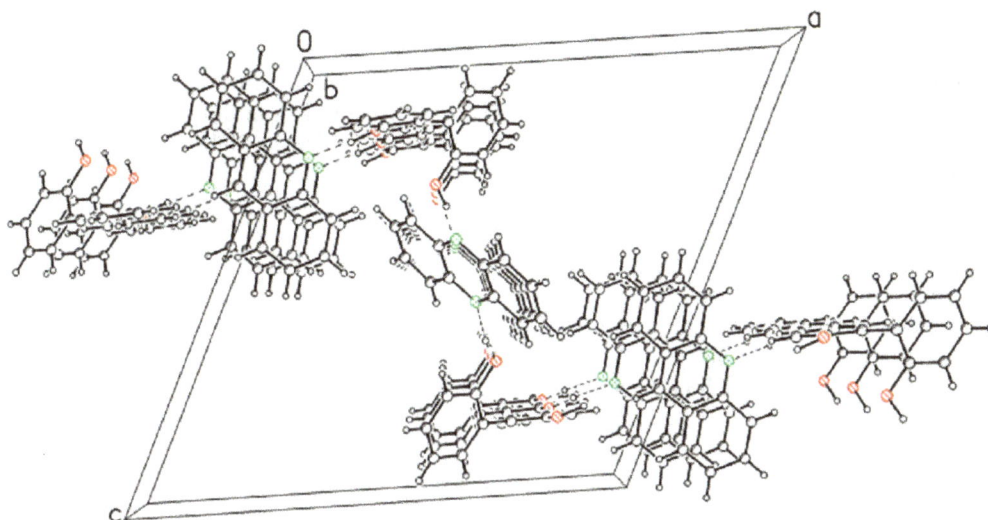

FIGURE 5.5 Extract of the structure of a tri-component cocrystals of phenazine, acridine, and 2,2′-dihydroxybiphenyl.

mill in an agate beaker with agate balls down to small particles with only diffuse reflections in the UV/Vis-spectra.

In the beginning, the cocrystals were only pulverised manually and measured laying on the potassium bromide matrix. Thereafter, the obtained cocrystals were also milled together with potassium bromide, wherein the cocrystals surprisingly remained being cocrystals, which could be checked by analysing the emerged photochromism. Furthermore, all cocrystals could be obtained through the melting of equimolar amounts of the components as well as through rubbing.

Later it was tested by the authoress if a certain amount of weight of both the initial substance and potassium bromide could also lead to the formation of cocrystals by milling. Surprisingly, even, the initial substances in a matrix of potassium bromide could be milled into "co-crystals" comprising particles with only diffuse reflections.

Therefore, before building up a comprehensive set of solutions of solvents and mixtures of initial substances to get a new cocrystal and trying to find the best solvent for the cocrystal formation of my selected initial substances, the authoress added a new first step before the crystallisation procedure. The first step of evaluation whether a cocrystal can be expected was an initial milling step of the initial substances in potassium bromide as a matrix together with a molar content of circa 10^{-3} to 10^{-4} mol. of the substances in the agate beaker with agate balls. Direct milling of the initial substance was not performed due to cost at that time. The results were analysed if cocrystal formation could be determined by infrared spectroscopy, e.g. all cocrystals showed a shift of the hydroxy bond OH– (vibrations) and C=N– (vibrations). Only then, a comprehensive set of solutions of solvents was prepared to raise crystals of the cocrystals for single-crystal structure analysis.

Unfortunately, this finding that the cocrystals can be obtained by weighting and milling of the educts in an amount of e.g. 10^{-2}, 10^{-3}, or 10^{-4} molar in an inorganic salt such KBr for a short period in the ball mill was not allowed to be published with authoress' Ph.D. work in 1999 and remained, therefore, knowhow since that time.

The above circumstance initially shows that even the inventor needs a certain time and distance to analyse a scientific situation in a more objective way. Here, also the term "cocrystals" fixes the mind of the potential inventor to perform solvent-assisted crystallisation to synthesise the cocrystals. But even if the inventor was able to draw the correct conclusions, in a scientific world a proof of concept will be ensured differently from a proof of concept in a patent application.

At that time and with the current knowledge of cocrystals and available databases, these findings were too new to be allowed by the supervisor to be published and the authoress had to delete all passages from her Ph.D. work.

More than 20 years later, with an incredible further development of the innovation transfer system and innovation management at universities and the force to acquire a third-party donor – probably a patent application would have been filed on a screening method for a cocrystal screening using a mechano-chemical method, e.g. based on friction, in particular, comprising of ball milling. Furthermore, for a scientific publication, additional research work might be done by a following Ph.D. student.

Why was this event mentioned in this chapter? It shows, very subjectively, the circumstances an invention may be faced with. An invention shall be novel, but the invention is sometimes made too early, and the time or the other responsible people need more time to decide on this topic or the inventor does not represent her own case properly. An attempt to resolve the issue 20 years later is not possible and would inevitably lead to a *post-ex facto analysis*.

The above issue shows that all inventions and here, in particular, mechanochemical synthesis of cocrystals, or also of new molecules in organic chemistry, may be bound in the past to scientific ideas the responsible scientists had. It may have taken some time to understand the underlying steps that even in a ball mill processed synthesis each molecule can leave its position in a crystal lattice and can move into a new position of a new crystal lattice.

Therefore, to avoid these circumstances in companies and universities, an innovation management was established requesting the inventors to report an invention and furnish assistance for checking the novelty, inventive step, and industrial applicability of a possible invention.

5.5 Innovations in Mechanochemistry in Different Industrial Areas and Patent Literature as a Springboard to Innovations

The above-mentioned historical patents to the current patents clearly show the wide range of disclosures of all technical fields that can be researched free of charge by everyone in patent databases. A comprehensive search over a broad range of many different technical disciplines is possible in only one patent database. Furthermore, as many databases possess English machine translations of Japanese, Chinese, and Korean patents or at least English abstracts, a worldwide search is possible with some restrictions. The database Espacenet[41] offers families of patents in different languages and in addition, machine translation in several other languages.

For example, the search for a ball mill may not only be performed for ball mills used in the chemical area but also in other fields of applications such as cellulose treatment in the paper industry, ball mills for inorganics such as porcelain industry or dispersers used in the construction chemistry. Furthermore, if the production is located in Europe, the United States, Japanese, or Chinese patent applications with no family member in Europe may give interesting information on milling apparatus or extruders for production in a patent-free country, here in this example in Europe or *vice versa*.

It is recommended to also use the patents in the different patent databases as a springboard for one's own developments. At least two different strategies may be considered. The first strategy may be based on evaluating lapsed patents and checking if the disclosed information may be used for further development.

A second and third strategy may encompass the development of technologies disclosed in pending patents or patent applications, wherein it has carefully been checked if the claimed inventions will not be infringed according to the relevant national law or the maturity date or lapse of the patents observed and a market launch takes place after the lapse of the patents.

5.5.1 Possibilities and Problems – Sustainable Industry

A general business concept for the development of sustainable processes for chemical compounds, NCEs (New Chemical Entity), or generic APIs (Active Pharmaceutical Ingredient) will be given below as a guideline for the development of a generic API. During patent and/or SPC term, the research, preparatory acts, manufacturing, or stock-pilling can be conducted under the research exemptions, Bolar exemptions, or SPC manufacturing wafer, wherein the concrete circumstances have to be checked for each issue and each step individually and carefully.

1.1 Define generic APIs for which substance patent(s) and SPC are lapsed and propose innovative manufacturing technology, with less energy, solvent consumption; or

1.2. Define APIs for which substance patent(s) are lapsed and SPC is pending and propose innovative manufacturing technology, with less energy, solvent consumption; or

1.3 Define APIs with pending patent and SPC and propose innovative manufacturing technology, with less energy and solvent consumption;

2.1 Try to propose new and inventive intermediates for the production of APIs;

2.2 Try to propose new and inventive solid-state modifications, such as hydrates, solvate, salts, APIs embedded in excipients with defined solubility as a product of the APIs;

3 Try to propose new and inventive reactors or combinations of reactors for the production of APIs, continuous reactors are much more preferred than batch-wise reactors;

- **4.** Try to propose innovative uses of digital transformation, AI (artificial intelligence), or robotics for competitive and scalable methods of production for the production.

The above very brief guideline gives pointers to several approaches for a holistic view on the development of new production processes of known or new substances on more sustainable processes. In addition, the coupling of the sustainable mechanochemical processes can be joined with a digital transformation to artificial intelligence-controlled and scalable green productions.

5.5.2 New Apparatuses, Systems, or Kits in Chemistry, Pharmacy, and Biotechnology

The application of mechanochemistry will lead to new emerging technologies for more sustainable chemical manufacturing with a reduction in CO_2 equivalents and a reduction in energy consumption.

With respect to biological material, novel and inventive processes may be developed through sustainable mechanochemistry methods which may lead to different kinds of isolated materials with deviating effects in comparison to solvent-assisted methods. Therefore, the isolated material from its natural environment or produced by means of this mechanochemical process can also be the subject of an invention even if it previously occurred in nature (section German patent law).[42]

Typically, with a new process, new reactors or process lines may also be needed. There are many possibilities to protect them. Commonly, the mechanical compound of a reactor or of a process line is described in patents, but the surface finish, such as the surface quality, texture, roughness, electrical properties, or coating of the inner parts may also be a relevant innovation which can, if it shall not remain know-how, be protected if the surface fulfils a certain technical purpose.

Many innovations deal with innovations that make the handling of apparatus easier, the process faster or safer, or reduce energy consumption or allow a kind of recycling of educts.

The focus on sustainable chemistry, pharmacy, and biotechnology using mechanochemical methods will lead to developments in many processes using extruders or ball mills for the synthesis of chemical compounds or solid-state forms, such as cocrystals.

A big challenge will be the development of systems for continuous solvent-free production with mechanochemical methods. The digital control system, as well as the feeding of the reactors or the processing of the mechanochemical manufactured reaction mixtures, will be an important playing field for many new innovations.

5.5.3 Computer Program Products and AI (Artificial Intelligence) in Mechanochemistry

As mentioned above in Section 5.2.4.1, according to Article 52 EPC[43] European patents shall be granted for any invention in all fields of technology, provided that they are new, involve an inventive step, and are susceptible to industrial applications with the exception of subject matter or activity as such on programs for computers, mathematical methods, presentation of information, schemes, rules, and methods for performing mental acts or doing business etc. Analogue rules are encompassed in most of the harmonised patent laws, such as in Section 1 of the German Patent law.[44]

Computer programs as such are excluded from patentability under Art. 52(2)(c) and (3). However, following the exclusion does not apply to computer programs having a *technical character*. It is pointed to the fact that a computer program and a corresponding computer-implemented method are distinct from each other, as the former refers to a sequence of computer-executable instructions specifying a method, whereas the computer-implemented method refers to a method being actually performed on a computer. Nevertheless, computer-implemented inventions (CII) are patentable if further requirements are met which will be discussed on the ruling of the EPC of the EPO.

According to the EPO:

> A computer-implemented invention (CII) is one which involves the use of a computer, computer network or other programmable apparatus, where one or more features are realised wholly or partly by means of a computer program.[44, 45]

It is established case law before the EPO that a claim on a computer-implemented invention requires the existence of a further technical effect of its own which goes beyond the normal effects of executing a program on the computer. Application of an industrial process possesses is a technical solution. Typical examples that confer a computer program a technical character are the control of the internal functioning of a computer itself, as well as the control of interfaces under the influence of the computer program. Examples may comprise an improvement of the efficiency or safety of a process. According to Article 52 EPC (2) and (3) EPC, claims directed to a computer-implemented method, a computer-readable storage medium, or a device will not be objected to for the reason that they involve the use of technical means such as a computer or any other technical means itself such as a computer or a computer-readable storage medium, which have a technical character.

Problems arise if a claim comprises a collection of technical and non-technical features as a patentable claim has to meet the following basic requirements: the claim must comprise (i) an "invention", belonging to any field of technology; the invention (ii) must be "susceptible of industrial application"; the invention (iii) must be "new"; and (iv) involve an "inventive step".[46]

As mentioned above, problems arise if a claim comprises a collection of technical and non-technical features as the patentability will be assessed only on technical features. Furthermore, the technical features have to meet the further requirements of clarity and sufficiency of disclosure for enabling third parties to rework the invention. The following very brief introduction to computer-implemented inventions has the purpose to give a short but holistic overview of the inventors of mechanochemistry-related inventions to be implemented as an industrial process in the chemical, biotechnological, or pharmaceutical industry.

According to the Guidelines of the EPO:

> In order to have a technical character, and thus not be excluded from patentability, a computer program must produce a *"further technical effect"* when run on a computer. A *"further technical effect"* is a technical effect going beyond the "normal" physical interactions between the program (software) and the computer (hardware) on which it is run. The normal physical effects of the execution of a program, e.g. the circulation of electrical currents in the computer, are not in themselves sufficient to confer technical character to a computer program (T 1173/97 and G 3/08).[47]

The technical effect of the computer-implemented invention has to be reflected in claimed features that cause a further technical effect (G 3/08). Further the computer-implemented method must solve a technical problem. Only if a claim passes the test for having technical character, the EPO will proceed to examine the novelty and inventive step.

If it has proven that the computer-implemented method involves at least one technical feature or causes a technical effect, the invention in addition has to be novel, inventive, and clear and must be disclosed in a manner that the skilled person can carry the invention out without undue burden, e.g. without being inventive.

If the claims comprise technical and non-technical features, only the technical features will be evaluated in the context of novelty and inventive steps. The assessment of the inventive step is then examined with the Comvik approach that was established in T 641/00 (OJ 2003, 352) to apply the problem–solution approach of the EPO only to those features that contribute to the technical character of the claims.[48]

Also, artificial intelligence (AI) and machine learning (ML) *per se* are excluded from the patentability according to Article 52 EPC as they are based on computational models and algorithms for classification, clustering, regression, and dimensionality reduction, such as neural networks, genetic algorithms, support vector machines, k-means, kernel regression, and discriminant analysis. The aforementioned models and algorithms belong to abstract mathematical considerations and therefore may as such not imply a technical means. But they can be used in various technological fields such as for the monitoring of mechanochemical processes, dosing of educts, temperature, or pressure control, for analytical monitoring of products and other steps that may be relevant for quality control of a process. These examples for the mentioned technical application furnish a technical contribution to such models. The examples for a technical purpose are not a complete list, but many further applications may contribute to a technical purpose. Also, the mere classification of abstract data records in a technical field without a specific technical purpose does not define technical use. The classification method may possess a technical purpose if the training set and training of the classifier contribute to a technical character of the invention and a technical purpose is achieved, e.g. a mechanochemical process is analysed in view of adverse by-products, particle size distribution of the reaction mixture or analyses a PXRD diffractogram on appearing polymorphs etc. In addition, for the computer-implemented method, at least one way of carrying out the invention must be disclosed in a manner sufficient that the skilled person can carry out the invention without undue burden.

Therefore, in addition to the above-mentioned possibilities of claiming new products, processes, and reactors, surface properties of reactors or instruments of new developments of a sustainable mechanochemistry, further interesting technical purposes may be protected by early innovators for amortisation of the investments, e.g. by a licensing model of the invention to third parties.

5.6 Conclusion and Forecast

The authoress would like to thank all readers for their interest in the above-described complex context and hopes that the reading was a little bit entertaining and exciting as well as that the readers received some useful suggestions that they can use in their future mechanochemical developments. It has to be mentioned that the chapter does not give any legal advice and the information is not to be understood to be compressive, but a very focused overview of the history and current intellectual property rights, such as patents in the context of mechanochemistry. It was an aim of the authoress to awaken the curiosity of the readers to consult the patent databases and to use them as a springboard for their own mechanochemical innovations.

Furthermore, it was an aim of the authoress to show that sometimes many other reasons than scientific reasons inhibit those innovations that are made, used in an industrial application, or be brought onto the market. In many cases imagination of concerned parties prevents real disruptive innovations, existing infrastructure, or marked conditions. The authoress also thanks all concerned parties and in particular the COST members for the possibility to join COST Action CA18112 "Mechanochemistry for Sustainable Industry", supported by the EU Framework Programme Horizon 2020, **which is a multidisciplinary network of European scientists, engineers, technologists, entrepreneurs, industrialists, and investors addressing** the exploitation of mechanical activation in **the production of chemicals through sustainable and economically** convenient **practices on the medium and large scales.**[49]

The time is ripe for "Mechanochemistry: a disruptive innovation for the industry of the future"[50, 51] as mechanochemistry is a very fast-developing scientific discipline establishing connections between chemistry, pharmacy, life sciences, biotechnology, material science, and environmental science. The above-exemplified mechanochemical reactions for the preparations of cocrystals impressively show the acceleration in screening processes performed by mechanochemical methods at ambient temperature instead of solvent-assisted crystallisations.

REFERENCES

1. IUPAC. Compendium of Chemical Terminology, 2nd ed. (the "Gold Book"). Compiled by A. D. McNaught and A. Wilkinson. Blackwell Scientific Publications, Oxford, 1997. Online version (2019-) created by S. J. Chalk. ISBN 0-9678550-9-8. https://doi.org/10.1351/goldbook.

2. Anastas, P. T., and Warner, J. C. *Green Chemistry: Theory and Practice.* New York: Oxford University Press, 1998, p. 30.

3. https://www.acs.org/content/acs/en/greenchemistry/principles/12-principles-of-green-chemistry.html, on 26.12.2021.

4. https://www.dpma.de/english/services/sme/what_is_ip/index.html.

5. https://www.wipo.int, on 26.12.2021.

6. https://www.epo.org/service-support/glossary.html#i, on 26.12.2021.

7. https://www.uspto.gov/learning-and-resources/glossary#sec-I, on 05.02.2022.

8. Yonge, C. D. The deipnosophists, or, banquet of the learned of Athenaeus: Henry G. Bohn. S. 835, 1854. https://e-courses.epo.org/wbts_int/litigation/History.pdf.

9. https://www.copyrighthistory.org/cam/index.php; http://www.copyrighthistory.org/cgi-bin/kleioc/0010/exec/showThumb/%2522i_1474%2522/start/%2522yes%2522# (from: Primary Sources on Copyright (1450–1900), eds L. Bently & M. Kretschmer, www.copyrighthistory.org), Wolfgang-Pfaller.de: Venediger Patentgesetz von 1474; http://www.wolfgang-pfaller.de/venedig.htm;Helmut Schippel: Die Anfänge des Erfinderschutzes in Venedig, in: Uta Lindgren (Hrsg.): Europäische Technik im Mittelalter. 800 bis 1400. Tradition und Innovation, 4. Auflage, Berlin 2001, S. 539–550.

10. https://e-courses.epo.org/wbts_int/litigation/History.pdf, on 26.01.2022.

11. https://www.jpo.go.jp/e/introduction/rekishi/seido-rekishi.html, on 26.01.2022.

12. https://www.jpo.go.jp/e/introduction/rekishi/seido-rekishi.html.

13. https://e-courses.epo.org/wbts_int/litigation/History.pdf; https://www.wipo.int/treaties/en/ip/paris/.

14. https://www.wipo.int/treaties/en/ip/paris/, on 26.01.2022.

15. https://www.wipo.int/pct/en/faqs/faqs.html, on 26.01.2022.

16. https://www.epo.org, on 26.01.2022.

17. https://www.wto.org/english/res_e/booksp_e/trips_agree_e/history_of_trips_nego_e.pdf, on 26.01.2022.

18. https://www.wto.org/english/tratop_e/trips_e/intel2_e.htm, on 26.01.2022.

19. https://www.uspto.gov/sites/default/files/documents/UK-SME-IP-Toolkit_FINAL.pdf, on 26.01.2022.

20. https://www.gesetze-im-internet.de/englisch_patg/englisch_patg.html#p0073, on 26.01.2022.

21. https://www.epo.org/law-practice/legal-texts/html/caselaw/2019/e/clr_i_a_2_4_2.htm, on 26.01.2022.

22. https://ec.europa.eu/growth/industry/strategy/intellectual-property/patent-protection-eu/supplementary-protection-certificates-pharmaceutical-and-plant-protection-products_de, on 26.01.2022; https://ec.europa.eu/docsroom/documents/43845?locale=en.

23. https://euipo.europa.eu/ohimportal/de/community-design-legal-texts; https://www.gesetze-im-internet.de, on 26.01.2022.

24. https://www.gesetze-im-internet.de, on 26.01.2022.

25. DFG. http://dingler.culture.hu-berlin.de/article/pj220/ar220110, on 26.01.2022.

26. https://www.chemie.de/news/1163866/giftige-luftschadstoffe-in-industriechemikalien-umwandeln.html, on 15.01.2022.

27. https://www.nobelprize.org/prizes/chemistry/1930/summary/, on 25.12.2021.

28. Lemmerer, A., Bernstein, J., Griesser, U. J., Kohlenberg, V., Többens, D. M., Lapidus, S. H., Stephens, P. W., and Esterhuysen, C. *Chem. Eur. J.* 2011, 17(48), 13445–13460.

29. Ling, A. R., and Baker, J. K. Halogen derivatives of quinone. Part III. Derivatives of quinhydrone. *J. Chem. Soc. Trans.* 1893, 63, 1314–1327.

30. Tanaka, K., and Toda, F. Solvent-free organic synthesis. *Chem. Rev.* 2000, 100, 1025–1074.

31. Trask, A. V., Motherwell, W. D. S., and Jones, W. Solvent-drop grinding: Green polymorph control of co-crystallization. *Chem. Commun.* 2004, 2004, 890–891.

32. Bailey Walsh, R. D., Bradner, M. W., Fleischman, S., Morales, L. A., Moulton, B., Rodriguez-Hornedo, N., and Zaworotko, M. J. Crystal engineering of the composition of pharmaceutical phases. *Chem. Commun.* 2003, 186–187.

33. Fleischman, S. G., Kuduva, S. S., McMahon, J. A., Moulton, B., Bailey Walsh, R. D., Rodriguez-Hornedo, N., and Zaworotko, M. J. Crystal engineering of the composition of pharmaceutical phases: Multi-component crystalline solids involving carbamazepine. *Cryst. Growth Des.* 2003, 909–919.

34. Childs, S. L., Chyall, L. J., Dunlap, J. T., Smolenskhaya, V. N., Stahly, B. C., and Stahly, G. P. Crystal engineering approach to forming cocrystals of amine hydrochlorides with organic acids. Molecular complexes of fluoxetine hydrochloride with benzoic, succinic, and fumaric acid, *J. Am. Chem. Soc.* 2004, 126, 13335–13342.

35. Remenar, J. F., Morissette, S. L., Peterson, M. L., Moulton, B., MacPhee, J. M., Guzman, H. R., and Almarsson, O. Crystal engineering of novel cocrystals of a triazole drug with 1,4-dicarboxylic acids. *J. Am. Chem. Soc.*, 2003, 125, 8456–8457.

36. Structure and properties of supramolecules form 1,4-Diaza-Compounds and bi-functional H-Bridge donors. Dr. Tanja Smolka, 1999, University Duisburg-Essen (Chemistry). https://duepublico2.uni-due.de/receive/duepublico_mods_00000460.

37. Smolka, T., Sustmann, R., and Boese, R. Phenazine and meso-1,2-Diphenyl-1,2-ethanediol – Partners in photochromic cocrystals. *J. Prakt. Chem.* 1999, 4, 378–383.

38. Smolka, T., Sustmann, R., and Boese, R. Design of a three component crystal based on the cocrystal of phenazine and 2,2′-dihydroxybiphenyl. *Struct. Chem.* 1999, 10, 429–431.

39. Felderhoff, M., Smolka, T., and Sustmann, R., Light-induced cooperative electron-proton transfer in hydrogen-bonded networks of diaryl substituted 1,4-bisimines and meso-1,2-diaryl-1,2-ethanediols. *J. Prakt. Chem.* 1999, 341, 639–648.

40. Smolka, T., Shaller, T., Sustmann, R., Bläser, D., and Boese, R. Structure and properties of cocrystals of phenazine and fumaric, 2,3-dihydroxyfumaric- and oxalic acid. *J. Prakt. Chem.*, 2000, 342(5), 465–472.

41. https://worldwide.espacenet.com/, on 12.02.2022.

42. https://www.gesetze-im-internet.de/englisch_patg/englisch_patg.html#p0014, on 13.02.2022.

43. https://www.epo.org/law-practice/legal-texts/epc.html, on 13.02.2022.

44. https://www.google.com/search?client=firefox-b-d&q=epo+guidelines, on 13.02.2022.

45. https://www.epo.org/law-practice/legal-texts/html/guidelines/e/j.htm, on 13.02.2022.

46. https://www.epo.org/law-practice/legal-texts/html/guidelines/e/g_i_1.htm.

47. https://www.epo.org/law-practice/legal-texts/guidelines.html, on 13.02.2022.

48. https://www.epo.org/law-practice/legal-texts/html/caselaw/2019/e/clr_i_d_9_1_3_b.htm, on 13.02.2022.

49. https://www.mechsustind.eu/.

50. Baláž, M., Vella-Zarb, L., Hernández, J. G., Halasz, I., Crawford, D. E., Krupička, M., André, V., Niidu, A., García, F., Maini, L., and Colacino, E.. Mechanochemistry: A disruptive innovation for the industry of the future. *Chem. Today*, 2019, 37, 32–34.

51. Hernández, J. G., Halasz, I., Crawford, D. E., Krupička, M., Baláž, M., André, V., Vella-Zarb, L., Niidu, A., García, F., Maini, L., and Colacino, E. European research in focus: Mechanochemistry for sustainable industry (MechSustInd) *Eur. J. Org. Chem.* 2020, 8–9.

Part 2

Solvent-Free Sustainable Technologies at Large Scale

6

Mechanochemistry and Industry: Process Intensification and Beyond

Valerio Isoni

CONTENTS

6.1 Introduction: Mechanochemistry and Industrial Adoption

Early documented examples of mechanochemistry using manual means such as mortar and pestle can be traced back to the 4th century BC (Takacs, 2000). In modern times, technological advancements led to the development of an array of equipment that could deliver high mechanical energy to let reactions take place in a solid state and/or in the presence of small quantities of liquids. Beside laboratory-scale instruments such as mixer mills, shaker mills and planetary mills designed and improved to address challenges of making reactions happen in the absence of solvents, examples of pilot-scale-sized equipment that can be operated in a semi-continuous manner have emerged. An example is the Simoloyer® CM100s, a preparative scale equipment that works on the principle of high-energy high-impact reactive milling in a horizontal rotary ball mill (Kaupp et al., 2002). A milder approach to mechanochemistry that had successful application in the manufacturing of Metal–Organic Frameworks (MOFs) is via reactive extrusion (Casaban et al., 2021). In analogy to solution-based synthetic chemistry, continuous reactive extrusion can be seen as the equivalent flow chemistry for solid/slurry reactions. The sustainable nature of mechanochemistry was recognised by the International Union of Pure and Applied Chemistry (IUPAC) as one of the top ten emerging technologies that would change our world (Gomollón-Bel, 2019). Moreover, mechanochemistry has the potential of enabling new, shorter synthetic strategies for complex molecules in a way that traditional solvent-based approaches might not be able to provide (Do & Friščić, 2016).

However, despite the good credentials and increase of scientific publications in the field in the past decade, industrial adoption of mechanochemistry for sustainable manufacturing has been rather slow. In the next sections a view will be provided on the major reasons for the slow industrial uptake, covering aspects related to the opportunities and intrinsic challenges during chemical process development and scale-up as seen through the lens of mechanochemistry.

DOI: 10.1201/9781003178187-8

6.2 Advantages of Mechanochemistry

Mechanochemistry in any of its many forms can be recognised as a form of process intensification in which the solvent is completely removed or drastically reduced, effectively lowering the score of Process Mass Intensity (PMI) and other green and process-related metrics, getting close to the ideal stoichiometry. The thorough mixing, shearing and/or grinding all contribute to a series of favourable outcomes such as overall faster kinetics, greater surface area of the reactants and reagents, and overcoming of mass transfer limitations such as solubility in certain solvents. For example, Liquid-Assisted Grinding (LAG) both in batch at the laboratory scale and continuously at the kilogram scale can be advantageous to overcome mass transfer of practically insoluble aromatic aldehydes in the aqueous reaction mixture for an exothermic carbonyl reduction (Isoni et al., 2017). When the reduction was performed in a traditional solution-based fed-batch mode, the starting material formed an immiscible solid layer that floated on the surface of the aqueous layer making the controlled addition and evolution of the reaction challenging. When the reaction was conducted in organic solvents that did not react with the reducing reagent or under neat conditions (one of the reactants in liquid state), the mass transfer limitation was observed to be associated with the ionic nature of $NaBH_4$. By switching to high shearing and kneading under LAG conditions, the researchers showed that the reaction could effectively take place in a matter of seconds (Figure 6.1). Besides improved mass transfer and good heat management in continuous reactive extrusion examples, mechanochemical transformations relying on higher energy, pulverisation and increased surface area of the reactants/reagents such as when using a mixer mill have been reported to perform Diels–Alder reaction in the solid state providing an estimated energy output of up to 111 kJ/mol (McKissic et al., 2014). Such an approach can be used to rethink of we look at some well-established industrial chemical processes that are particularly energy intensive and contribute to significant greenhouse gas emissions such as the century-old Haber–Bosch process. In this spirit, an international team has successfully shown that by means of ball milling and the use of the right catalyst, this is more than a mere possibility, synthesising ammonia in a laboratory at ambient pressure and a mild temperature of just 45 °C (Han et al., 2020). Besides energy considerations, these types of mechanochemical transformations allow access to molecular complexity that could not be achieved via traditional solution-based synthetic methods. An example is the unexpected alkyne [2 + 2 + 2 + 2] cycloaddition leading to cyclooctatetraenes obtained under milling conditions in the presence of nickel metal instead of the observed substituted benzenes obtained under solution-based conditions (Haley et al., 2016). Another example is a previously thought inaccessible tert-butyl-substituted adamantoid P4N6-phosphazane. The synthesis by conventional solution-based methods of this molecule was unsuccessfully attempted but proved to be readily accessible under milling conditions (Shi et al., 2016).

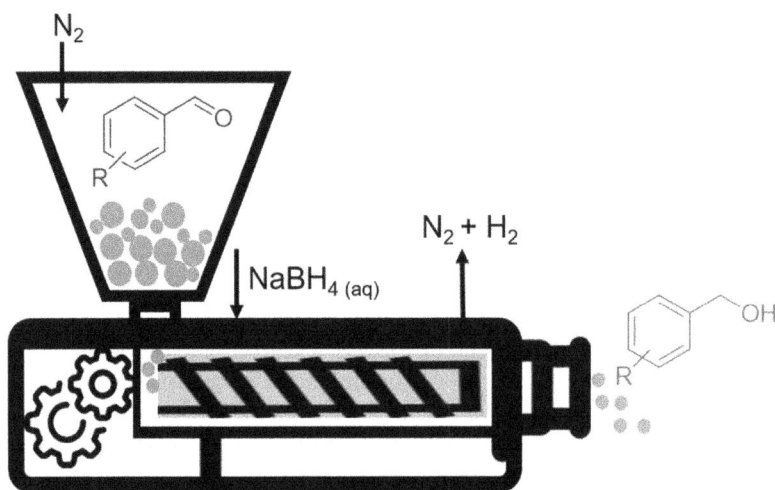

FIGURE 6.1 Example of exothermic large-scale reduction of insoluble aromatic aldehydes via reactive extrusion (based on Isoni et al., 2017).

Like the renewed industrial interest in photochemistry and electrochemistry, one of the most interesting selling points for the adoption of mechanochemistry lies in the opportunities to grow molecular diversity for better properties and provide at the same time a means to generate an intellectual property that is more difficult to be replicated by competitors less versed in non-traditional synthesis.

6.3 Disadvantages

Higher-energy mechanochemical processes can provide access to unique molecular structure, but this can come at the expense of parameters that are important for a confident uptake to an industrial scale such as heat management, material compatibility and metal contamination.

6.3.1 Heat Management

When in use, chemical reactors need to maintain a certain operating temperature in order to accommodate the endothermic/exothermic behaviour of chemical transformation or physical process in progress. Temperature control is important to achieve desired outcomes, such as yield and selectivity but also to avoid possible secondary reactions that can pose safety concerns (Stoessel, 2020). Put simply, for an exothermic reaction as the many examples that exist in the fine chemical industry, the heat generated during the chemical process (q) needs to be matched by a sufficient cooling capacity. The molar enthalpy (ΔH_r) and quantity of reactants (mol) can provide an idea of the energy generated during the reaction. The needed cooling will depend on several factors but most importantly the heat transfer coefficient of the reactor (U), the heat transfer area (A_s) and the difference in temperature (ΔT) between the heat exchanger Tj (e.g. a circulating fluid in the reactor's jacket) and the reaction contents (Tr).

Assuming no losses to the surroundings, under isothermal conditions:

$$q_{reactor} = -q_{reaction} \tag{6.1}$$

therefore, we can rewrite Equation (6.1) as:

$$U \times A_s \times \Delta T = -(\Delta Hr \times mol) \tag{6.2}$$

From Equation (6.2), it is possible to see that as the amount of heat generated increases, the surface area and/or the difference in temperature between the thermal fluid and the reaction should increase accordingly. However, this is not always possible for practical reasons. Similarly to what happens for liquid-phase chemical reactors, the geometry of batch mechanochemical reactors tends to have an unfavourable surface area-to-volume ratio that makes heat removal challenging as the size of the reactor and/or the exothermicity of the reaction increases (Figure 6.2). This can pose safety concerns for a mechanochemical process since the heat removal capacity needs to be sufficient to cope not only with the heat of the reaction but also with the additional heat generated by the mechanical action of the grinding media (e.g. milling balls) to avoid possible decomposition or secondary reactions that can trigger thermal runaways. At the laboratory scale, the quantity of heat generated is proportional to the quantity of reactants (relatively low at the gram scale) and it dissipates rather quickly (in the order of minutes). This allowed a rapid exploration of organic transformations to expand the chemist's toolbox using mechanochemistry (Tan & Friščić, 2017). However, despite ingenious approaches to improve the design of batch or semi-batch mechanochemical reactors to thermally control the outcome of the reaction (Cindro et al., 2019) or the use of cryogenic conditions to minimise the degradation of heat-sensitive materials (Huot et al., 2019), and to date, most examples remain confined to laboratory-scale preparations.

6.3.2 Material Compatibility

The material selection of chemical reactors and equipment is rather complex and it has to take into consideration several factors such as availability, chemical compatibility for the intended use, type of fabrication technology used, thermal properties and cost. In addition to those considerations, there are other

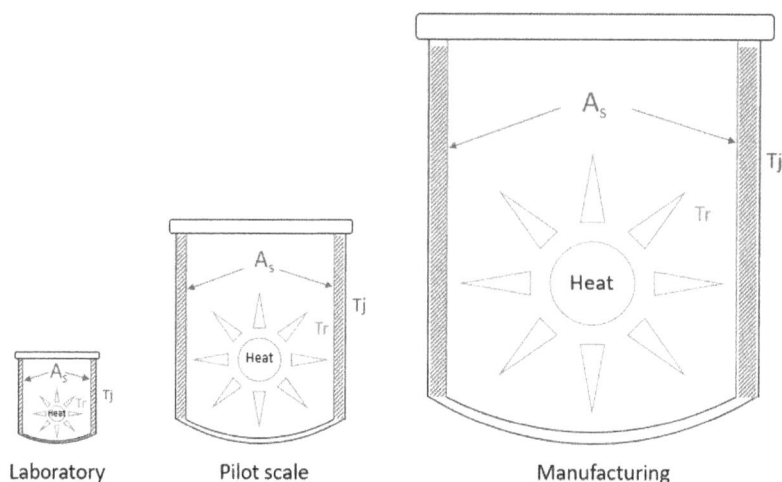

FIGURE 6.2   Unfavourable surface area-to-volume for batch reactors as size increases.

parameters that are specific to the scale and type of application that affect the decision-making process. Glass is arguably the most commonly used material at the laboratory scale followed by some alternative chemically resistant materials such as polytetrafluoroethylene (PTFE), but also ceramics such as Silicon Carbide (SiC), and metals such as stainless steel and higher metal alloys such as Hastelloy C-22. Due to its fragile nature, equipment completely made of glass are rarely used for manufacturing processes. The harder materials with good chemical compatibility are also generally mechanically resistant, but they fetch higher prices due to the more difficult manufacturing processes and the intrinsic higher cost of raw materials. Therefore, an alternative is to use processing equipment made up of a durable material with a reasonable cost and properties (e.g. stainless steel) and treated in order to have the internal parts in direct contact with chemicals lined with a more chemically resistant material such as glass or perfluorinated polymers. This approach is widely adopted in multipurpose batch plants for the manufacturing of both fine chemicals and pharmaceuticals as the best compromise among all the decision factors described above while handling liquid mixtures and slurries.

From a mechanochemistry equipment point of view, desirable properties of the materials such as toughness and hardness are critical to withstand shearing, kneading and/or impact forces. Imaginably, finding a good compromise between cost and mechanical properties is challenging and it gets even harder when considering an added dimension such as chemical compatibility. Materials that possess broader chemical resistance tend to be less mechanically durable (polymers, glass and ceramics vs metals) or much more expensive than commonly used alternatives (stainless steel vs higher alloys, titanium). The choice of material of construction for a mechanochemical process is also important as seen by numerous research papers where both the reactor and the grinding media were tuned to achieve the desired outcome for example using copper (Vogt et al., 2021), palladium (Vogt et al., 2019), zircon dioxide or stainless steel milling balls (Hwang et al., 2022). Because of the challenges highlighted in this section, to date, there is still work to be done to identify a generally accepted material of construction for mechanochemical processes that could be used for manufacturing purposes. Stainless steel is a good option in many cases, but its chemical compatibility *de facto* limits the possibilities to deal with more corrosive chemical environments due to potential damage of the equipment and/or the possible resulting metal contamination of the reaction mixture.

6.3.3 Metal Contamination

The mechanical forces in place in combination with the properties of the mechanochemical equipment can result in the incorporation of metals and/or material of construction into the reaction mixture and ultimately into the final product. This phenomenon has been described and even leveraged to achieve direct mechanocatalysis for cross-coupling reactions by carefully selecting the material for the construction of the milling medium and/or of the reactor (Amrute et al., 2021). However, in line with green chemical engineering principles, the material of constructions should be durable and should not interfere with the matrix of the reactive system to a point where contamination in the final product becomes a challenge to deal with further downstream in the chemical process. Metal contamination could also become an unexpected issue when the chemical compatibility of the material of construction of the mechanochemical equipment has not been fully tested such as in prolonged continuous processes where an incompatible chemical is generated as a by-product.

In addition to accidental contamination, metal catalysts can also be deliberately added to achieve desired reaction outcomes. As a result, those metal-catalysed reactions used in fine chemicals and pharmaceutical processes require stringent work-up and purification steps such as filtration, adsorption and extraction that rely on large volumes of solvents and metal-scavenging options to meet regulatory limits (Burke et al., 2018). In addition to the established solution-based protocols, several metal-catalysed mechanochemical transformations have been reported in recent years (Porcheddu et al., 2020). However, in terms of a comprehensive end-to-end mechanochemical process that starts with the reactive step and includes isolation of the pure product, there is still work to be done. The mostly solid nature of the reaction mixture in a mechanochemical transformation makes it unsuitable for many of the established purification methods that exist to deal with metal contamination. The solid-state palladium catalysed cross-coupling using a mixer mill (Kubota et al., 2019) can be used as a typical example: the reaction step was solventless, but the isolation and purification required column chromatography or recrystallisation in organic solvents.

Unfortunately, to date, metal-catalysed reactions remain an unsolved green chemistry challenge in terms of purification for both solution-based and mechanochemical approaches. It is expected that the situation will remain unchanged until part per million or better, lower levels of catalyst will be sufficient to drive the reaction to completion without the need for purification to really set mechanochemistry apart from solution-based approaches from a process development and scale-up perspective.

6.4 Process Development and Industry Adoption

The route towards commercialisation of new chemical entities from the initial idea stage, be them pharmaceutical intermediates, agrochemicals, monomers, polymers or functional materials, naturally goes through a series of research and development steps that could be loosely summarised in three major phases: (1) Discovery; (2) Process Development and (3) Manufacturing.

While these phases might have different durations (months to years) and/or requirements to meet regulatory demands (e.g. compliance with the Food and Drug Administration for pharmaceuticals) across the chemical industry, they reflect a difference in focus at each stage to enhance and accelerate the success rate towards commercialisation (Table 6.1).

At the discovery phase, the bulk of the effort is concentrated on screening reaction conditions that facilitate the rapid generation of libraries of molecular entities leveraging both well-established protocols as well as exploring newer technologies/protocols to access diverse and molecularly interesting motifs for the target application. At this R&D stage, new promising compounds with desired properties are synthesised in high purity and low quantity – typically up to gram scale for initial performance testing.

At the other end of the R&D spectrum, we find manufacturing processes, where kilograms, tonnes or thousands of tonnes of a said chemical product are produced yearly under rigorous protocols to ensure reproducible quality and performance. In this phase, the ability to sustain the business and be competitive in the market is of utmost importance. Hence, development efforts focus on maximising plant capacity, cleaning and plant turnaround, waste reduction and management, quality assurance and quality control.

Depending on the type subsector of the chemical industry, changes for potential technical improvement at this stage may require significant investments that need support by strong business cases. Even a change of solvent from a manufacturing process in a regulated environment such as pharma is unlikely to happen after pilot scale due to the rigid regulations and costly filing procedures in place. Hence, making

TABLE 6.1

Overview of the Three Main R&D Phases in the Chemical Industry and Their Different Focus towards Accelerated Entry to Market

	Quality	Efficiency	Safety	Environmental	Cost	Time to Market
Discovery	Selectivity, enantiomeric excess, crystallinity, metal content, impurities that compromise performance	Yield, temperature, pressure, synthetic step count	Chemical hazard evaluation	Solvent and reagent guides	Reagents and raw materials	Usually not too sensitive to it, focus on novelty
Process development	Analytical method development, *in situ* monitoring, alternative purification methods to chromatography	Process intensification, synthetic route (re)development, downstream processing, particle engineering	Calorimetry, heat transfer, mixing, exploring intrinsic safety, exposure limits, stability studies, dust explosion	Metrics, waste and emission, reuse and recycling of solvents	Reagents and raw material, energy use, new equipment/technology testing, kilo-lab and pilot plant studies	Technology/chemistry commercially available, speed of development
Manufacturing	Quality assurance, quality control, batch records, cleaning	Plant capacity, chemical inventory, energy integration, cycle time, plant turnaround	Hazard and operability study, procedural safety	Abatement system, waste disposal, fugitive emissions, local regulations	Operational and capital expenses, taxes, land, depreciation of assets, supply chain management	Speed of building a plant, speed of implementation, validation, filing with authorities

significant changes in this R&D phase is far from being a trivial matter and poses a challenge to meet a growing need for more sustainable manufacturing processes.

Multidisciplinary life cycle assessment studies on the manufacturing of APIs highlighted the importance of green principles commonly associated with mechanochemistry such as atom economy and less hazardous chemical synthesis (Isoni et al., 2022). Such an approach towards sustainability decision-making is regarded as more holistic than simply looking at individual metrics, but it can be very labour-intensive and unsuitable for both early discovery and/or process development. To bridge the gap, several quicker-to-use sustainability tools, metrics and pilot-scale demonstrations are available to help researchers design the best process in the shortest time with sustainability in mind, but the available information can quickly become overwhelming (Isoni et al., 2016). Solvent and reagent selection, maximising throughput, material cost, and safety are just a handful of aspects that need to be considered at the process development stage, where chemistry and chemical engineering need to work hand in hand in a relatively short period of time to generate a robust process for the making of kilograms of representative material. When time pressure is combined with high uncertainty of success and a yet-to-be-defined market demand for the product, a pattern tends to emerge: relying on established and well know protocols and technologies at the expense of potentially disruptive innovation.

Similarly to other alternative technologies and promising synthetic processes, mechanochemistry is currently discarded at the early process development phase in favour of more traditional approaches due to a lack of confidence to scale up the whole process at the manufacturing scale. As illustrated in Figure 6.3, the limited available literature show casing large-scale examples (>100 mmol) of synthetic mechanochemistry also contributes to the perpetuation of this vicious circle.

However, upscaling of chemical processes can look significantly different based on the type of operations needed and the scale of production. While bulk and certain speciality chemicals might be produced annually in the order of thousands of tonnes, not all products in the wider chemical industry need to meet such production volumes. Many low-volume high-value chemicals such as potent active pharmaceutical ingredients, catalysts and ligands, and/or custom compounds that cannot be synthesised in a conventional manner would naturally fall under this category (Colacino et al., 2021). A great example was reported by MOF technologies, a company based in the United Kingdom that leveraged the advantages of reactive extrusion for the manufacturing of MOFs for various applications (Casaban et al., 2021). Scaling up and manufacturing of hundreds to thousands of kilograms of such products a year is feasible at laboratory or kilo-lab scale by virtue of the extremely intensified nature of mechanochemistry that does not require solvents or sophisticated ancillary equipment to work. So a better question to ask is: what does it mean to scale up when your process is highly intensified?

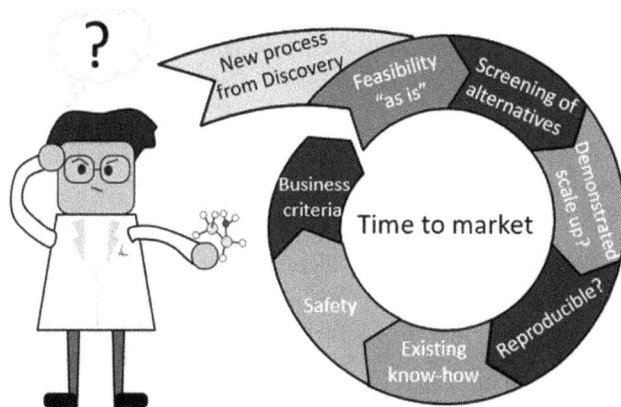

FIGURE 6.3 Technical and non-technical factors that affect technology selection at the process development stage.

6.5 Scaling-Up Does Not Always Mean Getting Larger

Mechanochemistry can be seen as an extreme form of process intensification, where minimal or better, no solvent is used for both the synthesis and ideally the work-up to yield the desired product. Extremely low Process Mass Intensity (PMI) values are usually associated with solventless processes since the solvent accounts for most of the material input that does not get incorporated into the final product. This is only true if no purification or work-up is needed after the reaction step. If a purification such as reslurrying, crystallisation or chromatography is needed, the PMI value of the overall process will be significantly higher and the benefit of going solventless could be questioned from a green chemistry perspective.

However, process intensification is not just about scoring better in terms of metrics, but it is a way to increase throughput by maintaining or significantly reducing the size of the equipment, towards cheaper, safer and more sustainable technologies (Boodhoo & Harvey, 2013). As a chemical process gets intensified, smaller equipment can be used, which means that they can be accommodated in laboratory-sized spaces or enclosed in fume hoods, enhancing the intrinsic safety aspect of having a large throughput in a safer-to-control environment. Similar considerations have been at the basis of the development and adoption of continuous manufacturing in pharmaceutical plants in recent years where compact rigs could be accommodated in walk-in fume hoods or ventilated enclosures without the need for notoriously costly explosion-proof components that are required for traditional manufacturing settings (Kleinebudde et al., 2017).

It is not hard to realise that if increasing the size of batch or semi-continuous mechanochemical reactor is challenging from a traditional scale-up point of view for the previously described reasons, the alternative to achieve higher throughput for manufacturing purposes is to scale out. Going bigger (scale-up) versus numbering up multiple smaller units (scale-out) is a common dilemma during the process development stage and the answer often depends less on technical feasibility and more on business and logistics angles. As shown in a cost analysis for the scale-up of standard unit operations in the chemical industry, the higher cost associated with an increased size of the equipment does not seem to follow a linear trend (Weber et al., 2020). For smaller- to medium-sized equipment, the cost variation vs size was less pronounced than for a significantly larger version of them, meaning that numbering up equipment for production can be more cost effective than scaling up for certain applications with lower throughput. By numbering up a series of batch higher mechanical energy reactors, it is conceivable to achieve low tonnage annual throughput of high-value compounds in a typical kilo-lab setting despite the less than ideal heat management. For reactive extrusion, the situation looks even more promising since the process can be performed in a continuous fashion for an extended period of time to meet demand. For example, a standard 18 mm hot melt extruder is capable of providing a product throughput of a few kilograms per hour. Running the equipment in a semi-continuous manner for a month will yield a few tonnes of material, which might be enough to satisfy the annual worldwide demand for high-value chemicals. Moreover, extrusion processes tend to rely on kneading, compression and conveying more than impact and commercial equipment that can control heat released during the reaction already exist (e.g. twin screw extruders, continuous granulators).

6.6 Beyond Process Intensification

The process intensification aspect of mechanochemistry is certainly attractive from an industrial perspective for manufacturing. The ability to use smaller equipment and/or increasing throughput can provide flexibility and lower entry barriers to new investments in equipment. Beyond process intensification, accessing "impossible" molecular complexity through mechanical activation modes can enable both lower energy processes than existing ones and/or exploring properties that simply could not be reached by traditional synthetic means. Besides the synthetic applications of mechanochemistry, many formulation and particle engineering-related works are available in the literature for the formation of co-crystals or organic salts of active pharmaceutical ingredients. For instance, drug–drug co-crystal research has been receiving attention from the industry in the last decade as the approach offers a low risk, but high-reward route to new and more effective medicines (Wang et al., 2020). An example of mechanochemistry in action in this space is the development of a new hybrid drug candidate for the treatment of malaria obtained via LAG of two antimalarial molecules, mefloquine and artesunate (do Prado et al., 2020).

Beyond providing new ways to access improved medicines, mechanochemistry has been used as a tool to tackle industrially relevant challenges such as the control of crystal forms and particle properties. Traditionally, depending on the type of solvents and processing condition chosen, different crystal forms of the same API can be obtained with resulting different properties that can negatively affect both the manufacturability and performance of the final product. This challenge in particle engineering was tested through the lens of mechanochemistry, showing that under LAG conditions the commonly accepted rule "one liquid for one specific polymorph" was simply incorrect (Hasa et al., 2016).

While drug–drug co-crystals and co-crystals, in general, are gaining momentum, most small-molecule APIs currently exist either in their free forms or salt forms. APIs can undergo salt formation reactions in a move to improve their biological and physicochemical properties or to overcome the non-desirable particle properties of free-form drugs, which may limit the API manufacturability during drug product processing. For instance, it has been observed that particle surfaces of API salt forms have a tendency to adsorb high levels of water resulting in improved electrostatic discharging upon mechanical processing (Karner & Anne Urbanetz, 2011). To understand the extent of how common this practice is in pharma, a study reported that approximately 50% of the approvals by the US FDA consist of APIs in their salt forms (Paulekuhn et al., 2007). Despite being less explored than the co-crystals, mechanochemistry has been reported as a valuable tool for medicinal chemists to form API salt forms with little to no solvent (Tan et al., 2016). From an industrial point of view, the salt formation step can be seen as the final synthetic step of a small-molecule API (primary manufacturing) and the beginning of a process to make the drug product (secondary manufacturing). In a global world, supply chains can be complex and it is common that an API is manufactured on one side of the world, while the corresponding drug product is made on the other side of the globe (Srai et al., 2015). Transportation between different manufacturing sites can occur in the order of weeks and the supply chain can be disrupted in both the short and long term in the event of changes in local regulations, demand or geopolitical reasons as seen for the COVID-19 pandemic (Ayati et al., 2020). For a more self-sufficient and resilient production of pharmaceuticals, the push for true end to end continuous manufacturing is clear despite the limited publications for this important multidisciplinary scientific topic (Domokos et al., 2021). Even the regulators recognised the need for a more modern pharmaceutical industry by embracing continuous manufacturing (FDA, 2019).

To bridge the gap between primary and secondary manufacturing of small-molecule drugs, mechanochemistry could play the intermediary role. Currently, primary manufacturing is predominantly by batch processing and solvent-laden steps such as purification and crystallisation. On the other hand, solid handling in a continuous manner is quite common for the production of drug products. While a multidisciplinary effort would be needed for the successful integration of primary and secondary manufacturing, it is already possible to envision a series of small continuous mechanochemical reactors that in the near future can perform synthesis and API salt formation with minimal to no solvent. Building on an example of API synthesis via twin screw extrusion (Crawford et al., 2020), in Figure 6.4 we can see how the muscle relaxer dantrolene sodium could be manufactured via two consecutive steps (API synthesis and salt formation) leveraging mechanochemistry and existing laboratory-size equipment. Besides the green credentials associated with an intensified process, integrating such a compact API facility into an established continuous secondary manufacturing line would be able to provide flexibility to changing market demands and better address supply chain challenges.

6.7 Conclusion

Mechanochemistry represents a mind shift in the way chemistry can be developed into more sustainable processes, accessing structurally intriguing molecules and as a way to bridge the gap between primary and secondary manufacturing in regulated environments such as pharma. The current challenges

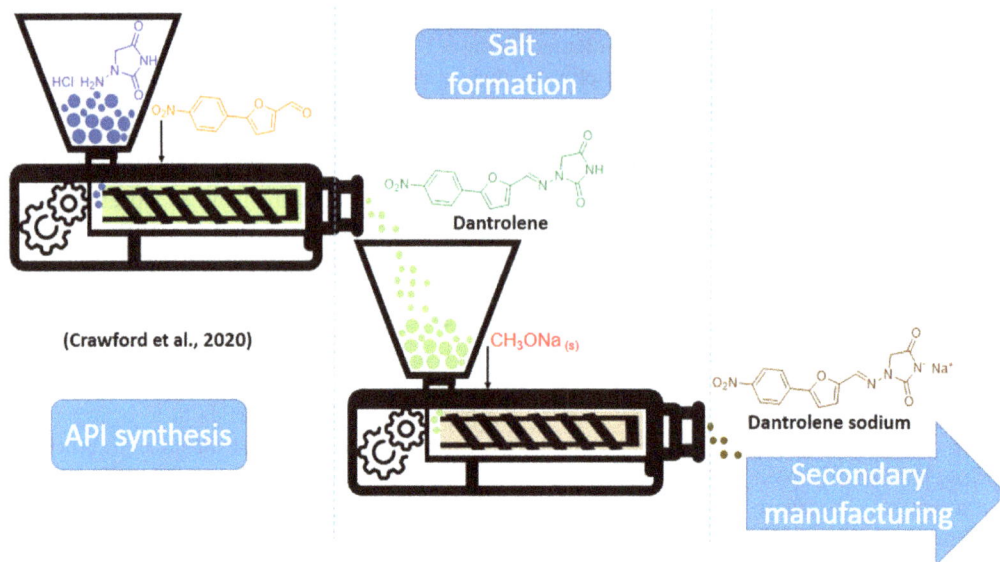

FIGURE 6.4   Reactor principles of different gasification processes (based on Crawford et al., 2020).

associated with traditional scale-up such as material compatibility, heat management and contamination might limit the opportunities in the short term for batch mechanochemistry to low-volume, high-value chemicals or the synthesis of small molecules in the gas phase. Another option is to bypass scale-up by going continuous. Reactive extrusion using continuous laboratory devices such as twin screw extruders can be leveraged to synthesise complex molecules, co-crystals, APIs and their salt forms with the possibility to integrate with secondary manufacturing.

Similarly to continuous/flow technology, mechanochemistry is still evolving and beyond the technological hurdles, economic, financial, organisational and educational barriers might be an even bigger barrier to adoption in the industry (Guenther & Fritzsche, 2013). The way forward is through close collaboration among manufacturers and academia to develop affordable equipment that can be independently tested and are fit for purpose, as well as large-scale demonstrations and transparency about lessons learned to bring awareness and build confidence that the process will work in a manufacturing setting.

REFERENCES

Amrute, A. P., De Bellis, J., Felderhoff, M., & Schüth, F. (2021). Mechanochemical synthesis of catalytic materials. *Chemistry – A European Journal*, *27*(23), 6819–6847. https://doi.org/10.1002/chem.202004583

Ayati, N., Saiyarsarai, P., & Nikfar, S. (2020). Short and long term impacts of COVID-19 on the pharmaceutical sector. *DARU: Journal of Pharmaceutical Sciences*, *28*(2), 799–805. https://doi.org/10.1007/s40199-020-00358-5

Boodhoo, K., & Harvey, A. (2013). *Process intensification technologies for green chemistry: Engineering solutions for sustainable chemical processing*. Wiley.

Burke, A. J., Marques, C. S., Turner, N. J., & Hermann, G. (2018). *Chapter 2. Active pharmaceutical ingredients in synthesis catalytic processes in research and development*. Wiley-VCH Verlag GmbH & Co. KG.

Casaban, J., Zhang, Y., Pacheco, R., Coney, C., Holmes, C., Sutherland, E., Hamill, C., Breen, J., James, S. L., Tufano, D., Wong, D., Stavrakakis, E., Annath, H., & Moore, A. (2021). Towards MOFs' mass market adoption: MOF technologies' efficient and versatile one-step extrusion of shaped MOFs directly from raw materials. *Faraday Discussions*, *231*, 312–325. https://doi.org/10.1039/d1fd00025j

Colacino, E., Isoni, V., Crawford, D., & García, F. (2021). Upscaling mechanochemistry: Challenges and opportunities for sustainable industry. *Trends in Chemistry*, *3*(5), 335–339. https://doi.org/10.1016/j.trechm.2026.02.008

Crawford, D. E., Porcheddu, A., McCalmont, A. S., Delogu, F., James, S. L., & Colacino, E. (2020). Solvent-free, continuous synthesis of hydrazone-based active pharmaceutical ingredients by twin-screw extrusion. *ACS Sustainable Chemistry & Engineering*, 8(32), 12230–12238. https://doi.org/10.1021/acssuschemeng.0c03816

do Prado, V. M., de Queiroz, T. B., Sá, P. M., Seiceira, R. C., Boechat, N., & Ferreira, F. F. (2020). Mechanochemistry for the production of a hybrid salt used in the treatment of malaria. *Green Chemistry*, 22(1), 54–66. https://doi.org/10.1039/c9gc02478f

Do, J.-L., & Friščić, T. (2016). Mechanochemistry: A force of synthesis. *ACS Central Science*, 3(1), 13–19. https://doi.org/10.1021/acscentsci.6b00277

Domokos, A., Nagy, B., Szilágyi, B., Marosi, G., & Nagy, Z. K. (2021). Integrated continuous pharmaceutical technologies—A review. *Organic Process Research & Development*, 25(4), 721–739. https://doi.org/10.1021/acs.oprd.0c00504

FDA. (2019). *FDA statement on FDA's modern approach to advanced pharmaceutical manufacturing.* U.S. Food and Drug Administration. Retrieved March 1, 2022, from https://www.fda.gov/news-events/press-announcements/fda-statement-fdas-modern-approach-advanced-pharmaceutical-manufacturing?platform=hootsuite

Gomollón-Bel, F. (2019). Ten chemical innovations that will change our world: IUPAC identifies emerging technologies in chemistry with potential to make our planet more sustainable. *Chemistry International*, 41(2), 12–17. https://doi.org/10.1515/ci-2019-0203

Guenther, A., & Fritzsche, E. M. (2013). Barriers to innovation challenges in the implementation of green chemistry and engineering. *CHEManager*, 12. https://www.chemanager-online.com/en/restricted-files/173508

Haley, R. A., Zellner, A. R., Krause, J. A., Guan, H., & Mack, J. (2016). Nickel catalysis in a high speed ball mill: A recyclable mechanochemical method for producing substituted cyclooctatetraene compounds. *ACS Sustainable Chemistry & Engineering*, 4(5), 2464–2469. https://doi.org/10.1021/acssuschemeng.6b00363

Han, G.-F., Li, F., Chen, Z.-W., Coppex, C., Kim, S.-J., Noh, H.-J., Fu, Z., Lu, Y., Singh, C. V., Siahrostami, S., Jiang, Q., & Baek, J.-B. (2020). Mechanochemistry for ammonia synthesis under mild conditions. *Nature Nanotechnology*, 16(3), 325–330. https://doi.org/10.1038/s41565-020-00809-9

Hasa, D., Miniussi, E., & Jones, W. (2016). Mechanochemical synthesis of multicomponent crystals: One liquid for one polymorph? A myth to dispel. *Crystal Growth & Design*, 16(8), 4582–4588. https://doi.org/10.1021/acs.cgd.6b00682

Huot, J., Cuevas, F., Deledda, S., Edalati, K., Filinchuk, Y., Grosdidier, T., Hauback, B. C., Heere, M., Jensen, T. R., Latroche, M., & Sartori, S. (2019). Mechanochemistry of metal hydrides: Recent advances. *Materials*, 12(17), 2778. https://doi.org/10.3390/ma12172778

Hwang, S., Grätz, S., & Borchardt, L. (2022). A guide to direct mechanocatalysis. *Chemical Communications*, 58(11), 1661–1676. https://doi.org/10.1039/d1cc05697b

Isoni, V., Khoo, H. H., & Ee, A. W. (2022). Green principles in active pharmaceutical ingredient manufacturing as seen through the lens of life cycle assessment. *Life Cycle Assessment*, 77–89. https://doi.org/10.1142/9789811245800_0005

Isoni, V., Mendoza, K., Lim, E., & Teoh, S. K. (2017). Screwing NaBH$_4$ through a barrel without a bang: A kneaded alternative to fed-batch carbonyl reductions. *Organic Process Research & Development*, 21(7), 992–1002. https://doi.org/10.1021/acs.oprd.7b00107

Isoni, V., Wong, L. L., Khoo, H. H., Halim, I., & Sharratt, P. (2016). Q-SA√ESS: A methodology to help solvent selection for pharmaceutical manufacture at the early process development stage. *Green Chemistry*, 18(24), 6564–6572. https://doi.org/10.1039/c6gc02440h

Karner, S., & Anne Urbanetz, N. (2011). The impact of electrostatic charge in pharmaceutical powders with specific focus on inhalation-powders. *Journal of Aerosol Science*, 42(6), 428–445. https://doi.org/10.1016/j.jaerosci.2016.02.010

Kaupp, G., Schmeyers, J., Naimi-Jamal, M. R., Zoz, H., & Ren, H. (2002). Reactive milling with the Simoloyer®: Environmentally benign quantitative reactions without solvents and wastes. *Chemical Engineering Science*, 57(5), 763–765. https://doi.org/10.1016/s0009-2509(01)00430-4

Kleinebudde, P., Khinast, J., & Rantanen, J. (2017). *Continuous manufacturing of pharmaceuticals.* Wiley.

Kubota, K., Seo, T., Koide, K., Hasegawa, Y., & Ito, H. (2019). Olefin-accelerated solid-state C–N cross-coupling reactions using mechanochemistry. *Nature Communications*, 10(1). https://doi.org/10.1038/s41467-018-08017-9

McKissic, K. S., Caruso, J. T., Blair, R. G., & Mack, J. (2014). Comparison of shaking versus baking: Further understanding the energetics of a mechanochemical reaction. *Green Chemistry*, *16*(3), 1628. https://doi .org/10.1039/c3gc41496e

Paulekuhn, G. S., Dressman, J. B., & Saal, C. (2007). Trends in active pharmaceutical ingredient salt selection based on analysis of the orange book database. *Journal of Medicinal Chemistry*, *50*(26), 6665–6672. https://doi.org/10.1021/jm701032y

Porcheddu, A., Colacino, E., De Luca, L., & Delogu, F. (2020). Metal-mediated and metal-catalyzed reactions under mechanochemical conditions. *ACS Catalysis*, *10*(15), 8344–8394. https://doi.org/10.1021/acscatal .0c00142

Shi, Y. X., Xu, K., Clegg, J. K., Ganguly, R., Hirao, H., Friščić, T., & García, F. (2016). The first synthesis of the sterically encumbered adamantoid phosphazane $p_4(N'Bu)_6$: Enabled by mechanochemistry. *Angewandte Chemie International Edition*, *55*(41), 12736–12740. https://doi.org/10.1002/anie.201605936

Srai, J. S., Harrington, T., Alinaghian, L., & Phillips, M. (2015). Evaluating the potential for the continuous processing of pharmaceutical products—A supply network perspective. *Chemical Engineering & Processing: Process Intensification*, *97*, 248–258. https://doi.org/10.1016/j.cep.2015.07.018

Stoessel, F. (2020). *Thermal safety of chemical processes: Risk assessment and process design*. Wiley-VCH.

Takacs, L. (2000). Quicksilver from cinnabar: The first documented mechanochemical reaction? *JOM*, *52*(1), 12–13. https://doi.org/10.1007/s11837-000-0106-0

Tan, D., & Friščić, T. (2017). Mechanochemistry for organic chemists: An update. *European Journal of Organic Chemistry*, *2018*(1), 18–33. https://doi.org/10.1002/ejoc.201700961

Tan, D., Loots, L., & Friščić, T. (2016). Towards medicinal mechanochemistry: Evolution of milling from pharmaceutical solid form screening to the synthesis of active pharmaceutical ingredients (APIs). *Chemical Communications*, *52*(50), 7760–7786. https://doi.org/10.1039/c6cc02015a

Vogt, C. G., Grätz, S., Lukin, S., Halasz, I., Etter, M., Evans, J. D., & Borchardt, L. (2019). Direct mechanocatalysis: Palladium as milling media and catalyst in the mechanochemical Suzuki polymerization. *Angewandte Chemie International Edition*, *58*(52), 18942–18947. https://doi.org/10.1002/anie .201911356

Vogt, C. G., Oltermann, M., Pickhardt, W., Grätz, S., & Borchardt, L. (2021). Bronze age of direct mechanocatalysis: How alloyed milling materials advance coupling in ball mills. *Advanced Energy & Sustainability Research*, *2*(5), 2100016. https://doi.org/10.1002/aesr.202100011

Wang, X., Du, S., Zhang, R., Jia, X., Yang, T., & Zhang, X. (2020). Drug-drug cocrystals: Opportunities and challenges. *Asian Journal of Pharmaceutical Sciences*. https://doi.org/10.1016/j.ajps.2020.06.004

Weber, R. S., Askander, J. A., & Barclay, J. A. (2020). The scaling economics of small unit operations. *Journal of Advanced Manufacturing & Processing*, *3*(1). https://doi.org/10.1002/amp2.10074

7

On the Theory and Recent Developments in "Batch Mechanochemical Synthesis – Scale-Up"

Steffen Reichle and Michael Felderhoff

CONTENTS

Mechanochemistry has undeniably become a well-established branch of chemistry and is studied and applied worldwide alongside classical thermochemistry as well as other alternative methods for inducing chemical reactions like photochemistry, electrochemistry, plasma chemistry or sonochemistry. The simplicity with which mechanochemistry can be realized – the easiest scenario is short milling of different reactants – is only one reason for its widespread use: The partly different reactivity (selectivity) in mechanochemical reactions but especially the fact that an enormous number of different substances (soft or hard matter, organic or inorganic) in the solid, liquid or gaseous state can be the object of mechanochemical investigations leads to various academic and industrial groups dealing with mechanochemistry. For industry, in particular, the Green Chemistry aspect of mechanochemistry can be of great interest with regard to the transition towards a more sustainable and resource-conserving mode of production. The reduction of solvents, the shortening of reaction times and synthesis stages, and the elimination of the need to heat chemical reactors should be mentioned as motivating factors. However, for a successful transfer from the laboratory to an applied process, any mechanochemical reaction must be superior to alternative methods, i.e., most of its advantages have to be retained even when scaled up from the laboratory. In academia, small-scale experiments naturally dominate and scaling up a mechanochemical reaction is not always the subject of research. In order to make this transfer to real-world application happen, more and more scientists are now also looking at synthesizing larger amounts of product through mechanochemistry. Here, a distinction can be made between two process modes, continuous and batch.

For continuously operated mechanochemical reactions, a higher product volume can be achieved simply by adjusting the run time or through parallelization of identical reactors. The great advantage here is that no further adaptions of the process parameters are necessary, and upscaling can be realized quickly

DOI: 10.1201/9781003178187-9

and easily in this way. However, the investigation of new mechanochemical reactions very often takes place in batch reactors, and, depending on the required energy input or material properties, a continuous mode may be less feasible or successful than performing the reaction in batch. Accordingly, it is also necessary to have the option of upscaling for mechanochemical reactions in batch. Simply enlarging the dimensions of the grinding bowl or mill can be technically demanding or can greatly increase the necessary energy requirement, so batch systems usually require a different type of mill as well as an adjustment of the various parameters (size and shape of grinding media, frequency etc.), which can cause time-consuming investigations. A simple numbering up of the small dimensions is not possible for a technical process because of the strongly increasing workload. In the following, an overview is given about the scale-up of mechanochemical systems in batch, first discussing potential milling devices for batch reactions and their working principles and characteristics, followed by recent research findings in this area.

7.1 Mills for Batch Processes and Their Scale-Up

In earlier times, people had a simple mortar and pestle at their disposal for grinding materials, and water- and windmills for larger quantities of products like grain, ores or paint pigments.

Today, a mechanochemist can rely on various automated grinding systems with a wide range of equipment, where grinding parameters such as frequency, time, temperature or the grinding medium and atmosphere can be easily adjusted. Motorized drives also allow higher energy densities, so that mechanochemical reactions can be easily realized in addition to the mere comminution of the material to be ground. Since, in addition to the magnitude, the type of mechanical stress (compression, shear, impact) also has a great influence on the type of chemistry, and it is of great advantage that different types of mills exist with different modes of operation. In addition to the pure utilization of gravity as in drum mills, the grinding media and the material to be ground can also be actively accelerated as in vibration, planetary or attrition mills. Especially in industry, further different types of mills exist for particle size reduction like hammer mills, jet mills or colloid mills. However, since these are always operated continuously, they will not be discussed here. For a detailed discussion of these mills, the reader is referred to a comprehensive review.[1]

For an effective mechanical stressing of the material and for force transmission, grinding media are added to the batch operation mills. On the laboratory scale, these are usually balls, and other shapes like rods or cylpebs can also be used on a pilot plant or technical scale.[2] Cylpebs are cylinders with equal diameter and length, which makes them one and a half times heavier compared to spheres of the same diameter. For this reason, they are preferentially used for fine grinding in technical applications. To have a first simple comparison between the different types of mills and larger dimensions, it helps to consult some simple key figures: The maximum energy that can be transferred to the material in the case of a completely elastic impact corresponds to the kinetic energy E of the grinding media with $E = 0.5mv^2$ (m and v being mass and velocity of the grinding object, respectively).

For a uniformly accelerated motion, the velocity depends on the square root of the acceleration a and the free path length, which can be taken as a first approximation with the grinding chamber diameter D as the maximum value, and, therefore, $v = \sqrt{2aD}$. Accordingly, high accelerations, large grinding chambers and heavy grinding media are to be targeted. Due to the complex movements of the grinding media in the various mills, which depend on the filling degree and other factors, these formulas should only be understood as a highly simplified starting point for comparison.

7.1.1 Mortar Grinder

Due to its working method based on the principle of manual grinding by means of mortar and pestle, the mortar grinder shall be briefly introduced here. The material is filled into the mortar grinder and is ground between a pestle and a rotating grinding bowl. Parameters that can be adjusted are the variable contact pressure of the pestle and the rotation frequency of the bowl. For the Pulverisette 2 of the Fritsch

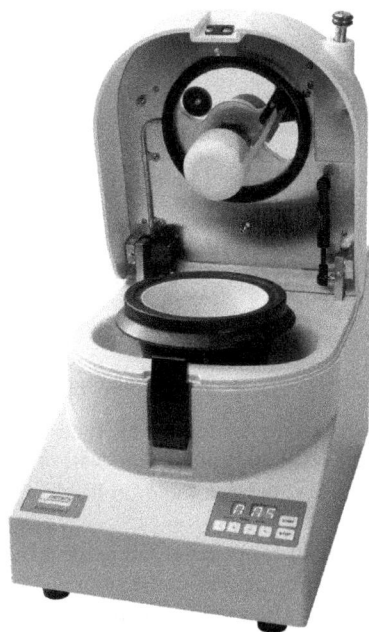

FIGURE 7.1   Pulverisette 2 mortar grinder by Fritsch GmbH. (With kind permission of Fritsch GmbH: www.fritsch.de.)

GmbH for example, it is possible to choose between a frequency of 70 or 80 rpm.[3] An efficient comminution is ensured by an out-of-center positioned pestle and a blade attached to the side, which mixes the powder and returns it to the pestle. Both the bowl and the pestle are available in different materials and the maximum material quantities to be ground are in the range of 100 g. Since some mechanochemical reactions also occur at mild conditions, the mortar grinder could potentially be suited for certain upscaling experiments in the laboratory.[4] In principle, however, it is only used for comminution and homogenization of solid samples or suspensions in the laboratory due to its low-energy input. Since more efficient technical mills exist for this task, the mortar grinder will remain confined to the laboratory (Figure 7.1).

7.1.2 Drum Mills

Drum mills are used industrially for more than 100 years for particle size reduction of large amounts of material. The fundamental working principle is simple: A drum that can be rotated freely around its horizontal axis, which is filled with milling media and the material to be ground. Depending on the rotational speed, several regimes can be distinguished. At low speeds, the cascade regime is established in which the grinding media, having been lifted by the rotation of the drum, roll back down over the grinding media located underneath. Accordingly, frictional forces predominate in this regime. If the grinding media receives more kinetic energy due to higher speeds, they lift off and parabolically fly back into the drum. Therefore, in this so-called cataract regime, impact forces dominate. If the centrifugal force on the grinding media corresponds to the gravitational force, the grinding media gets stuck to the wall of the drum and grinding no longer takes place. Figure 7.2 shows the three regimes schematically.

The critical centrifugal regime is established at a critical rotational speed of $n = 42.3/\sqrt{D}$. This, therefore, limits the rotational speeds which can be meaningfully applied on a technical scale. For industrial drum mills with volumes of up to several 100 m^3 having diameters in the upper meter range, this results in maximum rotational speeds of 10–30 rpm. For laboratory-scale drum mills with length dimensions in the range of 1 meter and below, rotational speeds of up to 100 rpm are possible. The TM500 by Retsch GmbH for example allows a maximum speed of 50 rpm.[5] For the most effective comminution of the material to be ground, in practical applications the rotational speed would have to be set in such a way that one works in a transitional range between the cascade and the cataract regime and the grinding

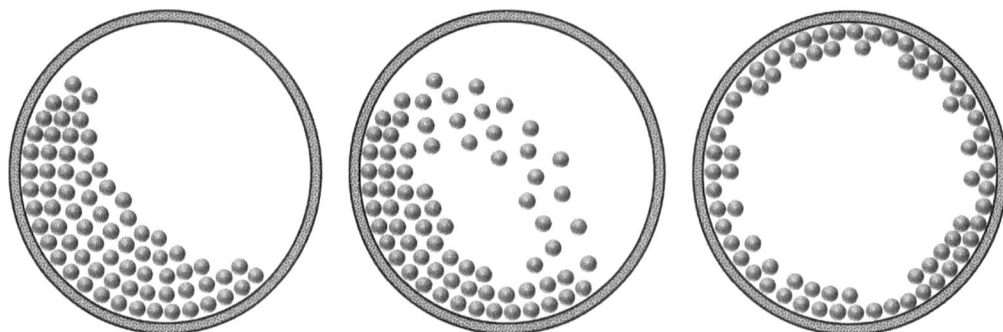

FIGURE 7.2   Cascade (left), cataract (middle) and centrifugal regime (right) in a drum mill.

media partly lift off but fall back onto the grinding media/material mixture, making use of both frictional and impact forces.[6] Attaching walking beams to the wall of the drum, which carry the grinding media along during rotation, can further increase the drop height of the milling media and thus their kinetic energy.

In industry, drum mills having a length-to-diameter ratio of less than 2 are usually operated in batch mode and are designed to be smaller, with volumes in the order of 10 m^3, unlike pipe mills with an L/D ratio greater than 2, which are designed for continuous operation.[7] For crushing and comminution operations, the latter reach throughputs of several 100 t/h. Pipe mills can be built as single-chamber mills or compartmented as multi-chamber mills. Compartmentalization allows the mill to be divided into sections with different functionalities, e.g. by performing only coarse pre-grinding in the first chamber and fine grinding in the subsequent chambers. The material to be ground can pass through the intermediate walls by sieves or classifiers, while the grinding media remain in the respective chambers.

Industrially, drum mills are well established for the comminution of materials like pigments, ores and quartz. However, for mechanochemical reactions they are severely limited, because the high kinetic energies of the milling media can only be realized by large free pathways and/or heavy weights, because the acceleration of the milling media is only determined by gravity and no additional acceleration takes place. Especially on the laboratory scale, therefore no mechanochemical reactions are possible that require high-energy impact in a short time. For bigger dimensions, a strong increase in reaction time can be expected despite the higher kinetic energies attainable due to the low rotational speeds. In addition to the fact that operation in such huge drum mills requires reactant quantities on a ton scale, this becomes one of the reasons for the lack of studies of mechanochemical reactions other than pure particle size reduction operations in industrial-size drum mills. However, reactions seem conceivable that proceed with low energy input or in which the special rolling motion of the grinding media is advantageous (binding and breaking of coordinative bonds, delamination of layered materials). The latter has been successfully demonstrated, for example, for the exfoliation of graphite by using rods in a laboratory-scale drum mill.[8]

7.1.3 Vibration Mills

In vibration mills, the grinding vessel with the grinding media and material inside is actively moved back and forth, with the grinding vessel performing horizontal, vertical, elliptical, circular, arcuate, 8-shaped motions or combinations thereof. In this energetic movement, the material is stressed mainly by impact, but also friction. High accelerations enable high-energy inputs, which is why vibratory mills are perfectly suited for carrying out mechanochemical reactions.

For each mill, in addition to the form of motion, the achievable accelerations depend on the amplitude and the variable frequency. For lab-scale vibration mills typical values of the amplitude range from a few millimeters up to several centimeters and frequencies up to 30 Hz and beyond are possible. This results in peak accelerations of several 10 *g* (an italic *g* is used throughout the text when referring to g-forces to avoid confusion with the SI unit for mass) with milling media velocities in the range of 1 m/s, for

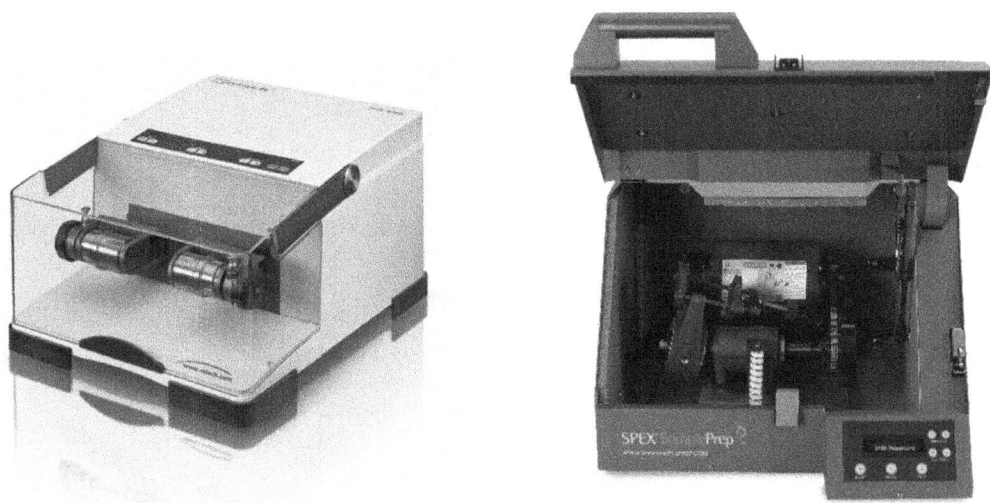

FIGURE 7.3   MM400 (Retsch GmbH) and SPEX 8000M (SPEX SamplePrep, LLC). (With kind permission of the respective companies: www.retsch.de and www.spexsampleprep.com.)

example, 36 *g* for the mixer mill MM400 from Retsch GmbH (performing an arcuate movement with an amplitude A ≈ 1 cm at a frequency of 30 Hz and $a_{max} = A\omega^2$) or more than 40 *g* for the SPEX 8000 mixer mill.[9, 10] Both mills are shown in Figure 7.3.

Due to the high kinetic energies present in vibration mills, they are perfectly suited for lab-scale mechanochemistry. Depending on the material's density, amounts on the order of 10 g per jar can be processed. A wide range of different jar designs and materials (steel, Al_2O_3, ZrO_2, WC, PTFE etc.) are available. A direct scale-up of the dimensions of such shaker mills is technically not feasible due to the direct change of motion, which would lead to severe stresses on the motor and the mechanics. A simple way to increase the material amount a little at least for laboratory purposes is the possibility of numbering up the number of jars used per sample holder: For example, the MM500 vario from Retsch GmbH can hold up to six milling jars, as shown in Figure 7.4.[11] Larger amounts on a multi-gram scale are therefore easy to synthesize. Nonetheless, this principle will be limited to the lab scale, as discussed earlier.

Industrial vibration mills do not perform a shaking motion, they are large tubular grinding chambers mounted on springs (steel, rubber) which are excited to circular vibrations by a motorized drive. The motor is flexibly connected to an unbalanced weight mounted horizontally around the mill's center of gravity, forcing the weight to rotate and causing the entire structure to vibrate. It can reach rotational speeds of up to 3000 rpm, achieving vibration radii of up to 10 mm and more. Accelerations that can be achieved are in the range of 10 *g*. The material to be ground and the grinding media are circulated in the opposite direction to the direction of rotation of the drive, with the grinding media being partially spun through the drum. The material to be ground is thus subjected to impact, friction and shear.

The pipe diameter of these vibratory mills is in the order of several 10 cm, and it must not be too large to ensure effective momentum propagation in the bulk, which becomes less energetic as the distance from the wall increases. Accordingly no larger pipes are used to increase the capacity of the vibratory pipe mills, but several pipes are arranged around the central axis of rotation. Each of the pipes can have a volume of up to 2000 L. Such multi-tube vibratory pipe mills include up to six tubes, with arrangements of two or three tubes being the most common. Figure 7.5 shows a vibratory pipe mill with two tubes.

In the case of multi-tube vibratory pipe mills, the individual tubes can also be interconnected, allowing compartmentalization, in which, as in the case of drum mills, each grinding chamber can be filled with different grinding media as required. Vibratory pipe mills are preferably used in continuous mode and achieve throughputs of several tons per hour, but they are equally suitable for experiments in batch. In order to avoid a strong rise in temperature, the pipes can be designed as double-shell vessels so that cooling can be provided by means of cooling water during the grinding process. For special applications, pipes with ceramic or polymer coating are available in addition to the standard steel pipes.

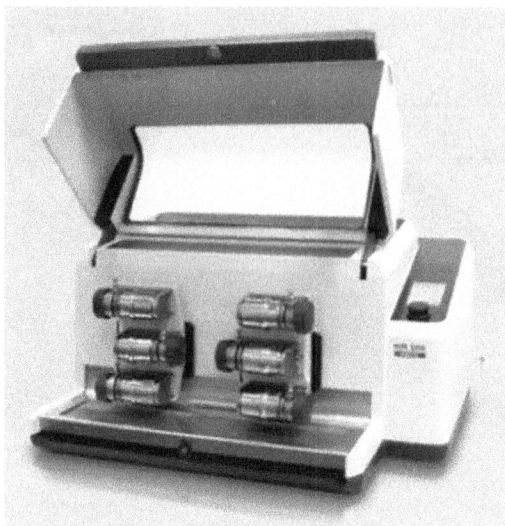

FIGURE 7.4   MM500 vario (Retsch GmbH). (With kind permission of Retsch GmbH: www.retsch.de.)

FIGURE 7.5   Vibratory pipe mill Palla™ 65U with two tubes. (Reproduced without further adaptions from K. Andres, F. Haude, *J. S. Afr. Inst. Min. Metall.* **2010**, *110*, 125-131.)

A further development of the vibratory pipe mill is the so-called eccentric vibratory mill, which emerged from a cooperation between the Technical University Clausthal and the machine manufacturing company Siebtechnik GmbH.[12] In this case, the drive is not located in the center of gravity of the mill, but an unbalanced weight is placed directly to the side of the grinding chamber and driven by a motor. A counterweight is flanged to the opposite side of the chamber to compensate for the imbalance. A schematic diagram of the eccentric vibratory mill and a commercially available mill are shown in Figure 7.6. Volumes of the grinding chambers range from around 10 L to almost 1000 L.

Due to its special design, the mill does not perform circular oscillating movements but inhomogeneous vibrations consisting of elliptical, circular and linear oscillations. This allows amplitudes of up to 20 mm, resulting in a sharp increase in grinding media acceleration and high kinetic energies.

It also allows the maximum tube diameter to be increased up to 1 m, surpassing that of classical vibratory pipe mills. According to the manufacturer, this results in up to a doubling of throughput with energy consumption reduced by half. The possibility of cooling due to the double-shell design, different materials as lining and the modular connection of several chambers remains. They can also be operated continuously or in batch mode.

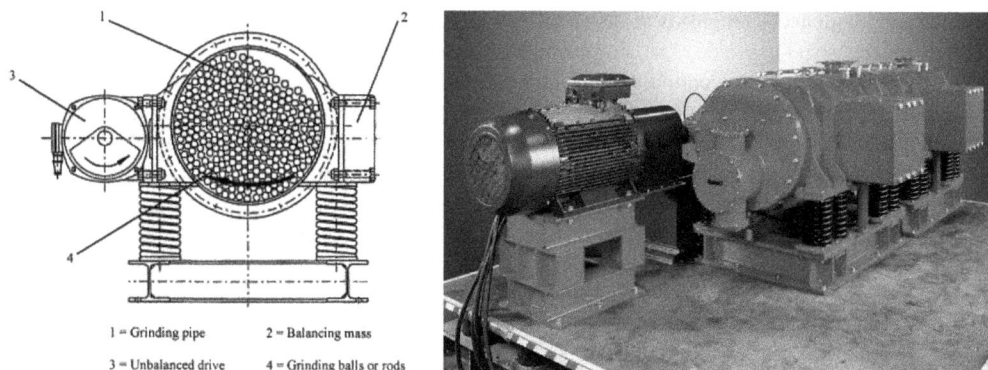

| 1 = Grinding pipe | 2 = Balancing mass |
| 3 = Unbalanced drive | 4 = Grinding balls or rods |

FIGURE 7.6 Diagram of an eccentric vibratory mill[12] and image of the ESM 856-2ks (Siebtechnik GmbH). (With kind permission of Siebtechnik GmbH: www.siebtechnik-tema.de.)

Another advantage due to the inhomogeneous vibration is that the main wear zone in the grinding chamber is larger than in the vibratory pipe mills and extends to the lower 180° instead of 90°, which has a positive effect on the service life of a chamber.[13] Instead of a counterweight, a second unbalanced weight can also be used, enabling two vibration radii to be switched between depending on the desired forces in the mill (e.g. preferably friction or impact).[14] Due to all these properties, the mill is excellently suited for mechanochemical reactions and their upscaling and is one of the mill types for which this has already been tested many times.

7.1.4 Planetary Ball Mills

Planetary ball mills are based on another special operating principle: One or more grinding bowls are located on a rotating so-called sun disk and rotate around their own axis, this direction of rotation being opposite to that of the sun disk. This special movement causes centrifugal acceleration relative to the axes of the sun disk and the grinding vessel as well as Coriolis force, which are mutually superimposed. Depending on the selected rotational speeds, different regimes of the grinding media movement are created, similar to the regimes observed in drum mills.

In the sliding regime, the grinding media is pressed against the inner wall of the bowl and moves in the direction of the rotary motion of the bowl to a maximum point and then rolls off over the grinding media packing. In this regime, the stress on the material takes place almost exclusively by friction. An increase in acceleration results in the throwing regime, with the grinding media no longer rolling over the remaining bulk, but being thrown to the opposite side. In addition to friction and shearing, impact forces are also transmitted to the material in this regime. With a further increase in acceleration, turbulent motion sets in before the centrifugal force becomes so strong that the grinding media are pressed firmly against the wall (circle regime) and no milling of the material takes place. The speed of the sun disk is often the only parameter in a planetary ball mill that can be varied by the experimenter, since the transmission ratio between the sun disk and the grinding bowls is usually fixed (most often roughly 1:2). Typical values for the rotational speed of the sun disk are in the range of up to 1000 rpm, sometimes even higher. However, there are also mills where the sun disk and the grinding bowls have individual drives and the respective speeds can be varied separately. Depending on their size, enormous accelerations of up to 100 g can be achieved in laboratory-scale mills (for larger devices rather than 20–30 g), which is why they are excellently suited for mechanochemical reactions. Depending on the choice of the mill, grinding bowls from a few ml volume up to 500 mL can be selected, making it easy to carry out reactions on a scale of several grams even on a laboratory scale. The high-energy density in planetary mills can lead to a strong temperature increase in the grinding bowl, which is why the mills usually have air cooling. In addition, pauses between the individual runs can also be set in advance.

In the past, it was assumed that planetary mills were hardly suitable for upscaling due to their high-energy density, as this would place too great a demand on cooling and especially the drive.[15]

Even if such high accelerations as in the small-dimension laboratory mills are not achieved, planetary ball mills for industrial usage do exist. For example, the Russian company TTD St. Petersburg produces, among other mills, horizontal continuous planetary ball mills for crushing operations with grinding chamber volumes of up to 300 L (four such chambers per mill).[16] Another variant has eight grinding chambers, each with a volume of 50 L. These mills have a feed spout through which the material flows into a distribution chamber and is directed outward into the individual milling drums by centrifugal forces. Even with such dimensions, accelerations of up to 35 *g* can be achieved with these mills. Throughputs of several tons per hour are possible for crushing operations. It remains to be seen whether it can also be used for batch reactions, as problems, especially with thermally sensitive materials, are to be expected. Other large-scale planetary ball mills are manufactured and distributed by the Changsha Tianchuang Powder Technology Company.[17] Both horizontal and vertical mills with up to 100 L total volume (four jars with 25 L) are available for batch operation. Even though all these mills are intended to be used for particle size reduction, scaling up mechanochemical reactions in planetary ball mills nonetheless seems quite realistic, although this must always be assessed on a case-by-case basis and is certainly not possible for every type of material.

7.1.5 Attrition Mills

In vibration and planetary mills, the grinding chamber is always moved in addition to the grinding media and the material. Attrition mills, on the other hand, consist of a stationary grinding vessel in which the material to be ground is actively moved by a rotor, so that no energy has to be expended for the movement of the grinding vessel.

The rotor consists of an axle with several blades, which can have different geometries – disks, optionally perforated with different shapes, rods, propellers or screws are common. These blades set the grinding media and the material to be ground in motion during rotation. In addition to acceleration in the tangential direction, the grinding media is also moved axially and radially, which also ensures good mixing.[18] Due to the gradients in the velocity profiles, the grinding media exerts compressive forces that lead to effective comminution of the material. In addition to compression, impact and frictional forces also prevail in attrition mills. With rotational speeds in the range of 1500 rpm and rotational velocities of 30 m/s and more, large kinetic energy inputs can be achieved in attrition mills, which is why they are well suited for mechanochemical applications in addition to pure size reduction tasks. A double-shell design prevents excessive temperature rise with the aid of a cooling liquid. At the same time, with grinding chamber volumes of up to 1000 L, they are ideally suited for large-scale batch experiments up to the 100 kg range. A typical attrition mill can be seen in Figure 7.7.

Attrition mills can not only be operated batch-wise, but continuous process operation is also possible. An impressive example of this is the Isamill, a gigantic horizontal disk attrition mill in which continuous fine grinding (of slurries) can be carried out using beads as grinding media.[19] The largest version of the Isamill includes a drum volume of 50,000 L and can process over 1000 m^3/h of material.

When using powders in vertically arranged attrition mills, process-related difficulties can arise: Accumulation of powder at the bottom of the grinding drum can lead to problems during the start or, by making mixing more difficult, prevent uniform mechanical stressing of the entire material to be ground. For this reason, horizontal attrition mills have been extensively studied for dry grinding. Commercial devices are built and marketed under the name Simoloyer® by the Zoz GmbH.[20]

These mills are available for high-energy grinding applications both for laboratory scale with grinding chambers in the range of 1 L with rotational speeds up to 1800 rpm and for industrial applications with up to 900 L and 325 rpm. Cooling of the grinding chamber is also possible here. Figure 7.8 shows the Simoloyer® CM100 as an example.

The productivity in batch operation for Simoloyer® mills is in the range of 100 kg/h – with all working steps (charging, discharging) already being included. Continuous operation is also possible with the Simoloyer®. Furthermore, various materials are also available as coatings or linings for the grinding chamber (Al_2O_3, WC). When other materials are selected, the rotor with its blades is still made of steel for stability reasons, but the ends of the blades can be provided with ceramic tips or the entire rotor can be plasma-coated. However, for this, potential wear can pose a problem for the stability and durability of a technical process.

FIGURE 7.7 SD-50 attrition mill (Union Process, Inc.). (With kind permission of Union Process, Inc: www.unionprocess.com.)

FIGURE 7.8 Simoloyer® CM100 (Zoz GmbH). (With kind permission of Zoz GmbH: www.gmbh.zoz.de.)

7.2 Inorganic Chemistry

Investigations into the upscaling of mechanochemical reactions towards the development of industrial processes already go back many decades. The increased reactivity, which is caused by a grinding process in addition to the pure particle size reduction, has been exploited for large-scale processing of inorganic substances since the middle of the 20th century. Baláž gives a very good overview of previous, but also current developments in applied mechanochemistry in his summaries.[21, 22] Examples from the field of inorganic chemistry include the treatment of minerals, e.g. improving the flotation properties of sulfidic and oxidic ores, increasing the solubility of phosphates for subsequent use as fertilizers, or improving the extraction properties of various ores. Furthermore, examples of waste treatment (e.g. degradation of asbestos) and cement production can be found. Since the focus was on applications in mechanochemistry, there are also many continuous processes discussed in addition to batch processes in drum, vibration and attrition mills. In the following, only additional new developments in scaling up batch reactions will be discussed.

7.2.1 Particle Size Reduction

Some investigations into batch reactions, mainly aiming at particle size reduction, were carried out by Zoz GmbH using the horizontal attrition mills (Simoloyer®). The production of enameled components requires fine enamel powders, which are applied as a slurry to a material made of metal or glass and subsequently burnt. Enamels are conventionally produced in drum mills, which requires long processing times due to the low kinetic energy. For this reason, grinding media made of steel cannot be used and the grinding chambers must be lined with ceramic materials to avoid impurities that would show up in the burnt enamel as unwanted color changes.

It was shown that the high-energy input in a Simoloyer® reduced the enamel processing time to less than 2 minutes.[23] As a proof of concept, 200 g enamel chunks from various industrial manufacturers were ground in a Simoloyer® CM01 with a 2 L grinding chamber at 1500 rpm, reducing the average particle size from the millimeter range down to a few micrometers. However, the problem with the batch mode was that the influence of the discharging time of the grinding chamber – which must, without external assistance, take place while the mill is running in order to transport the powder out – became significant due to the short processing time. If the duration of the discharging process is of the same order of magnitude as the grinding process itself, the material will also change during discharging and the powder that comes out of the grinding chamber last will be different from the fraction that was discharged at the beginning, rendering the production of a homogeneous product impossible. The batch process here, therefore, served as the experimental basis for the development of a (semi)-continuous process, in which the powder is actively conveyed out of the chamber by means of gas flow and separated by a cyclone filter. The chosen method used here, referred to as depression mode, is implemented by applying a vacuum to the cyclone filter. The results of batch screening were successfully reproduced and d_{50} values below 30 µm were obtained after only 2.5 min of processing, while in a comparison with a drum mill, particles larger than 1 mm were predominantly obtained even after 5 h. The short grinding time with high-energy input also allowed processing in a steel chamber as well as the use of steel balls, since the short contact reduced impurities to a minimum and thus optically no (at 1.5 min grinding time) or hardly any (after 2.5 min) changes occurred in the enamel coatings.

The Simoloyer® has also been successfully used to produce metal flakes. Flakes of Al, Ag, Cu, brass, Au, Ti, Fe and Zn could be produced in a Simoloyer® CM01 with a 2.0 L grinding chamber.[24] For this purpose, generally 200 g of a mixture of the appropriate metal powder and 0.5 wt.-% stearic acid as process control agent (to suppress agglomeration) were ground under air or argon for about 30 min on average. This treatment resulted in the production of flakes which had a diameter-to-thickness ratio of generally more than 100, with lower values only obtained for Ti and Fe due to their comparatively high hardness. This process is thus a suitable alternative to the production of metal flakes by means of stamping machines. Further upscaling with a productivity of more than 1 ton per day was possible by using the Simoloyer® CM100 equipped with a 100 L grinding chamber.[25] In this case, 20 kg of a mixture of copper and 0.5 wt.-% stearic acid were milled at 430 rpm. Flakes of sufficient quality were obtained after only 3 min. Together with the time for discharge, this resulted in a productivity of 60 kg/h. Here, too, the batch test served as the basis for the development of a semi-continuous process in which productivity could be increased by a further 33 %.

7.2.2 Cementitious Materials

Classic Portland cement is produced by burning a mixture of predominantly lime together with clay and sand or other aggregates. During this process, various calcium silicates are formed, which form calcium silicate hydrate phases during the subsequent setting of this so-called cement clinker, which interlock with each other and thus give the cement its strength. The cement industry is extremely energy-intensive, consuming up to 2 % of the global energy consumption.[26] Most of this energy is required as thermal energy for the firing process, and the remaining electrical energy is mainly needed to grind the whole starting materials.[27]

Here, optimized mechanochemical processes could not only improve the energy consumption of grinding but, in the case of mechanochemical activation, also shorten the subsequent firing process, which could result in energy and cost reductions. In addition to the mechanochemical treatment of classic cement clinker, this method also opens up the possibility of increasing the content of aggregates whose reactivity would otherwise be too slow or of introducing higher proportions of waste materials from other processes such as fly ash or granulated blast furnace slag without loss of quality. In this way, the material properties of the resulting cement (setting time, strength) can also be influenced, and CO_2 emissions can be reduced through the lower consumption of Portland cement. Correspondingly, mechanochemical batch processes are ideally suited for producing larger quantities of novel cement formulations and investigating their properties.

For example, the changes in the grinding of kaolin clay, a potential aggregate, were studied in a conventional drum mill.[28] Here, 5 kg of kaolin clay was milled per batch between 10 and 1200 min at 46 rpm. In addition to the reduction of the mean particle size from 23 µm to just under 6 µm, a strong change in chemical reactivity was also observed. Kaolin clay, which mainly consists of kaolinite and quartz, shows dehydroxylation behavior upon heating in addition to the loss of absorbed water. This dehydroxylation, which partially already occurred during the grinding process, also started at lower temperatures in the ground products, the longer the processing time was (reduction from 503 °C to 466 °C). In addition, the reactive silica content also increased from 18 % to 33 % (after 1200 min) and was thus higher than the silica content in kaolinite, suggesting concomitant reactions with the quartz present in the clay. The suitability of the ground kaolin clay as a cementitious material was demonstrated by the production of a mortar that exhibited an almost fivefold higher compressive strength when using the 1200 min ground kaolin compared to the unground material. Similar results, but after much shorter milling times, were found when a planetary mill was used.[29] A total of 1.04 kg (260 g per 500 ml steel jar) of proclay kaolinite were ground for 1, 3 or 9 h at 400 rpm. In addition to a very rapid crystallite size reduction and increase of the strain, it could be shown by infrared and thermal analyses that the dehydroxylation progressively takes place with longer grinding times up to completion (no thermally induced dehydroxylation detected after 9 h grinding).

Cement can also be directly mechanochemically activated. Generally, an optimum balance would have to be found for the grinding process with regard to an industrial process, since no significant improvements in the properties of the cement can be expected if the stress is too low, and if the stress is too high, a disproportionately large amount of energy has to be expended with hardly any improvement in the properties. Therefore, the influence of the rotational speed was investigated in an attrition mill.[30] Portland cement was ground in an attritor (grinding chamber and rotating disks made of Al_2O_3) at a scale of about 100 g. Up to a speed of 5 m/s, the physicochemical changes were primarily particle size reduction effects towards a mean size of 6 µm. A further increase above this speed did not lead to any further reduction of the mean size – instead, predominantly mechanochemical transformations then set in (amorphization), which became visible on the basis of the weakly pronounced reflections of the calcium silicate hydrate phases in the X-ray diffraction pattern. No measurements of the setting time or compressive strength were carried out, so unfortunately no statements can be made about an optimum rotational speed in this attritor. Nevertheless, the attritor showed a clear energy efficiency compared to a laboratory-scale drum mill. Studies on the grinding of cement in attritors can also be found for the horizontal Simoloyer®.[31] In a Simoloyer® CM01 with a 2 L grinding chamber and Si_3N_4 lining and blades, up to 115 g of Portland cement were ground at 900 and 500 rpm (alternating 4 and 1 min) for 1 h. In this study as well, a strong degradation of the crystalline phases was shown. Due to the high kinetic energy in the Simoloyer®, the average particle size was reduced to 2 µm (from 30 to 60 µm without grinding) and a strongly homogeneous material was obtained, with only a very small proportion of particles of different sizes compared to the strongly heterogeneous starting material. This fineness of the material reduced the setting time to just 2–3 min. The compressive strength of the cement produced in this way was almost twice as high after just 1 day as that of the unprocessed cement after 1 month. However, it was also found that small cracks appeared in the grinding chamber lining after prolonged use of the Simoloyer®, so further improvements would be necessary with regard to this.

Another intensively investigated possibility to produce new cementitious materials is the addition of various aggregates or waste materials to reduce the content of Portland cement in the final product. Ball milling helps to obtain cement of high quality despite these otherwise low-reactive aggregates. For example, the content of Portland cement could be reduced to 45 %.[32] A further 45 % blast furnace slag and either 10 % silica fume or a reactive silica-based admixture were used as aggregates. Here, the grinding process was carried out in a drum mill with a 5 kg batch for 10, 20 and 30 min each. Within this short time, the particles were crushed to an average size of up to 8 μm and, especially for the mixture with the reactive silica admixture, comparable values of setting time as well as compressive strength could be achieved with respect to the pure Portland cement as reference. At the optimum of 20 min grinding, the compressive strength could even be increased by 62 % at an age of 28 days. Even the use of high proportions (50 %) of quartz or fly ash together with Portland cement resulted in high-quality cement after a short processing time.[33] This was impressively demonstrated by comparing the mechanochemically treated materials with the untreated blends. In each case, 2 kg of material was ground for 30 min in a vibratory pipe mill at 150 rpm. Compared to the pure blends, a strong hydration reaction was observed for both aggregates (quartz or fly ash) immediately after the addition of water. This was reflected in the increased compressive strengths of the cement produced, which from the first day on were significantly higher than those of the untreated blends (after 28 days almost four times as high for quartz and one and a half times for fly ash). This resulted from both the increased reactivity of the ground products and the finer microstructure of the mixtures after the grinding process, where the mean particle sizes were in the range of reference cement (pure Portland cement or rapid cement), resulting in a homogeneous microstructure even after setting. The low content of Portland cement in such mechanochemically modified cement offers great potential for energy savings in the burning process. For a content of 50 % fly ash and 50 % Portland cement, an energy consumption of only 54 % was calculated compared to the production of conventional cement.[14] Accordingly, such types of cement were already produced on a large scale and used for construction works (e.g. paving), with several stages in vibratory or attrition mills being used to process the materials.[14] For example, CemPozz, a cementitious material with a high fly ash content, was marketed and used in Texas.[34, 35] Zoz GmbH also sells a Portland cement reduced by high admixtures of blast furnace slag that has been mechanochemically activated under the name FuturZement.[36]

Furthermore, studies on the scale-up of completely Portland cement-free, alkali-activated cementitious materials were also made.[37] Here, laboratory-scale production was first carried out on a 1 kg scale in a drum mill (volume 5.6 L).[38] The cement produced had a higher compressive strength compared to Portland cement and can be made entirely without a burning process from fly ash, blast furnace slag, albite, sodium silicate and hydroxide, calcium oxide and borax. The scale-up was performed in a steel drum mill with 225 L volume, where the gas atmosphere consisted of flue gas in order to store additional CO_2 in the cementitious material. Tests were carried out from 30 min to 4 h grinding time. Already after 30 min a significant reduction of the particle size was found, which after 4 h showed an average value of 6.5 μm. Infrared spectroscopy showed an uptake of CO_2, which was as high as 3 % after 4 h. However, the uptake in the form of carbonates, which increased with grinding, also resulted in the compressive strength of the cement prepared with this material being highest for the product ground for only 30 min. The reason was the impairment of hydration reactions due to the lower alkalinity of the longer-milled products. For the production of a total of 630 kg of this cementitious material for the construction of a sidewalk, the grinding time was therefore shortened to 18 min and batches of 10 kg were produced in the pilot plant.

Novel formulations of cementitious materials have also already been tested in the eccentric vibrating mill. Two kilograms of a 1:1 mixture of quartz and hydrothermally synthesized calcium silicate hydrate were ground for 4 h and the product was then treated at various temperatures up to 500 °C, which is almost 1000 °C lower than in classical cement production.[39] Calorimetric measurements after the addition of water showed high hydraulic reactivity in the cementitious materials, which in some cases were significantly higher than for known calcium silicates. Grinding and subsequent thermal treatment also increased the degree of silicate polymerization to up to 27 % dimeric species. However, no properties of set cement made from these materials were investigated.

The examples discussed impressively demonstrate the many ways in which mechanochemistry can assist in the production of novel cement formulations. The additional examples of industrial usage that have already taken place also underline the potential of mechanochemical batch processes, although it remains unknown to what extent these materials and processes will be used in larger projects.

7.2.3 Mechanical Alloying and Activation

Mechanical alloying describes a mechanochemical process in which new, homogeneous composite materials are obtained by treating blends of powdered starting materials. It originated in 1966 in the preparation of oxide-strengthened nickel alloys and has since developed into an industrial method in which materials are produced on a ton scale.[40] Since past results and historical developments have been discussed extensively elsewhere, only more recent examples will be given here. For an excellent overview with many examples, the reader is referred to the review by Suryanarayana.[40]

Hüller et. al. investigated the mechanical alloying of an aluminum alloy with NiTi or Al_2O_3 (10 vol.-% each) in different batch planetary mills with acceleration factors of 6.5 (Pulverisette 6, Fritsch), 28 and 50 (MPP-1 and MPP-1-1, both ttd-spb).[41] The last two mills were intended to serve as a reference for the continuously operating large-scale planetary ball mills with similar acceleration factors, in order to get an indication of the potential upscaling in those. Initial trials of grinding the pure aluminum alloy showed that in the mill with a g-factor of 28, there was a significant particle size reduction to an average size of around 7.5 µm in a very short time (30 min) accompanied by a strong degree of amorphization compared to the mill with a g-factor of 6.5, in which instead there was an increase in the average particle size due to cold welding. Also in the planetary mill with a g-factor of 50, significant comminution was seen after 20 min, but to a lower extent than with a g-factor of 28. The subsequent mechanical alloying tests with (pre-milled) starting materials were therefore carried out for a maximum of 15 min in the MPP-1 planetary mill. This showed that alloying could be carried out successfully: Strong reduction of the crystallite size as observed via X-ray diffraction as well as electron micrographs of the Al + NiTi compound showed that the formation of the nanocrystalline metal matrix composite was possible in this planetary mill in a very short time, so that further scaling up to the continuously operating planetary mills also appears possible. Unfortunately, no exact data on the quantities used were provided, but due to the size of the planetary mills, a range of around 100 g can be assumed.

Similar quantities were also used in studies by Zoz GmbH on mechanical alloying in Simoloyer® attrition mills. For a 100 g blend of Ti and Ni powders in a Simoloyer® CM01 with 0.5 L grinding chamber under argon, grinding was performed at different frequencies (1500 and 900 rpm, 4 and 1 min, respectively) to suppress the adhesion of the material to the grinding chamber and rotor due to its ductility.[42] This procedure is referred to by the authors as cycle operation and is intended to prevent cold welding or attachment of the powder on the machine components by different proportions of deformation, fracture and welding in the corresponding grinding regimes and thus represents an alternative to the use of a process control agent. Grinding was carried out in several intervals of up to 10 h. It could be shown that still in the first 6 h an enlargement of the particles up to 400 µm took place and only gradually decreased in the further course of the grinding process to approx. 20 µm, which indicated a homogenization of the phases. Partial, but not complete, amorphization could be detected by X-ray diffraction. A major problem that could not be completely solved by the cycle operation remained the adhesion of the powder to machine components, so that only 70 % of the material could be recovered. Similar results were obtained in the mechanical alloying of blends of Ti and a pre-alloyed Al–Nb compound.[43] First, a total of 130 g powder was dry-milled in a Simoloyer® CM01 with a 0.5 L grinding chamber at 1300 rpm under argon. Progressive alloying could be followed by powder diffraction patterns, with complete amorphization taking place after 15 h. Since no cycle operation was carried out and the yield was very small, a further test was carried out with alternating rotation frequencies (1300 and 900 rpm) for 15 h. As a result, 80 % of the product was subsequently recovered during discharging. A larger variant of the Simoloyer® was successfully used to produce oxide dispersion-strengthened ferritic alloys.[44] In a CM20 Simoloyer® with a 20 L grinding chamber, 3 kg of a mixture of the ferritic alloy and Y_2O_3 powder were ground under argon. Again, a variation of the rotor frequency (200 and 450 rpm in 30 s intervals) was applied and milling was performed for 4, 20 and 40 h. Electron micrographs showed an inhomogeneous distribution of the oxide inclusions for a grinding time of 4 h, changing to a homogeneous distribution at 40 h. Correspondingly, an increase in yield strength with increasing grinding time was found, which was explained not only by the oxidic inclusions but also by further changes in the microstructure (increased grain boundary density, dislocation hardening).

In addition to the synthesis of materials by mechanical alloying, materials or mixtures can also be activated for subsequent steps. For example, in the synthesis of Si_3N_4 ceramics, it was found that mechanochemical pretreatment of silicon with NH_4Cl for a period of 8 or 12 h, within which the degree of amorphization and formation of nanocrystalline silicon increased, led to a decrease in N_2 pressure in the subsequent combustion synthesis to Si_3N_4.[45] Unfortunately, no exact information was provided on the mill or the amount of material used, but 400 g of material was used in the ignition, so a larger batch approach to milling can be assumed.

In a Simoloyer®, the mechanical treatment allowed electric arc furnace dust to be activated for subsequent extraction.[46] First, 200 g of the dust was pretreated for 20 min at 1300 rpm in a Simoloyer® CM01 with a 2 L grinding chamber, and the subsequent extraction was also carried out in the mill after the addition of a leaching solution – but in a vertical position. Due to the mechanical treatment during the leaching process, the leaching duration could be reduced to 10 min. All leachable Zn and Pb species could be extracted, while the non-leachable residues of zinc ferrite and magnetite could be easily separated with hydrochloric acid due to the previous mechanical treatment. Compared to the untreated dust, where complete leaching did not occur even after more than 1 h, there was thus a clear time advantage for the mechanochemical process.

The disintegration of mineral materials was also performed in an eccentric vibration mill.[47] In a batch of 400 g, a laterite ore was ground with sulfur for 1 h and then digested with sulfuric acid. The formation of $NiSO_4$ and additional sulfidic phases significantly increased the rate of leaching of Ni, Co and Fe compared to pure mechanochemical activation without sulfur.[48] Simoloyer® mills have furthermore been used to pre-treat and activate potential hydrogen storage materials, even on a kilogram scale.[49] Three kilograms of FeTiMn alloy were treated under argon in a Simoloyer® CM100 with 100 L grinding chamber for only 10 min. The aim was to make the otherwise high temperatures required to activate the material unnecessary by the mechanochemical pretreatment. In fact, the treatment was shown to produce even stronger reactivity (faster kinetics) and 50 % higher hydrogen capacity in the material than conventional thermal activation. This was attributed to the strong change in microstructure (particle and crystallite size reduction, increased number of grain boundaries and defects). However, the high-energy input of the Simoloyer® had a detrimental effect on the amount of powder recovered. Of the 6 kg used, only 2 kg of material could be recovered, because the powder was strongly attached to the grinding chamber and the balls. This shows once again that there is still a need for optimization of dry large-scale mechanochemical processes.

The scaled-up production of Mg-based hydrides for solid-state hydrogen storage was demonstrated in quantities of 0.5 up to 1 kg of material.[50] In this process a batch of 500 g MgH_2 and a Zr–Ni alloy as catalysts (weight ratio 9:1) was ball-milled in an attritor-type mill for 116 h. The whole milling and handling of the material were done under inert gas conditions to protect the material from oxidation. X-ray diffraction patterns after milling indicate the prevalence of β-MgH_2 with a significant peak broadening. This indicates reduced particle size and/or crystal disordering as a result of the long milling times. The storable hydrogen amount reaches values between 5.3 and 5.6 wt.%. Hydrogenation and dehydrogenation times at 300 °C are in the order of a few minutes, which is comparable to milled MgH_2 materials in lab-scale quantities. The first step in developing and optimizing mechanochemical processes of industrial size is to conduct corresponding experiments with smaller quantities of materials on a laboratory scale. Subsequently, the implementation in the industrial production process takes place based on these results. One example is the mechanochemical processing of MgH_2 for hydrogen storage.[51, 52] A systematic study of milling MgH_2 with different transition metals as catalysts was performed as a first step. The experiments were done in a planetary ball mill (ball-to-powder ratio 50:1) with MgH_2 and transition metals in different ratios for up to 20 h of milling. The behavior of hydrogen uptake and release and cycle stability were tested for these samples. Based on these results, the McPhy company prepared for several years commercial quantities of ball-milled MgH_2 for the storage of hydrogen. MgH_2 with additives was nanostructured through ball milling at an industrial scale using attritor mills for a few hours under inert gas conditions. The activated MgH_2 was afterward mixed with 10 wt.% expanded natural graphite and compacted under high pressure to disks. Expanded graphite increases the thermal conductivity of the material for rapid heat release and heat input during the hydrogenation and dehydrogenation processes.[53] In these disks, the MgH_2 is protected from oxidation and these can then be stacked to larger units under ambient conditions. Additionally, graphite facilitates the ball-milling process by functioning as a process control agent, reducing powder loss by adhering to the reactor walls, as demonstrated by milling 500 g of MgH_2 with 1 % graphite.[54]

Complex aluminum hydrides, f.i. $NaAlH_4$ or Na_3AlH_6, are among the most advanced and intensely investigated solid-state hydrogen storage materials in the last 20 years. $NaAlH_4$ can reach a storage capacity of more than 5 wt.% at around 100 °C. By suitable doping with catalysts, the re- and de-hydrogenation kinetics can be dramatically improved so that re-hydrogenation times on the order of minutes can be achieved. Mechanochemical processes are often used for this activation process and subsequently several methods for the production of activated $NaAlH_4$ in kg-scales were developed. During mechanochemical treatment, $NaAlH_4$ reacts with the dopant $TiCl_3$ according to Equation (9.1) to produce finely divided small Ti particles that act as catalysts in hydrogenation and dehydrogenation processes.

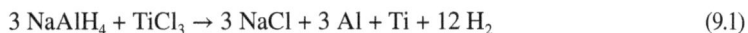

$$3\ NaAlH_4 + TiCl_3 \rightarrow 3\ NaCl + 3\ Al + Ti + 12\ H_2 \qquad (9.1)$$

The preparation of 8.8 kg of the $NaAlH_4$ storage material was carried out by ball milling of $NaAlH_4$/$TiCl_3$ mixtures (2.5–3.0 mol% Ti) in 2.5–3.0 kg batches.[55] Milling was performed in an ESM 234 vibratory mill from Siebtechnik (Mülheim an der Ruhr). The cylindric milling vessel (19 L) was filled with 55.5 kg milling balls, each ball of 12 mm, 6.5 g. The resulting ball-to-powder ratio was 20:1. The evolved hydrogen gas during the milling was measured. The hydrogen storage capacity of the non-optimized material was 3.0–3.3 wt.% H_2.

The synthesis of activated $NaAlH_4$ can also be carried out starting from the decomposition products NaH and Al metal with subsequent hydrogenation Equation (9.2).

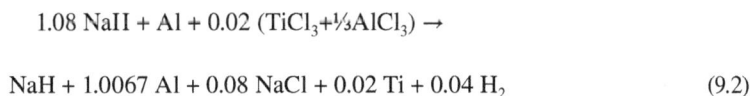

$$1.08\ NaII + Al + 0.02\ (TiCl_3 + \tfrac{1}{3}AlCl_3) \rightarrow$$

$$NaH + 1.0067\ Al + 0.08\ NaCl + 0.02\ Ti + 0.04\ H_2 \qquad (9.2)$$

This was demonstrated up to 1 kg batches for the synthesis of 8 kg of material.[56] Milling was carried out under inert gas conditions using a modified ESM 236 vibratory tube mill from Siebtechnik (Mülheim an der Ruhr, Germany) with a ball-to-powder ratio of 140:1. In order to avoid agglomeration, 100 ml of cyclohexane was added. After 2 h of milling the material was dried through evacuation of the mill. In a similar way an amount of 21.5 kg of Ti-activated $NaAlH_4$ hydrogen storage material was synthesized starting from NaH, Al and $TiCl_3$.[57] After the activation process, the material was mixed with expanded natural graphite for better dispersion and heat conductivity. Several runs with 125 g batches were run in a Fritsch P5 mill at 280 rpm for 4 h (4 cycles a 1 h per cycle). The material was used for a hydrogen storage tank demonstration project. For another demonstration project, an amount of 1.9 kg of Ti-activated Na_3AlH_6 was synthesized by ball milling.[58] The starting mixture was $NaAlH_4$ + 2NaH, and $TiCl_3$, Al metal and activated carbon were added. In several runs of 25 g each, the total amount of material was prepared under mechanochemical conditions in an inert gas atmosphere. Grinding was performed with a planetary ball mill (Fritsch Pulversisette 5) using 500 ml stainless steel grinding jars and grinding balls (diameter 20 mm) at 4.71 s^{-1}, with a ball-to-powder ratio of 32:1.

7.2.4 Reactive Milling for the Synthesis of Materials

Many studies concerning the synthesis of larger amounts of functional inorganic materials were done with the cooperation of the Slovakian Institute of Geotechnics and Eberhard Gock using the eccentric vibration mill developed by him. Here, a main distinction can be made between the synthesis of binary compounds from their elements and ternary or quaternary compounds from the elements or other precursors. The aim has always been to produce important compound classes such as semiconductors for potential applications as thermoelectric, photovoltaics or optoelectronics. Mechanochemistry is particularly interesting with regard to the synthesis of semiconductors, since the crystalline size-dependent band gap in mechanochemically produced products can have different values than in thermally produced products. The batch sizes in most cases were 100 g of starting material, which was ground in an eccentric vibration mill ESM 654 usually under argon in a 5 L grinding chamber made of steel or WC. The synthesis of PbSe took place after only 20 min with a conversion of almost 100 %.[59, 60] A comparison with a laboratory mill (Pulverisette 6) showed that this high conversion required only 10 min longer in the scaled-up experiment. In an analogous study for the synthesis of ZnSe, it was found that the degree of conversion was lower (approx. 80 % after 60 min for both the industrial mill and the laboratory mill).[61] However, optimization of the milling conditions to potentially quantitative conversion was not undertaken. In contrast, the synthesis of CuSe was almost quantitatively achieved after 5 min.[62] With direct band gaps of 1.6 eV for the product produced in a laboratory planetary mill and 1.8 eV for the material from the eccentric vibration mill, the values were blue-shifted compared to bulk CuSe, but several tenths of eV red-shifted compared to alternative synthesis methods and were thus in an optimal range for potential solar cell materials. The fact that the eccentric vibration mill can also be used successfully for larger preparations than 100 g was demonstrated in the synthesis of CuS.[63] In initial investigations with 300 g batches, it became clear that the synthesis from the elements takes place in several stages: A very fast reaction leading to a CuS/Cu_2S composite followed by further consumption of unreacted sulfur leading to a phase composition mainly consisting of CuS with less than 1 % free sulfur after 40 min. This synthesis could also be reproduced in an even larger batch of 7.5 kg in a 50 L milling chamber (60 min milling time) to demonstrate the possibility of large-scale operation. The synthesis of ternary sulfidic compounds was also tested in the eccentric vibration mill. The synthesis of Cu_2SnS_3 from CuS and SnS was carried out with a conversion of over 85 % after 240 min, with a small proportion of impurity phases (e.g. unreacted CuS).[64] A comparison with a laboratory planetary mill shows that this allowed a comparable conversion after about 20 min and an almost complete conversion after 1 h. Cu_3SbS_4 was successfully produced from the elements, also requiring a slightly longer processing time of 180 min compared to 120 min in the laboratory mill for complete conversion.[65] In addition, contamination of WC by abrasion was evident. A detailed study of the reaction between Cu, Sb and S, in this study milled in a steel grinding chamber, showed the consecutive mechanochemical reaction sequence starting with the formation of Cu_2S and CuS, which are consumed in the further course and Cu_3SbS_3, Cu_3SbS_4 and $Cu_{12}Sb_4S_{13}$ are formed, with the proportion of the tetrahedrite phase $Cu_{12}Sb_4S_{13}$ increasing with increasing grinding time.[66] The products ground for different times were then thermally treated by spark plasma sintering (SPS), and it was found that even 1 h of grinding was sufficient to subsequently obtain predominantly the tetrahedrite phase. In the case of longer-stressed samples, an increased proportion of famatinite Cu_3SbS_4 was found after SPS. The method of mechanochemical synthesis followed by SPS was additionally used to dope the tetrahedrite $Cu_{12}Sb_4S_{13}$ with bismuth.[67] The elements were again milled for 1 h and then thermally treated. However, an increased proportion of bismuth in the structure resulted in a lower proportion of the tetrahedrite phase at the expense of Cu_3SbS_4 and Cu_3SbS_3 and thus in a deterioration of the thermoelectric properties of the products. Another report, in addition to the synthesis of the tetrahedrite $Cu_{12}Sb_4S_{13}$, includes the syntheses of $CuFeS_2$, Cu_2SnS_3 and the quaternary compounds $Cu_6Fe_2SnS_8$ and Cu_2ZnSnS_4 from their elements by grinding in an eccentric vibration mill followed by SPS.[68] For all compounds, a successful synthesis could be carried out and the materials could be produced in high purity (partly, however, with low amounts of side phases). The thermoelectric performance (electrical resistivity, Seebach coefficient and thermal conductivity) of the products prepared in the eccentric vibration mill was nevertheless comparable to that of compounds prepared mechanochemically on a laboratory scale. In addition to Cu_2ZnSnS_4, the synthesis of the stannite Cu_2FeSNS_4 was also the subject of another study.[69] Here, the Cu_2FeSnS_4 was synthesized not only from the elemental precursors, but also from the mineral precursors $CuFeS_2$, CuS and SnS, as well as from a combination of mineral and elemental compounds (CuS, SnS, Fe, S). Since the binary sulfides occur as intermediate products anyway, the alternative synthesis routes were also successful, although small amounts of unreacted starting materials were also found here due to un-optimized grinding conditions. Also for Cu_2ZnSnS_4 a possible synthesis starting from mineral precursors (CuS and SnS together with Zn and S) could be carried out.[70] Residues of copper sulfides detected by Raman spectroscopy in the product could be removed by further thermal treatment. In the Cu–Fe–Sn–S system, in addition to the known stannite Cu_2FeSnS_4, the rhodostannite $Cu_2FeSn_3S_8$ could also be produced from the elements.[71] Due to the complex reaction sequence, in which CuS and Cu_2SnS_3 are formed intermediately, and in order to obtain the highest possible conversion to the desired rhodostannite, a grinding time of 6 h was necessary, with small amounts of unreacted iron still present, which could not be removed even after another 4 h of grinding. By SPS, a composite material of mawsonite $Cu_6Fe_2SnS_8$ and the stannite Cu_2FeSnS_4 was accessible.[72] For this purpose, the elemental precursors were ground in an eccentric vibration mill. As expected, first binary and then ternary sulfides were formed until after 120 min the mawsonite $Cu_6Fe_2Sn_2S_8$ was formed as the main crystalline phase. Subsequent thermal treatment formed the composite material consisting of over 50 % mawsonite and 40 % stannite. The thermoelectric $Cu_{26}V_2Sn_6S_{32}$ was also successfully formed in the Cu–V–Sn–S system.[73] After the initial formation of the copper sulfides and Cu_2SnS_3, the crystallinity of the product was not good enough to safely assume the formation of $Cu_{26}V_2Sn_6S_{32}$ in the further course due to the similar reflection positions with Cu_2SnS_3 at higher diffraction angles. However, subsequent SPS improved the crystallinity to such an extent that it could be shown that a phase-pure product is obtained in this way. Using ^{119}Sn Mössbauer spectroscopy and electron microscopy studies, it was shown that longer milling times (over 6 h) are necessary for this, since vanadium-rich core-shell structures form before this point and Sn can be found on more than one crystallographic site. The thermoelectric properties of the products from the synthesis in the eccentric vibration mill were comparable to materials produced in a laboratory-scale planetary mill.

An alternative method for the preparation of metal sulfides by metathesis reaction between the corresponding metal acetates and Na_2S was also investigated in the eccentric vibration mill. The reaction with hydrated copper acetate gave nanocrystalline CuS after only 6 min, with the product being contaminated with $CuSO_4 \cdot 5 H_2O$.[74] The formation of the sulfate was explained by the high reactivity of the mechanochemically prepared CuS, resulting in a reaction with air and crystal water. A thermal analysis showed that the thermal stability of the product was lower than usually observed for the bulk material, since the thermal effects were shifted up to 200 °C to lower temperatures. This route could also be used to synthesize the sulfides of Pb, Zn and Cd in another study.[75] Moreover, these sulfides could also be prepared in a time of 6 min without additional oxidation products. In addition to the syntheses mentioned above, further experiments were carried out on the reduction of various metal sulfides (Cu_2S, PbS and Sb_2S_3) with iron in the industrial eccentric vibration mill.[76] In these experiments, the consumption of Fe was followed magnetometrically with the milling time and then the rate constants for the respective reactions were determined from this data. It was found that the product formation rate was the highest for Cu, followed by Sb and Pb. In the case of PbS, after the maximum applied grinding time of 120 min, PbS starting material could still be found compared to the other two reactions. In another study, in addition to Cu_2S, the reduction of CuS with Fe was also investigated.[77] While the reduction of Cu_2S again proceeded completely, significant amounts of Fe were still found in the case of CuS even after 480 min, which was attributed to the different reduction rates in the two systems due to the different oxidation states of copper.

In addition to the numerous investigations in the eccentric vibration mill, a chemical reaction was also carried out in a Simoloyer® by dry grinding the reactants on a multi-gram scale.[78] The reaction of a silver bronze Ag_3Sn with silver oxide Ag_2O resulted in the formation of an Ag/SnO_2 composite material. The reaction was first carried out in a batch of 350 g in a 30 L drum mill, but no complete conversion was found even after 36 h due to the low kinetic energy input. The long processing time also led to severe abrasion of the Al_2O_3 balls used, resulting in high levels of impurity in the milled powder. In further investigations in a Simoloyer® CM01 with 2 L grinding chamber and up to 200 g starting materials, the influence of the temperature was also investigated by designing and applying a special heating jacket. Before this, however, the optimum reaction temperature was determined to be 100 °C by means of a vibrating frame grinder. The subsequent synthesis in the Simoloyer® could be fully realized under optimal conditions already after about 30 min. The strong reaction acceleration also reduced the influence of the gas atmosphere (air or argon), as a strong difference was found for this reaction at slow kinetics when grinding in the presence of air or under argon/vacuum in the processing time required. This was explained both by a different mechanism of the reaction under air due to the involvement of molecular oxygen and potential passivation by CO_2 with the formation of surface silver carbonate.[79]

In the field of catalyst synthesis, He et al. were able to prepare precursors of the subsequently supported catalyst on a kilogram scale in a planetary mill.[80] By milling Pd(acac)$_2$ and Zn(acac)$_2$ for 10 h, such a good distribution of Pd(acac)$_2$ was achieved that aggregation was strongly suppressed in the subsequent calcination and reduction steps and single-atom catalysts could be obtained. Not only atomically distributed Pd on ZnO could be prepared using this approach, but this method was also transferable to the synthesis of Rh/ZnO and Cu/ZnO single-atom catalysts. The activity and stability of the prepared Pd/ZnO catalyst were subsequently successfully demonstrated in the hydrogenation of phenylacetylene and CO oxidation. Using a similar preparation method, ceria-supported gold single-atom catalysts were obtained also on the order of 1 kg by first grinding the corresponding acetates.[81]

7.3 Organic Materials

7.3.1 Co-crystals

The upscaling of quantitative co-crystal formation of APIs (active pharmaceutical ingredients) is of great importance. Especially in cases where reactions can be carried out close to 100 % yield without work-up of the products or removal of catalysts or other by-products. The development of co-crystals can improve several properties like dissolution rate, physical and chemical stability, processability or bioavailability.

The 1:1 co-properties like the crystal of α-(D)-Glucose and urea was produced in quantitative yields of 200 g batches after only 5 min of milling time.[82] The milling was performed in a 2 L stainless steel horizontal rotary-ball mill (Simoloyer©). Two kilograms of steel ball with a diameter of 5 mm were used. The mill was water cooled and had a temperature of 15 °C at the wall and a maximum temperature of 10 °C in the center of the mill. The rotor was run at 900 min^{-1} (610 W power) for 5 min to achieve quantitative yields. The resulting co-crystals were finally characterized with IR spectroscopy.

Under the trade name Faban©, a co-crystal consisting of dithianone (1) and pyrimethanil (2) is marketed as an ingredient in crop protection products. Because the co-crystal has a significantly different solubility than the individual ingredients, it can be used to protect plants over a much longer period of time while significantly reducing the amount of material used (Figure 7.9).

In addition to the synthesis methods that take place in solution, this 1:1 co-crystal can also be obtained through mechanical–chemical processing. For this purpose, a variety of mills are used, which are available in sizes for commercial production.[83] Similarly, a co-crystal with dithianone can be obtained by mechanochemical processes in which pyrimethanil is replaced by the structurally very similar cyprodinil.[84]

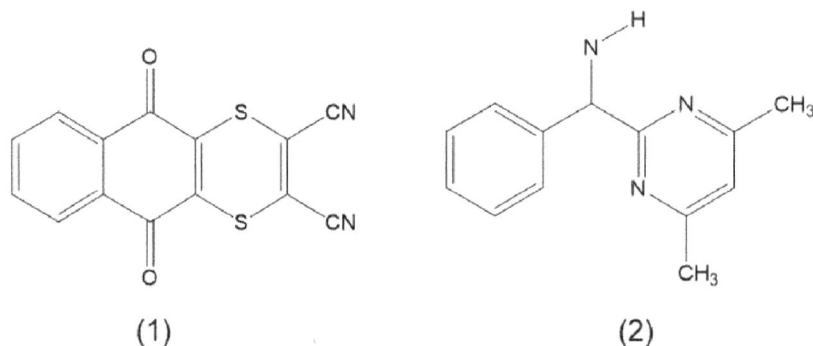

FIGURE 7.9   Dithianone (1) and Pyrimethanil (2).

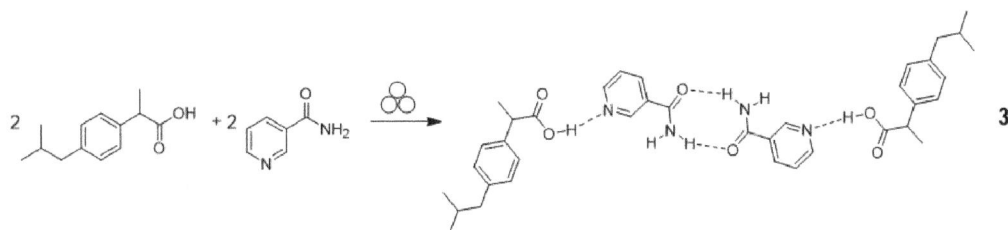

FIGURE 7.10   Mechanochemical formation of an Ibuprofen-Nicotinamide co-crystal.

Recently a study was published focusing on the use of mechanochemical processes to prepare the Active Pharmaceutical Ingredients (APIs) ibuprofen–nicotinamide co-crystal (Equation 9.3). The co-crystal formation was upscaled to kilogram scale in an eccentric vibration mill (Figure 7.10).[85]

One experiment was performed in an ESM23G-1bs eccentric vibrating mill with a grinding vessel and balls made of steel. The diameter of the steel balls was 40 mm (260 g per ball). An overall weight of 76 kg of balls (ca. 15 L) was used, resulting in a free volume of 5 L inside the grinding vessel. For the experiment, an amount of 2.8 kg of a stoichiometric 1:1 ibuprofen–nicotinamide mixture was filled into the mill. The second experiment was carried out in an ESM-234-1bk eccentric vibrating mill. In this case, the grinding vessel and the balls were made of Al_2O_3. The diameter of the Al_2O_3 balls was also 40 mm (77 g per ball). The overall weight of the balls was 15 kg (6.5 L) and the free volume inside the milling chamber 2.5 L. For this experiment, an amount of 1.4 kg of a stochiometric 1:1 ibuprofen–nicotinamide mixture was filled into the mill.

In both cases, the horizontal vibration width was 11–12 mm and the vertical 7–8 mm. The total grinding time was 150 min. The reaction in the steel grinding plant was completed after about 1 hour. The reaction with Al_2O_3 spheres requires slightly more time for completion. The products were characterized by X-ray diffraction and DSC measurements. No starting materials could be observed after the indicated milling times.

7.3.2 Other Organic Reactions

Only a few examples of mechanochemically organic reactions on a kg scale exist.

Under similar reaction conditions (200 g reactants, 15 min milling) as for the preparation of the glucose–urea cocrystal, a quantitative reaction of 4-hydroxy-benzaldehyde with 4-aminobenzoic acid to give the corresponding imine can also be carried out (Equation 9.4).[82] The water formed during the condensation reaction is incorporated into the crystal structure as crystal water and therefore does not interfere with the mechanochemical conversion. If necessary, the crystal water can be completely removed in vacuo at a slightly elevated temperature. The progress of the reaction can be monitored by IR spectroscopy. The carbonyl vibrations at 1667 and 1668 cm^{-1} partially disappeared and were replaced by new vibrations of the resulting imine compound at 1687 cm^{-1} (Figure 7.11).

The depolymerization of cellulose under the mechanocatalytic process using different types of acidic clays was demonstrated in kg-scale experiments.[86] Milling alone without any catalyst was not sufficient to hydrolyze the glycosylic bond in cellulose. Different acidic solid materials like kaolinite ($Al_2Si_2O_7 \cdot 2\,H_2O$), alumina ($Al_2O_3$), Y-type zeolite or bentonite ($Al_2Si_4O_{11} \cdot H_2O$) showed a good catalytic ability. Dehydration products like 5-hydroxy-methyl-furfural (HMF) are of interest for use as a fuel or chemical feedstock. Several experiments were performed on a kg scale using a small Union Process attritor. Also the mechanocatalytic depolymerization of H_2SO_4-impregnated lignocellulose was shown to produce fully water-soluble products on the scale of 1 kg.[87] A comparison between small-scale experiments using a planetary ball mill and the gram scale (Simoloyer® CM01) and 1 kg batch experiments (Simoloyer® CM20) has additionally shown that the energy efficiency of the process improved upon upscaling.

It could be shown that the mechanocatalytic processing of materials has significant advantages over current methods, and the depolymerization of other biopolymers such as chitin or synthetic polymers like nylon or polyethylene terephthalate may be possible.

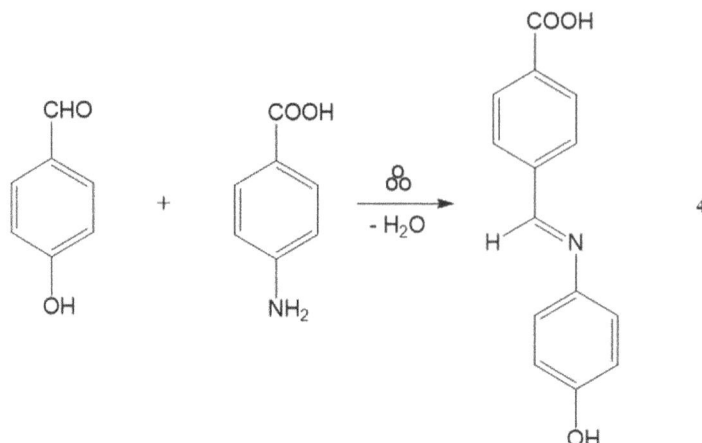

FIGURE 7.11   Imine formation between 4-hydroxy-benzaldehyde and 4-aminobenzoic acid.

REFERENCES

1. S. Bernotat, K. Schönert. Size reduction. In: *Ullmann's Encyclopedia of Industrial Chemistry*, 6th ed., 40v. Weinheim, Germany: John Wiley & Sons, Inc., (2000). (https://doi.org/10.1002/14356007.b02_05).
2. F. Shi. *Miner. Eng.* 2004, *17*(11–12), 1259–1268.
3. https://www.fritsch-international.com/sample-preparation/milling/mortar-grinder/details/product/pulverisette-2/.
4. V. Štrukil, M. D. Igrc, L. Fábián, M. Eckert-Maskić, S. L. Childs, D. G. Reid, M. J. Duer, I. Halasz, C. Mottillo, T. Friščić. *Green Chem.* 2012, *14*(9), 2462–2473.
5. https://www.retsch.com/products/milling/ball-mills/tm-500/function-features/.
6. H. Zoz, R. Reichardt, J. S. Kim. *Keram. Z.* 2001, *53*, 384–392.
7. K. Höffl. *Zerkleinerungs- und Kassiermaschinen*. Berlin, Heidelberg, New York: Springer-Verlag, 1986.
8. G. Zhumagalieva, V. Pershin, A. Tkachev, A. Vorobiev, A. Pasko, E. Galunin. *AIP Conf. Proc.* 2018, *2041*, 020010.
9. https://www.retsch.com/products/milling/ball-mills/mixer-mill-mm-400/function-features/.
10. A. R. Aktaş, G. Aktaş, M. Gürü. *Gu J. Sci* 2017, *30*, 187–194.
11. https://www.retsch.com/products/milling/ball-mills/mm-500-vario/function-features/.
12. E. Gock, K.-E. Kurrer. *Powder Technol.* 1999, *105*(1–3), 302–310.
13. E. Gock, V. Vogt, R. Florescu. Neue Entwicklungen rotationsschwingender Feinzerkleinerungsmaschinen. *TU Contact* 2001, *8*, 45–50.
14. E. Gock, V. Vogt, D. E. Kaufmann, J. C. Namyslo. Synthese Schlag auf Schlag – Die Schwingmühle als chemischer Reaktor. *TU Contact* 2006, *18*, 34–38.
15. G. Heinicke. *Tribochemistry*. Berlin: Akademie-Verlag, 1984.
16. http://www.ttd.spb.ru.
17. http://www.lab-mills.com/about.html.
18. A. Kwade, L. Blecher, J. Schwedes. *Chem. Ing. Tech.* 1997, *69*(6), 836–839.
19. https://www.glencoretechnology.com/en/technologies/isamill.
20. https://gmbh.zoz.de/?page_id=1238.
21. P. Baláž. Applied mechanochemistry. In: *Mechanochemistry in Nanoscience and Minerals Engineering*. Berlin, Heidelberg: Springer-Verlag, 2008.
22. P. Baláž, E. Dutková. *Miner. Eng.* 2009, *22*(7–8), 681–694.
23. H. Zoz, H. U. Benz, G. Schäfer, M. Dannehl, J. Krüll, F. Kaup, H. Ren, R. Reichardt. *High Kinetic Processing of Enamel, Part I*, 2022. Can be found under https://gmbh.zoz.de/?page_id=1155.
24. H. Zoz, H. Ren, R. Reichardt, H. U. Benz, A. Nadkarni, G. Wagner. *Ductile Metal Flakes based on [Au], [Ag], [Al], [Cu], [Ti], [Zn] and [Fe] Materials by High Energy Milling Part I*, 2022. Can be found under https://gmbh.zoz.de/?page_id=1155.

25. H. Zoz, D. Ernst, R. Reichert, H. Ren, T. Mizutani, M. Nishida, H. Okouchi. *Simoloyer® CM100s, Semi-Continuous Mechanical Alloying on a Production Scale using Cycle Operation – Part II*, 2022. Can be found under https://gmbh.zoz.de/?page_id=1155.

26. E. Worrell, L. Price, N. Martin, C. Hendriks, L. O. Meida. *Annu. Rev. Energy Environ.* 2001, *26*(1), 303–329.

27. N. A. Madlool, R. Saidur, N. A. Rahim, M. Kamalisarvestani. *Renew. Sustain. Energy Rev.* 2013, *19*, 18–29.

28. A. Mitrović, M. Zdujić. *Int. J. Miner. Process.* 2014, *132*, 59–66.

29. R. Hamzaoui, F. Muslim, S. Guessasma, A. Bennabi, J. Guillin. *Powder Technol.* 2015, *271*, 228–237.

30. G. Mucsi, A. Rácz, V. Mádai. *Powder Technol.* 2013, *235*, 163–172.

31. H. Zoz, D. Jaramillo V, Z. Tian, B. Trindade, H. Ren, O. Chimal-V, S. Diaz de la Torre. *High Performance Cements and Advanced Ordinary Portland Cement Manufacturing by HEM-Refinement and Activation*, 2022. Can be found under https://gmbh.zoz.de/?page_id=1155.

32. K. Sobolev. *Cem. Concr. Compos.* 2005, *27*(7–8), 848–853.

33. H. Justnes, L. Elfgren, V. Ronin. *Cem. Concr. Res.* 2005, *35*(2), 315–323.

34. H. Justnes, V. Ronin, J.-E. Jonasson, L. Elfgren. Mechanochemical technology: Synthesis of energetically modified cements (EMC) with high volume fly ash content. *Proceedings of the 12th International Congress on the Chemistry of Cement (ICCC)*, 2007, Montreal, Canada, paper 84 (TH3-14.4) 12 pp.

35. V. Ronin, L. Elfgren. An industrially-proven solution for sustainable pavements of high volume pozzolan concrete—Using energetically modified cement, 2022. Available at: www.emccement.com/wiki/pdfs/emc_sustainable_pavements_with_high_volume_natural_pozzolans.pdf.

36. B. Funk, R. Trettin, H. Zoz. *Mater. Today: Proc.* 2017, *4*, 81–86.

37. F. Matalkah, A. H. Alomari, P. Soroushian. *Case Stud. Constr. Mater.* 2020, *13*, e00463.

38. F. Matalkah, P. Soroushian. *Constr. Build. Mater.* 2018, *158*, 42–49.

39. K. Garbev, G. Beuchle, U. Schweike, D. Merz, O. Dregert, P. Stemmermann. *J. Am. Ceram. Soc.* 2014, *97*(7), 2298–2307.

40. C. Suryanarayana. *Prog. Mater. Sci.* 2001, *46*(1–2), 1–184.

41. M. Hüller, G. G. Chernik, E. L. Fokina, N. I. Budim. *Rev. Adv. Mater. Sci.* 2008, *18*, 366–374.

42. H. Zoz, D. Ernst, I. S. Ahn, W. H. Kwon. *Mechanical Alloying of Ti-Ni based Materials using the Simoloyer®*, 2022. Can be found under https://gmbh.zoz.de/?page_id=1155.

43. II. Zoz, D. Ernst, II. Weiss, M. Magini, C. Powell, C. Suryanarayana, F. H. Froes. *Mechanical Alloying of Ti-24Al-11Nb using the Simoloyer®*, 2022. Can be found under https://gmbh.zoz.de/?page_id=1155.

44. R. Didomizio, S. Huang, L. Dial, J. Ilavsky, M. Larsen. *Metall. Mater. Trans. A* 2014, *45*(12), 5409–5418.

45. J.-T. Li, Y. Yang, H.-B. Jin. *Key Eng. Mater.* 2007, *336*, 911–915.

46. H. Zoz, G. Kaupp, H. Ren, K. Goepel, M. R. Naimi-Jamal. *Recycling of EAF Dust by Semi-Continuous High Kinetic Process*, 2022. Can be found under https://gmbh.zoz.de/?page_id=1155.

47. H. Basturkcu, M. Achimovičová, M. Kaňuchová, N. Acarkan. *Hydrometallurgy* 2018, *181*, 43–52.

48. H. Basturkcu, N. Acarkan, E. Gock. *Int. J. Miner. Process.* 2017, *163*, 1–8.

49. J. M. Bellosta von Colbe, J. Puszkiel, G. Capurso, A. Franz, H. U. Benz, H. Zoz, T. Klassen, M. Dornheim. *Int. J. Hydrogen Energy* 2019, *44*(55), 29282–29290.

50. B. Molinas, A. A. Ghilarducci, M. Melnichuk, H. L. Corso, H. A. Peretti, F. Agresti, A. Bianchin, S. Lo Russo, A. Maddalena, G. Principi. *Int. J. Hydrogen Energy* 2009, *34*(10), 4597–4601.

51. P. de Rango, A. Chaise, J. Charbonnier, D. Fruchart, M. Jehan, Ph. Marty, S. Miraglia, S. Rivoirard, N. Skyabina. *J. Alloys Compd.* 2007, *446–447*, 52–57.

52. M. Jehan, D. Fruchart. *J. Alloys Compd.* 2013, *580*, S343–S348.

53. A. Chaise, P. de Rango, Ph. Marty, D. Fruchart, S. Miraglia, R. Olivès, S. Garrier. *Int. J. Hydrogen Energy* 2009, *34*(20), 8589–8596.

54. M. Verga, F. Armanasco, G. Guardamagna, C. Valli, A. Bianchin, F. Agresti, S. Lo Russo, A. Maddalena, G. Principi. *Int. J. Hydrogen Energy* 2009, *34*(10), 4602–4610.

55. A. Pommerin, M. Felderhoff. Unpublished results.

56. G. A. Lozano, N. Eigen, C. Keller, M. Dornheim, R. Bormann. *Int. J. Hydrog. Energy* 2009, *34*(4), 1896–1903.

57. T. A. Johnson, S. W. Jorgensen, D. E. Dedrick. *Faraday Discuss.* 2011, *151*, 327–352.

58. R. Urbanczyk, K. Peinecke, M. Meggough, P. Minne, S. Peil, D. Bathen, M. Felderhoff. *J. Power Sources* 2016, *324*, 589–597.

59. M. Achimovičová, P. Baláž, J. Ďurišin, A. Rečnik, J. Kováč, A. Šatka, A. Feldhoff, E. Gock. Mechanochemical synthesis of nanocrystalline lead selenide. In: *VI International Conference on Mechanochemistry and Mechanical Alloying (INCOME 2008)*, 2008.

60. M. Achimovičová, P. Baláž, J. Ďurišin, N. Daneu, J. Kováč, A. Šatka, A. Feldhoff, E. Gock. *Int. J. Mater. Res.* 2011, *102*(4), 441–445.

61. M. Achimovičová, P. Baláž, T. Ohtani, N. Kostova, G. Tyuliev, A. Feldhoff, V. Šepelák. *Solid State Ion* 2011, *192*(1), 632–637.

62. M. Achimovičová, M. Baláž, V. Girman, J. Kurimský, J. Briančin, E. Dutková, K. Gáborová. *Nanomaterials* 2020, *10*(10), 2038.

63. M. Achimovičová, E. Dutková, E. Tóthová, Z. Bujňáková, J. Briančin, S. Kitazono. *Front. Chem. Sci. Eng.* 2019, *13*(1), 164–170.

64. M. Hegedüs, M. Baláž, M. Tešínský, M. J. Sayagués, P. Siffalovic, M. Krulaková, M. Kaňuchová, J. Briančin, M. Fabián, P. Baláž. *J. Alloys Compd.* 2018, *768*, 1006–1015.

65. E. Dutková, M. J. Sayagués, M. Fabián, M. Baláž, M. Achimovičová. *Mater. Lett.* 2021, *291*, 129566.

66. P. Baláž, E. Guilmeau, N. Daneu, O. Dobrozhan, M. Baláž, M. Hegedus, T. Barbier, M. Achimovičová, M. Kaňuchová, J. Briančin. *J. Eur. Ceram. Soc.* 2020, *40*(5), 1922–1930.

67. P. Baláž, E. Guilmeau, M. Achimovičová, M. Baláž, N. Daneu, O. Dobrozhan, M. Kaňuchová. *Nanomaterials* 2021, *11*(6), 1386.

68. P. Baláž, M. Achimovičová, M. Baláž, K. Chen, O. Dobrozhan, E. Guilmeau, J. Heitmánek, K. Knížek, L. Kubíčková, P. Levinský, V. Puchý, M. J. Reece, P. Varga, R. Zhang. *ACS Sustain. Chem. Eng.* 2021, *9*(5), 2003–2016.

69. P. Baláž, M. Hegedüs, M. Achimovičová, M. Baláž, M. Tešínský, E. Dutková, M. Kaňuchová, J. Briančin. *ACS Sustain. Chem. Eng.* 2018, *6*(2), 2132–2141.

70. P. Baláž, M. Hegedus, M. Baláž, N. Daneu, P. Siffalovic, Z. Bujňáková, E. Tóthová, M. Tešínský, M. Achimovičová, J. Briančin, E. Dutková, M. Kaňuchová, M. Fabián, S. Kitazono, O. Dobrozhan. *Prog. Photovolt. Res. Appl.* 2019, *27*(9), 798–811.

71. M. Baláž, O. Dobrozhan, M. Tešínský, R.-Z. Zhang, R. Džunda, E. Dutková, M. Rajňák, K. Chen, M. J. Reece, P. Baláž. *Powder Technol.* 2021, *388*, 192–200.

72. P. Baláž, M. Hegedüs, M. Reece, R.-Z. Zhang, T. Su, I. Škorvánek, J. Briančin, M. Baláž, M. Mihálik, M. Tešínský, M. Achimovičová. *J. Electron. Mater.* 2019, *48*(4), 1846–1856.

73. M. Hegedüs, M. Achimovičová, H. Hui, G. Guélou, P. Lemoine, I. Fourati, J. Juraszek, B. Malaman, P. Baláž, E. Guilmeau. *Dalton Trans.* 2020, *49*(44), 15828–15836.

74. E. Godočíková, P. Baláž, J. M. Criado, C. Real, E. Gock. *Thermochim. Acta* 2006, *440*(1), 19–22.

75. E. Godočíková, P. Baláž, E. Gock, W. S. Choi, B. S. Kim. *Powder Technol.* 2006, *164*(3), 147–152.

76. P. Baláž, E. Dutková, I. Škorvánek, E. Gock, J. Kovac, A. Šatka. *J. Alloys Compd.* 2009, *483*(1–2), 484–487.

77. M. Baláž, M. Tešínský, J. Marquardt, M. Škrobian, N. Daneu, M. Rajňák, P. Baláž. *Adv. Powder Technol.* 2020, *31*(2), 782–791.

78. H. Zoz, H. Ren, N. Späth. *Improved Ag-SnO$_2$ Electrical Contact Material Produced by Mechanical Alloying*, 2022. Can be found under https://gmbh.zoz.de/?page_id=1155.

79. N. Lorrain, L. Chaffron, C. Carry, P. Delcroix, G. Le Gaër. *Mater. Sci. Eng. A* 2004, *367*(1–2), 1–8.

80. X. He, Y. Deng, Y. Zhang, Q. He, D. Xiao, M. Peng, Y. Zhao, H. Zhang, R. Luo, T. Gan, H. Ji, D. Ma. *Cell Rep. Phys. Sci.* 2020, *1*, 100004.

81. T. Gan, Q. He, H. Zhang, H. Xiao, Y. Liu, Y. Zhang, X. He, H. Ji. *Chem. Eng. J.* 2020, *389*, 124490.

82. G. Kaupp, J. Schmeyers, M. R. Naimi-Jamal, H. Zoz, H. Ren. *Chem. Eng. Sci.* 2002, *57*(5), 763–765.

83. US 8324233 B2.

84. WO 2013/030777 A1.

85. E. Colacino, M. Felderhoff, to be published.

86. S. M. Hick, C. Griebel, D. T. Restrepo, J. H. Truitt, E. J. Buiker, C. Bylda, R. C. Blair. *Green Chem.* 2010, *12*(3), 468–474.

87. M. D. Kaufman Rechulski, M. Käldström, U. Richter, F. Schüth, R. Rinaldi. *Ind. Eng. Chem. Res.* 2015, *54*(16), 4581–4592.

8

Acoustic Synthesis (Solvent-Free) and Resonant Acoustic Mixing (RAM)

Maria Elena Rivas

CONTENTS

8.1 Introduction: Background

Resonant acoustic mixing (RAM) is a technology developed by ResoDyn Corporation which provides a low-energy contactless mixing system [1]. By applying this novel mixing system, acoustic wave energy is transferred to mechanical movement under resonant conditions. In resonant acoustic mixing (RAM), the sample vessel is placed on a plate connected to a bed of springs and forced to oscillate at the fixed mechanical resonance frequency of the device reference. This motion is transferred to the powder, introducing intense, local mixing zones. Figure 8.1a and b describes how the RAM technique can be applied as a method for mechanochemical processes under significantly more gentle conditions than those experienced during ball milling. In general, the vibration action created during RAM (resonant acoustic mixing) can be used to complete various processes, not limited to coating, sieving, and mixing of materials having a significant impact on the development of more environmentally friendly products for different industries [2].

Investigations into the manufacturing of energetic materials via RAM technology have highlighted many potential advantages [3]. These include shorter time scales, improved mix homogeneity, reduced waste output due to flexibility of load/batch size, and the absence of moving parts (a potential ignition source). In general, resonant acoustic mixing (RAM) is a new technology designed for the intensive mixing of powders that offers the capability to process powders with minimal damage to particles. In addition. Additionally, RAM can process higher-viscosity products, giving the opportunity for the development of new families of energetics. Examples include powder metallurgy food preparation, propellant manufacturing, and pharmaceuticals [4–7]. In all those applications homogeneity of the powder mixture is the goal. It should be also highlighted that RAM represents a potentially disruptive technology in the pharmaceutical industry, where solid dosage forms (which include tablets and capsules) are usually produced from a pre-blended powder. That means that the effective dose of an Active Pharmaceutical Ingredient (API) in a tablet or capsule is not only dependent on the correct weight of the API and all excipients in the bulk but also on the homogeneity after mixing.

This book chapter examines how the resonant acoustic mixing (RAM) technology can be applied to the processing to produce pharmaceuticals materials, characterisation techniques, and some insights into the potential of RAM for powder mixing.

DOI: 10.1201/9781003178187-10

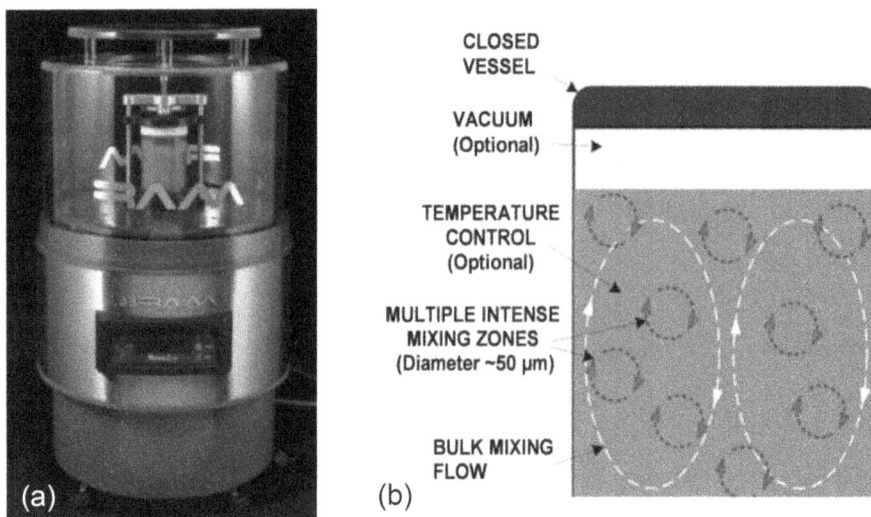

FIGURE 8.1   (a) A lab RAM bench-top resonant acoustic mixer. Source: ResoDyn Corporation. (b) Diagrammatic representation of microscopic mixing zones induced during the mix cycle.

8.2 RAM-Assisted Co-Crystallisation Examples

Fusing solid-state reactions with crystallisation techniques is an attractive way of adapting "Green Chemistry" to industrial manufacturing as one of the 12 principles of Green Chemistry is the reduction and/or possible removal of solvents from a chemical reaction. Solvent-free crystallisation reduces waste, cost, environmental, and health impacts, as well as batch-to-batch variations due to solvents sourced from different suppliers with different impurity profiles [8–12]. Traditionally, cocrystals have been prepared by using a solution or slurry crystallisation. The use of solution crystallisation represents some challenges as the components of the cocrystal may have different solubilities in the solvent used for crystallisation, therefore, making it more challenging to ensure the production of the right form of the cocrystal [13]. At a small scale, mechanochemical methods such as dry grinding and liquid-assisted grinding using ball mills have been demonstrated as efficient methods to generate cocrystals and have been shown to circumvent the issues seen with solution crystallisation [14–16]. However, the ball milling process has been proven to be more difficult to scale up. More recently scale-up of co-crystallisation by mechanochemical means was demonstrated using twin-screw extrusion (TSE) making cocrystals at large scales [17–22]. TSE offers a unique opportunity not found in grinding or milling approaches because they are difficult to scale up to manufacture large amounts of the desired cocrystal. In fact, one of the benefits of TSE is that the process of crystallisation can be carried out under solventless conditions. Daurio demonstrated that TSE can be an important method of preparation of cocrystals at a large scale by studying four different cocrystal systems. The authors emphasise that temperature and extent of mixing are the main parameters for a successful conversion in the process. TSE equipment consists basically of two co-/counter-rotating screws in a single barrel. By mechanical action of screwing, the components are mixed along the length of the barrel forming cocrystals. Although TSE has provided a means to carry out mechanochemical synthesis in a continuous, large scale, and efficient fashion, which is adaptable to a manufacturing process. A clear advantage of RAM can be seen over ball milling and twin-screw extrusion (TSE) processes and is the lower shear approach where low-frequency high-intensity acoustic energy is used to agitate samples. Additionally, in RAM the energy is uniformly distributed across the sample and the mixing elements do not meet the sample. As mentioned before, RAM is an emerging technology for pharmaceutical applications, and therefore there is not vast literature in this area. However, in this chapter a review of the main work conducted on the development of cocrystals by resonant acoustic mixing will be described.

Nagapudi et al. [23] demonstrated a high throughput 96-well plate cocrystal screening using RAM for the screening of caffeine, carbamazepine, and theophylline as model cocrystal formers. In this work, the following cocrystal formers were used in the screening: Oxalic acid, glutaric acid, malic acid, maleic acid, succinic acid, malonic acid, benzoic acid, adipic acid, and citric acid. Scale-up experiments were also conducted at a 400 mg scale using RAM with a theophylline–citric acid cocrystal system. Five different conditions were tested: (i) Solids with no beads, (ii) solids with water, (iii) solids with bead sand water, (iv) solids with ethanol, and (v) solids with beads and ethanol. The materials were subjected to acoustic mixing at an acceleration of 60 G. The authors report the best conditions obtained when the solvent was added during mixing and no positive effect was observed with the additions of beads during the preparations (Figure 8.2).

The authors claimed a high success rate in terms of cocrystal, demonstrated for three APIs: Caffeine, theophylline, and carbamazepine. The application of RAM in the production of both anhydrated and hydrated forms of theophylline–citric acid cocrystals was demonstrated, including scalability of theophylline–oxalic acid cocrystal at 80 g scale. Acceleration and mixing time were identified as the key parameters to be optimised for the successful production of cocrystals. The use of beads along with liquid-assisted acoustic mixing was introduced in this work to accelerate the kinetics of cocrystal formation. RAM is thus established as a scalable and environmentally friendly mechanochemical technique to produce cocrystals (Figure 8.3).

Zack Guo et al. [24] evaluated a diverse set of drug and polymer combinations by acoustic fusion. In this work, amorphous solid dispersions (ASD) on the mg scale were formed, indicating that this approach can be a simple procedure for ASD drug formulations. Specifically, the researchers demonstrated the effectiveness of this acoustic fusion process by producing amorphous solid dispersions of various BCS class 2 and 4 drug candidates, including torcetrapib, itraconazole, and lopinavir, with a variety of polymer systems, such as copovidone, Soluplus®, PEG1500, Vitamin-E TPGS, Kolliphor EL, and Eudragit. Formulations of these ASD drug products demonstrated higher solubility of the drug substance compared to the solubility of the crystalline form of the drug. The model drugs torcetrapib in either HPMCAS-LF, copovidone + Vitamin-E TPGS, or Soluplus® showed improved supersaturation solubility in an aqueous buffer *in vitro* compared to the drug in crystalline form, suggesting that the acoustic fusion process resulted in an amorphous solid dispersion state like those formed in spray drying (SD) or hot melt extrusion (HME) processes (Figure 8.4).

On the other hand, Gonnet et al. [25] introduced catalytic organic synthesis by resonant acoustic mixing (RAM) as a mechanochemical methodology that does not require bulk solvent or milling media. The researchers used as model reactions ruthenium-catalysed ring-closing metathesis, Enyne metathesis, and copper-catalysed sulphonamide–isocyanate coupling. This emerging process has proven to be faster and simpler than conventional ball milling. Moreover, the method can be readily scaled up, as demonstrated

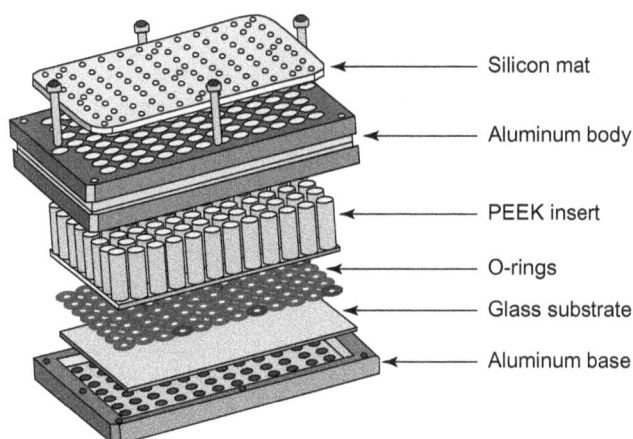

FIGURE 8.2 A 96-well high throughput assembly.

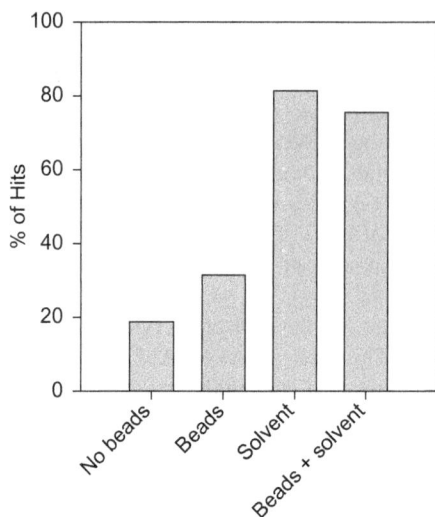

FIGURE 8.3   Outcome of HTS screening reported as the percentage of hits obtained from the screen under different mixing conditions.

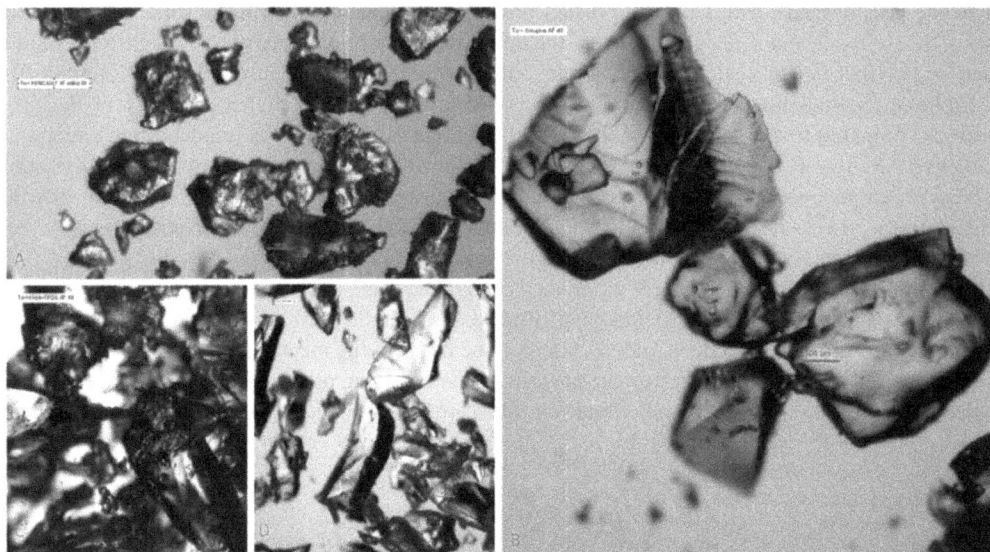

FIGURE 8.4   Microscope images of acoustic fusion products of torcetrapib: torcetrapib/HPMCAS-LF (A, top left), torcetrapib/Soluplus® (B, right), torcetrapib/VA64/TPGS (C, bottom, first from left), Torcetrapib/VA64 (D, bottom, second from left).

by the straightforward catalytic synthesis of the antidiabetic drug Tolbutamide from hundreds of milligrams to at least 10 grams, without any significant changes in reaction conditions.

Karagianni et al. [26] applied resonant acoustic technology (RAM) for the formation of different complexes and multiphase systems. The researchers studied the synthesis of carbamazepine–nicotinamide cocrystals by adding different solvents such as CH_3Cl, H_2O, DMF, DMSO, and MeOH, for 100 mg of the cocrystal formation. The experiments were conducted in a laboratory RAM equipment operated for 2 hours at 90% intensity. Results generated in this research confirmed the use of resonant acoustic mixing for high-purity co-crystalline cocrystal production at a large scale (Figure 8.5).

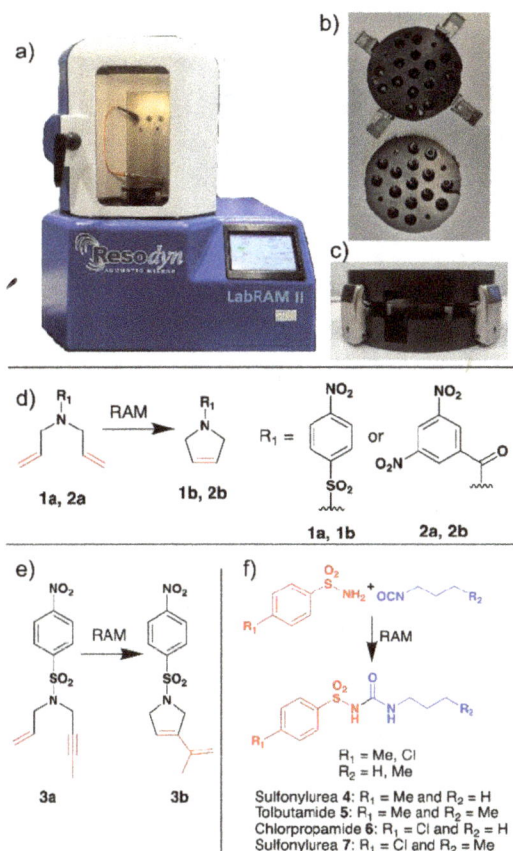

FIGURE 8.5 (a) The LabRAM II instrument. Different views of the custom-designed RAM sample holder: (b) Top view and (c) side view, with reaction vials mounted. Reaction schemes for explored reactions: (d) Ruthenium-catalysed ring-closing metathesis (RCM), (e) ruthenium-catalysed ring-closing ene-yne metathesis (RCEYM), and (f) copper-catalysed coupling to produce sulfonylureas, including APIs Tolbutamide and Chlorpropamide.

8.3 In Situ Characterisation of Resonant Acoustic Mixing

In the mechanochemical process, reactions are normally conducted in confined and rapidly moving reaction vessels, making the in situ analysis a very challenging task. In the last year, in situ analysis of the mechanochemical process, mainly for organic reactions, has been developed and improved. The analysis is being focused on powder X-ray diffraction (PXRD) and Raman spectroscopy [27, 28].

A similar approach has been recently implemented to the resonant acoustic mixing reaction (RAM). In marked contrast to ball milling techniques, the lack of milling bodies in the RAM experiment does not hinder the co-crystallisation of the two starting materials, which occurred readily and was independent of the frequency of oscillation. For example, Adam A et al. [29] reported the first in situ real-time monitoring of a RAM process using synchrotron X-ray radiation. The co-crystallisation was monitored at two acceleration settings of 50 G and 100 G. Both possible products (CBZDH and CBZ.NIC) display characteristic Bragg peaks in the region above a d-spacing of 7 Å, and so a qualitative indication of the relative rates of the co-crystallisation process could be assessed by monitoring. Results indicated an effect of the mixing frequency (G) on the conversion to the final material. Moreover, when the mixture was subjected to specific 50 G, an initial induction period was observed. This observation helped to support the hypothesis of no reaction between the two starting materials prior to the RAM mixing process (Figure 8.6). Within 10 seconds, the first indications of product peaks were observed.

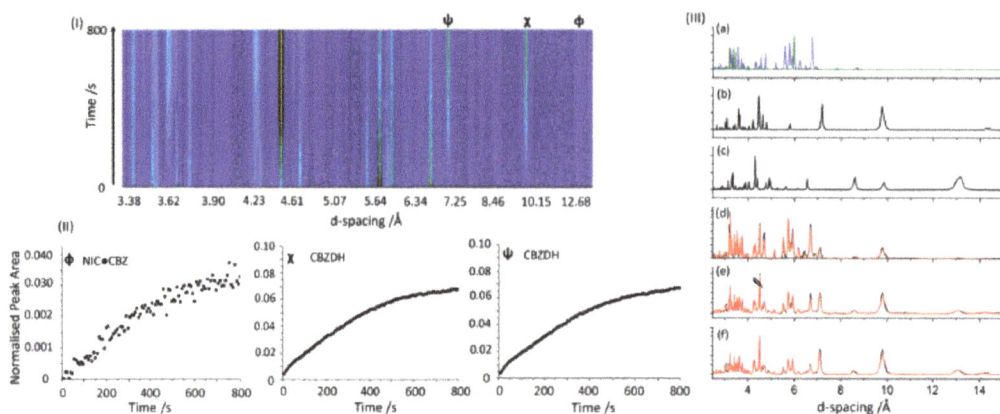

FIGURE 8.6   RAM treatment of 1:1 mixture of CBZ:NIC at 50 G. (I) Time-resolved XRPD profile of RAM treatment. (II) Normalised integrated peak areas of product peaks at d-spacing (C) 7.20 Å, (w) 9.78 Å, and (f) 13.10 Å. (III) Simulated XRPD patterns for (a) CBZ III (blue) and NIC I (green), (b) CBZDH, (c) CBZNIC, and Rietveld refined profiles at (d) 112 s, (e) 400 s, and (f) 800 s. Experimental (black) and refined profiles (red) are shown.

It was found in this research that the formation of CBZDH occurred at a nearly continuous rate before reaching a final plateau. The continuous and consistent growth of CBZDH before this point, however, suggests that the formation of reactive contacts occurred readily, despite the absence of any milling bodies. On mixing at 50 G, the major product peaks do not correspond to the thermodynamically most favourable phase (CBZ.NIC) but instead indicate the formation of CBZDH as the kinetically controlled product. Rietveld refinement showed that no more than 10 mol% of the 1:1 CBZ.NIC cocrystal was formed over the period of the experiment. This in situ analysis provided clear insights into the reaction mechanism of the formation of a CBZ.NIC cocrystal via Resodyn Lab RAM in the absence of milling bodies.

8.4 Resonant Acoustic Mixing (RAM): A New Mixing Technique

Over the years several techniques have been investigated for achieving the required mixing quality, as well as to prevent segregation after mixing pharmaceutical powders. Segregation can either be prevented by granulation or by designing a powder mixture and a process which minimise segregation potential if a direct compression process shall be realised. Kottlan et al. [30] conducted research focused on the acoustic mixing (blending) step of this process and especially on the quantification of the mixing quality. The authors investigated the influence of fill level, mixing time, and acceleration magnitude on the degree of mixing. As the aim was to develop a technique to assess mixing quality on a visual basis, no APIs were used; microcrystalline celluloses (MCC), i.e., Emcocel 90 M (E-90) and Avicel PH-102 (A-102), were used instead. Additionally, the level of mixing of the powders was assessed by an image analysis developed by the Quantitative Image Group, TU Delft, Netherlands and EzyFit [31]. Figure 8.7 shows intense-mixing zones within the powder bed exhibiting a lower density and higher particle velocities.

In general, these mixing zones lead to homogenisation by strongly enhancing diffusive mixing and by the convection induced in denser regions of the powder bed. Regardless of the frequency, we observed that the formation of dilute intense-mixing zones starts at the top of the bed. In these intense-mixing zones, the powder behaves like a granular gas, as it is observed for shallow granular beds at lower accelerations. For 100 Hz, a mixing zone can be observed that starts to form at the top and then in the course of time spreads over about 70% of the container, leading to fast mixing. Together with at least one other mixing zone, it produces an additional convective flow, conveying powder upwards in the middle of the container. The combination of these two phenomena is responsible for mixing at this frequency. From this research work, it was confirmed how resonant acoustic missing can be applied to generate a

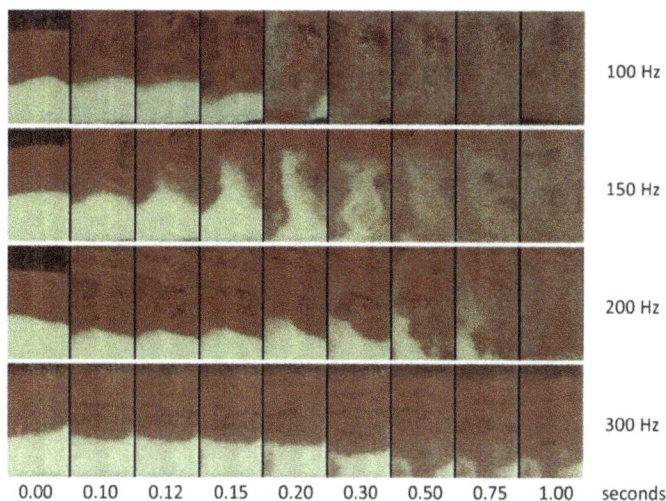

FIGURE 8.7 Images recorded during the mixing process showing the progress of mixing. (For interpretation of the references to colour in this figure legend, the reader is referred to the web version of this article.)

single-tablet-scale mixing process. Mixing performance was investigated for a variety of frequencies and amplitudes using a coloured tracer, high-speed video recordings, and image analysis. Results provided insights into the rate and quality of the mixing process.

REFERENCES

1. ResoDyn Acoustic Mixers; ResonantAcoustic® Mixing Technical White Paper, 2010.
2. A. Vandenberg, K. Wille; *Construction & Building Materials* 164 (2018) 716–730.
3. K. S. Hope, H. J. Lloyd, D. Ward, A. A. L. Michalchuk; Resonant Acoustic Mixing and Its Applications to Energetic Materials. *NRTEM* (2015).
4. D. J. A. Ende, S. R. Anderson, J. S. Salan; Development and Scale-Up of Cocrystals Using Resonant Acoustic Mixing. *Organic Process Research & Development* 18 (2014), 331–341.
5. R. Tanaka, K. Ashizawa, Y. Nakamura, N. Takahashi, Y. Hattori, M. Otsuka; Verification of the mixing processes of the active pharmaceutical ingredient, excipient and lubricant in a pharmaceutical formulation using a resonant acoustic mixing technology. *RSC Advances* 6 (2016), 87049–87057.
6. K. Zheng, K. Kunnath, Z. Ling, L. Chen, R. Davé; Influence of guest and host particle sizes on dry coating effectiveness: When not to use high mixing intensity. *Powder Technology* 366 (2020), 150–163.
7. D. I. A. Millar, W. G. Marshall, I. D. H. Oswald, C. R. Pulham; High-pressure structural studies of energetic materials. *Crystallography Reviews* 16 (2010), 115–132.
8. A. Bak, A. Gore, E. Yanez, M. Stanton, S. Tufekcic, R. Syed, A. Akrami, M. Rose, S. Surapaneni, T. Bostick, A. King, S. Neervannan, D. Ostovic, A. Koparkar; *Journal of Pharmaceutical Sciences* 97(9) (2008), 3942–3956.
9. D. P. McNamara, S. L. Childs, J. Giordano, A. Iarriccio, J. Cassidy, M. S. Shet, R. Mannion, E. O'Donnel, A. Park; *Pharmaceutical Research* 23(8) (2006), 1888–1897.
10. K. Stavropoulos, S. C. Johnston, Y. Zhang, B. G. Rao, M. Hurrey, P. Hurter, E. M. Topp, I. Kadiyala; *Journal of Pharmaceutical Sciences* 104(10) (2015), 3343–3350.
11. N. Variankaval, R. Wenslow, J. Murry, R. Hartman, R. Helmy, E. Kwong, S. D. Clas, C. Dalton, I. Santos; *Crystal Growth & Design* 6(3) (2006), 690–700.
12. Y. Wang, Z. Yang, H. Li, X. Zhou, Q. Zhang, J. Wang, Y. Liu; *Propellants, Explosives, Pyrotechnics* 39(4) (2014), 590–596.
13. R. A. Chiarella, R. J. Davey, M. L. Peterson; *Crystal Growth & Design* 7(7) (2007), 1223–1226.
14. T. Friščić, L. Fábián, J. C. Burley, W. Jones, W. D. S. Motherwell; *Chemical Communications* 48 (2006), 5009–5011.
15. E. Colacino, M. Carta, G. Pia, A. Porcheddu, P. C. Ricci, F. Delogu; *ACS Omega* 3(8) (2018), 9196–9209.

16. E. Colacino, G. Dayaker, A. Morère, T. Friščić; *Journal of Chemical Education* 96(4) (2019), 766–771.
17. D. Daurio, C. Medina, R. Saw, K. Nagapudi, F. Alvarez-Núñez; *Pharmaceutics* 3(3) (2011), 582–600.
18. D. Daurio, K. Nagapudi, L. Li, P. Quan, F. Alvarez-Núñez; *Faraday Discussions* 170 (2014), 235–249.
19. R. S. Dhumal, A. L. Kelly, P. York, P. D. Coates, A. Paradkar; *Pharmaceutical Research* 27(12) (2010), 2725–2733.
20. C. Kulkarni, C. Wood, A. L. Kelly, T. Gough, N. Blagden, A. Paradkar; *Crystal Growth & Design* 15(12) (2015), 5648–5651.
21. S. Li, T. Yu, Y. Tian, C. P. McCoy, D. S. Jones, G. P. Andrews; *Molecular Pharmaceutics* 13(9) (2016), 3054–3068.
22. D. E. Crawford, C. K. G. Miskimmin, A. B. Albadarin, G. Walker, S. L. James; *Green Chemistry* 19(6) (2017), 1507–1518.
23. K. Nagapudi, E. Y. Umanzor, C. Masui; *International Journal of Pharmaceutics* 521(1–2) (2017), 337–345.
24. Z. Guo, C. Boyce, T. Rhodes, L. Liu, G. M. Salituro, K. Lee, A. Bak, D. H. Leung; *International Journal of Pharmaceutics* 592 (2021), Article 120026.
25. L. Gonnet, C. B. Lennox, J. Do, I. Malvestiti, S. G. Koenig, K. Nagapudi, T. Friščić; *Angewandte Chemie International Edition* 61(13) (2022), Article e202115030.
26. A. Karagianni, M. Malamatari, K. Kachrimanis; *Pharmaceutics* 10(1) (2018 March), 18.
27. I. Sović, S. Lukin, E. Meštrović, I. Halasz, A. Porcheddu, F. Delogu, P. C. Ricci, F. Caron, T. Perilli, A. Dogan, E. Colacino; *ACS Omega* 5(44) (2020), 28663–28672.
28. M. Frenette, G. Cosa, T. Friščić; *CrystEngComm* 15 (2013), 5100–5106.
29. A. A. L. Michalchuk, K. S. Hope, S. R. Kennedy, M. V. Blanco, E. V. Boldyrev, C. R. Pulham; *Chemical Communications* 54 (2018), 4033.
30. A. A. L. Michalchuk, K. S. Hope, S. R. Kennedy, M. V. Blanco, E. V. Boldyreva, C. R. Pulham; *Chemical Communications* 54(32) (2018), 4033–4036.
31. A. Kottlan, B. J. Glasser, J. G. Khinast; *Powder Technology* 387 (2021), 385–395.

Part 3

Case Studies/Perspective

9

Sustainable Large-Scale Synthesis of Fine Chemicals, Active Pharmaceutical Ingredients and Biologically Active Compounds by Non-traditional Activation Methods and Biocatalysis

Manisha Mishra and Béla Török

CONTENTS

9.1 Introduction

The concept of sustainable methodologies in the chemical industry has been developed to meet the scientific challenges by eliminating but at the least limiting environmental hazards and pollution, which were the result of traditional chemical production, was first brought by Paul Anastas and Roger Garrett in 1991 and later termed "green chemistry (GC)" by Joe Breen.[1] This green chemistry concept became increasingly popular since 1995 when the US Presidential Green Chemistry Challenge (PGCC) has been established to highlight and endorse hazard-free, environmentally conscious and economically competitive synthesis in the chemical and pharmaceutical industries in the United States. The foundation of the field became stronger and more clear with the introduction of the 12 principles of green chemistry by Anastas and Warner.[2] A recent report predicted that the global market of green production is increasing exponentially, as it hit \$98.5 billion by the end of 2020.[3] There has been an increasing focus on the adoption of principles of green chemistry (GC) for attaining more sustainable manufacturing of pharmaceuticals[4–6] and fine chemicals[7] on a large scale. The synthesis of raw materials, intermediates and active pharmaceutical ingredients (APIs) generates substantial waste byproducts and hazardous contaminants, toxic solvents, reagents and contaminated gas. The practice of sustainable methods is now being emphasized more to address these issues.[8] Multiple emerging areas of research that include the development of green solvents,[9] use of water as the reaction medium[10–12] and solvent-free or no-media-involved reactions,[13, 14] reaction media in the form of aqueous biphasic catalysis,[15] the application of sustainable raw materials[16] as well as the use of non-traditional activation techniques[17, 18] have been developed and became favored over the traditional processes to synthesize various fine chemicals and APIs.[19] The application of different types of catalysis was also broadly investigated for the same purpose.[20–23] Considering the importance of the employment of sustainable approaches in the chemical industry, here, in this chapter, we will be reviewing the preparative to large-scale synthesis of various fine chemicals and APIs with

DOI: 10.1201/9781003178187-12

the aid of alternatives or non-traditional energy input systems such as microwave heating, ultrasounds, mechanochemical mixing, photochemical synthesis and application of biocatalysts.

9.2 Microwave-Assisted Synthesis of Fine Chemicals and APIs

High-speed microwave-assisted syntheses with modern and innovative techniques[24–27] are very common in contemporary organic chemistry research. Microwave-mediated reactions reduce reaction times, increase product yields and enhance product purities compared to conventionally heated experiments, and thus microwave protocols are used extensively in pharmaceutical, agrochemical and related industries as a frontline methodology.[28–32] While many reactions have been performed on a small scale using microwave heating, few reactions have been further developed into preparative-scale synthesis at least to the gram scale as a batch format using multimode microwave instrumentation.[33–36] Since microwave irradiation is able to penetrate only a few centimeters into the reaction mixture, the maximum volume that can be irradiated is about 1 L[37] in one batch. Therefore, it is necessary to run multiple batches for a large-scale production. An alternative option to the batch process is to create a batch (or stop)-flow arrangement where the reactor is set up inside the microwave reactor with ports and the reaction mixture is pumped through it to a receiving container. This technique has rapidly found its place in laboratories because it appears as the solution to scale up reactions to a larger scale. In the continuous flow technique, the reaction mixture is continuously pumped through a microwave reactor.[38] This continuous operation of a small reactor is capable of making large quantities of product in a short time.[39] Synthetic protocols that are aimed at large-scale synthesis have been recently reviewed,[40] providing a comprehensive picture of MW-assisted synthesis. Current research has focused more on performing microwave chemistry under continuous flow conditions.[41–46] Herein, we will review the preparative to large-scale synthesis of fine chemicals and APIs under batch and continuous flow process in microwave-assisted reactions.

A bench-top microwave batch reactor that can accommodate a single 1 L reaction vessel has been explored in the large-scale synthesis of various molecules. A master-wave BTR-manufactured microwave instrument served for these batch-type scale-up reactions in a 250–750 mL scale following the one-vessel-at-a-time concept. Depending on the specific chemistry, a daily output of several hundred grams of product per day was attainable with sequential batch processing. Benzimidazoles, an important class of heterocycles which are found in numerous biologically active substances,[47] were synthesized in the large batch reactor, that was heated using the full 1700 W magnetron power. The 630 mL reaction mixture (*o*-phenylenediamine and acetic acid) was heated for 7 min leading to a total reaction time of 34 min, including cooling to 60 °C. By using this multimode bench-top reactor, 300 g of benzimidazole product was isolated[48] (Scheme 9.1). Several other processes, such as the Newman–Kwart and Diels–Alder reactions, Pd-catalyzed cross-couplings and a Knoevenagel condensation, were also investigated.

Cyclic acetals and ketals, in particular 1,3-dioxolane and its derivatives, are substances of interest as they serve as reagents or solvents in a variety of organic syntheses.[49] They are also used in the production of various materials.[50] Hamelin and co-workers[51] developed a scaled-up technique for the synthesis of dioxolanes, dithiolanes and oxathiolanes which was rapid and environmentally friendly. In these two-step reactions mediated by montmorillonite K10 under microwave irradiation (ketalization and transketalization) as a batch process, dioxolanes, oxathiazoles and dithiolanes were isolated (Scheme 9.2) with >75% yield. The reactions were scaled up to as high as 2 moles.

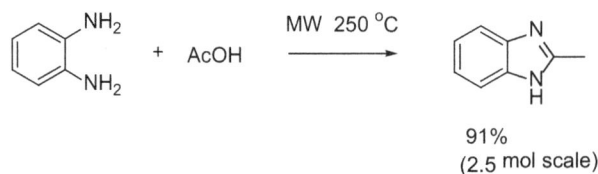

91%
(2.5 mol scale)

SCHEME 9.1   Microwave-assisted synthesis of 2-methylbenzimidazole.

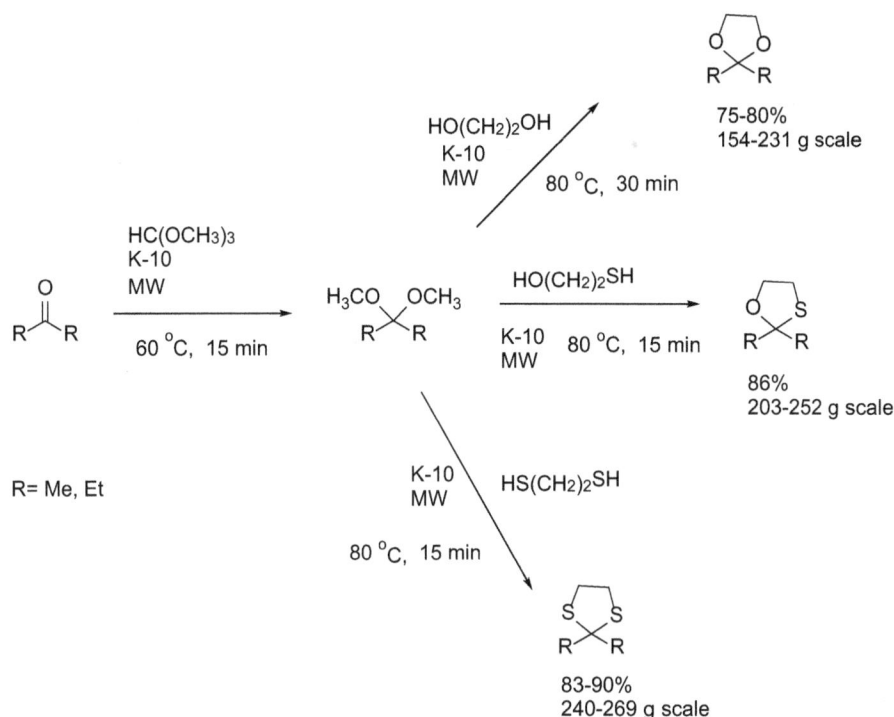

SCHEME 9.2   Synthesis of dioxolanes, oxathiazoles and dithiolanes by a microwave-assisted K-10 catalyzed reaction.

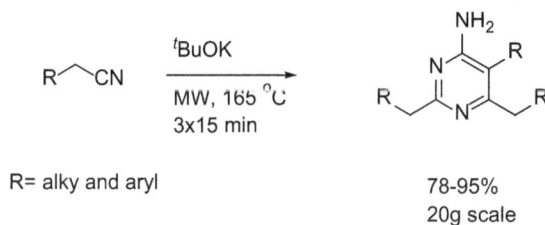

SCHEME 9.3   Microwave-assisted synthesis of 4-aminopyrimidines.

4-Aminopyrimidine is an important key heterocyclic component in many commercial antiseptics like Acsulfiso, Aristoplo (Aristoplomb), Agsulfiso, Sulfiona (Aristamid), Sulfisomi and Forsulfis (traded under Aristamid, Belfarosan, Cibazol, Dimedon and Domian).[52, 53] A simple and scalable method was developed[54] for the synthesis of 4-aminopyrimidines by trimerization of nitriles in 20 g scale in a single-mode batch process. The treatment of a neat sample of various nitriles in the presence of a catalytic quantity (3 mol %) of potassium *tert*-butoxide at 165 °C for 3 × 15 min under microwave irradiation provided the pure pyrimidine products (Scheme 9.3).

Various scaled-up (100 mmol) organic transformations involving different solvents and reaction conditions were studied by Stadler et al.[55] in larger multimode systems. The transformations included multi-component Biginelli dihydropyrimidine and Kindler thioamide syntheses, Heck and Negishi reactions, solid-phase organic synthesis and Diels–Alder cycloaddition reactions using gaseous reagents in pre-pressurized reaction vessels. The reactions were carried out in a parallel batch processing technique in the microwave reactor (1400 W multimode) equipped with magnetic stirring, on-line temperature and pressure control. The results of these scale-up experiments are summarized in Scheme 9.4. Depending on the type of reaction and the applied scale, 10–100 g of product could be isolated from a single run.

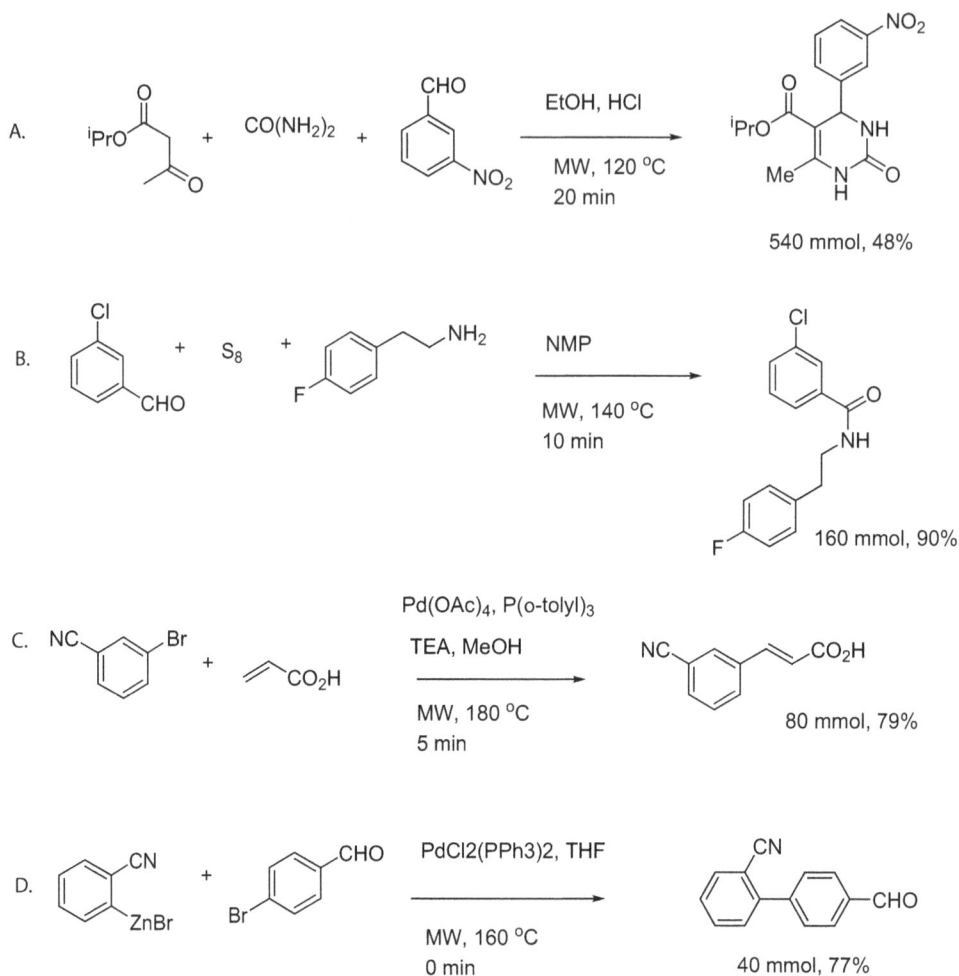

SCHEME 9.4   C–C bond forming reactions by microwave activation.

Acacetin (5,7-dihydroxy-4′-methoxyflavone), a constituent of flavone naturally present in plants, has anti-cancer and anti-inflammatory activities. Acacetin inhibits cell growth and cell cycle progression[56] and induces apoptosis in gastric carcinoma cells,[57] breast cancer MCF-7 cells,[58] prostate cancer DU145 cells,[59] and lung cancer AS49 cells via activating the caspase cascade and regulating other molecular factors related to apoptosis.[60] There are various methods available in the literature describing the synthesis of this flavone derivative on small scale.[61] Radoiu and group[62] have reported a green, simple and cost-effective system for the production of acacetin in bulk scale from phloroglucinol and excess ethyl -3-(4-methoxyphenyl)-3-oxopropionate in the presence of PEG1000 as a solvent via microwave irradiation at 915 MHz (Scheme 9.5). In the large scale-up, 50 g of phloroglucinol, 149 g ketoester and 180 mL PEG1000 were taken in a three-neck flask and heated with microwaves to 250 °C. The reaction was run for 70 min. After extraction with ethanol and removing the solvent, 79 g crude product (70% yield) was isolated.

Eptifibatide, a disulfide-bridged cyclo-heptapeptide drug, is marketed in the United States under the trademark Integrilin (Figure 9.1). This drug is used as an antithrombotic agent to treat patients with acute coronary syndrome and/or acute myocardial infarction.[63] A microwave-assisted solid-phase peptide synthesis (MW-SPPS) scalable to kilogram-scale manufacturing, for the preparation of eptifibatide in a single reactor process, was described by Papini's group.[64] The linear peptide-resin chain was developed in the first step under microwave conditions. After several operations that include amino acid couplings,

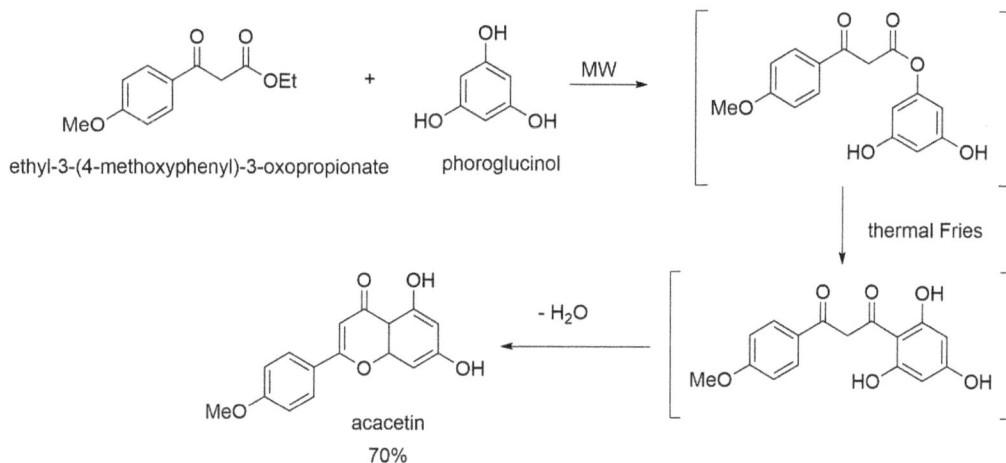

SCHEME 9.5 Microwave-assisted synthesis of acacetin.

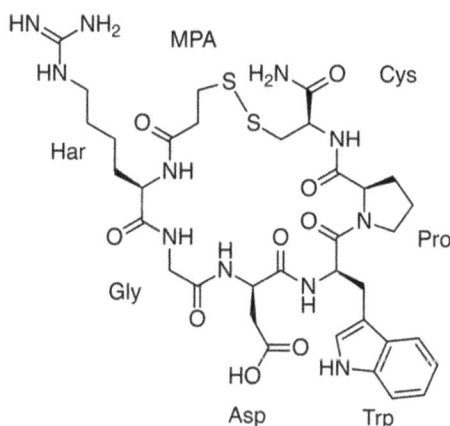

FIGURE 9.1 The structure of eptifibatide: N^6-(aminoiminomethyl)-N_2-(3-mercapto-1-oxopropyl)-L-lysylglycyl-L-α-aspartyl-L-tryptophyl-L-prolyl-L-cysteinamide, cyclic (1 → 6)-disulfide.

orthogonal side chain protecting groups' removal and final head to MPA on cysteine side chain cyclization, eptifibatide was obtained in 61% yield. The overall formation of this molecule has been depicted in Scheme 9.6.

1-Arylpyrazole, an important pharmacophore, is present in several marketed drugs.[65] A batch microwave scale-up synthesis of the pyrazole unit was performed by Kappe and group[66] in a Synthos 3000 multimode instrument. The condensation of acetylacetone with phenylhydrazine in ethanol under acidic conditions was studied. In the scale-up experiments, the best result ultimately was achieved by using 3 mol/L phenylhydrazine solution in ethanol, 1.1 equivalent of acetylacetone and 1 mol % HCl as an additive (Scheme 9.7). The reaction was processed in 43 min providing 95% yield (468 g).

Vitamin E analogs are important natural antioxidants that protect cells by scavenging excess free radicals. Vitamin E derivatives with shorter isoprenoid side chains present higher antioxidant capacity and improved hydrophilic character as documented by Pentland et al. in their study on cellular arachidonic acid metabolism.[67] Since the demand for vitamin E derivatives has significantly increased in recent years, it is desirable to increase the production of vitamin E and its derivatives on large scale. The vitamin E family is comprised of eight chromanol analogs that fall into two subfamilies: tocopherols (TPs) and tocotrienols (T-3s). These families are characterized by the hydrocarbon chain at chromanol-C2.

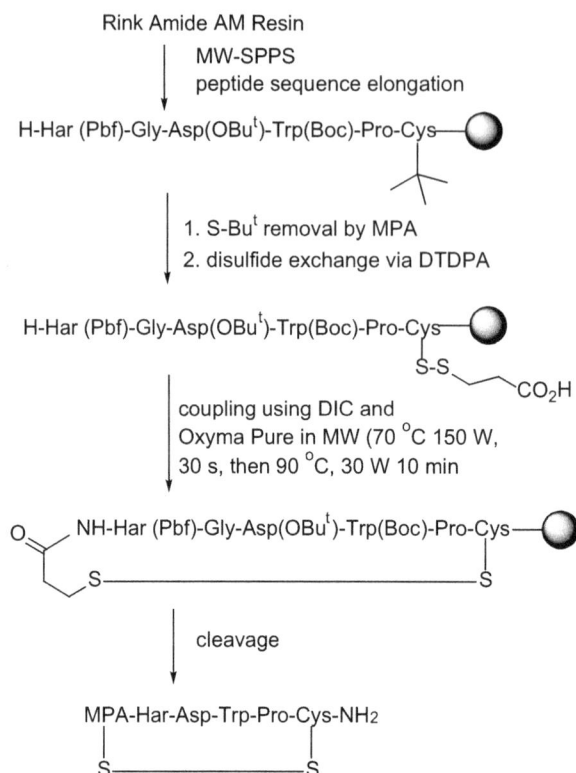

SCHEME 9.6   Microwave-assisted solid-phase synthesis of eptifibatide.

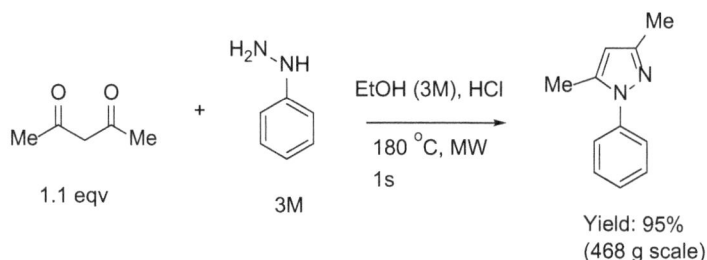

SCHEME 9.7   Pyrazole synthesis under microwave-assisted conditions.

Whereas T-3s have a C2 isoprenoid side chain with three non-conjugated double bonds (C3′–C4′, C7′–C8′, C11′–C12′). The TPs have an identical but fully saturated chain. The four naturally occurring TP and T-3 analogs (α-, β-, γ-, δ-) differ from each other only through variations in methyl substitutions at C5, C7 and C8 on the aromatic chromanol ring. The scale-up of MW-assisted Friedel–Crafts alkylation was investigated by Cravotto et al.[68] up to a 100 g scale for the synthesis of vitamin E and other derivatives. A suspension of TMHQ and $Sc(OTf)_3$ in toluene was heated at 110 °C (400 Watt) under vacuum (200 mbar) inside the MW rotating reaction vessel (200 rpm). The high-yielding protocol is potentially suitable for industrial applications (Scheme 9.8).

Even though the large laboratory MW units have been able to produce fine chemicals and various APIs to at the gram or kilogram scale in batch processes, these systems have a few inherent limitations. Recently, numerous studies on microwave-assisted synthesis have revealed the possibility of transferring the methodology from the batch mode to a conventionally heated or microwave-heated flow mode. Continuous flow (CF) systems developed in microwave instruments have attracted significant interest

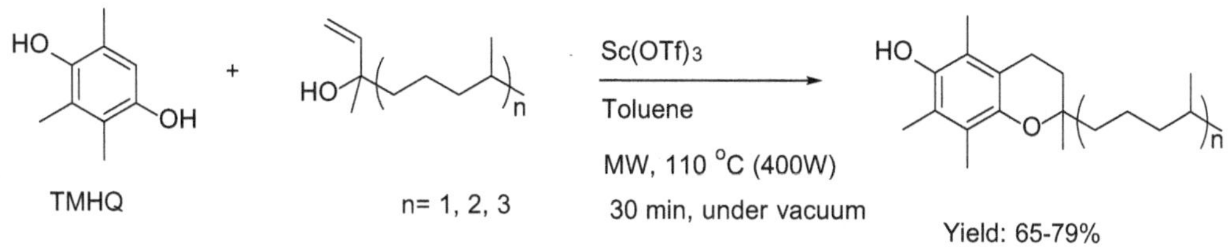

SCHEME 9.8 Microwave-assisted synthesis of vitamin E derivatives.

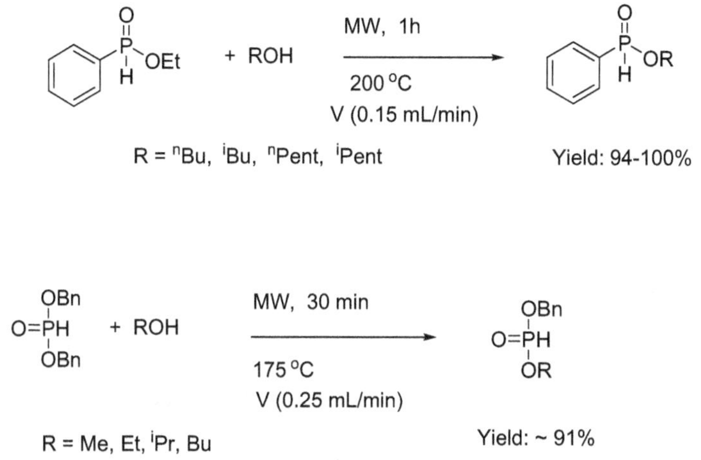

SCHEME 9.9 Microwave-assisted flow synthesis of phosphonates and phosphinates.

due to the advantages in safety, handling and scale-up possibilities.[38, 69] Microwave-assisted CF synthesis methods have been used successfully for many different organic transformations.[70–74]

H-Phosphinates and H-phosphonates are typical starting materials for the Hirao P–C coupling reactions and the Kabachnik–Fields condensations, resulting in the formation of aryl-phosphinates/phosphonates and α-aminophosphonates, respectively. Syntheses of H-phosphinates and H-phosphonates into flow processes under microwave conditions with ionic liquid (IL) additives were investigated by Keglevich's group.[75] Both esterification and trans-esterification reactions were studied (Scheme 9.9). In the esterifications, the flow preparation afforded ca. 0.75 g ester/30 min. The continuous flow reactions were performed in a system using a CEM® Discover (300 W) focused MW reactor equipped with a CEM® 10-mL Flow Cell Accessory continuous flow unit (irradiated volume 7 mL).

Indene-C_{60} monoadduct ($IC_{60}MA$) and indene-C_{60} bisadduct ($IC_{60}BA$) were identified as important components in the making of polymer-based solar cells due to their higher LUMO energy levels which result in higher open cell voltages.[76] Indene-C_{60} monoadduct ($IC_{60}MA$) and indene-C_{60} bisadduct ($IC_{60}BA$) were identified as superior to PCBM as PSC components.[77] $IC_{60}MA$ was recently employed in perovskite solar cells as well.[78]

A scalable, safe, and efficient continuous synthesis of photovoltaic cell components $IC_{60}MA$ and $IC_{60}BA$ was demonstrated[79] (Scheme 9.10) using a non-chlorinated solvent by means of a commercially available microwave flow reactor system. A key factor behind the successful preparation of $IC_{60}MA$ and $IC_{60}BA$ is their enhanced miscibility in polymeric donors. The Diels–Alder reaction has been successfully studied in the chemical derivatization of graphene and fullerene. Indene-C_{60} monoadduct and an indene-C_{60} bisadduct were synthesized on a gram scale (1.07 g/h), maintaining the flow rate at 11.6 mL/min.

Scale-up synthesis of benzimidazoles was also explored by Kappe's group[74] in a continuous flow microwave system in the kilogram scale operating at high temperature/high pressure processing 20 L/h. Organ et al. established another example of this reaction carried out in a microwave-assisted continuous

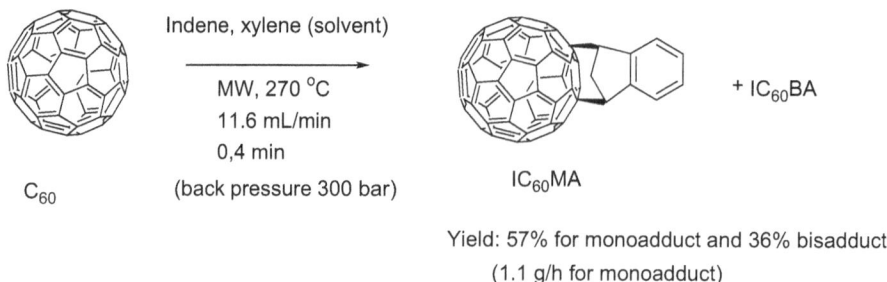

Indene, xylene (solvent)

MW, 270 °C
11.6 mL/min
0,4 min
(back pressure 300 bar)

C_{60}

$IC_{60}MA$

+ $IC_{60}BA$

Yield: 57% for monoadduct and 36% bisadduct
(1.1 g/h for monoadduct)

SCHEME 9.10   Synthesis of $IC_{60}MA$ and $IC_{60}BA$.

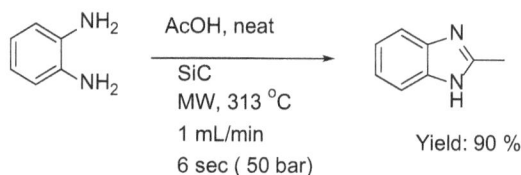

AcOH, neat

SiC
MW, 313 °C
1 mL/min
6 sec (50 bar)

Yield: 90 %

SCHEME 9.11   Synthesis of 2-methylbenzimidazole in a microwave-assisted continuous flow system.

flow reactor. This system was equipped with a SiC reactor tube and back-pressure regulators that guaranteed no pressure fluctuations during the reaction.[80]

The continuous multitubular millireactor/heat exchanger (MTMR) assembly of a commercial microwave instrument has been developed for the scale-up of fine chemical synthesis. The inner walls of the tubular millireactors were coated with copper. This copper coating can act as a microwave-absorbing material as well as a catalyst.[81] The copper-coated MTMR was tested to run a multicomponent reaction (MCR) to produce 1,3-diphenyl-2-propynyl piperidine starting from benzaldehyde, piperidine and phenyl acetate (Scheme 9.12). The production of propargyl amine was commercially successful on a scale of 1 kg/day.[82] A production rate of 333 ± 11 $kg_{prod}/(kg_{cat}\cdot h)$ was achieved in a single microwave cavity at 373 ± 5 K and at a total reactant flow rate of 1.66×10^{-9} m³/s.

9.3 Ultrasound-Assisted Synthesis

Ultrasounds had been applied in chemical reactions for nearly a century,[83] starting with Loomis' experiments in 1927.[84, 85] The first mechanism explaining its effect on reactions was published in 1945.[86] Its use is extensive now, with the widespread synthetic applications and the non-invasive medical diagnostics started in the 1980s, and by now, the use of ultrasonics became a mainstream technique in both laboratories and industries,[87, 88] significantly contributing to green synthetic efforts as well.[89–93] The typical frequency range for sonochemical synthesis is 20–100 kHz, also called power ultrasounds.[94] The higher frequency ranges are called high-frequency ultrasounds (100 kHz to 2 MHz),[95] and diagnostic ultrasound for medical use (5–10 MHz).[96]

In contrast to MAOS, sonochemical synthesis is limited to liquids. Theoretical calculations indicated that energy released during cavitation generates high temperature (1000–3000 K) and pressure (~10^5 kPa)[97] In the early 2000s Suslick et al. provided experimental evidence, observing 4000–5000 K temperatures.[98] The best interpretation of the experimental observations is the *hot spot* model,[99] but other theories (e.g. electric model)[100] have also been proposed. The experimental variables significantly affect sonochemical synthesis.[101] Frequency, power, temperature, external pressure and the protecting gas or solvent are all important variables. As an additional benefit, the ultrasonic treatment also provides extremely effective mixing and powerful surface cleaning that is particularly important when using solid catalysts or reagents (e.g. Grignard reagent preparation). Herein, we will illustrate the preparative to large-scale synthesis of fine chemicals and APIs under sonochemical conditions.

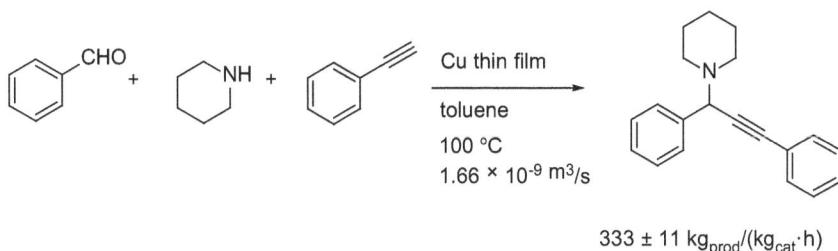

SCHEME 9.12 Microwave-assisted propargyl amine synthesis catalyzed by Cu.

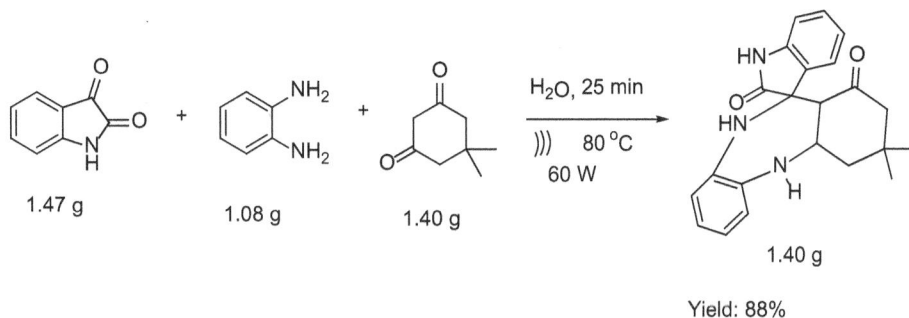

SCHEME 9.13 Gram-scale synthesis of benzodiazepines under ultrasound irradiation.

Benzodiazepines are a class of nitrogen heterocycles in which the benzene ring is connected with the diazepine ring. Benzodiazepines are considered the most vital class of heterocyclic compounds due to their widespread biological and pharmacological activities.[102] A green protocol via ultrasonic-assisted synthesis of benzodiazepine ring was developed by Singh and coworkers.[103] The multicomponent reaction was carried out by using various substituted isatin, diphenylamine and 1,3-diketone derivatives as reactants in an aqueous medium with excellent yield (95%). In the gram-scale synthesis of benzodiazepines, a mixture of isatin (10 mmol, 1.47 g) 1,2-phenylenediamine (10 mmol, 1.08 g) and 5,5-dimethyl cyclohexane-1,3-dione (10 mmol, 1.40 g) in water was stirred and heated at 80 °C for 25 min under ultrasonic irradiation. The product was isolated as a solid precipitate when cold water was added to the reaction mixture after the desired time (Scheme 9.13).

Another important nitrogen heterocycle towards many vital biologically important molecules, such as the cholesterol-lowering drug atorvastatin[104] and the anti-inflammatory analgesic tolmetin,[105] is pyrrole. In the traditional method, pyrroles are derived through the condensation reaction of α-aminocarbonyl compounds with 1,3-dicarbonyls in the Knorr condensation reaction.[106] Glacial acetic acid is used as the solvent in this reaction. Wang et al. investigated the possibility of using lactic acid instead of glacial acetic acid as a bio-based green solvent for the scalable sonochemical synthesis of pyrrole derivatives using the Knorr condensation.[107] This ecofriendly protocol successfully produced the desired pyrrole derivatives on a large scale with excellent yield (up to 92%) under low-power ultrasonic irradiation which were provided by an ultrasonic cleaner at 40 kHz, 500 W (Scheme 9.14). This synthesis uses lactic acid which is inexpensive, safe and easy to handle and hence it could be very useful in the industrial preparation of pyrroles.

Chalcones are naturally occurring molecules consisting of two aromatic rings connected by an α,β-unsaturated ketone. Due to their natural fluorescence, potential pharmaceutical properties and antiproliferative activity against leukemic cells, the synthesis of chalcones has gained interest in the pharmaceutical industry.[108] The most conventional strategies for the synthesis of chalcones in the laboratory setting are to perform Claisen–Schmidt condensation in either basic or acidic media under homogeneous conditions followed by in situ dehydration. Arafa[109] has reported a sustainable Claisen–Schmidt-type reaction to synthesize a series of chalcone derivatives under ultrasound-assisted conditions in water using novel IL,

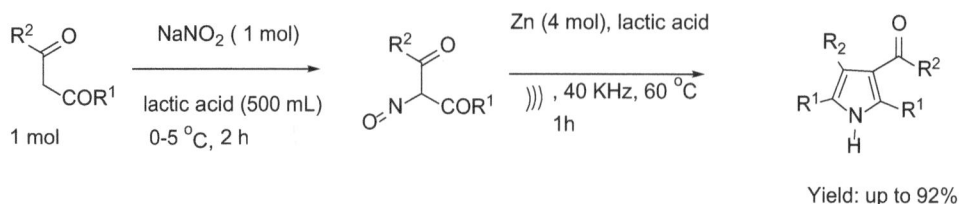

SCHEME 9.14   Scale-up synthesis of pyrroles under ultrasound irradiation.

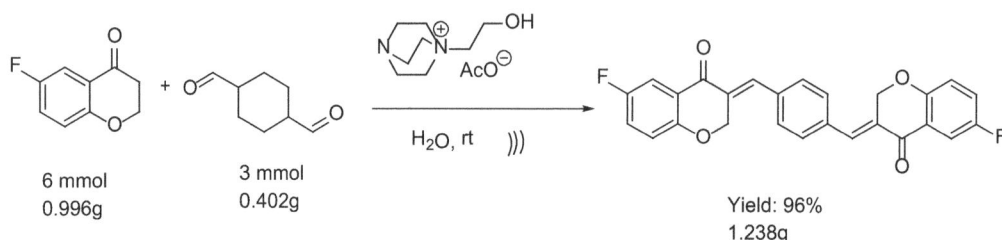

SCHEME 9.15   Scaled-up synthesis of chalcones in an ultrasound-activated protocol.

[DABCO-EtOH][AcO]. This protocol is ecofriendly green and sustainable. In a scale-up experiment, 6-fluorochroman-4-one (6 mmol, 0.996 g) and terephthalaldehyde (3 mmol, 0.402 g) were chosen using 15 mol% of [DABCO-EtOH][AcO]. The reaction was run under ultrasonic irradiation at room temperature for 10 min, which in turn provided 96% (1.238 g) yield in a 3 h reaction (Scheme 9.15).

Imines and their derivatives are very useful and common intermediates in various organic syntheses, for the preparation of heterocycles and β-aminoacids.[110] Stefani et al.[111] have developed a mild, convenient and improved protocol for the preparation of imines by sonochemical activation (Scheme 9.16). The imines were synthesized by the reaction of aldehydes and amines in the presence of aluminum silica as the promoter in ethanol as the solvent. In the scale-up condition, starting aldehydes and imines were taken on a 50 mmol scale. The reaction afforded the desired imines in a very high yield (85–96%).

Polyvinyl alcohol (PVA) is a linear synthetic polymer produced via partial or full hydrolysis of polyvinyl acetate to remove the acetate groups. The amount of hydroxylation determines the physical characteristics, chemical properties and mechanical properties of the PVA. PVA is used in several medical devices that are implanted in the body. Particulate PVA has been used to treat vascular embolisms,[112] to prepare hydrophilic coatings to improve neurologic regeneration,[113] and as tissue adhesion barriers.[114] An interesting aspect of PVA is its flexibility to be cross-linked with several cross-linking reagents, such as phytic acid, which improves the thermal stability and conductivity of PVA. The porous structure of PVA has been explored well in bioengineered tissue scaffolds. On the other hand, phytic acid being a bulky molecule will create extra free volume for gas adsorption due to its inefficient packing in the solid state. PVA/PA polymer films show mesoporosity, and thus, they are promising materials for gas/vapor sorption.[115] An ultrasound-assisted synthesis of polyvinyl alcohol/phytic acid polymer (PVA/PA polymer) from PVA and PA via the esterification reaction was developed by Niu and coworkers.[116] In the experimental procedure, 2.20 mL of PA was dissolved in 5.00 mL of distilled water first followed by 1 min of ultrasound radiation. Afterwards, a 60.00 mL aqueous solution of PVA (8g PVA) was added into the PA solution and irradiated for 3h, at around 90 °C to obtain the PVA-PA polymer (Scheme 9.17).

9.4 Application of Photochemistry in the Scalable Synthesis of Fine Chemicals and APIs

Light has long been recognized as a valuable tool for performing organic reactions.[117] It acts as a traceless reagent: adding energy to a chemical system without generating waste in the process.[118] Historically, UV light has found most application in synthesis; however, recent studies have shown that visible light can

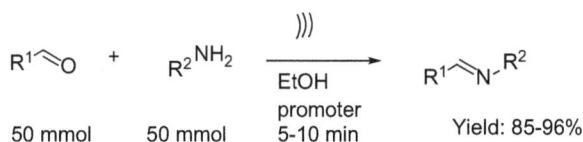

SCHEME 9.16 Sonochemical synthesis of imines.

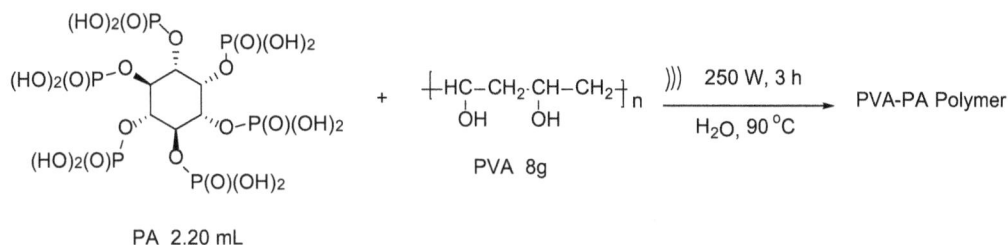

SCHEME 9.17 Synthesis of PVA-PA polymer by ultrasound-assisted conditions.

affect a wide range of useful transformations as well.[119] Applications of photochemistry towards the synthesis of APIs are still not very common. It is probably due to the large size of reaction vessels where the light cannot penetrate properly into the reaction mixture. The application of flow chemistry can provide a more efficient method to access photochemical transformations in a range of scales with inexpensive laboratory equipment.[120] In the flow process, the reaction mixture is carried out through transparent polymer tubing or a transparent chip microreactor which is irradiated with a light source.

New and practical continuous flow photochemical reactors have been designed by the Booker-Milburn group[121] using UV-transparent FEP tubing in conjunction with established UV lamp technology. These reactors have reported, for the first time, that preparative synthetic organic photochemistry can be carried out continuously in various scales by using readily available components in a standard laboratory fume hood. For example, the intramolecular [5 + 2] photocycloaddition of 3,4-dimethyl-1-pent-4-enylpyrrole-2,5-dione to form the bicyclic azepine was studied in this flow reactor. In a scale-up procedure, the starting material was irradiated at 8 mL/min flow rate, giving the bicyclic product in 80% an isolated yield. This is a highly significant protocol producing the azepine in 178 g in a 24 h period (Scheme 9.18).

A free radical tandem cyclization reaction was carried out with an N-substituted pyrrole derivative catalyzed by Ru-(bpy)$_3$Cl$_2$ in a simple flow reactor assembled with blue LEDs (5.88 W) irradiating a coil of PFA tubing[122] (Scheme 9.19). The reaction mixture was passed through this tube and an output of 2.88 mmol/h was obtained in 1 min residence time. When the same reaction was performed in a batch process with a 15 W fluorescent lamp, the output was 0.012 mmol/h.

A unique and interesting application of a nanocrystalline suspension flow method for the enantiospecific photodecarbonylation of a hexasubstituted homochiral ketone was developed by Hernandez-Linares et al.[123] The synthesis, crystallization and solid-state photochemistry of acyclic, chiral, hexasubstituted (+)-(2R,4S)-2-carbomethoxy-4-cyano-2,4-diphenyl-3-pentanone suspended in water was chosen to demonstrate this solid-state photodecarbonylation. The reaction affords (+)-(2R,3R)-2-carbomethoxy-3-cyano-2,3-diphenyl-butane with two adjacent stereogenic, all-carbon substituted quaternary, centers in quantitative chemical yield and 100% diastereoselectivity and enantiomeric excess (Scheme 9.20). The reaction does not require further extraction, removal of reagents or purification. The scale-up reaction was carried out using two photoreactors that were connected in a series equipped with 450 W medium pressure Hg arc lamp.

As substituted benzyl bromides represent common photochemically accessible building blocks for many molecules that are important in the pharmaceutical and medicinal field, their scalable synthesis is important in the chemical industry.[124] Kappe et al.[125] have developed a process for performing photochemical benzylic bromination, utilizing in situ bromine generation from HBr and NaBrO$_3$, from lab to pilot scale. 2,6-Dichlorotoluene (DCT) was selected as a suitable model substrate for this case study

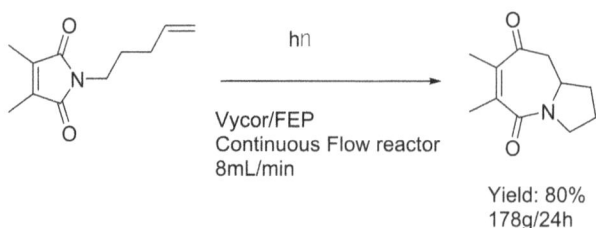

SCHEME 9.18   Synthesis of bicyclic azepine via a light-initiated [5+2]-cycloaddition reaction.

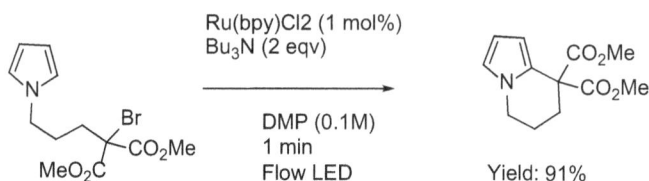

SCHEME 9.19   Visible light-initiated cyclization of an N-substituted pyrrole.

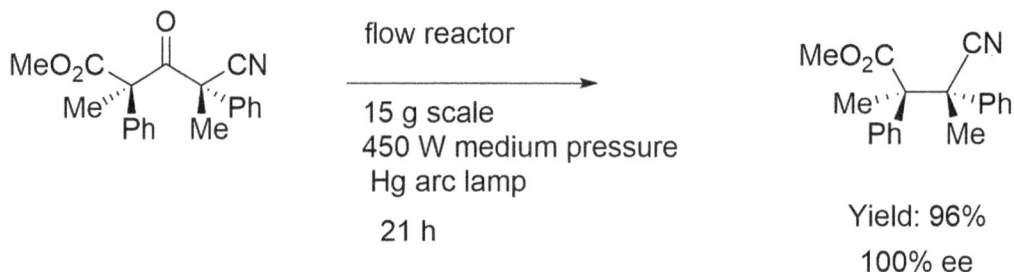

SCHEME 9.20   A flow method for the enantiospecific photodecarbonylation of a hexasubstituted homochiral ketone.

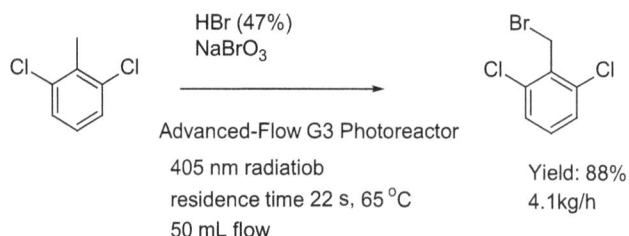

SCHEME 9.21   A large-scale photochemical benzylic bromination in a flow system.

(Scheme 9.21). This work is a very significant scale-up of a flow photochemical process, which demonstrates progress towards the widespread implementation of large-scale flow photochemistry in the fine chemical industry.

Since photocycloaddition has been a well-established way of accessing strained ring scaffolds, it can be a starting point for drug discovery. Based on this objective, Booker-Milburn et al. described the preparation of tricyclic aziridines in multigram quantities by a flow photocycloaddition protocol. These strained products can undergo different ring-opening/annulation reactions to yield a broad range of fused polyheterocycles. Starting with aziridine, the alkaloid (±)-3-demethoxyerythratidinone was obtained in 69% yield after a five-step sequence.[126] To synthesize the aziridine scaffold, Rutjes et al.[127] constructed a higher throughput system, built around a larger commercially available lamp. This system consists of a Philips PL-L 55 W UV–C (254 nm) lamp and FEP tubing (id 2.7 mm, reactor volume 105 mL wound

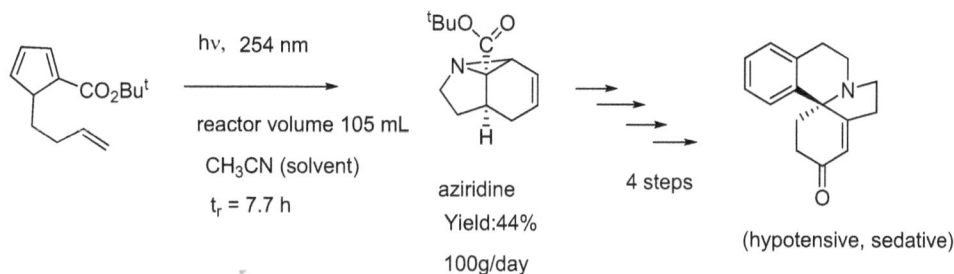

SCHEME 9.22 Photochemical flow cyclization to yield an aziridine precursor for the synthesis of (±)-3-demethoxyerythratidinone.

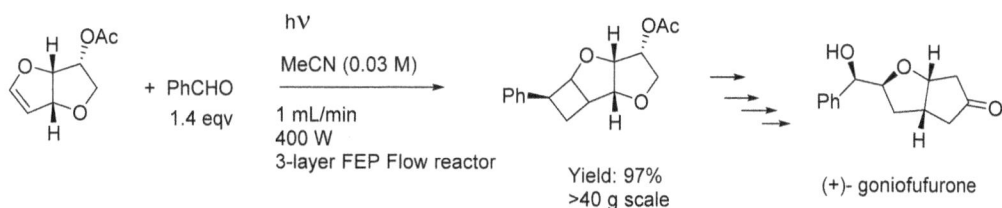

SCHEME 9.23 A photosynthetic flow protocol for the synthesis of a (+)-goniofufurone precursor.

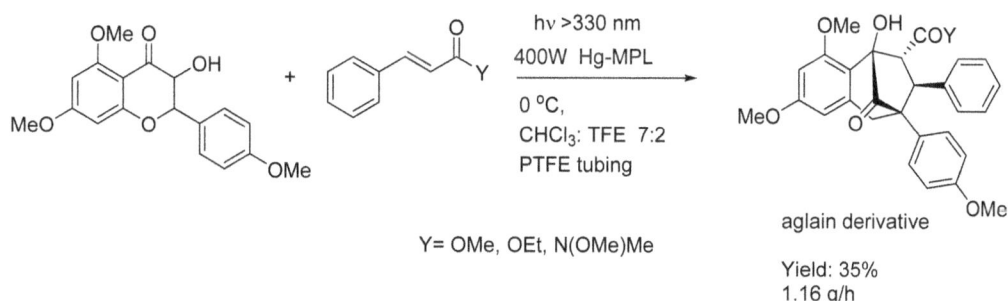

SCHEME 9.24 A photochemical flow [3 + s2]-cycloaddition reaction towards the synthesis of aglain.

around the lamp). The tubing is externally cooled with water and contained in a metal jacket to protect the surrounding environment from the UV light. This system reaches a significantly high throughput and can produce >100 g/day of aziridine via the cycloaddition reaction (Scheme 9.22).

A natural product (+)-goniofufurone, found in goniothalamus and isolated from *G. arvensis* trees of the plant family *Annonaceae*, exhibits promising antitumor properties.[128] The key precursor oxetane analog was synthesized involving the Paterno-Büchi [2 + 2] photocycloaddition in a flow reactor (Scheme 9.23). This method enables a very effective scaling up of the reaction (e.g., 150 g/24 h, 97% isolated yield) in a short time.[129]

The rocaglates are complex secondary metabolites isolated from *Aglaia* genus plants. There are several members of this family such as rocaglamide, silvestrol and rocaglaol that exhibit potent cytotoxicity for several cancer cell lines.[130] Beeler's research group[131] has developed a series of simple continuous flow photoreactors for carrying out the ESIPT [3+2]-photocycloaddition, generating aglain derivatives on a large scale (Scheme 9.24). It was observed that the reactors can be effectively numbered up to increase the reaction throughput without significant change in the overall reaction outcome. In this method a triple reactor system was lined up to synthesize 12.1 g of the critical aglain in 10.5 h.

Scharf and coworkers have used concentrated sunlight as an alternative and "freely available" sustainable light source for the solar production of selected fine chemicals.[132] Visible light-assisted facile

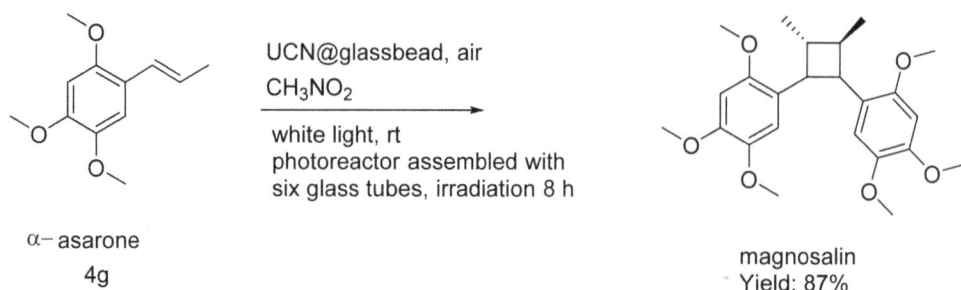

α− asarone
4g

magnosalin
Yield: 87%

SCHEME 9.25   A photocatalytic [2 + 2] dimerization of α-asarone catalyzed by UCN under white light.

production of high-value compounds in a continuous flow reactor in the presence of heterogeneous photoredox catalysis was also reported by Wang's research group.[133] Stable, inexpensive and sustainable polymeric carbon nitrides (UCN) were used as heterogeneous photocatalysts in the flow reactor system to produce the cyclobutane analogs via [2+2]-cycloaddition reaction of α-asarone catalyzed by UCN under visible light. In a continuous flow photoreactor with high light penetration, substituted cyclobutanes can be synthesized on a gram scale in a high yield of 81% (Scheme 9.25).

9.5 Mechanochemical Synthesis

Mechanochemical methods for the synthesis of chemicals is an environmentally friendly approach that can provide efficient, clean and high-yielding processes for modern synthetic chemistry.[134] It is widely used for the preparation of various substances, including small organic molecules, pharmaceutical cocrystals, metal–organic frameworks, nanomaterials and nanocomposites.[135] In mechanochemical synthesis, the materials are grounded in a mechanical ball mill or in hand grinding. Ball-Milling techniques are common and represent a powerful tool in which the reactions take place either in a solvent-free condition or using a small amount of solvent assisting grinding. During the last two decades, numerous protocols have been published using the method of ball milling for synthesis in many areas of organic chemistry.[136]

Spiroimidazolines are prevalent building blocks of numerous biologically active compounds such as a nicotinic acetylcholine receptor ligand,[137] α-adrenergic agonist6 and β-site amyloid precursor protein cleaving enzyme 1 (BACE1) inhibitors GNE-629 and GNE-892. A gram-scale synthesis of spiroimidazolines via N-iodosuccinimide promoted intermolecular cyclization of alkenes with amidines under solvent-free ball-milling conditions was reported by Xu et al. (Scheme 9.26).[138] This protocol is green, solvent-free and high yielding (up to 96%). The reactions were performed with amidines and alkenes along with N-iodosuccinimide (NIS) within a stainless steel ball (10 mm in diameter) in a Retsch MM400 mixer mill.

A very straightforward and efficient synthesis protocol was developed by Brahmachari et al.[139] for the synthesis of pharmaceutically interesting diversely functionalized (*E*)-3-(2-arylhydrazono)chro mane-2,4-diones. This one-pot three-component reaction was carried out between 4-hydroxycoumarins, primary aromatic amines and *tert*-butyl nitrite under ball-milling conditions without the aid of any catalysts/additives. In the gram-scale experiment, 4-hydroxycoumarin, *p*-toluidine and *tert*-butyl nitrite were chosen as reactants and ball-milled for 6 min at 500 rpm. The product was isolated with 94% yield (Scheme 9.27).

Another simple and convenient ball-milling protocol for the synthesis of 2,3-dihydroquinazolin-4(1H)-one moiety was described by Saha's group.[140] In this solvent-free synthesis of dihydroquinazolinone, anthranilamide and various aryl aldehydes were reacted under mechanochemical grinding conditions. The reaction was catalyzed by *p*-TSA and completed within a very short time. The products were isolated in high yield. In a scale-up experiment, 22.0 mmol of anthranilamide, 24.2 mmol of *p*-nitrobenzaldehyde and 10 mol % *p*-TSA were milled in a ball mill without any solvent. Ball milling was carried out in a tumbler ball-milling instrument having a frequency of 40 rpm. The product 2-(4-nitrophenyl)-2,3-dih

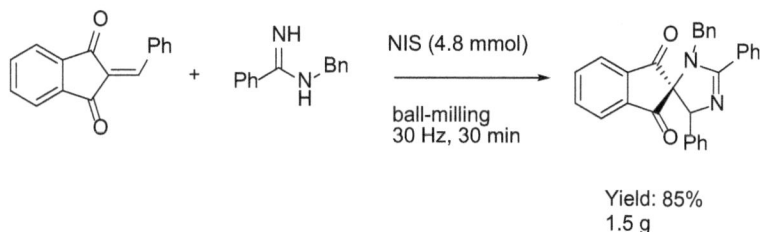

SCHEME 9.26 N-Iodosuccinamide (NIS) promoted mechanochemical synthesis of spiroimidazoline.

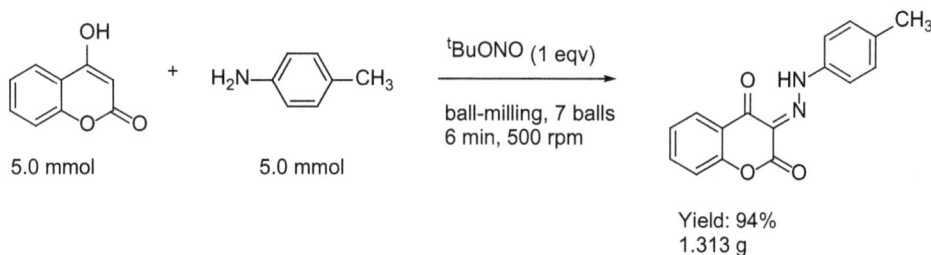

SCHEME 9.27 Mechanochemical synthesis of (*E*)-3-(2-(*p*-tolyl)hydrazono)chromane-2,4-dione.

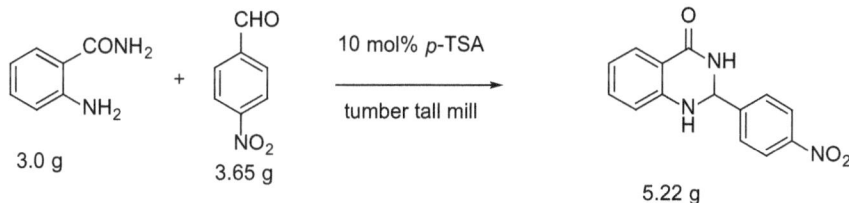

SCHEME 9.28 Mechanochemical coupling and cyclization of anthranilaminde and *p*-nitrobenzaldehyde.

ydroquinazolin-4(1H)-one (5.22 g) could be synthesized within 240 min (Scheme 9.28). It is worth mentioning that dihydroquinazolin-4(1H)-ones have extensive applications in the pharmaceutical industry due to their distinct biological behaviors such as anti-cancer, diuretic, anti-inflammatory, anti-convulsant and anti-hypertensive activities.[141]

Twin-screw extrusion (TSE) is being used extensively in the food, polymer and pharmaceutical industries.[142] This mechanochemical technique has been gaining interest from green chemists[143] as it has the potential for the efficient, continuous mechanochemical synthesis of varieties of chemicals including cocrystals and metal–organic frameworks. Several mechanochemical condensation reactions such as the Knoevenagel condensation and aldol reaction were performed successfully (Scheme 9.29).[144] James' group[145] investigated different condensation reactions (the Knoevenagel condensation, imine formation and the Michael addition) successfully on a large scale by TSE. In these processes, the molecules were synthesized without any added solvent.

Cocrystallization has been a convenient way of improving chemical and physical stability,[146] and bioavailability[147] of APIs in the industry. Vitamin C is common as a pharmaceutical and food supplement in the industry. One of the first cocrystals approved for human consumption by the FDA (21CFR172.315) is a cocrystal of vitamin C (or L-(+)-ascorbicacid) and nicotinamide (a form of vitamin B3) as a formulation for inclusion into multivitamin preparations. This cocrystal has antiscorbutic properties (prevention of scurvy) and certain activity against tuberculosis.[148] A mechanochemical cocrystallization procedure was reported by Stolar et al. (Scheme 9.30)[149] to demonstrate the reaction between ascorbic acid and nicotineamide in various scales. The large-scale cocrystallization was achieved by both ball-milling (10 g) and TSE (100 g) techniques. Using MeOH or EtOH as additives, this technique resulted in the quantitative

Knooevenagel Condensation

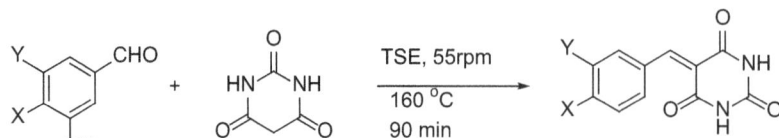

X= OH, Y= H
X= OMe, Y= H
X= OMe, Y= Br

TSE, 55rpm
160 °C
90 min

Yield: upto 90%

0.11 kg/h

Imine Formation

Ball Milling
TSE
30 min
55rpm, 120 oC

Yield: 99%

0.03 kg/h

Michael Addition

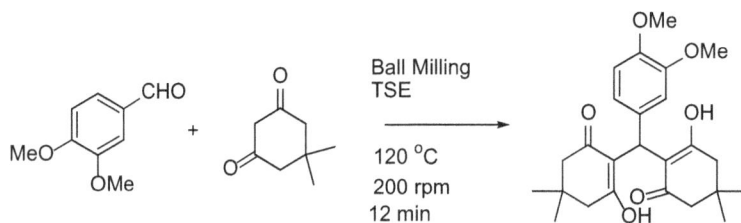

Ball Milling
TSE
120 °C
200 rpm
12 min

Yield: 98%

0.07 kg/h

SCHEME 9.29 Organic synthesis by twin-screw extrusion (TSE).

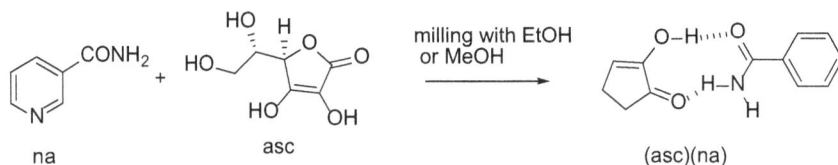

milling with EtOH
or MeOH

na   asc   (asc)(na)

SCHEME 9.30 Mechanochemical cocrystallization of ascorbic acid (Vit C) and nicotinamide.

formation of target polymorph form after 60 min of milling. In the TSE method at the screw speed of 180 rpm, the reaction throughput was in the range of 21–23 g/min for each extrusion cycle.

Air-flow impacting is a common method for the reduction of particle size on the large scale. The mechanism of air-flow impacting is different from conventional mechanical ball milling, hand grinding and TSE. In air-flow impacting, a gaseous medium is introduced, and impacting takes place by colliding particles at high speed. Air-flow impacting exhibits various advantages such as the absence of cross-contamination in conventional mechanochemical synthesis methods, the ability to fragmentize heat-sensitive substances and large-scale preparation at several kg/min rates. In addition, it can achieve sustained response and is environmentally friendly since it avoids the use of various organic solvents. The application of this method was demonstrated first in a chemical synthesis developed

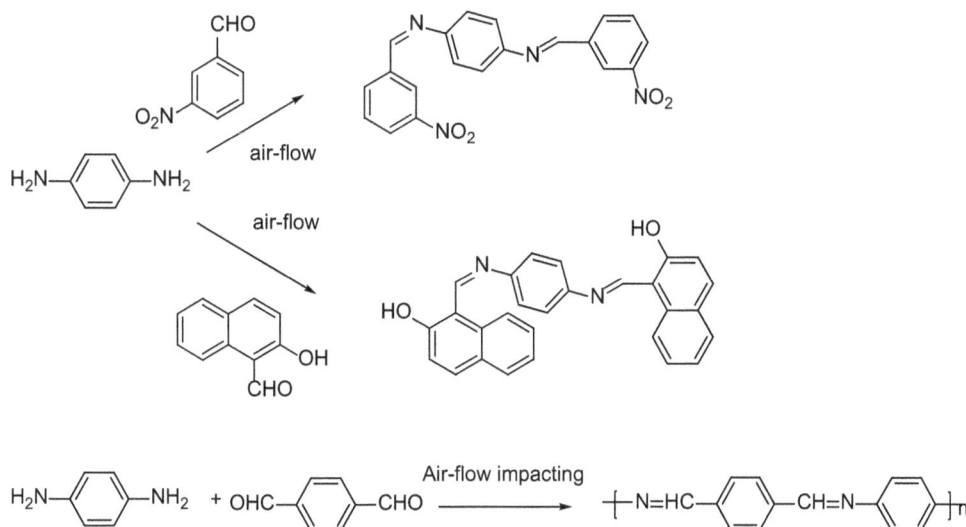

SCHEME 9.31 Synthesis of Schiff bases and a polymeric Schiff base by air-flow method.

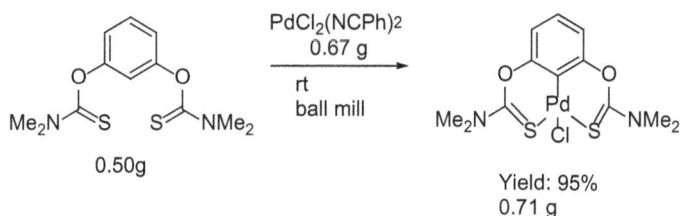

SCHEME 9.32 Mechanochemical synthesis of an organolpalladium compound.

by Zuo and coworkers.[150] N,N′-bis(*m*-nitrobenzylidene)-*p*-phenylenediamine, N,N′-bis(2-hydroxy-1-naphthylmethylene)-*p*-phenylene-diamine and polymeric Schiff bases were prepared by air-flow impact technology. A very high throughput rate of 90 kg/h was readily achieved by this method (Scheme 9.31).

Pincer complexes with monoanionic tridentate ligand framework possess a unique class of organometallic compounds that can be applied as catalysts from organic synthesis to materials science and medicinal chemistry.[151] First ever synthesis of an organometallic Pd(II) pincer complex was derived via C–H bond activation of the bis(thiocarbamate) ligand with PdCl$_2$(NCPh)$_2$ under mechanochemical conditions.[152] Mechanochemical synthesis of this organometallic compound was performed in an electrically powered ball mill in which the starting ligand (0.50 g, 1.76 mmol) and PdCl$_2$(NCPh)$_2$ (0.67 g, 1.76 mmol) were shaken in a stainless steel jar with two grinding balls using a Narva DDR GM 9458 vibration mill. The complex product was isolated in an excellent yield of 95% (0.71 g) just after 2 min of mechanochemical activation and a simple workup procedure (Scheme 9.32).

9.5.1 Biocatalysts in Fine Chemical and API Synthesis

Biocatalysis generally refers to the use of enzymes, or enzyme-containing cells, in chemical transformations. Organic synthesis with the aid of enzymes has been known since the early 20th century. Scientists proved that components of living cells could be applied to useful chemical transformations. For example, (*R*)-mandelonitrile was synthesized from benzaldehyde and hydrogen cyanide using a plant extract.[153] A revolutionary experiment, reported by Zaksand Klibanov,[154] showed that porcine pancreatic lipase-catalyzed transesterification reaction between tributyrin and various primary and secondary alcohols can take place in a 99% organic medium. It was also demonstrated that the enzyme was of good heat stability even at 100°C for many hours and exhibited a high catalytic activity at that temperature. It led to the realization among organic chemists that biocatalysis could have a much broader scope in organic synthesis. Since enzymes are usually highly enantioselective catalysts, biocatalysis was widely used in the manufacturing of pharmaceutical intermediates and fine chemicals. Examples include the lipase-catalyzed resolution of chiral precursors for the synthesis of diltiazem (a blood pressure drug), the hydroxynitrile-lyase-catalyzed synthesis of intermediates for herbicides,[155] carbonyl-reductase-catalyzed syntheses of enantiopure alcohols for cholesterol-lowering drugs, statins, lipase-catalyzed synthesis of wax esters such as myristyl myristate or cetyl ricinoleate for cosmetics,[156] nitrile-hydratase-catalyzed hydration of acrylonitrile to acrylamide for polymers[157] and many more. Because of the efficient catalytic properties of enzymes shown in the synthesis of various chemicals, biocatalysis has no longer remained confined to academic laboratories but evolved to industrial technology for the enantioselective synthesis of various APIs.[158] The numerous biocatalytic routes, scaled up for pharmaceutical manufacturing, demonstrate their competitiveness with traditional chemical processes.[159] In the last two decades, protein engineering techniques, such as directed (in vitro) evolution,[160] have engineered desired enzymes so that they can pursue pre-defined properties with regard to substrate specificity, activity, selectivity, stability and pH optimum in a particular reaction. Lastly, the development of effective immobilization techniques has improved the operational stability of enzymes and cost effectiveness of the processes.[161] Herein, we will be focusing on applications of enzymes in various syntheses of small molecules that are important in the pharmaceutical industry. For example, 6-aminopenicillanic acid (6-APA) is the key intermediate for the manufacturing of semi-synthetic penicillins. In a conventional method, 6-PPA is produced by the cleavage of a side of penicillin G (Scheme 9.33). In the enzymatic process, by employing commercially available penicillin G amidase, the same reaction was carried out in the water at 37 °C.[162] This method has enabled the cost reduction of the production of 6-APA and as an additional benefit, the catalyst was reusable.

The introduction of semi-synthetic β-lactam antibiotics transformed the 6-aminopenicillanic acid (6-APA) into a major pharmaceutical intermediates namely areampicillin and amoxicillin, in which the side chain of 6-APA is D-phenylglycine and 4-hydroxyphenylglycine, respectively.[163] In 1960, the discovery of penicillin G acylase (penicillin amidohydrolase) was reported independently by four industrial research groups.[164] Over the years, penicillin G acylase variants with improved stability were obtained by screening and by employing recombinant DNA technology. These new enzymes, combined with efficient immobilization methods,[165] made the recycling of the biocatalyst possible, generating a more efficient production of β-lactams and dramatic reductions in enzyme cost. The industrial production of β-lactam antibiotics occurs using two different penicillin G acylases, one designed for the hydrolysis of benzylpenicillin (HydPGA) to give the β-lactam nucleus 6-APA and the

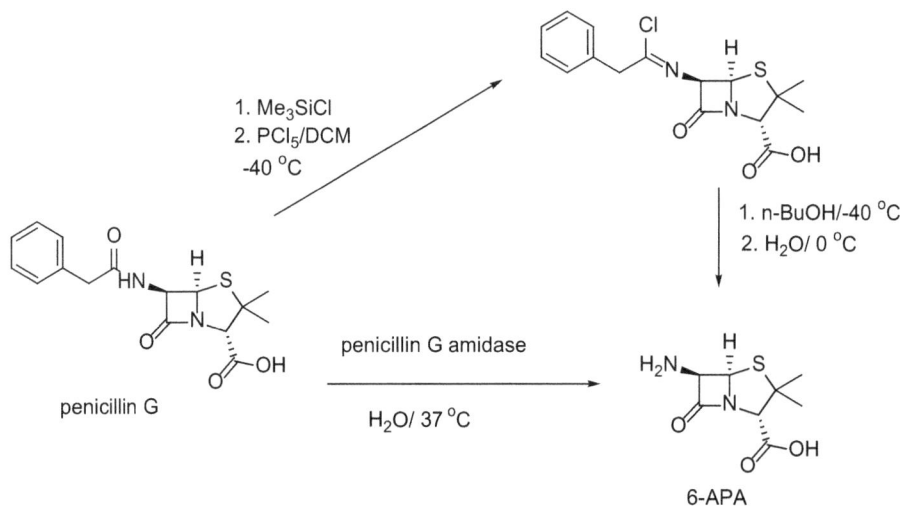

SCHEME 9.33 Enzymatic vs chemical process for the synthesis of 6-APA.

other designed for the synthesis (SynPGA) of semi-synthetic β-lactams as ampicillin or amoxicillin (Scheme 9.34).[166]

Omapatrilat (generic name Vanlev), an anti-hypertensive drug, inhibits both angiotensin-converting enzymes (ACEs) and neutral endopeptidase (NEP). One of the important building blocks for the synthesis of omapatrilat is (*S*)-2-amino-5-(1,3-dioxolan-2-yl)-pentanoic acid [(*S*)-allysine ethylene acetal]. This molecule can be obtained by reductive amination of a ketoacid acetal, using phenylalanine dehydrogenase (PDH) from *thermoactinomyces intermedius* ATCC 33205[167] (Scheme 9.35). The reaction required NADH and ammonia, and NAD was produced during the reaction. The NAD can be recycled to NADH by the oxidation of formate to CO_2, using formate dehydrogenase (FDH). *T. intermedius* PDH was cloned and expressed in *Escherichia coli*. Recombinant *E. coli* as a source of PDH and heat-dried *Candida boidinii* as a source of FDH, 197 kg of the product was produced in three batches with an average yield of 91% and 98%ee.

During the last two decades, directed strategies have been rigorously studied in the improvement of the catalytic efficiency of enzymes. In contrast to the traditional approach where substrates are chosen to fit into the catalyst, recently the process for a desired synthesis is designed with pre-defined parameters, and enzymes are being engineered based on the requirements. This strategy has been very useful and applied in pharmaceutical industries for preparing active pharmaceutical ingredients. For example, the key intermediate, (*R*)-4-cyano-3-hydroxybutyrate (HN), to atorvastatin (the active ingredient in Lipitor, a cholesterol-lowering drug) was synthesized in two-step three-enzyme process that was carried out by employing pre-defined process idea.[168] After many rounds of DNA shuffling and screening, the process was standardized and the best variant of HHDH was used where the overall volumetric productivity per mass catalyst load of the cyanation process was improved ~2500-fold, comprising a 14-fold reduction in

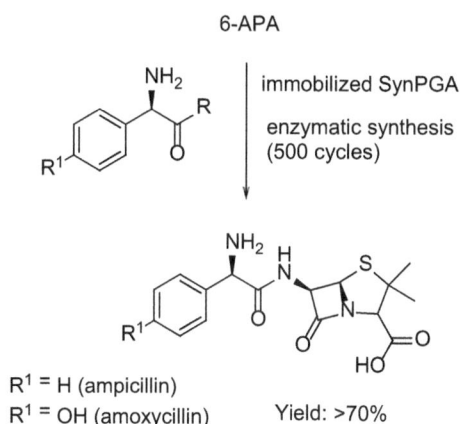

SCHEME 9.34 Biocatalytic synthesis of β-lactams.

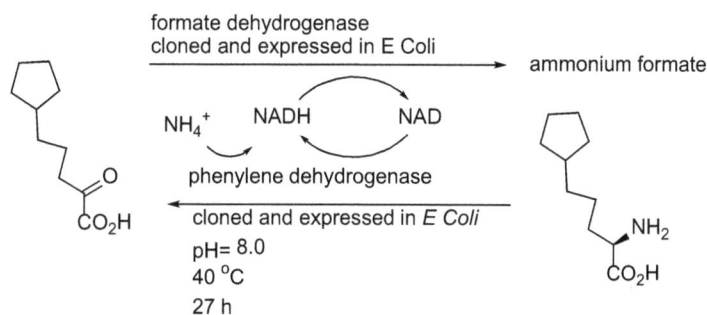

SCHEME 9.35 Enzymatic synthesis of allysine ethylene acetal.

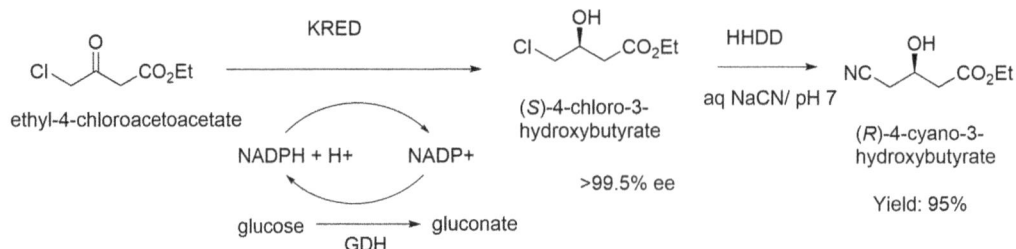

SCHEME 9.36 Enzymatic synthesis of (*R*)-4-cyano-3-hydroxybutyrate.

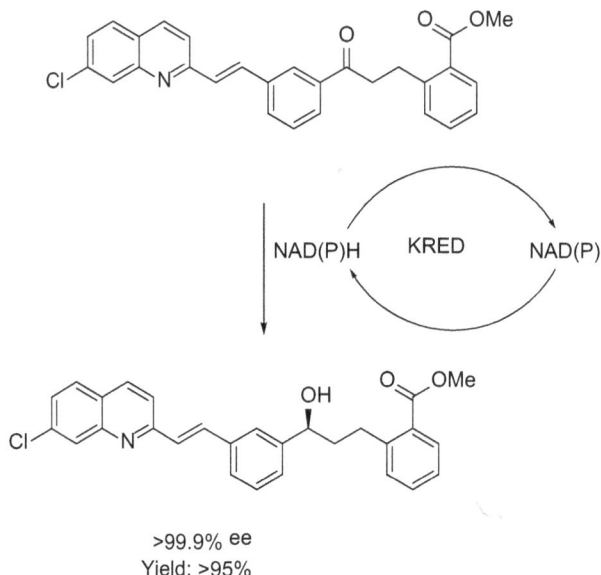

SCHEME 9.37 Synthesis of a montelukast intermediate by KRED reduction.

reaction time (72–5 h), a 7-fold increase in substrate loading (20–140 g/L), a 25-fold reduction in enzyme use (30–1.2 g/L) and a 50% improvement in isolated yield (Scheme 9.36).

In another example (Scheme 9.37), engineered KRED has been successfully used to catalyze the asymmetric reduction of (*E*)-methyl-2-(3-(3-(2-(7-chloroquinolin-2-yl)vinyl)phenyl)-3-oxopropyl)benzoate to the corresponding (*S*)-alcohol, a key intermediate in the synthesis of montelukast (Singulair[169]).[170] After optimizing the reaction condition, the activity of the biocatalyst was improved by subjecting variants to directed evolution technologies under the prescribed process parameters. The resulting biocatalytic process is economical and robust. The product is obtained in >95% yield in >99.9% ee and >98.5% chemical purity in 200+ kg scale.

The application of biocatalysis for the resolution of racemic chiral amines with the aid of lipase was reported in the early 1990s.[171] However, the process is not operable on an industrial scale as the reactions were very slow. A protein engineering tool was developed[172] to synthesize a chiral amine from the carbonyl group employing processed transaminase. The active intermediate for sitagliptin, which is the active API of the antidiabetic agent Januvia, was synthesized using (*R*)-selective transaminase. The transaminase catalyst was designed so that it can possess the necessary active site to perform a desired transformation (Scheme 9.38). This efficient biocatalytic method can replace the rhodium-catalyzed synthesis for the large-scale manufacturing of sitagliptin.

1,4-Butanediol (BDO) is a key raw material used in the production of over 2.5 million tons per annum of plastics and fibers. A sustainable commercial-scale process for the production of 1,4-butanediol from carbohydrate feedstock was developed by Yim's group[173] by employing an engineered *E. coli* strain.

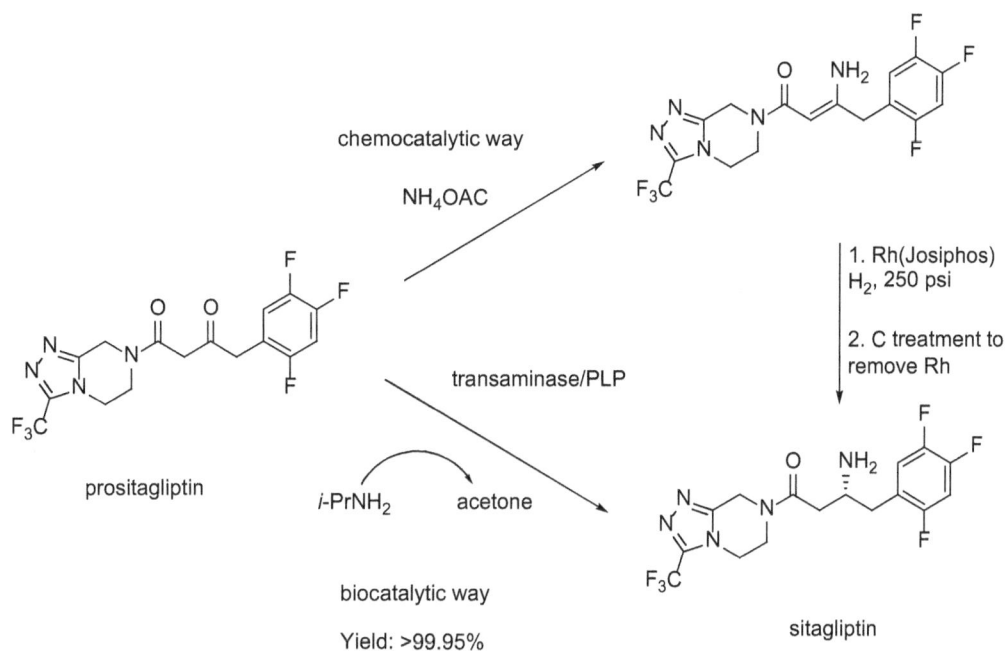

SCHEME 9.38 Synthesis of sitagliptin by enzymatic transamination.

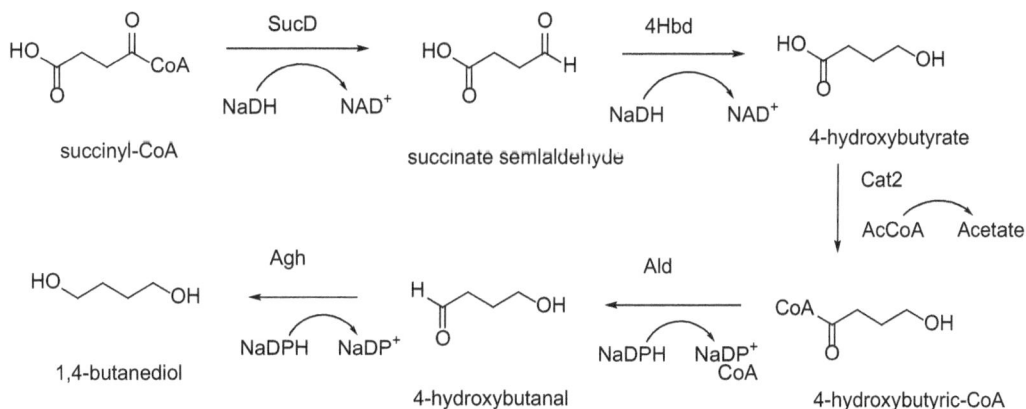

SCHEME 9.39 Selected enzymatic route to 1,4-butanediol.

Bio-BDO produced in this method meets industry specifications and performance requirements for all major existing applications. The authors claim that the bio-BDO can be produced from dextrose with up to 83% lower total CO_2-equivalent emissions/kg BDO and 67% lower fossil energy usage relative to the petrochemical process. The selected pathway for the production of BDO has been described in Scheme 9.39. This selected route comes off from the citrate cycle to 2-oxoglutarate that is decarboxylated to succinyl-CoA.

9.6 Conclusions

Even though the fundamental idea of green chemistry and its guiding principles have been known for decades, the real implementation of those rules in drug design and chemical synthesis on a large scale is still limited. The application of non-traditional activation methods and biocatalysis has already revolutionized laboratory-scale organic synthesis and they are making their way into industrial production of chemicals. This chapter summarized the recent development made in these fields providing representative examples from the application of microwave-, ultrasound- and light-activated synthesis as well as the preparation of chemicals by mechanochemistry. As an additional green tool biocatalysis has also made significant advances and was also discussed in this chapter. It is worth noting that all methods that suffered a significant drawback regarding the penetration depth of the energy form into the reaction mixtures (microwaves, ultrasounds and photochemistry) have been aided by the introduction of the tools of flow chemistry to ensure that by the use of specific reactors and the appropriate selection of the flow rate and residence time now, the use of these activation methods can be readily be extended to large-scale operations. Given the advantages that the combination of these tools present, it is safe to predict that flow-based non-traditional activations will make further advances in the future. Advances in biotechnology, computational approaches and advanced genome sequencing have made biocatalytic synthetic methods more powerful than ever. More engineered enzymes are now becoming available with lower cost through genetic modifications coupled with recombinant expressions. Also directed engineering is playing an important role in biomass-based production of fine chemicals.

REFERENCES

1. Anastas, P. T.; Williamson, T. C. *Green Chemistry: Frontiers in Benign Chemical Syntheses and Processes*. Oxford University Press, Oxford, 1998.
2. Anastas, P. T.; Warner, J. C. *Green Chemistry: Theory and Practice*. Oxford University Press, New York, 1998.
3. Ratti, R. Industrial applications of green chemistry: Status, challenges and prospects. *SN Appl. Sci.* 2020, 2(2), 263.
4. Kopack, E. Sustainability: A foundation for pharma, generic and government partnerships? *Curr. Opin. Green Sustain. Chem.* 2018, 11, 54–57.
5. Kar, S.; Sanderson, H.; Roy, K.; Benfenati, E.; Leszczynski, J. Green chemistry in the synthesis of pharmaceuticals. *Chem. Rev.* 2022, 122(3), 3637–3710.
6. Vlocskó, R. B.; Apaydın, S.; Török, B.; Török, M. The multitarget approach as a green tool in medicinal chemistry. In *Contemporary Chemical Approaches for Green and Sustainable Drugs* (Török, M. Ed.), Elsevier, Amsterdam, Oxford, Cambridge, 2022.
7. Xie, G.; Vlocskó, R. B.; Török, B. Green synthetic methods in drug discovery and development. In *Contemporary Chemical Approaches for Green and Sustainable Drugs* (Török, M. Ed.), Elsevier, Amsterdam, Oxford, Cambridge, 2022.
8. Török, B. Sustainable synthesis. In *Green Chemistry: An Inclusive Approach* (Török, B.; Dransfield, T. Eds.), Elsevier, 2018, Chp. 2.2, pp. 49–89.
9. (a) Clarke, C. J.; Tu, W.-C.; Levers, O.; Bröhl, A.; Hallett, J. P. Green and sustainable solvents in chemical processes. *Chem. Rev.* 2018, 118(2), 747–800. (b) Abou-Shehada, S.; Clark, J. H.; Paggiola, G.; Sherwood, J. Tunable solvents: Shades of green. *Chem. Eng. Process.* 2016, 99, 88–96. (c) Alder, C. M.; Hayler, J. D.; Henderson, R. K.; Redman, A. M.; Shukla, L.; Shuster, L. E.; Sneddon, H. F. Updating and further expanding GSK's solvent sustainability guide. *Green Chem.* 2016, 18(13), 3879–3890.
10. Draye, M.; Chatel, G.; Duwald, R. Ultrasound for drug synthesis: A green approach. *Pharmaceuticals* 2020, 13(2), 23–57.
11. Banik, B. K.; Sahoo, B. M. Reactions in water: Synthesis of biologically active compounds. In *Green Approaches: Medicinal Chemistry for Sustainable Drug Design* (Banik, B. Ed.), Elsevier, Amsterdam, the Netherlands, 2020.
12. Maleki, A.; Rahimi, J.; Demchuk, O. M.; Wilczewska, A. Z.; Jasinski, R. Green in water Sonochemical synthesis of tetrazolopyrimidine derivatives by a novel core-shell magnetic nanostructure catalyst. *Ultrason. Sonochem.* 2018, 43, 262–271.
13. Senthilkumar, G.; Neelakandan, K.; Manikandan, H. A convenient, green, solvent free synthesis and characterization of novel fluoro chalcones under grind-stone chemistry. *Chem. Sin.* 2014, 5, 106–113.
14. Thirunarayanan, G.; Vijayakumar, S. Solvent-free synthesis and antimicrobial potential of some (2E)-4-methoxyphenyl chalcones. *Pharm. Chem.* 2018, 10, 43–47.

15. Santi, C.; Jacob, R. G.; Monti, B.; Bagnoli, L.; Sancineto, L.; Lenardao, E. J. Water and aqueous mixtures as convenient alternative media for organoselenium chemistry. *Molecules* 2016, 21(11), 1482.

16. Kokel, A.; Török, B. Sustainable production of fine chemicals and materials using non-toxic renewable sources. *Toxicol. Sci.* 2018, 161(2), 214–224.

17. Shahid, A.; Ahmed, N.; Saleh, T. S.; Mokhtar, M.; Basahel, S.; Schwieger, W.; Mokhtar, M. Solvent free Biginelli reactions catalyzed by hierarchical zeolite utilizing a ball mill technique: A green sustainable process. *Catalysts* 2017, 7(12), 84.

18. Török, B.; Schäfer, C. (Eds.) *Non-traditional Activation Methods in Green and Sustainable Applications: Microwaves, Ultrasounds, Photo, Electro and Mechanochemistry and High Hydrostatic Pressure.* Elsevier, Cambridge, Oxford, 2021.

19. Török, B.; Dranfield, T. (Eds.) *Green Chemistry: An Inclusive Approach.* Elsevier, Cambridge, Oxford, 2018.

20. Török, B.; Schäfer, C.; Kokel, A. *Heterogeneous Catalysis in Sustainable Synthesis*; Elsevier, Cambridge, Oxford, 2021.

21. Sheldon, R. A.; Woodley, J. M. Role of biocatalysis in sustainable chemistry. *Chem. Rev.* 2018, 118(2), 801–838.

22. Sharma, S.; Das, J.; Braje, W. M.; Dash, A. K.; Handa, S. A glimpse into green chemistry practices in the pharmaceutical industry. *ChemSusChem* 2020, 13(11), 2859–2875.

23. Bandichhor, R.; Bhattacharya, A.; Diorazio, L.; Dunn, P.; Fraunhoffer, K.; Gallou, F.; Hayler, J.; Hickey, M.; Hinkley, B.; Humphreys, L.; Kaptein, B.; Mathew, S.; Richardson, P.; White, T.; Wuyts, S.; Yin, J. Green chemistry articles of interest to the pharmaceutical industry. *Org. Process Res. Dev.* 2014, 18(12), 1602–1613.

24. (a) Kappe, C. O.; Dallinger, D. Controlled microwave heating in modern organic synthesis: Highlights from the 2004–2008 literature. *Mol. Divers.* 2009, 13(2), 71–193. (b) Caddick, S.; Fitzmaurice, R. Microwave enhanced synthesis. *Tetrahedron* 2009, 65(17), 3325–3355.

25. Lidström, P.; Tierney, J. P. (Eds.) *Microwave-Assisted Organic Synthesis.* Blackwell Publishing, Oxford, 2005.

26. Loupy, A. (Ed.) *Microwaves in Organic Synthesis*, 2nd ed. Wiley-VCH, Weinheim, 2006.

27. Larhed, M.; Olofsson, K. (Eds.) *Microwave Methods in Organic Synthesis.* Springer, Berlin, 2006.

28. Larhed, M.; Hallberg, A. Microwave-assisted high-speed chemistry: A new technique in drug discovery. *Drug Discov. Today* 2001, 6(8), 406–415.

29. Kappe, C. O.; Stadler, A. *Microwaves in Organic and Medicinal Chemistry.* Wiley-VCH, Weinheim, 2005.

30. Kappe, C. O.; Dallinger, D. The impact of microwave synthesis on drug discovery. *Nat. Rev. Drug Discov.* 2006, 5(1), 51–63.

31. Daştan, A.; Kulkarni, A.; Török, B. Environmentally benign synthesis of heterocyclic compounds by combined microwave-assisted heterogeneous catalytic approaches †. *Green Chem.* 2012, 14(1), 17–37.

32. Kokel, A.; Schäfer, C.; Török, B. Microwave-assisted reactions in green chemistry. In *Encyclopedia of Sustainable Science and Technology* (Meyers, R. A. Ed.), Springer-Nature, 2018. https://doi.org/10.1007/978-1-4939-2493-6_1008-1.

33. Strauss, C. R. On scale up of organic reactions in closed vessel microwave systems. *Org. Process Res. Dev.* 2009, 13(5), 915–923.

34. Kremsner, J. M.; Stadler, A.; Kappe, C. O. The scale-up of microwave-assisted organic synthesis. *Top. Curr. Chem.* 2006, 266, 233–278.

35. (a) Moseley, J. D.; Lenden, P.; Lockwood, M.; Ruda, K.; Sherlock, J.-P.; Thomson, A. D.; Gilday, J. P. A comparison of commercial microwave reactors for scale-up within process chemistry. *Org. Process Res. Dev.* 2008, 12(1), 30–41. (b) Bowman, M. D.; Holcomb, J. L.; Kormos, C. M.; Leadbeater, N. E.; Williams, V. A. Approaches for scale-up of microwave-promoted reactions. *Org. Process Res. Dev.* 2008, 12(1), 41–57. (c) Moseley, J. D.; Woodman, E. K. Scaling-out pharmaceutical reactions in an automated stop-flow microwave reactor. *Org. Process Res. Dev.* 2008, 12(5), 967–981. (d) Bowman, M. D.; Schmink, J. R.; McGowan, C. M.; Kormos, C. M.; Leadbeater, N. E. Scale-up of microwave-promoted reactions to the multigram level using a sealed-vessel microwave apparatus. *Org. Process Res. Dev.* 2008, 12(6), 1078–1088. (e) Ianelli, M.; Bergamelli, F.; Kormos, C. M.; Paravisi, S.; Leadbeater, N. E. Application of a batch microwave unit for scale-up of Alkoxycarbonylation reactions using a near-stoichiometric loading of carbon monoxide. *Org. Process Res. Dev.* 2009, 13(3), 634–637.

36. Kokel, A.; Schäfer, C.; Török, B. Application of microwave-assisted heterogeneous catalysis in sustainable synthesis design. *Green Chem.* 2017, 19(16), 3729–3751.
37. Kappe, C. O. Controlled microwave heating in modern organic synthesis. *Angew. Chem. Int. Ed.* 2004, 43(46), 6250–6284.
38. Mooney, T.; Török, B. Microwave-assisted flow systems in the green production of fine chemicals. In *Nontraditional Activation Methods in Green and Sustainable Applications: Microwaves; Ultrasounds; Photo-, Electro- and Mechanochemistry and High Hydrostatic Pressure* (Török, B.; Schäfer, C. Eds.), Elsevier, Oxford, Cambridge, 2021, Chp 3. pp. 101–136.
39. Anderson, N. G. Practical use of continuous processing in developing and scaling up laboratory processes. *Org. Process Res. Dev.* 2001, 5(6), 613–621.
40. Martina, K.; Cravotto, G.; Verma, R. Impact of microwaves on organic synthesis and strategies toward flow processes and scaling up. *J. Org. Chem.* 2021, 86(20), 13857–13872.
41. Baxendale, I. R.; Hayward, J. J.; Ley, S. V. High throughput screening. *Comb. Chem.* 2007, 10, 802–836.
42. Glasnov, T. N.; Kappe, C. O. Microwave-assisted synthesis under continuous-flow conditions. *Macromol. Rapid Commun.* 2007, 28(4), 395–410.
43. Benaskar, F.; Hessel, V.; Krtschil, U.; Löb, P.; Stark, A. Intensification of the capillary-based Kolbe-Schmitt synthesis from resorcinol by reactive ionic liquids, microwave heating, or a combination thereof. *Org. Process Res. Dev.* 2009, 13, 970–982.
44. Dressen, M. H. C. L.; Van de Kruijs, B. H. P.; Meduldijk, J.; Vekemans, J. A. J. M.; Hulshof, L. A. From batch to flow processing: Racemization of N-acetylamino acids under microwave heating. *Org. Process Res. Dev.* 2009, 13(5), 888–895.
45. Smith, C. J.; Iglesias-Sigüenza, F. J.; Baxendale, I. R.; Ley, S. V. Flow and batch mode focused microwave synthesis of 5-amino-4-cyanopyrazoles and their further conversion to 4-aminopyrazolopyrimidines. *Org. Biomol. Chem.* 2007, 5(17), 2758–2761.
46. Shore, G.; Morin, S.; Mallik, D.; Organ, M. G. Pd PEPPSI-IPr-mediated reactions in metal-coated capillaries under MACOS: The synthesis of indoles by sequential aryl amination/Heck coupling. *Chem. Eur. J.* 2008, 14(4), 1351–1356.
47. Alamgir, M.; Black, D. St. C.; Kumar, N. Synthesis, reactivity and biological activity of benzimidazoles. In *Bioactive Heterocycles III* (Khan, M. T. H. Ed.), Topics in Heterocyclic Chemistry; Springer, Berlin, Germany, 2007, 9, pp. 87–118.
48. Dallinger, D.; Lehmann, H. C.; Moseley, J. D.; Stadler, A.; Kappe, C. O. Scale-up of microwave-assisted reactions in a multimode bench-top reactor. *Org. Process Res. Dev.* 2011, 15(4), 841–854.
49. (a) Delfort, B.; Durand, I.; Jaecker, A.; Lacome, T.; Montagne, X.; Paille, F. Diesel fuel compounds containing glycerol acetals, US Patent 2003/0163949 A1 (4 September 2003). (b) Mushrush, G. W.; Beal, E. J.; Hardy, D. R.; Hughes, J. M.; Cummings, J. C. Jet fuel system icing inhibitors: Synthesis and characterization. *Ind. Eng. Chem. Res.* 1999, 2497–2502.
50. Bekemeier, H.; Koester, D. The pharmacology of some 1,3-dioxolanes, 1,3-oxathiolanes and 1,3-dithiolanes. *Pharmazie* 1976, 31(5), 317–723.
51. Perio, B.; Dozlas, M.-J.; Hamelin, J. Ecofriendly fast batch synthesis of dioxolanes, dithiolanes, and oxathiolanes without solvent under microwave irradiation. *Org. Process Res. Dev.* 1998, 2(6), 428–430.
52. Hong, Y.-L.; Hossler, P. A.; Calhoun, D. H.; Meshnick, S. R. Inhibition of recombinant pneumocystis carinii dihydropteroate synthetase by sulfa drugs. *Antimicrob. Agents Chemother.* 1995, 39(8), 1756–1763.
53. Ehlert, C.; Strunz, H.; Visser, K.; Wiesse, M.; Seydel, J. K. J. Inhibition of the conjugation of PABA with glycine in vitro by sulfamoyl benzoic acids, sulfonamides, and penicillins and its relation to tubular secretion. *J. Pharm. Sci.* 1998, 87(1), 101–108.
54. Baxendale, I. R.; Ley, S. V. Formation of 4-aminopyrimidines via the trimerization of nitriles using focused microwave heating. *J. Comb. Chem.* 2005, 7(3), 483–489.
55. Stadler, A.; Yousefi, B. H.; Dallinger, D.; Walla, P.; Van der Eycken, E.; Kaval, N.; Kappe, C. O. Scalability of microwave-assisted organic synthesis. From single-mode to multimode parallel batch reactors. *Org. Process Res. Dev.* 2003, 7, 707–716.
56. Hsu, Y. L.; Kuo, P. L.; Lin, C. C. Acacetin inhibits the proliferation of Hep G2 by blocking cell cycle progression and inducing apoptosis. *Biochem. Pharmacol.* 2004, 67(5), 823–829.
57. Pan, M. H.; Lai, C. S.; Hsu, P. C.; Wang, Y. J. Acacetin induces apoptosis in human gastric carcinoma cells accompanied by activation of caspase cascades and production of reactive oxygen species. *J. Argric Food Chem.* 2005, 53(3), 620–630.

58. Shim, H. Y.; Park, J. H.; Paik, H. D.; Nah, S. Y.; Kim, D. S.; Han, Y. S. Acacetin induced apoptosis of human breast cancer MCF-7 cells involves caspase cascade, mitochondria-mediated death signaling and SAPK/JNK1/2-c-jun activation. *Mol. Cells* 2007, 24(1), 95–104.

59. Shen, K. H.; Hung, S. H.; Yin, L. T.; Huang, C. S.; Chao, C. H.; Liu, C. L.; Shih, Y. W. Acacetin, a flavonoid, inhibits the invasion and migration of human prostate cancer DU145 cells via inactivation of the p38 MAPK signaling pathway. *Mol. Cell. Biochem.* 2010, 333(1–2), 279–291.

60. Hsu, Y. L.; Kuo, P. L.; Liu, C. F.; Lin, C. C. Acacetin induced cell cycle arrest and apoptosis in human non-small cell lung cancer A549 cells. *Cancer Lett.* 2004, 212(1), 53–60.

61. (a) Kabalka, G. W.; Mereddy, A. R. Microwave-assisted synthesis of functionalized flavones and-chromones. *Tetrahedron Lett.* 2005, 46(37), 6315–6317. (b) Sarda, S. R.; Pathan, M. Y.; Paike, V. V.; Pachmase, P. R.; Jadhav, W. N.; Pawar, R. P. A facile synthesis of flavones using recyclable ionic liquid under microwave irradiation. *Arkivoc*, 2006, 43–48; (c) Moghaddam, F. M.; Ghaffarzadeh, M.; Abdi-Oskoui, S. H. Solvent-free synthesis of functionalized flavones under microwave irradiation. *J. Chem. Res. Synop.* 1999, 9, 574–575.

62. Radoiu, M.; Chantreux, D.; Marchiori, B. Scale-up of one-step synthesis of acacetin and apigenin using 915 MHz microwaves. *Chem. Eng. Process.* 2017, 114, 39–45.

63. Lau, J. L.; Dunn, M. K. Therapeutic peptides: Historical perspectives, current development trends, and future directions. *Bioorg. Med. Chem.* 2018, 26(10), 2700–2707.

64. Sabatino, G.; D'Ercole, A.; Pacini, L.; Zini, M.; Ribecai, A.; Paio, A.; Rovero, P.; Papini, A. M. An optimized scalable fully automated solid-phase microwave assisted cGMP-ready process for the preparation of eptifibatide. *Org. Process Res. Dev.* 2021, 25(3), 552–563.

65. McDonald, E.; Jones, K.; Brough, P. M.; Drysdale, M. J.; Workman, P. Discovery and development of pyrazole-scaffold Hsp90 inhibitors. *Curr. Top. Med. Chem.* 2006, 6(11), 1193–1203.

66. Damm, M.; Glasnov, T. N.; Kappe, C. O. Translating high-temperature microwave chemistry to scalable continuous flow processes. *Org. Process Res. Dev.* 2010, 14(1), 215–224.

67. Pentland, A. P.; Morrison, A. R.; Jacobs, S. C.; Hruza, L. L.; Hebert, J. S.; Packer, L. Tocopherol analogs suppress arachidonic acid metabolism via phospholipase inhibition. *J. Biol. Chem.* 1992, 267(22), 15578–15584.

68. Rotolo, L.; Gaudino, E. C.; Carnaroglio, D.; Barge, A.; Tagliapietra, S.; Cravottoa, G. Fast multigram scale microwave-assisted synthesis of vitamin E and C10-, C15- analogues under vacuum. *RSC Adv.* 2016, 6(68), 63515–63518.

69. Strauss, C. R.; Trainor, R. W. Developments in microwave-assisted organic chemistry. *Aust. J. Chem.* 1995, 48(10), 1665–1692.

70. Smith, C. J.; Iglesias-Sigüenza, F. J.; Baxendale, I. R.; Ley, S. V. Flow and batch mode focused microwave synthesis of 5-amino-4-cyanopyrazoles and their further conversion to 4-aminopyrazolopyrimidines. *Org. Biomol. Chem.* 2007, 5(17), 2758–2761.

71. Glasnov, T. N.; Kappe, C. O. Microwave-assisted synthesis under continuous-flow conditions. *Macromol. Rapid Commun.* 2007, 28(4), 395–410.

72. Bremner, W. S.; Organ, M. G. Multicomponent reactions to form heterocycles by microwave-assisted continuous flow organic synthesis. *J. Comb. Chem.* 2007, 9(1), 14–16.

73. Bowman, M. D.; Holcomb, J. L.; Kormos, C. M.; Leadbeater, N. E.; Williams, V. A. Approaches for scale-up of microwave-promoted reactions. *Org. Process Res. Dev.* 2008, 12(1), 41–57.

74. Morschhauser, R.; Krull, M.; Kayser, C.; Boberski, C.; Bierbaum, R.; Püschner, P. A.; Glasnov, T. N.; Kappe, C. O. Microwave-assisted continuous flow synthesis on industrial scale. *Green Process. Synth.* 2012, 1(3), 281–290.

75. Kiss, N. Z.; Henyecz, R.; Keglevich, G. Continuous flow esterification of a H-phosphinic acid, and transesterification of Hphosphinates and H-phosphonates under microwave conditions. *Molecules* 2020, 25(3), 719–735.

76. (a) He, Y.; Chen, H.-Y.; Hou, J.; Li, Y. Indene-C(60) Bisadduct: A new acceptor for high-performance polymer solar cells. *J. Am. Chem. Soc.* 2010, 132(4), 1377–1382. (b) Kang, H.; Cho, C.-H.; Cho, H.-H.; Kang, T. E.; Kim, H. J.; Kim, K.-H.; Yoon, S. C.; Kim, B. J. Controlling number of indene solubilizing groups in multiadduct fullerenes for tuning optoelectronic properties and open-circuit voltage in organic solar cells. *ACS Appl. Mater. Interfaces* 2012, 4(1), 110–116.

77. Cho, H.-H.; Cho, C.-H.; Kang, H.; Yu, H.; Oh, J. H.; Kim, B. J. *Korean J. Chem. Eng.* 2015, 32(2), 261–267.

78. Kegelmann, L.; Wolff, C. M.; Awino, C.; Lang, F.; Unger, E. L.; Korte, L.; Dittrich, T.; Neher, D.; Rech, B.; Albrecht, S. *ACS Appl. Mater. Interfaces* 2017, 9(20), 17245–17255.

79. Barham, J. P.; Tanaka, S.; Koyama, E.; Ohneda, N.; Okamoto, T.; Odajima, H.; Sugiyama, J.-I.; Norikane, Y. Selective, scalable synthesis of C60-Fullerene/Indene monoadducts using a microwave flow applicator. *J. Org. Chem.* 2018, 83(8), 4348–4354.

80. Sauks, J. M.; Mallik, D.; Lawryshyn, Y.; Bender, T.; Organ, M. A. A continuous-flow microwave reactor for conducting high-temperature and high-pressure chemical reactions. *Org. Process Res. Dev.* 2014, 18(11), 1310–1314.

81. Shore, G.; Yoo, W.-J.; Li, C.-J.; Organ, M. G. Propargyl amine synthesis catalysed by gold and copper thin films by using microwave-assisted continuous-flow organic synthesis (MAPOS). *Chem. Eur. J.* 2010, 16(1), 126–133.

82. Patil, N. G.; Benaskar, F.; Rebrov, E. V.; Meuldijk, J.; Hulshof, L. A.; Hessel, V.; Schouten, J. C. Continuous multitubular millireactor with a Cu thin film for microwave assisted fine chemical synthesis. *Ind. Eng. Chem. Res.* 2012, 51(44), 14344–14354.

83. Mason, T. J.; Lorimer, J. P. *Applied Sonochemistry – The Uses of Power Ultrasound in Chemistry and Processing.* Wiley-VCH, Weinheim, 2002.

84. Wood, R. W.; Loomis, A. L. Physical and biologic effects of high frequency sound waves of great intensity. *Phil. Mag.* 1927, 4, 414–436.

85. Richards, W. T.; Loomis, A. L. The chemical effects of high frequency sound waves I. Preliminary survey. *J. Am. Chem. Soc.* 1927, 49(12), 3086–3100.

86. Grieser, F.; Choi, P. K.; Enomoto, N.; Harada, H.; Okitsu, K.; Yasui, K. (Eds.) *Sonochemistry and the Acoustic Bubble.* Elsevier, Amsterdam, 2015.

87. Boudjouk, P. *Ultrasounds, Its Chemical, Physical, and Biological Effects* (Suslick, K. S. Ed.), VCH, Weinheim, 1988.

88. Suslick, K. S. Sonocatalysis. In *Handbook of Heterogeneous Catalysis* (Ertl, G.; Knozinger, H.; Weitkamp, J. Eds.), Wiley-VCH, New York, 1997.

89. Draye, M.; Kardos, N. Advances in green organic sonochemistry. *Top. Curr. Chem.* 2016, 374(5), 1–29.

90. Chatel, G. How sonochemistry contributes to green chemistry? *Ultrason. Sonochem.* 2018, 40(B), 117–122.

91. Cintas, P. Ultrasound and green chemistry - Further comments. *Ultrason. Sonochem.* 2016, 28, 257–258.

92. Mason, T. J. Sonochemistry and the environment - Providing a "green" link between chemistry, physics and engineering. *Ultrason. Sonochem.* 2007, 14(4), 476–483.

93. Ellstrom, C. J.; Török, B. Application of sonochemical activation in green synthesis. In *Green Chemistry: An Inclusive Approach* (Török, B.; Dransfield, T. Eds.), Elsevier, Oxford, Cambridge, 2018, Chp. 3.19., pp. 673–693.

94. Luche, J.-L. (Ed.) *Synthetic Organic Sonochemistry.* Springer, New York, 1998.

95. Cousin, T.; Chatel, G.; Kardos, N.; Andrioletti, B.; Draye, M. High frequency ultrasound as a tool for elucidating mechanistic elements of cis-cyclooctene epoxidation with aqueous hydrogen peroxide. *Ultrason. Sonochem.* 2019, 53, 120–125.

96. Lockwood, G. R.; Turnball, D. H.; Christopher, D. A.; Foster, F. S. Beyond 30 MHz [applications of high-frequency ultrasound imaging]. *IEEE Eng. Med. Biol. Mag.* 1996, 15(6), 60–71.

97. Tudela, I.; Sáez, V.; Esclapez, M. D.; Díez-García, M. I.; Bonete, P.; González-García, J. Simulation of the spatial distribution of the acoustic pressure in sonochemical reactors with numerical methods: A review. *Ultrason. Sonochem.* 2014, 21(3), 909–919.

98. Didenko, Y. T.; Suslick, K. S. The energy efficiency of formation of photons, radicals and ions during single-bubble cavitation. *Nature* 2002, 418(6896), 394–397.

99. Suslick, K. S.; Hammerton, D. A.; Cline, R. E. Sonochemical hot spot. *J. Am. Chem. Soc.* 1986, 108(18), 5641–5642.

100. Margulis, M. A. Sonoluminescence and sonochemical reactions in cavitation fields: A review. *Ultrasonics* 1985, 23(4), 157–169.

101. Török, B.; Balázsik, K.; Török, M.; Felföldi, K.; Bartók, M. Asymmetric reactions in Sonochemistry. *Ultrason. Sonochem.* 2001, 8(3), 191–200.

102. (a) Jiang, K.; Jia, Z.-J.; Yin, X.; Wu, L.; Chen, Y.-C. Asymmetric quadruple aminocatalytic domino reactions to fused carbocycles incorporating a spirooxindole motif. *Org. Lett.* 2010, 12(12), 2766–2769. (b) Schimer, J.; Cígler, P.; Veselý, J.; Grantz Šašková, K.; Lepsik, M.; Brynda, J.; Řezáčová,

P.; Kožíšek, M.; Císařová, I.; Oberwinkler, H.; Kraeusslich, H. G.; Konvalinka, J. Structure-aided design of novel inhibitors of HIV protease based on a benzodiazepine scaffold. *J. Med. Chem.* 2012, 55(22), 10130–10135. (c) Clevenger, K. D.; Ye, R.; Bok, J. W.; Thomas, P. M.; Islam, M. N.; Miley, G. P.; Robey, M. T.; Chen, C.; Yang, K.; Swyers, M.; Wu, E.; Gao, P.; Wu, C. C.; Keller, N. P.; Kelleher, N. L. Interrogation of benzomalvin biosynthesis using fungal artificial chromosomes with metabolomic scoring (FAC-MS): Discovery of a benzodiazepine synthase activity. *Biochemistry* 2018, 57(23), 3237–3243.

103. Maury, S. K.; Kumar, D.; Kamal, A.; Singh, H. K.; Kumari, S.; Singh, S. A facile and efficient multicomponent ultrasound assisted "on water" synthesis of benzodiazepine ring. *Mol. Divers.* 2021, 25(1), 131–142.

104. Simsek, S.; Schalkwijk, C. G.; Wolffenbuttel, B. H. R. Effects of rosuvastatin and atorvastatin on glycaemic control in type 2 diabetes-the CORALL study. *Diabet. Med.* 2012, 29(5), 628–631.

105. (a) Reddy, L. A.; Chakraborty, S.; Swapna, R.; Bhalerao, D.; Malakondaiah, G. C.; Ravikumar, M.; Kumar, A.; Reddy, G. S.; Naram, J.; Dwivedi, N.; Roy, A.; Himabindu, V.; Babu, B.; Bhattacharya, A.; Bandichhor, R. Synthesis and process optimization of amtolmetin: An antiinflammatory agent. *Org. Process Res. Dev.* 2010, 14(2), 362–368. (b) Olsen, J.; Li, C.; Skonberg, C.; Bjørnsdottir, I.; Sidenius, U.; Benet, L. Z.; Hansen, S. H. Studies on the metabolism of tolmetin to the chemically reactive acyl-coenzyme: A thioester intermediate in rats. *Drug Metab. Dispos.* 2007, 35(5), 758–764.

106. Magnus, N.; Staszak, M. A.; Udodong, U. E.; Wepsiec, J. P. Synthesis of 4-cyanopyrroles via mild Knorr reactions with β-ketonitriles. *Org. Process Res. Dev.* 2006, 10(5), 899–904.

107. Wang, S.-F.; Guo, C.-L.; Cui, K.-K.; Zhu, Y.-T.; Ding, J.-X.; Zou, X.-Y.; Li, Y.-H. Lactic acid as an invaluable green solvent for ultrasound-assisted scalable synthesis of pyrrole derivatives. *Ultrason. Sonochem.* 2015, 26, 81–86.

108. Jacques, A. V.; Stefanes, N. M.; Walter, L. O.; Perondi, D. M.; Efe, F.; de Souza, L. F. S.; Sens, L.; Syracuse, S. M.; de Moraes, A. C. R.; de Oliveira, A. S.; Martins, C. T.; Magalhaes, L. G.; Andricopulo, A. D.; Silva, L. O.; Nunes, R. J.; Santos-Silva, M. C. Synthesis of chalcones derived from 1-naphthylacetophenone and evaluation of their cytotoxic and apoptotic effects in acute leukemia cell lines. *Bioorg. Chem.* 2021, 116, 105315.

109. Arafa, W. A. A. Sustainable catalytic process with a high eco-scale score for the synthesis of novel series of Bischalcones through Claisen–Schmidt condensation. *J. Heterocycl. Chem.* 2018, 55(2), 456–464.

110. (a) Hou, X. L.; Wu, J.; Fan, R. H.; Ding, C. H.; Luo, Z. B.; Dai, L. X. Towards reaction selectivities of imines and aziridines. *Synlett* 2006, 181–193. (b) Silveira, C. C.; Vieira, A. S.; Braga, A. L.; Russowsky, D. Stereoselective Mannich-type reaction of chlorotitanium α-phenylseleno esters enolates with aromatic aldimines. *Tetrahedron* 2005, 61(39), 9312–9318.

111. Karla, P. G.; Alexandre, S. G.; Órfão, A. T. G.; Rodrigo, C.; Claudio, M. P. P.; Hélio, A. S. Eco-friendly synthesis of imines by ultrasound irradiation. *Tetrahedron Lett.* 2007, 48(10), 1845–1848.

112. (a) Covey, A. M.; Tuorto, S.; Brody, L. A.; Sofocleous, C. T.; Schubert, J.; von Tengg-Kobligk, H.; Getrajdman, G. I.; Schwartz, L. H.; Fong, Y.; Brown, K. T. Safety and efficacy of preoperative portal vein embolization with polyvinyl alcohol in 58 patients with liver metastases. *AJR Am. J. Roentgenol.* 2005, 185(6), 1620–1626. (b) Tadavarthy, S. M.; Moller, J. H.; Amplatz, K. Polyvinyl alcohol (Ivalon)—A new embolic material. *Am. J. Roentgenol. Radium Ther. Nucl. Med.* 1975, 125(3), 609–616.

113. Maquet, V.; Martin, D.; Malgrange, B.; Franzen, R.; Schoenen, J.; Moonen, G.; Jerome, R. Peripheral nerve regeneration using bioresorbable macroporous polylactide scaffolds. *J. Biomed. Mater. Res.* 2000, 52(4), 639–651.

114. Weis, C.; Odermatt, E. K.; Kressler, J.; Funke, Z.; Wehner, T.; Freytag, D. Poly(vinyl alcohol) membranes for adhesion prevention. *J. Biomed. Mater. Res. B Appl. Biomater.* 2004, 70, 191–202.

115. Othayoth, A. K.; Paul, S.; Muralidharan, K. Polyvinyl alcohol-phytic acid polymer films as promising gas/vapor sorption materials. *J. Polym. Res.* 2021, 28(7), 249–259.

116. Lia, Y.; Songa, Y.; Lia, J.; Lib, Y.; Lic, N.; Niu, S. A scalable ultrasonic-assisted and foaming combination method preparation polyvinyl alcohol/phytic acid polymer sponge with thermal stability and conductive capability. *Ultrason. Sonochem.* 2018, 42, 18–25.

117. Evans, R. C.; Douglas, P.; Burrows, H. D. (Eds.) *Applied Photochemistry.* Springer, Dodrecht, 2013.

118. Bonfield, H. E.; Knauber, T.; Levesque, F.; Moschetta, E. G.; Susanne, F.; Edwards, L. J. Photons as a 21st century reagent. *Nat. Commun.* 2020, 11(1), 804.

119. (a) Narayanam, J. M. R.; Stephenson, C. R. J. Visible light photoredox catalysis: Applications in organic synthesis. *Chem. Soc. Rev.* 2011, 40(1), 102–113. (b) Xuan, J.; Xiao, W.-J.; Photoredox, visible-light visible-light photoredox catalysis. *Angew. Chem. Int. Ed.* 2012, 51(28), 6828–6838. (c) Ciriminna, R.; Delisi, R.; Xu, Y.-J.; Pagliaro, M. Toward the waste-free synthesis of fine chemicals with visible light. *Org. Process Res. Dev.* 2016, 20(2), 403–408.

120. (a) Knowles, J. P.; Elliott, L. D.; Booker-Milburn, K. I. Flow photochemistry: Old light through new windows. *Beilstein J. Org. Chem.* 2012, 8, 2025–2052. (b) Oelgemoeller, M. Highlights of photochemical reactions in microflow reactors. *Chem. Eng. Technol.* 2012, 35(7), 1144–1152.

121. Hook, B. D. A.; Dohle, W.; Hirst, P. R.; Pickworth, M.; Berry, M. B.; Booker-Milburn, K. I. A practical flow reactor for continuous organic photochemistry. *J. Org. Chem.* 2005, 70(19), 7558–7564.

122. Tucker, J. W.; Zhang, Y.; Jamison, T. F.; Stephenson, C. R. J. Visible-light photoredox catalysis in flow. *Angew. Chem. Int. Ed.* 2012, 51(17), 4144–4147.

123. Hernandez-Linares, M. G.; Guerrero-Luna, G.; Pérez-Estrada, S.; Ellison, M.; Ortin, M.-M.; Garcia-Garibay, M. A. Large-scale green chemical synthesis of adjacent quaternary chiral centers by continuous flow photodecarbonylation of aqueous suspensions of nanocrystalline ketones. *J. Am. Chem. Soc.* 2015, 137(4), 1679–1684.

124. Saikia, I.; Borah, A. J.; Phukan, P. Use of bromine and bromo-organic compounds in organic synthesis. *Chem. Rev.* 2016, 116(12), 6837–7042.

125. Steiner, A.; Roth, P. M. C.; Strauss, F. J.; Gauron, G.; Tekautz, G.; Winter, M.; Williams, J. D.; Kappe, C. O. Multikilogram per hour continuous photochemical benzylic brominations applying a smart dimensioning scale-up strategy. *Org. Process Res. Dev.* 2020, 24(10), 2208–2216.

126. Blackham, E. E.; Booker-Milburn, K. I. A short synthesis of (±)-3-Demethoxyerythratidinone by ligand-controlled selective heck cyclization of equilibrating enamines. *Angew. Chem. Int. Ed.* 2017, 56(23), 6613–6616.

127. Blanco-Ania, D.; Gawade, S. A.; Zwinkels, L. J. L.; Maartense, L.; Bolster, M. G.; Benningshof, J. C. J.; Rutjes, F. P. J. T. Rapid and scalable access into strained scaffolds through continuous flow photochemistry. *Org. Process Res. Dev.* 2016, 20(2), 409–413.

128. Popsavin, V.; Sreco, B.; Benedeković, G.; Francuz, J.; Popsaviń, M.; Kojic, V.; Bogdanović, G. Design, synthesis and antiproliferative activity of styryl lactones related to (+)-Goniofufurone. *Eur. J. Med. Chem.* 2010, 45(7), 2876–2883.

129. Ralph, M.; Ng, S.; Booker-Milburn, K. I. Short flow photochemistry enabled synthesis of the cytotoxic lactone (+)-Goniofufurone. *Org. Lett.* 2016, 18(5), 968–971.

130. An, F. L.; Wang, X. B.; Wang, H.; Li, Z. R.; Yang, M. H.; Luo, J.; Kong, L. Y. Cytotoxic Rocaglate derivatives from leaves of Aglaia Perviridis. *Sci. Rep.* 2016, 6, 6–17.

131. Yueh, H.; Gao, Q.; Porco, J. A.; Beeler, A. B. A photochemical flow reactor for large scale syntheses of Aglain and Rocaglate natural product analogues. *Bioorg. Med. Chem.* 2017, 25(23), 6197–6202.

132. (a) Pohlmann, B.; Scharf, H.-D.; Jarolimek, U.; Mauermann, P. Photochemical production of fine chemicals with concentrated sunlight. *Sol. Energy* 1997, 61(3), 159–168. (b) Esser, P.; Pohlmann, B.; Scharf, H.-D. The photochemical synthesis of fine chemicals with sunlight. *Angew. Chem. Int. Ed. Engl.* 1994, 33(20), 2009–2023.

133. Yang, C.; Li, R.; Zhang, K. A. I.; Lin, W.; Landfester, K.; Wang, X. Heterogeneous photoredox flow chemistry for the scalable organosynthesis of fine chemicals. *Nat. Commun.* 2020, 11(1), 1239–1246.

134. (a) Wang, G.-W. Mechanochemical organic synthesis. *Chem. Soc. Rev.* 2013, 42(18), 7668–7700. (b) Boldyreva, E. Mechanochemistry of inorganic and organic systems: What is similar, what is different? *Chem. Soc. Rev.* 2013, 42(18), 7719–7738. (c) James, S. L.; Adams, C. J.; Bolm, C.; Braga, D.; Collier, P.; Friščič, T.; Grepioni, F.; Harris, K. D. M.; Hyett, G.; Jones, W.; Krebs, A.; Mack, J.; Maini, L.; Orpen, A. G.; Parkin, I. P.; Shearouse, W. C.; Steed, J. W.; Waddell, C. Mechanochemistry: Opportunities for new and cleaner synthesis. *Chem. Soc. Rev.* 2012, 41(1), 413–447.

135. (a) Wang, Y.; Wang, H.; Jiang, Y.; Zhang, C.; Shao, J.; Xu, D. Fast, solvent-free and highly enantioselective fluorination of β-keto esters catalyzed by chiral copper complexes in a ball mill. *Green Chem.* 2017, 19(7), 1674–1677; (b) Hermann, G. N.; Jung, C. L.; Bolm, C. Mechanochemical indole synthesis by rhodium-catalysed oxidative coupling of acetanilides and alkynes under solventless conditions in a ball mill. *Green Chem.* 2017, 19(11), 2520–2523; (c) Howard, J. L.; Sagatov, Y.; Repusseau, L.; Schotten, C.; Browne, D. L. Controlling reactivity through liquid assisted grinding: The curious case of mechanochemical fluorination. *Green Chem.* 2017, 19(12), 2798–2802; (d) Vadivelu, M.; Sugirdha, S.;

Dheenkumar, P.; Arun, Y.; Karthikeyan, K.; Praveen, C. Solvent-free implementation of two dissimilar reactions using recyclable CuO nanoparticles under ball-milling conditions: Synthesis of new oxindole-triazole pharmacophores. *Green Chem.* 2017, 19(15), 3601–3610; (e) Tan, D.; Friščić, T. Carbodiimide insertion into sulfonimides: One-step route to azepine derivatives via a two-atom saccharin ring expansion. *Chem. Commun. (Camb)* 2017, 53(5), 901–904.

136. Stolle, A.; Szuppa, T.; Leonhardt, S. E. S.; Ondruschk, B. Ball milling in organic synthesis: Solutions and challenges. *Chem. Soc. Rev.* 2011, 40(5), 2317–2329.

137. (a) Fan, G.-P.; Liu, Z.; Wang, G.-W. Efficient ZnBr2-catalyzed reactions of allylic alcohols with indoles, Sulfamides and anilines under high-speed vibration milling conditions. *Green Chem.* 2013, 15(6), 1659–1664. (b) Thorwirth, R.; Stolle, A.; Ondruschka, B.; Wild, A.; Schubert, U. S. Fast, ligand-and solvent-free copper-catalyzed click reactions in a ball mill. *Chem. Commun. (Camb)* 2011, 47(15), 4370–4372. (c) Fischer, F.; Scholz, G.; Batzdorf, L.; Wilke, M.; Emmerling, F. Synthesis, structure determination, and formation of a theobromine: Oxalic acid 2:1 cocrystal. *Cryst. Eng. Comm.* 2015, 17, 824–829. (d) Biswal, B. P.; Chandra, S.; Kandambeth, S.; Lukose, B.; Heine, T.; Banerjee, R. Mechanochemical synthesis of chemically stable isoreticular covalent organic frameworks. *J. Am. Chem. Soc.* 2013, 135(14), 5328–5331. (d) Dekamin, M. G.; Eslami, M. Highly efficient organocatalytic synthesis of diverse and densely functionalized 2-amino-3-cyano-4H-pyrans under mechanochemical ball milling. *Green Chem.* 2014, 16(12), 4914–4921.

138. Xu, H.; Chen, K.; Liu, H.-W.; Wang, G.-W. Solvent-free N-iodosuccinimide-promoted synthesis of spiroimidazolines from alkenes and amidines under Ball Milling Conditions. *Org. Chem. Front.* 2018, 5(19), 2864–2869.

139. Brahmachari, G.; Karmakar, I.; Karmakar, P. Catalyst- and solvent-free Csp2–H functionalization of 4-hydroxycoumarins via C-3 dehydrogenative aza-coupling under Ball-Milling. *Green Chem.* 2021, 23(13), 4762–4770.

140. Yashwantrao, G.; Jejurkar, V. P.; Kshatriya, R.; Saha, S. Solvent-free, mechanochemically scalable synthesis of 2,3-Dihydroquinazolin-4(1H)-one using Brønsted acid catalyst. *ACS Sust. Chem. Eng.* 2019, 7(15), 13551–13558.

141. (a) Jafari, E.; Khajouei, M. R.; Hassanzadeh, F.; Hakimelahi, G. H.; Khodarahmi, G. A. Quinazolinone and quinazoline derivatives: Recent structures with potent antimicrobial and cytotoxic activities. *Res. Pharm. Sci.* 2016, 11(1), 1–14. (b) Rakesh, K. P.; Darshini, N.; Shubhavathi, T.; Mallesha, N. Biological applications of quinazolinones analogues: A review. *Org. Med. Chem. J.* 2017, 2, 555–585. (c) He, L.; Li, H.; Chen, J.; Wu, X.-F. Recent advances in 4(3H)-quinazolinone syntheses. *RSC Adv.* 2014, 4(24), 12065–12077.

142. Crawford, D. E.; Casaban, J. Recent developments in mechanochemical materials synthesis by extrusion. *Adv. Mater.* 2016, 28(27), 5747–5754.

143. Giles, H. F.; Wagner, J. R.; Mount, E. M. *Extrusion: The Definitive Processing Guide and Handbook.* William Andrew Elsevier, Amsterdam, 2014.

144. (a) Stolle, A.; Schmidt, R.; Jacob, K. Scale-up of organic reactions in ball mills: Process intensification with regard to energy efficiency and economy of scale. *Faraday Discuss.* 2014, 170, 267–286. (b) Burmeister, C. F.; Stolle, A.; Schmidt, R.; Jacob, K.; Breitung-Faes, S.; Kwade, A. Experimental and computational investigation of Knoevenagel condensation in planetary ball mills. *Chem. Eng. Technol.* 2014, 37(5), 857–864. (c) Kaupp, G.; Rezami-Jamal, M.; Schmeyers, J. Solvent-free Knoevenagel condensations and Michael additions in the solid state and in the melt with quantitative yield. *Tetrahedron* 2003, 59(21), 3753–3760. (d) Balema, V. P.; Wiench, J. W.; Pruski, M.; Pecharsky, V. K. Mechanically induced solid-state generation of phosphorus ylides and the solvent-free Wittig reaction. *J. Am. Chem. Soc.* 2002, 124(22), 6244–6245.

145. Crawford, D. E.; Miskimmin, C. K. G.; Albadarin, A. B.; Walker, G.; James, S. L. Organic synthesis by twin screw extrusion (TSE): Continuous, scalable and solvent-free. *Green Chem.* 2017, 19(6), 1507–1518.

146. (a) Trask, A. V.; Motherwell, W. D. S.; Jones, W. Pharmaceutical cocrystallization: Engineering a remedy for caffeine hydration. *Cryst. Growth Des.* 2005, 5(3), 1013–1021. (b) Trask, A. V.; Motherwell, W. D. S.; Jones, W. Physical stability enhancement of theophylline via cocrystallization. *Int. J. Pharm.* 2006, 320(1–2), 114–123. (c) Jones, W.; Motherwell, W.; Trask, A. V. Pharmaceutical cocrystals: An emerging approach to physical property enhancement. *MRS Bull.* 2006, 31(11), 875–879.

147. (a) Shiraki, K.; Takata, N.; Takano, R.; Hayashi, Y.; Terada, K. Dissolution improvement and the mecha- nism of the improvement from cocrystallization of poorly water-soluble compounds. *Pharm. Res.* 2008, 25(11), 2581–2592. (b) McNamara, D. P.; Childs, S. L.; Giordano, J.; Iarriccio, A.; Cassidy, J.; Shet, M. S.; Mannion, R.; O'Donnell, E.; Park, A. Use of a glutaric acid cocrystal to improve oral bioavailabil- ity of a low solubility API. *Pharm. Res.* 2006, 23(8), 1888–1897. (c) Good, D. J.; Rodríguez-Hornedo, N. Solubility advantage of pharmaceutical cocrystals. *Cryst. Growth Des.* 2009, 9(5), 2252–2264. (d) Babu, N. J.; Nangia, A. Solubility advantage of amorphous drugs and pharmaceutical cocrystals. *Cryst. Growth Des.* 2011, 11(7), 2662–2679.

148. (a) Buu-Hoï, N. P.; Ratsimamanga, A. R.; Xuong, N. D.; Nigeon-Dureuil, M. Study of nicastubine, an addition compound of vitamin PP and vitamin C. *Bull. Soc. Chim. Biol. (Paris)* 1953, 35(3–4), 326–333. (b) Stahly, G. P. A survey of cocrystals reported prior to 2000. *Cryst. Growth Des.* 2009, 9(10), 4212–4229.

149. Stolar, T.; Lukin, S.; Tireli, M.; Sović, I.; Karadeniz, B.; Kereković, I.; Matijasič, G.; Gretić, M.; Katancič, Z.; Dejanović, I.; di Michiel, M.; Halasz, I.; Uzarevic, K. Control of pharmaceutical cocrystal polymorphism on various scales by mechanochemistry: Transfer from the laboratory batch to the large- scale extrusion processing. *ACS Sust. Chem. Eng.* 2019, 7(7), 7102–7110.

150. Sun, B.; He, Y.; Peng, R.; Chu, S.; Zuo, J. Air-flow impacting for continuous, highly efficient, large-scale mechanochemical synthesis: A proof-of-concept study. *ACS Sust. Chem. Eng.* 2016, 4(4), 2122–2128.

151. (a) Morales-Morales, D.; Jensen, C. M. (Eds.) *The Chemistry of Pincer Compounds.* Elsevier, New York, 2007; (b) Organometallic pincer chemistry. In *Top. Organomet. Chem* (van Koten, G.; Milstein, D. Eds.), 40, Springer, 2013; (c) Szabo, K.; Wendt, O. F. (Eds.) *Pincer and Pincer-Type Complexes: Applications in Organic Synthesis and Catalysis.* Wiley-VCH, Weinheim, 2014. (d) Selander, N.; Szabó, K. J. Catalysis by palladium pincer complexes. *Chem. Rev.* 2011, 111(3), 2048–2076; (e) Niu, J.-L.; Hao, X.-Q.; Gong, J.-F.; Song, M.-P. Symmetrical and unsymmetrical pincer complexes with group 10 metals: Synthesis via aryl C–H activation and some catalytic applications. *Dalton Trans.* 2011, 40(19), 5135– 5150; (f) van Koten, G. *J. Organomet. Chem.* 2013, 730, 156–164; (g) Assay, M.; Morales-Morales, D. *Dalton Trans.* 2015, 44(40), 17432–17447.

152. Aleksanyan, D. V.; Churusova, S. G.; Aysin, R. R.; Klemenkova, Z. S.; Nelyubina, Y. V.; Kozlov, V. A. The first example of mechanochemical synthesis of organometallic pincer complexes. *Inorg. Chem. Commun.* 2017, 76, 33–35.

153. Rosenthaler, L. Durch enzyme bewirkte asymmetrische synthese. *Biochem. Z.* 1908, 14, 238–253.

154. Zaks, A.; Klibanov, A. M. Enzymic catalysis in organic media at 100°C. *Science* 1984, 224(4654), 1249–1251.

155. Griengl, H.; Schwab, H.; Fechter, M. The synthesis of chiral cyanohydrins by oxynitrilases. *Trends Biotechnol.* 2000, 18(6), 252–256.

156. Hills, G. Industrial use of lipases to produce fatty acid esters. *Eur. J. Lipid Sci. Technol.* 2003, 105(10), 601–607.

157. Nagasawa, T.; Nakamura, T.; Yamada, H. Production of acrylic acid and methacrylic acid using Rhodococcus rhodochrous J1 nitrilase. *Appl. Microbiol. Biotechnol.* 1990, 34(3), 322–324.

158. Bommarius, A. S. Biocatalysis: A status report. *Chem. Biomol. Eng.* 2015, 6, 319–345.

159. Bornscheuer, U. T.; Huisman, G. W.; Kazlauskas, R. J.; Lutz, S.; Moore, J. C.; Robins, K. Engineering the third wave of biocatalysis. *Nature* 2012, 485(7397), 185–194.

160. Reetz, M. T. Biocatalysis in organic chemistry and biotechnology: Past, present, and future. *J. Am. Chem. Soc.* 2013, 135(34), 12480–12496.

161. Sheldon, R. A.; van Pelt, S. Enzyme immobilization in biocatalysis: Why, what and how. *Chem. Soc. Rev.* 2013, 42(15), 6223–6235.

162. Sheldon, R. A.; van Rantwijk, F. Biocatalysis for sustainable organic synthesis. *Aust. J. Chem.* 2004, 57(4), 281–289.

163. (a) Wegman, M. A.; Janssen, M. H. A.; van Rantwijk, F.; Sheldon, R. A. Towards biocatalytic synthesis of β-lactam antibiotics. *Adv. Synth. Catal.* 2001, 343(6–7), 559–577.

164. (a) Batchelor, F. R.; Doyle, F. P.; Nayler, J. H. C.; Rolinson, G. N. Synthesis of penicillin: 6-aminopenicil- lanic acid in penicillin fermentations. *Nature* 1959, 183(4656), 257–258. (b) Rolinson, G. N.; Batchelor, F. R.; Butterworth, D.; Cameron-Wood, J.; Cole, M.; Eustace, G. C.; Hart, M. V.; Richards, M.; Chain, E. B. Formation of 6-aminopeni-cillanic acid from penicillin by enzymatic hydrolysis. *Nature* 1960, 187,

236–237. (c) Claridge, C. A.; Gourevitch, A.; Lein, J. Bacterial penicillin amidase. *Nature* 1960, 187, 237–238. (d) Huang, H. T.; English, A. R.; Seto, T. A.; Shull, G. M.; Sobin, B. A. Enzymatic hydrolysis of the side chain of penicillins. *J. Am. Chem. Soc.* 1960, 82(14), 3790–3791.

165. Kallenberg, A. I.; van Rantwijk, F.; Sheldon, R. A. Immobilization of penicillin Gacylase: The key to optimum performance. *Adv. Synth. Catal.* 2005, 347(7–8), 905–926.

166. Bruggink, A.; Roos, E. C.; de Vroom, E. Penicillin acylase in the industrial production of β-lactam antibiotics. *Org. Process Res. Dev.* 1998, 2(2), 128–133.

167. Hanson, R. L.; Howell, J.; LaPorte, T.; Donovan, M.; Cazzulino, D.; Zannella, V.; Montana, M.; Nanduri, V.; Schwarz, S.; Eiring, R. et al. Synthesis of Allylsine ethylene acetal using phenylalanine dehydrogenase from Thermoactinomyces intermedius. *Enzyme Microb. Technol.* 2000, 26(5–6), 348–358.

168. Ma, S. K.; Gruber, J.; Davis, C.; Newman, L.; Gray, D.; Wang, A.; Grate, J.; Huisman, G. W.; Sheldon, R. A. A green-by-design biocatalytic process for atorvastatin intermediate. *Green Chem.* 2010, 12(1), 81–86.

169. Belley, M. L.; Leger, S.; Labelle, M.; Roy, P.; Xiang, Y. B.; Guay, D. U.S. Patent 5,565,473, 1996.

170. Liang, J.; Lalonde, J.; Borup, B.; Mitchell, V.; Mundorff, E.; Trinh, N.; Kochrekar, D. A.; Cherat, R. N.; Pai, G. G. Development of a biocatalytic process as an alternative to the (-)-DIP-Cl-Mediated asymmetric reduction of a key intermediate of montelukast. *Org. Process Res. Dev.* 2010, 14(1), 193–198.

171. Sheldon, R. A.; van Rantwijk, F. Enantioselective acylation of chiral amines catalyzed by serine hydrolases. *Tetrahedron* 2004, 60(3), 501–519.

172. Saville, C. K.; Maney, J. M.; Mundorff, E. C.; Moore, J. C.; Tam, S.; Jarvis, W. R.; Colbeck, J. C.; Krebber, A.; Fleitz, F. J.; Brands, J.; Devine, P. N.; Huisman, G. W.; Hughes, G. J. Biocatalytic asymmetric synthesis of chiral amines from ketones applied to sitagliptin manufacture. *Science* 2010, 329(5989), 305–309.

173. Burgard, A.; Burk, M. J.; Osterhout, R.; Van Dien, S.; Yim, H. Development of a commercial scale process for production of 1,4-butanediol from sugar. *Curr. Opin. Biotechnol.* 2016, 42, 118–125.

10

Eccentric Vibration Milling for Mechanochemical Synthesis

A Straightforward Route to Advanced Materials

Peter Baláž, Marcela Achimovičová, Erika Dutková, and Matej Baláž

CONTENTS

10.1 Introduction

Mechanochemistry invokes changes in solids due to the application of mechanical energy. This energy is supplied by high-energy milling which leads to the production of various surface and bulk defects. Besides these fundamental defects in the solid phase, a plethora of further effects can be selectively created depending on the mill type and milling conditions applied. When no chemical changes are

DOI: 10.1201/9781003178187-13

documented, the process is called *mechanical activation*. However, if the impact of milling is performed at higher energy levels, a new chemical species can be formed. This process is called *mechanochemical activation (mechanochemical synthesis, mechanochemical reduction, mechanochemical exchange, mechanical alloying, etc.)*. The term *reactive milling* is also used in the literature. In technology, the possibilities of mechanochemistry can be utilized in cases where the economic demand, high temperatures as well as wet processes are under consideration. The accompanying environmental aspects are particularly attractive (Baláž 2000, 2008; Baláž et al. 2013; Baláž 2021).

The great potential of mechanochemistry to produce well-applicable chalcogenides will be illustrated in this chapter. As milling equipment, an eccentric vibratory mill (ESM) has been applied and the properties of the products have been compared with those prepared in a laboratory mill to document the applicability and effectivity of the milling process on a large scale.

10.2 Eccentric Vibratory Mill (ESM)

The concept of ESM mill has been developed at the Institute of Mineral and Waste Processing, Waste Disposal and Geomechanics at the Clausthal University of Technology in close co-operation with Siebtechnik, Mülheim (both in Germany). Pioneering theoretical works on the operation of the mill were elaborated by Professor Gock and co-authors (Gock and Kurrer 1996). They verified the application of this type of mill in several branches of industry including recycling technology (Gock and Kurrer 1999).

The ESM is a single tube mill of modular design. Figure 10.1 shows a diagram of such a module. On one side of the spring-mounted grinding tube (1) is the exciter unit (3) consisting of a bearing block and unbalanced masses, which are connected to the motor by means of the drive shaft. For the purpose of mass compensation, a counterweight (2) is located on the opposite side of the exciter unit. Unlike conventional vibratory tube mills with a center drive, this mill produces elliptic, circular, and linear vibrations, which lead to unequaled amplitudes of vibration up to 20 mm and a high degree of disaggregation of the grinding media (4). As a consequence, more intensive milling is produced because a broader zone of ball contact with the material is created. The specific energy consumption was reduced

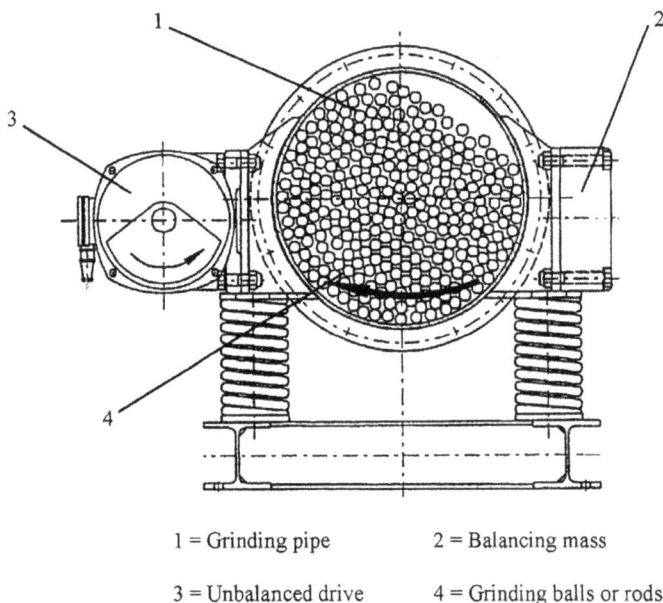

1 = Grinding pipe	2 = Balancing mass
3 = Unbalanced drive	4 = Grinding balls or rods

FIGURE 10.1   Schematic view of the ESM. Reprinted with permission from ref. (Gock and Kurrer 1999). Copyright 1999 Elsevier.

FIGURE 10.2 ESM pilot scale (type ESM 654-10.5ks), half module, grinding tube diameter 650 mm (the photograph is courtesy of the authors).

FIGURE 10.3 "Satellite" version of ESM: satellite tube in a steel-gray color is mounted on the corpus of the mill in blue color (operator manipulates with flexible pipe for purge and filling of satellite with argon) (the photograph is courtesy of the authors).

by 50%, while the throughput was increased by a factor of 1.5–2.0 when compared with conventional vibratory mills.

Figure 10.2 shows a pilot ESM in the plant at Clausthal University. The machine is equipped with two imbalanced motors of different frequencies, directly flanged to the grinding tube. The motor can run at 1500 min^{-1} or 1000 min^{-1} at different oscillating circuit diameters. The unused motor serves as the counterweight. In general, with this configuration, it is possible to focus on frictional stress at higher frequencies and impact stress at lower frequencies.

In the pilot plant practice, so-called "satellite" version of ESM has been applied as a preparatory step before milling in the whole volume of the mill (Figure 10.3). In this configuration of milling, up to hundreds of grams of mass can be treated. The applied milling parameters for this configuration are in harmony with the parameters applied for milling in the whole volume of the mill where in principle tens of kilograms of solids can be treated (remark: the results presented in this chapter were exclusively obtained in the "satellite" version of ESM).

10.3 Mechanochemical Synthesis in Materials Science

There is a big demand for energy-related applications of metal–chalcogenide compounds nowadays. Their application in batteries, supercapacitors, and hydrogen-related applications is intensively promoted (Liu et al. 2016). The nanostructuring phenomenon, which is a characteristic feature of mechanochemical synthesis, has been found to further boost the application potential (Rui, Tan, and Yan 2014). In Table 10.1 selected applications of chalcogenides are summarized.

10.3.1 Binary Systems

The mechanochemical synthesis of binary chalcogenides in the ESM mill will be described in this chapter. The simple scheme of the process is presented in Figure 10.4. The dry mode is more simple and nanostructured products can be obtained in a straightforward manner just by mixing and milling of corresponding elements. However, produced nanosized particles are frequently agglomerated into microsized composites which require further classification (Figure 10.4). Besides the principle of dry mode synthesis (a), also a scheme of the process has been proposed by a computer simulation (b).

The wet mode of sulfide synthesis developed by our group in 2003 in the so-called "acetate route" (Baláž et al. 2003) overcomes the problem of agglomeration; however, the synthesis is more complicated because additional operations are needed (Figure 10.5).

10.3.1.1 MeS Sulfides (Me = Pb, Zn, Cd, Cu)

According to Table 10.1, binary sulfides are well-applicable in various fields. Several binary sulfides were synthesized in the ESM mill via a wet mode combining corresponding acetates with sodium sulfide according to Equation 10.1:

$$(CH_3COO)_2 Me.xH_2O + Na_2S \rightarrow MeS + 2CH_3COONa + xH_2O \tag{10.1}$$

where $x = 3, 2, 2,$ and 4 for Pb, Zn, Cd, and Cu acetate, respectively (Godočíková et al. 2006).

The XRD patterns of corresponding binary sulfides are summarized in Figure 10.6. The crystalline nature of the products is clearly seen, and no wear from millings balls was detected. The average size of MeS particles determined from XRD data is 4–14 nanometers for powders milled for 6 min.

TABLE 10.1

Chalcogenides in Energy-Related Applications

Chalcogenide	Application
$CuInS_2$, $CuInSe_2$, Cu_2ZnSnS_4, $CuIn_xGa_{(1-x)}Se_2$	Solar energy conversion
ZnS, PbS, Ag_2S, CdSe	Infrared windows and detectors
PbS, CdS, $CuInS_2$	Light-emitting diodes
FeS, Co_9S_8, MoS_2, WS_2, CoS_2	Hydrogen evolution and storage
$Cu_{12}Sb_4S_{13}$, CuS, CuSe, Cr_2S_3	Thermoelectricity
TiS_2, MoS_2, WS_2, NiS_2, GeS	Lithium- and sodium-ion batteries
ZnS:Mn, SrS:Ce, $BaAl_2S_4$:Eu	Luminescence
Ag_2S, PbS	Ion-selective electrodes
Cr_5S_6, $Fe_xMn_{1-x}S$, $Co_xMn_{1-x}S$	Giant magnetoresistors
CdS, CdSe, CdTe, Ag_2S	Diagnostics
MoS_2, Sb_2S_3, SnS, WS_2, FeS	Wear resistance
MoS_2, FeS, ZnS, RuS_2, WS_2	Catalysts

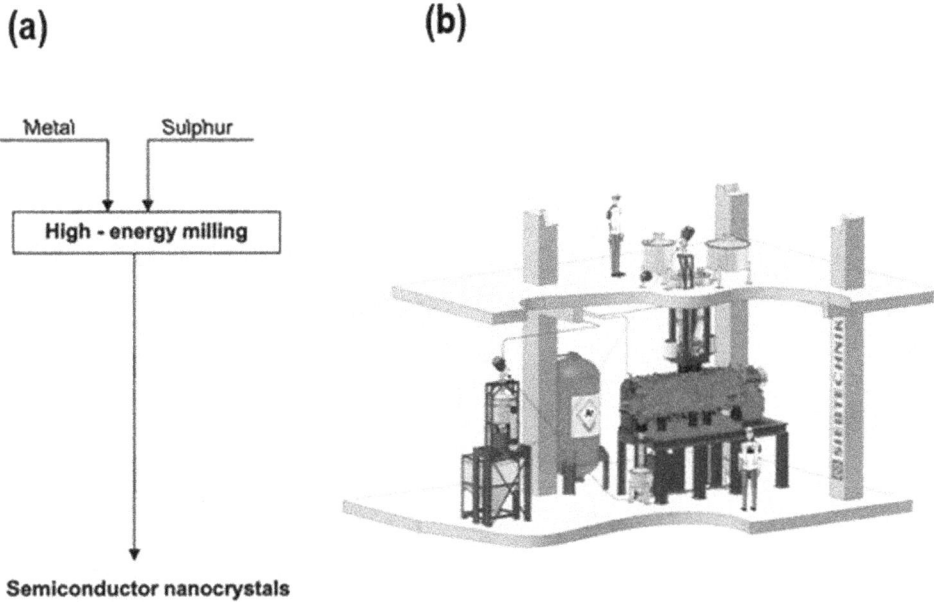

FIGURE 10.4 (a) Mechanochemical synthesis of semiconductor nanocrystals and (b) plant for mechanochemical synthesis in the ESM mill under an inert atmosphere (computer simulation). Modified and reprinted with permission from ref. Baláž et al. (2017). Copyright 2017 Springer Nature.

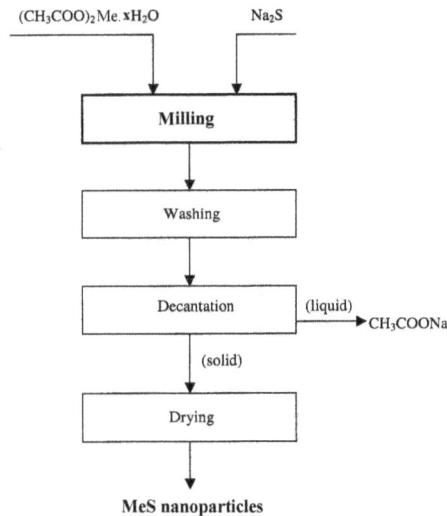

FIGURE 10.5 Flowchart of the mechanochemical synthesis of MeS nanoparticles in an industrial mill (wet mode). Reprinted with permission from ref. Godočíková et al. (2006). Copyright 2006 Elsevier.

10.3.1.2 MeSe Selenides (Me = Pb, Zn, Cu)

Binary metal selenides were synthesized by a dry mode rapid one-pot mechanochemical synthesis in the ESM mill, as well as in order to verify the feasibility of industrial nanocrystalline semiconductors production (Achimovičová, Baláž, Ďurišin, et al. 2011; Achimovičová, Baláž, Ohtani, et al. 2011; Achimovičová et al. 2020). These semiconductors are used in various optoelectronic applications, thermoelectrics, and in battery research as electrode materials (Yu et al. 2004; Kumar and Singh 2009; Zyoud et al. 2018). The XRD patterns of the synthesized binary selenides PbSe, ZnSe, and CuSe with

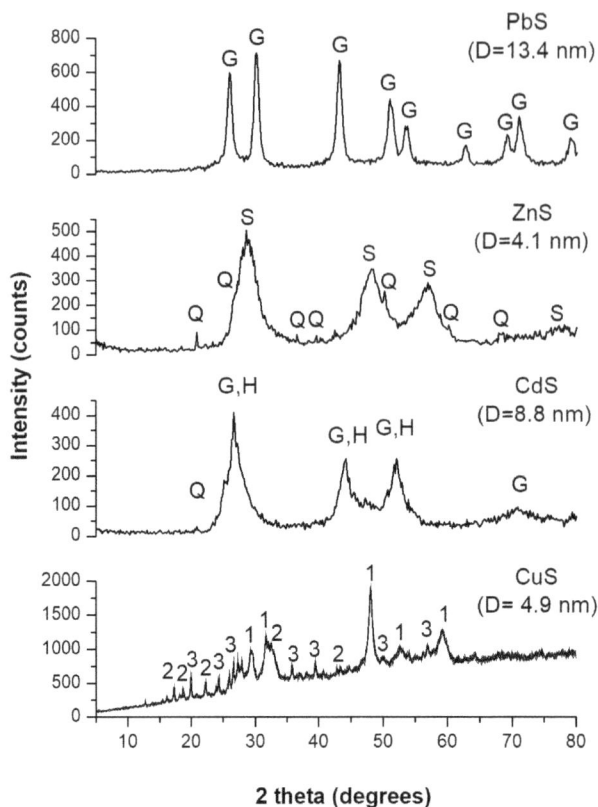

FIGURE 10.6   XRD patterns of mechanochemically synthesized copper sulfide CuS (1 – covellite, 2 – chalcanthite, 3 – bonattite), cadmium sulfide CdS (G – greenockite, H – hawleyite, Q – quartz), zinc sulfide ZnS (S – sphalerite, Q – quartz), and lead sulfide PbS (G – galena). Modified and reprinted with permission from ref. Godočíková et al. (2006). Copyright 2006 Elsevier.

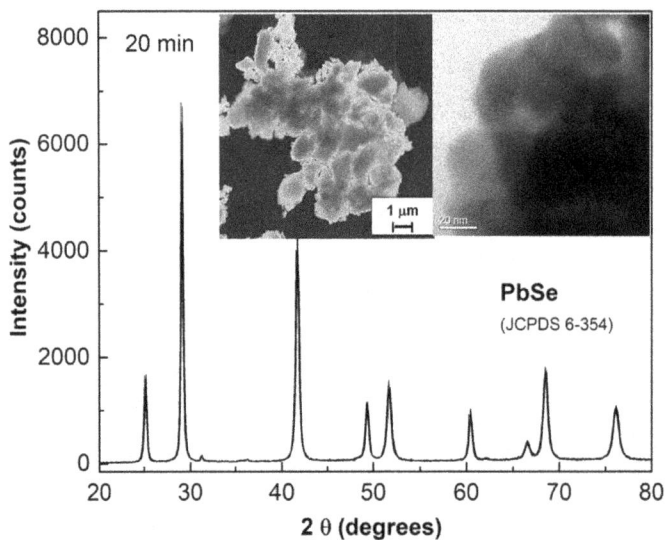

FIGURE 10.7   XRD pattern of mechanochemically synthesized lead selenide PbSe for 20 min. TEM/HRTEM images of agglomerated nanocrystals (inset). "Mechanochemical synthesis of nanocrystalline lead selenide: industrial approach" by Marcela Achimovičová et al. is licensed under CC by 4.0.

the TEM/HRTEM images of their agglomerated nanocrystals are shown in Figures 10.7–10.9. The calculated average crystallite size was 10 and 64 nm for ZnSe and CuSe, respectively. Agglomerated nanoparticles of prepared metal selenide semiconductors form clusters with a size of about 10.2–50 μm.

10.3.2 Ternary Systems

Polymetal sulfides represent the variability of composition and to some extent resemble high-entropy alloys produced from several elements by mechanical alloying (Vaidya, Muralikrishna, and Murty 2019).

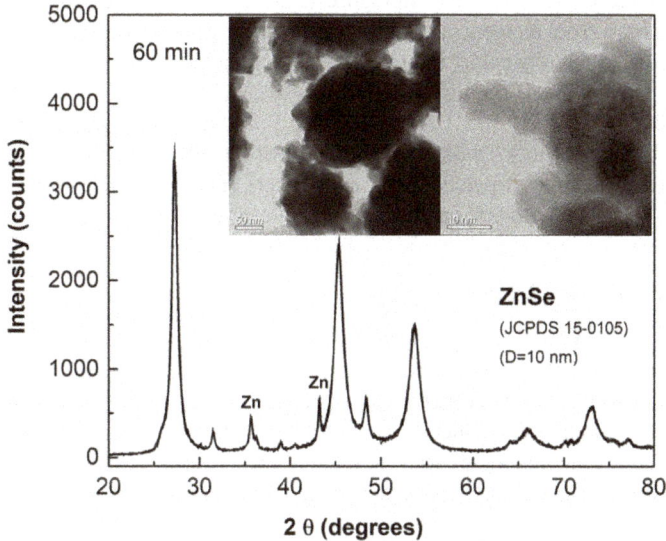

FIGURE 10.8 XRD pattern of mechanochemically synthesized zinc selenide ZnSe for 60 min. TEM/HRTEM images of agglomerated nanocrystals (inset). Modified and reprinted with permission from ref. Achimovičová, Baláž, Ohtani, et al. (2011). Copyright 2011 Elsevier.

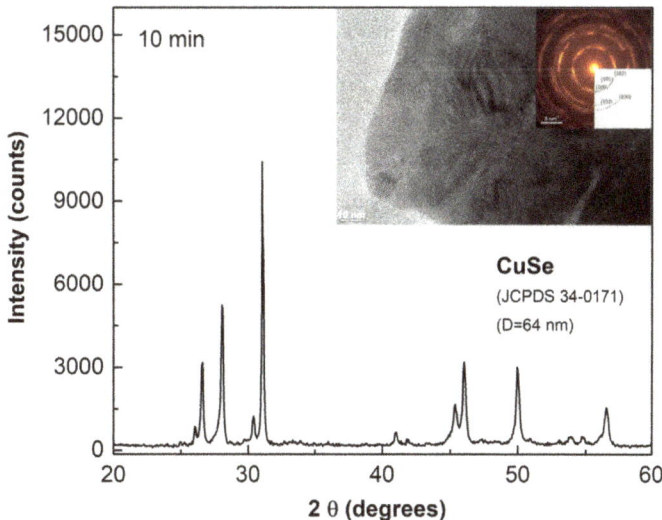

FIGURE 10.9 XRD pattern of mechanochemically synthesized copper selenide CuSe for 10 min. HRTEM image of agglomerated nanocrystals and their structure, corresponding to SAED pattern (inset). "Comparative Study of Nanostructured CuSe Semiconductor Synthesized in a Planetary and Vibratory Mill" by Marcela Achimovičová et al. is licensed under CC by 4.0

10.3.2.1 Mohite Cu₂SnS₃

Mohite belongs to the group of semiconductor materials suitable for application in photoelectric devices and solar cells (Lokhande et al. 2016). Its utilization is also known in lithium batteries and perspective thermoelectric materials. This sulfide is abundant in nature as a mineral mohite. To obtain a synthetic mohite, there are multiple ways of its preparation, including the solid-state mechanochemical approach applied on a laboratory scale (Vanalakar et al. 2015; Baláž, Daneu, et al. 2018). Mohite also represents an intermediate in the synthesis of the corresponding quaternary sulfides; see Part 10.3.3.

Recently mohite was prepared using elemental precursors in an ESM mill (Baláž, Achimovičová, et al. 2021). The XRD patterns are illustrated in Figure 10.10 for milling times 60, 180, and 240 min. The straightforward synthesis from elements is complicated by the possible presence of several mohite polymorphs in the product, concretely cubic, monoclinic, and tetragonal ones.

Another strategy to synthesize mohite was to apply pre-milled CuS and SnS precursors and their subsequent co-milling in ESM (Hegedüs et al. 2018). The product was obtained after 240 min of milling (Figure 10.11). The results showed the formation of nearly stoichiometric mohite nanoparticles with an average particle size of approximately 10–15 nanometers. The material with its physical properties seemed to be suitable as an absorber for photovoltaic applications.

10.3.2.2 Tetrahedrite Cu₁₂Sb₄S₁₃

Tetrahedrite $Cu_{12}Sb_4S_{13}$ in nature belongs to the most common sulfominerals. However, except Cu, Sb, and S, it can also contain a plethora of other elements, and its proper formula is $(Cu,Ag)_{10}(Cu,Zn,Fe,Cd,Hg)_2(Sb,Bi,As)_4S_{13}$ where the individual elements can be in the M^+, M^{2+}, and M^{3+} oxidation states (Pattrick and Hall 1983). In extractive metallurgy, tetrahedrite is exploited mainly for copper, antimony, and silver content. Several technologies were developed for its metallurgical processing including the MELT process in Slovakia (applying mechanochemical leaching) and the SUNSHINE process in the USA (Baláž 2008).

Synthetic tetrahedrite fulfills the criteria recommended for the thermoelectric application of energy materials. Namely, its high symmetry crystal structure, a small electronegativity difference among

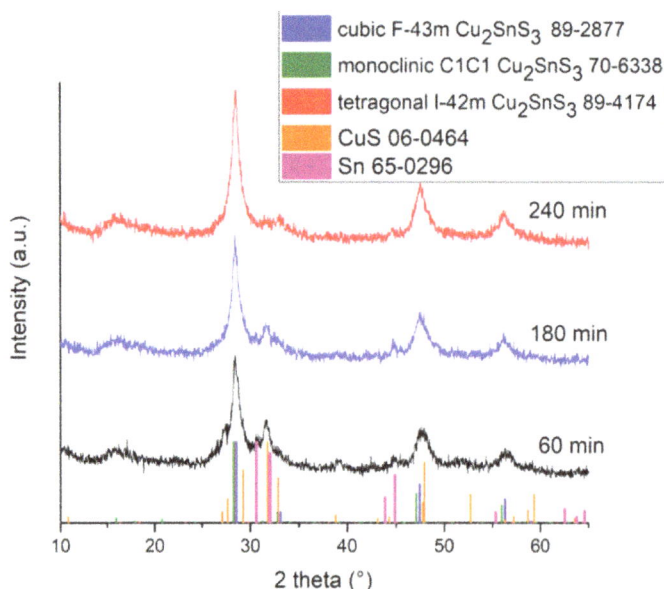

FIGURE 10.10 XRD patterns of the Cu–Sn–S mixture milled for 60, 180, and 240 min in an eccentric vibratory mill. Reprinted and adapted with permission from ref. Baláž, Achimovičová, et al. (2021). Copyright 2021 American Chemical Society.

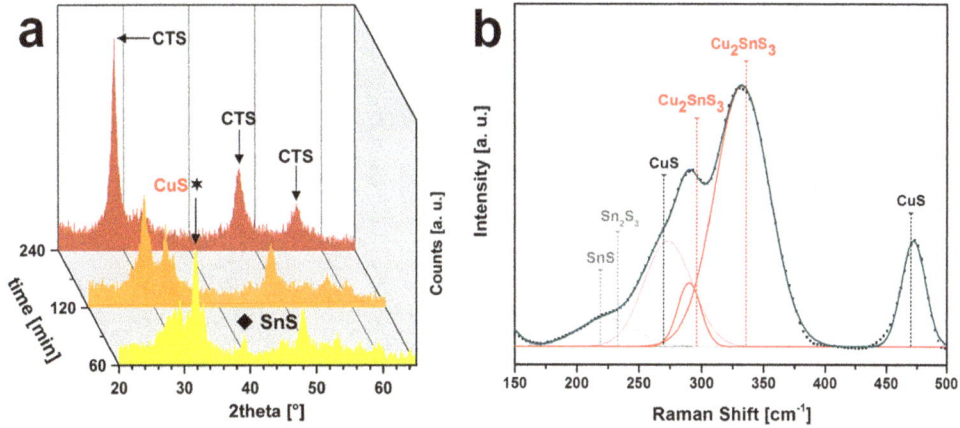

FIGURE 10.11 Reaction progress of mohite mechanochemical synthesis in ESM monitored via X-ray diffraction patterns (a). The fitted Raman spectra for the tetragonal phase are shown in (b) with expected Raman shifts. Reprinted and adapted with permission from ref. Hegedüs et al. (2018). Copyright 2018 Elsevier.

constituent elements yielding covalent chemical bonds, and large complex unit cells are promising parameters (He and Tritt 2017). Indeed, in 2012, its ultra-low thermal conductivity was revealed (Suekuni et al. 2012) and a plethora of papers appeared to synthesize this sulfide, as reported in the review in Baláž, Achimovičová, et al. (2021).

Recently, a scalable synthesis of tetrahedrite by the application of ESM has been performed (Baláž, Achimovičová, et al. 2021). Milling time for 1–6 hours was applied. Several phases were formed by milling, starting from binary copper sulfides, which subsequently reacted with antimony to form several ternary systems like skinnerite Cu_3SbS_3, famatinite Cu_3SbS_4, and tetrahedrite $Cu_{12}Sb_4S_{13}$ (Table 10.2).

After densification using SPS technology, the samples were subjected again to XRD analysis. In this case, the highest ratio of tetrahedrite/famatinite was obtained for the sample milled for 1 hour (Figure 10.12). This composition has a positive impact on thermoelectric performance as the presence of famatinite is undesirable in thermoelectrics. Indeed, the highest figure-of-merit ZT (efficiency) and the lowest thermal conductivity κ were obtained for the sample milled for 1 hour (Figure 10.13).

In the second paper on ESM synthesis of tetrahedrite, bismuth was applied as a dopant by its co-milling with Cu, Sb, and S during the synthesis step (Baláž, Guilmeau, et al. 2021). The idea was to elucidate the appearance and behavior of ternary sulfides which accompany tetrahedrite. Bi presence suppressed the production of tetrahedrite $Cu_{12}Sb_4S_{13}$ and promoted the formation of famatinite Cu_3SbS_4 and skinnerite Cu_3SbS_3 (Figure 10.14).

10.3.2.3 Chalcopyrite CuFeS₂

Chalcopyrite $CuFeS_2$ is an earth-abundant semiconductor serving in extractive metallurgy as a copper source. It has unusual magnetic and electrical properties. Depending on the applied stress (supplied by, e.g., high-energy milling), the naturally occurring ordered tetragonal phase (α-$CuFeS_2$) can show the structural and magnetic transitions. These changes influence fundamental physical properties like optical band gap, electrical conductivity, and thermoelectric performance (Baláž, Dutková, et al. 2020). Thus this mineral can also find applications as an energy material.

Mechanochemical synthesis of chalcopyrite $CuFeS_2$ was performed in ESM according to the reaction:

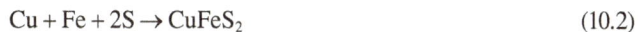

$$Cu + Fe + 2S \rightarrow CuFeS_2 \qquad (10.2)$$

In Figure 10.15 XRD patterns of chalcopyrite prepared by laboratory milling in a planetary mill (A) and by industrial milling in ESM (B) are given for comparison. Small amounts of admixtures such as

TABLE 10.2

XRD Qualitative Analysis of Mechanochemically Synthesized Powders (L Corresponds to Low Content of Phase)

Milling Time (h)	Identified Phase							
	Cu	Sb	S	Cu_2S	CuS	Cu_3SbS_3	Cu_3SbS_4	$Cu_{12}Sb_4S_{13}$
0	+	+	+	+	+			
1					+	+	+	+
2					+	+	+	+
3					L	+	+	+
4						L	+	+
6							+	+

Reprinted with permission from ref. Baláž, Guilmeau, et al. (2020). Copyright 2020 Elsevier.

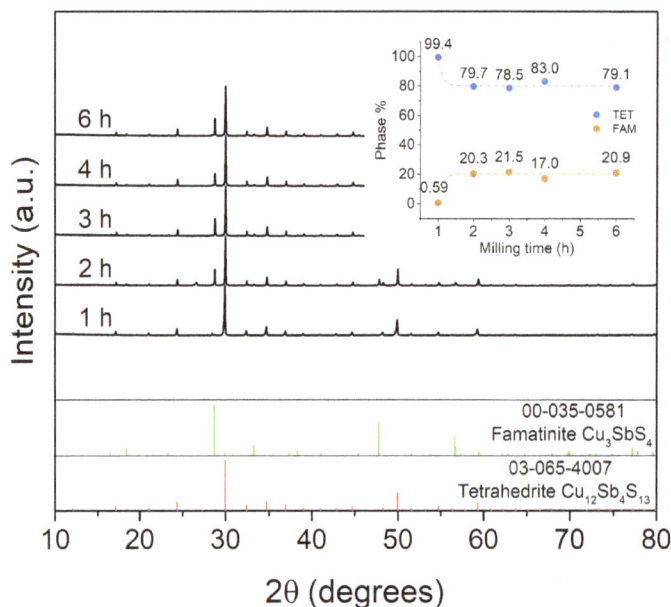

FIGURE 10.12 XRD patterns of densified samples after SPS versus milling time (milling time in hours shown on patterns). Inset: relative ratio of tetrahedrite to famatinite determined from Rietveld refinement. Reprinted with permission from ref. Baláž, Guilmeau, et al. (2020). Copyright 2020 Elsevier.

pyrite FeS_2 and bornite Cu_5FeS_4 next to chalcopyrite were recorded for laboratory and industrial milling, respectively. In the latter case, WC was also found as a consequence of wear from milling balls.

In Figure 10.16 the thermoelectric performance of chalcopyrite $CuFeS_2$ is illustrated for the products obtained in both mills. The discussion about individual thermoelectric characteristics is out of the scope of this chapter and can be followed in Baláž, Achimovičová, et al. (2021). However, a significantly better figure-of-merit ZT (efficiency) values for the products prepared in the industrial mill is worth mentioning.

10.3.2.4 Famatinite Cu_3SbS_4

Famatinite Cu_3SbS_4 is a promising p-type semiconductor and belongs among ternary chalcogenides, which are in research focus today. Cu_3SbS_4 has a direct band gap (0.9–1.8 eV) depending on the crystal structure and high absorption coefficient 10^4–10^5 cm^{-1} (Kavinchan et al. 2018). This material is known to have several applications, specifically in solar photovoltaic (Ramasamy et al. 2014) as well as in thermoelectric devices (van Embden and Tachibana 2012). Until now, famatinite has been fabricated by

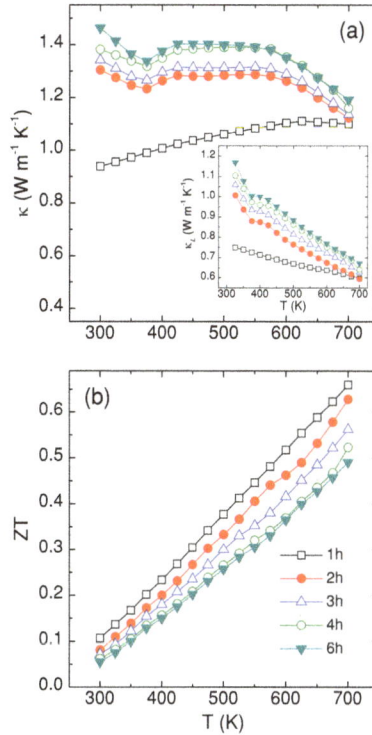

FIGURE 10.13 Temperature dependence of (a) thermal conductivity κ and (b) figure-of-merit ZT of densified tetrahedrite samples after SPS for different milling times. Inset in (a) displays the temperature dependence of the lattice thermal conductivity. Reprinted with permission from ref. Baláž, Guilmeau, et al. (2020). Copyright 2020 Elsevier.

several procedures involving cyclic microwave radiation, colloidal hot-injection methods, hydrothermal route, sputtering, solvothermal chemical route, chemical bath deposition technique, etc. Many of these methods are environmentally harmful. In this context, the ball milling method represents a green and environmentally friendly alternative (Baláž et al. 2013).

Recently, famatinite was synthesized using elemental precursors, namely copper, antimony, and sulfur in a stoichiometric ratio of 3:1:4 in the ESM mill with the utilization of a protective atmosphere (Dutková et al. 2021). The progress of the mechanochemical synthesis of Cu_3SbS_4 for milling times 60–180 minutes is illustrated in Figure 10.17.

The presence of famatinite structure after 180 minutes of milling with a crystallite size of 10 nm was confirmed by both LeBail refinement of the X-ray powder diffraction data and transmission electron microscopy. The tungsten carbide (WC) was also present as a consequence of the abrasion of milling balls.

The optical properties of the Cu_3SbS_4 obtained were studied by UV-Vis spectroscopy (Figure 10.18). The band gap is determined from the intercept of the extrapolated linear fit to the experimental data of the Tauc plot (inset in Figure 10.18). The determined direct optical band gap 1.24 eV is blue-shifted relatively to the bulk Cu_3SbS_4. The observed blue shift could be attributed to the existence of very small nanocrystallites.

10.3.3 Quaternary Systems

Conventional co-milling of individual elements and the sequential co-milling (as in the case of stannite Cu_2ZnSnS_4, where first binary compounds CuS and SnS were synthesized and then co-milled with Zn and S) strategies were applied for the mechanochemical synthesis of quaternary chalcogenides.

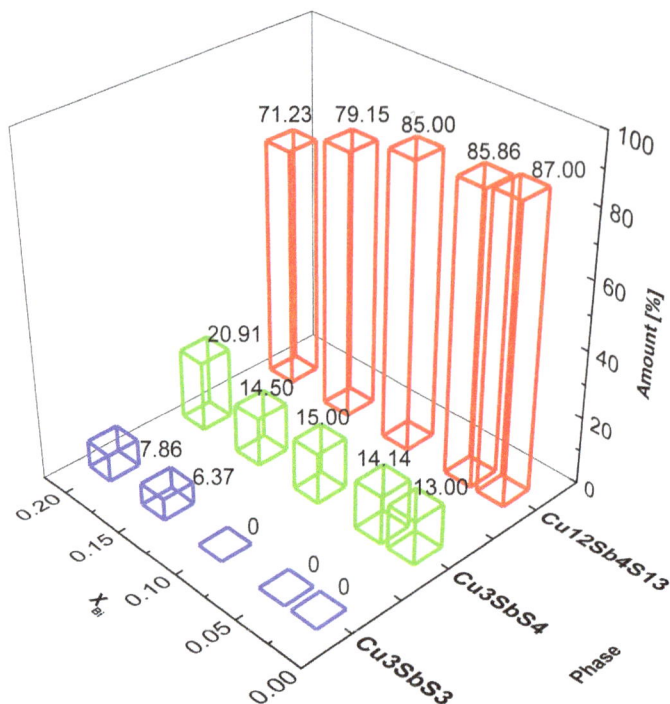

FIGURE 10.14   Phase distribution of tetrahedrite $Cu_{12}Sb_4S_{13}$, famatinite Cu_3SbS_4, and skinnerite Cu_3SbS_3 for densified samples after SPS as a function of the nominal Bi content, X_{Bi}. "Bismuth Doping in Nanostructured Tetrahedrite: Scalable Synthesis and Thermoelectric Performance" by Peter Baláž et al. is licensed under CC by 4.0.

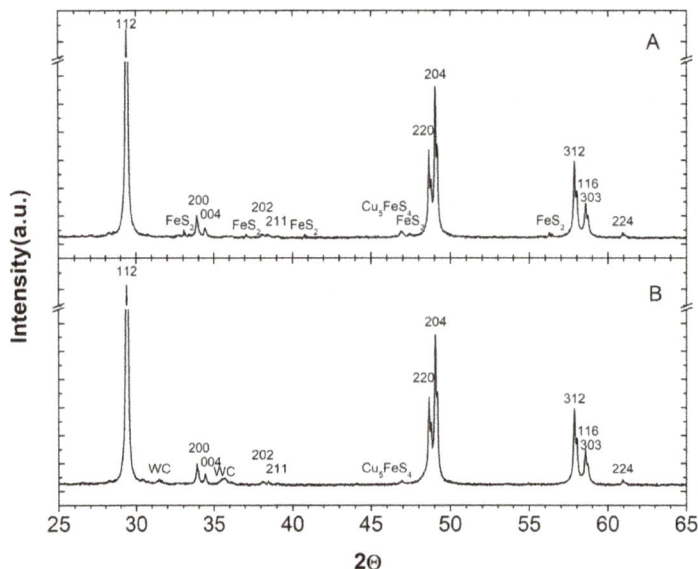

FIGURE 10.15   XRD patterns of chalcopyrite $CuFeS_2$. Milling time: (A) 60 min (laboratory mill) and (B) 720 min (industrial mill). Reprinted with permission from ref. Baláž, Achimovičová, et al. (2021). Copyright 2021 American Chemical Society.

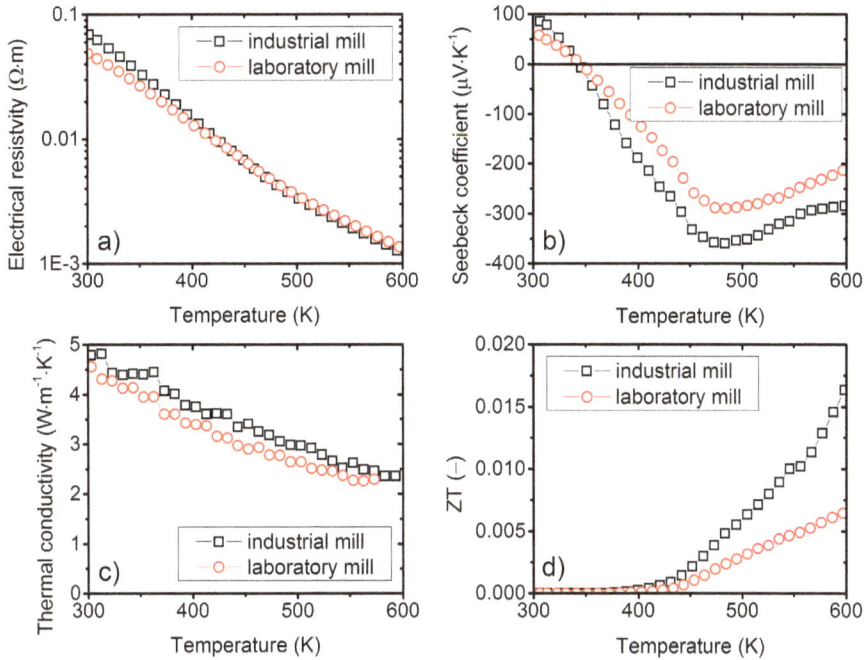

FIGURE 10.16 Thermoelectric performance of chalcopyrite $CuFeS_2$ as a function of temperature: (a) electrical resistivity, (b) Seebeck coefficient, (c) thermal conductivity, and (d) figure-of-merit ZT. Milling time: laboratory mill, 60 min; industrial mill, 720 min. Reprinted with permission from ref. Baláž, Achimovičová, et al. (2021). Copyright 2021 American Chemical Society.

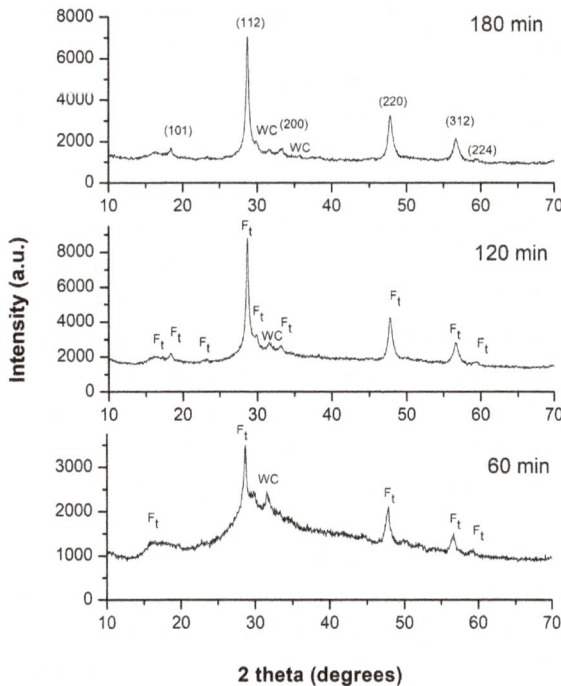

FIGURE 10.17 XRD patterns of Cu_3SbS_4 synthesized for different milling times. The identified phases are marked as follows: Cu_3SbS_4-tetragonal famatinite (F_t) and tungsten carbide (WC). Reprinted with permission from ref. Reprinted with permission from ref. Dutková et al. (2021). Copyright 2021, Elsevier.

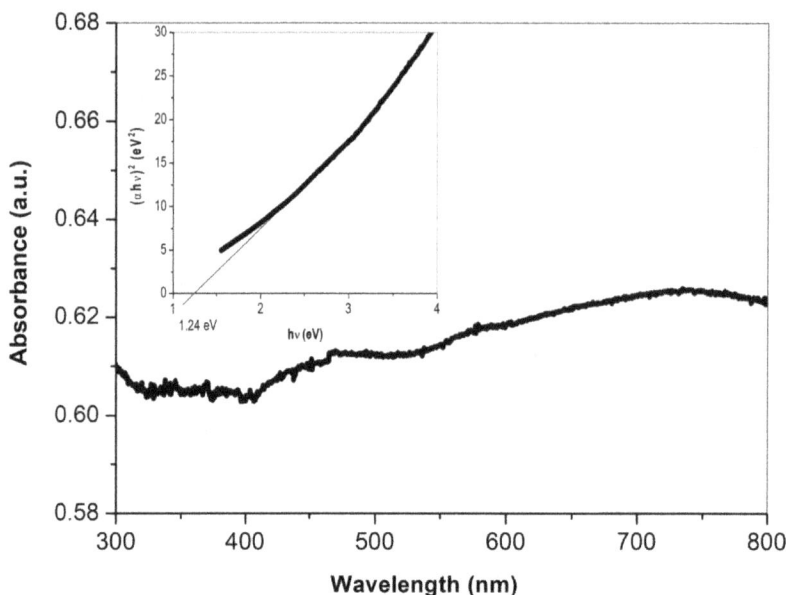

FIGURE 10.18   UV-Vis spectrum of Cu_3SbS_4 and Tauc plot (inset). Reprinted with permission from ref. Dutková et al. (2021). Copyright 2021, Elsevier.

For quaternary sulfides, the choice of reaction precursors was directed at environmentally friendly and earth-abundant elements like copper, iron, zinc, tin, and sulfur. Especially tin chalcogenides represent prospective thermoelectric materials in materials science. As stated in Cai et al. (2019), tin-based chalcogenides were the most intensively researched topic in the field during 2014–2018.

10.3.3.1 Kesterite Cu_2ZnSnS_4

Quaternary sulfide kesterite Cu_2ZnSnS_4 (CZTS) possesses many advantageous characteristics for photovoltaic applications, such as suitable band gap, high absorption coefficient, and high radiation stability. In contrast to the more examined CIGS ($CuIn_xGa_{1-x}Se_2$), kesterite provides a promising alternative because of its environmental advantage (S instead of toxic Se), low cost, and availability (abundant and common Zn and Sn instead of scarce In and Ga) (Siebentritt and Schorr 2012). Kesterite has been prepared by several techniques, such as solution-based synthesis, hot-injection synthesis, electrochemical deposition, and microwave irradiation (Ratz et al. 2019). However, these techniques are complex and time-consuming and need high temperatures and toxic organic solvents. In this study, we demonstrate the use of elemental precursors (Cu, Zn, Sn, and S) to obtain Cu_2ZnSnS_4 by a solid-state process at ambient temperature without using solvents. Two alternatives following Equations (10.3) and (10.4) were tested:

$$2Cu + Zn + Sn + 4S \rightarrow Cu_2ZnSnS_4 \tag{10.3}$$

$$2CuS + SnS + Zn + S \rightarrow Cu_2ZnSnS_4 \tag{10.4}$$

In both cases the pure kesterite phase was obtained after 360 minutes of milling in ESM (Figure 10.19) with an average crystallite size of 15 nm for both mixtures (Baláž et al. 2019).

The applied mechanochemical approach enabled us to obtain kesterite in ESM. The results confirm the possibility of up-scaling the mechanochemical one-pot synthesis process of kesterite which represents a key perspective absorber for future solar cells.

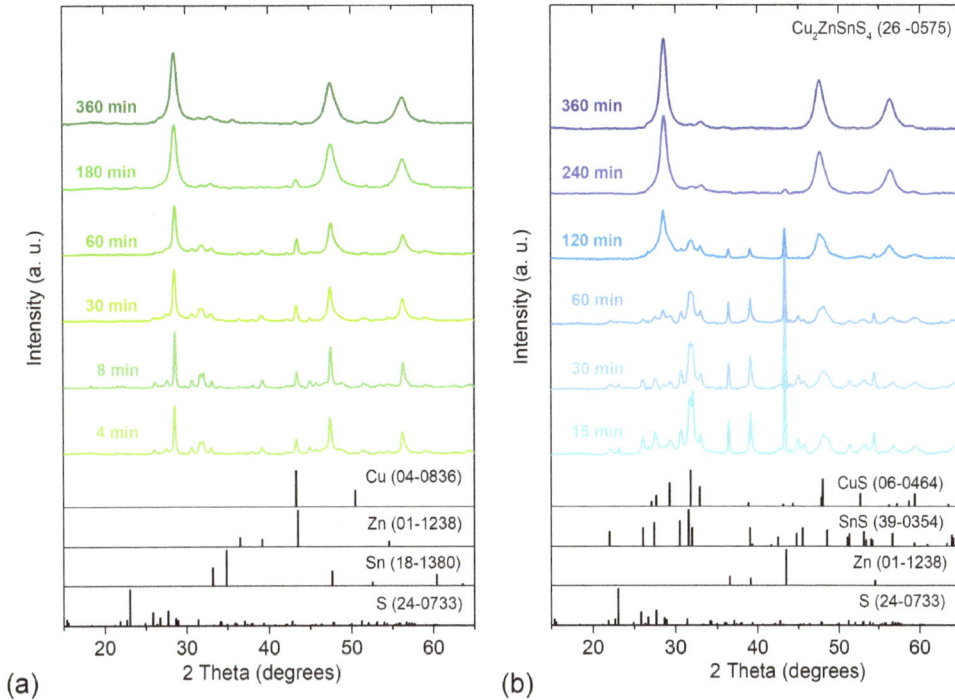

FIGURE 10.19 X-ray diffractometry patterns of stoichiometric mixtures milled according to Equation (10.3) (a) and Equation (10.4) (b) for various times. Reprinted with permission from ref. Baláž et al. (2019). Copyright 2019 John Wiley and Sons.

10.3.3.2 Stannite Cu₂FeSnS₄

Quaternary sulfide stannite Cu_2FeSnS_4, together with kesterite, represents perspective material in solar cell technology. Using ESM, stannite was prepared according to Equation (10.5):

$$2Cu + Fe + Sn + 4S \rightarrow Cu_2FeSnS_4 \quad (10.5)$$

Results of the synthesis are presented in Figure 10.20 where the creation of a new phase can be documented with respect to milling time. Already at the beginning of the synthetic process, stannite is present. However, the other intermediates like CuS, FeS, and non-consumed Fe can be also traced. At the end of the mechanochemical synthesis, rhodostannite $Cu_2FeSn_3S_8$ was identified as a product of stannite transformation (Baláž, Hegedüs, et al. 2018). Its synthesis is described in more detail in Section 10.3.3.3.

A new idea for stannite synthesis is outlined in Equation (10.6):

$$CuFeS_2 + CuS + SnS \rightarrow Cu_2FeSnS_4 \quad (10.6)$$

Natural chalcopyrite $CuFeS_2$ from Kazakhstan as a reaction precursor was used. In this case, a natural mineral represents "a motive" for the preparation of a synthetic analog. This concept was recently verified for the mechanochemical synthesis of chalcopyrite $CuFeS_2$ (Baláž, Dutková, et al. 2020).

The results of this synthesis are exemplified by Figure 10.20, where the conversion degree for stannite synthesis is shown for various milling times (this is calculated based on the consumption of elemental sulfur). Starting from 120 minutes of milling, a conversion degree over 94% can be obtained. However, as previously reported (Baláž, Hegedüs, et al. 2018), stannite is not the only product. Besides stannite, the non-consumed chalcopyrite $CuFeS_2$, covellite CuS, herzenbergite SnS, and rhodostannite $Cu_2FeSn_3S_8$ are also present.

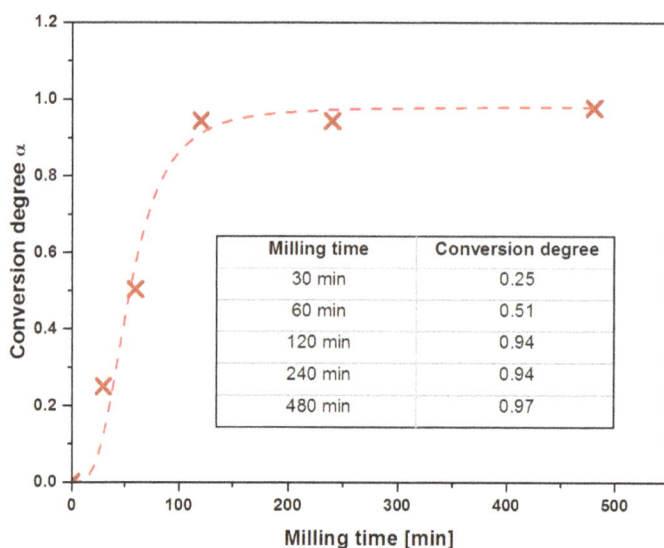

FIGURE 10.20   Conversion degree α vs. milling time for stannite synthesis calculated from the data of Soxhlet's extraction determining the content of non-reacted sulfur. Reprinted with permission from ref. Baláž, Hegedüs, et al. (2018). Copyright 2018 American Chemical Society.

10.3.3.3 Rhodostannite $Cu_2FeSn_3S_8$

Rhodostannite $Cu_2FeSn_3S_8$ as a stannite-free product was synthesized in ESM (Baláž, Dobrozhan, et al. 2021). Using methods of XRD and magnetometry, a successful reaction progress until 6 hours of milling was confirmed. The crystallite size of rhodostannite around 12 nm was determined. As magnetic iron was used as a reaction precursor, the method of magnetometry for tracking the synthetic progress was applied (Figure 10.21). The values of magnetization decrease with milling time, which confirms the embedding of magnetic Fe into a less magnetic sulfide product. For samples milled for 360 and 600 min, the saturation magnetization is almost the same, and thus the maximum possible conversion was most probably reached. From the residual saturation magnetization value, we can deduce that only a slight amount of non-consumed iron is present.

10.3.3.4 Mawsonite $Cu_6Fe_2SnS_8$

Elemental precursors (Cu, Fe, Sn, and S) were used for obtaining mawsonite $Cu_6Fe_2SnS_8$ using a solid-state process at ambient temperature in ESM. The milling up to 240 min at ambient temperature and in argon over-pressure was applied. However, in the sample where the highest conversion was obtained, 12–13% of stannite was also present together with small amounts of chalcopyrite and bornite (Baláž et al. 2019). The mentioned sulfides represent good thermoelectric candidates among energy materials. Therefore a comparison of thermoelectric performance for this synthetic composite with pure mawsonite obtained by laboratory milling (Zhang et al. 2017) was performed (Figure 10.22). The calculated parameter figure-of-merit ZT as a measure of thermoelectric efficiency was even higher for industrially milled samples. The time required to obtain the product was also significantly reduced when using ESM, namely 4 hours were required in ESM, whereas 96 hours were necessary for a laboratory-scale synthesis.

10.3.3.5 Colusite $Cu_{26}V_2Sn_6S_{32}$

Colusites $Cu_{26}T_2Sn_6S_{32}$ represent a big group of ternary sulfides where T = V, Nb, Ta, Cr, Mo, and W (Hegedüs et al. 2020), which possess degenerated semiconducting properties suitable for energy applications. A pristine colusite $Cu_{26}V_2Sn_6S_{32}$ was synthesized in ESM by milling from 1 up to 12 hours. The

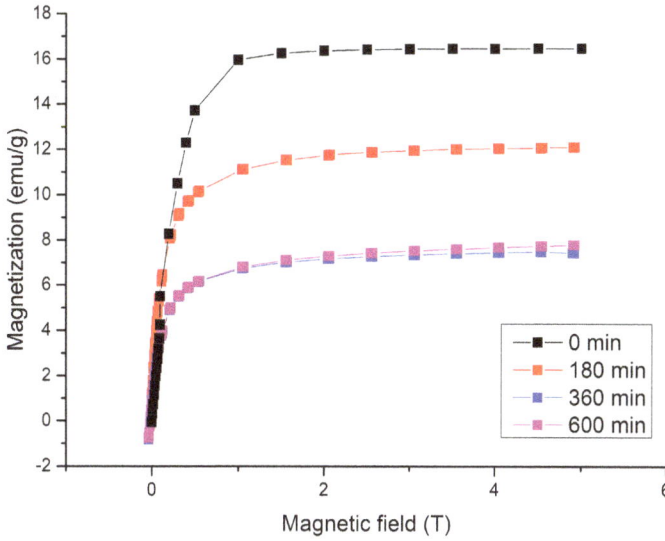

FIGURE 10.21 The first quadrant of magnetization curves for the samples milled for different times. Reprinted with permission from ref. Baláž, Dobrozhan, et al. (2021). Copyright 2021 Elsevier.

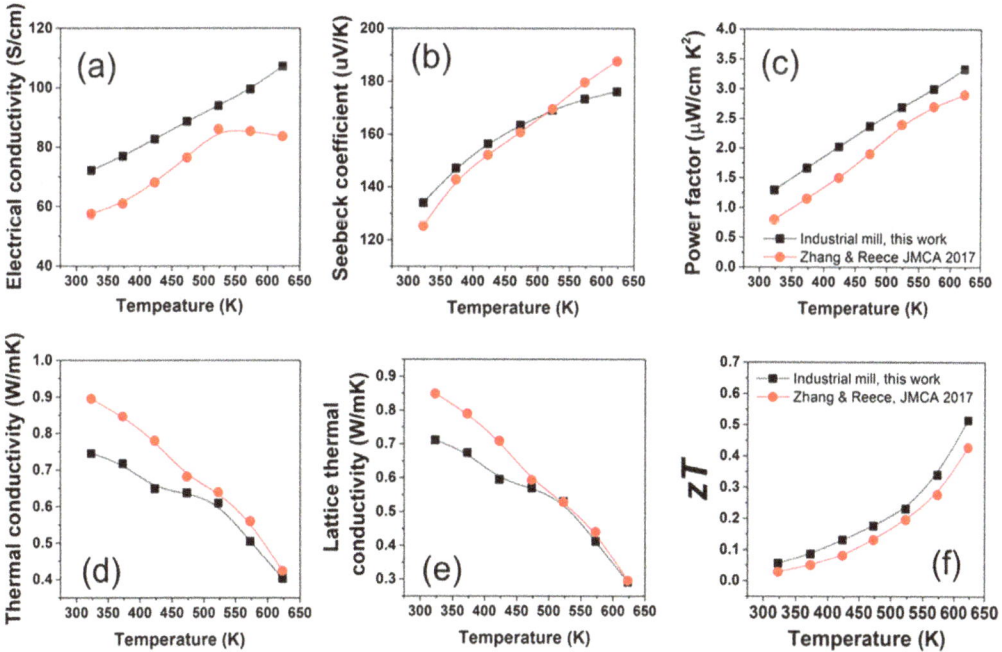

FIGURE 10.22 Thermoelectric performance for the samples milled for 4 h in an industrial mill (black points) and 96 h in a laboratory mill (red points). Reprinted with permission from ref. Baláž, Achimovičová, et al. (2021). Copyright 2021 American Chemical Society.

samples were investigated by X-ray powder diffraction. For the sample milled for 12 hours, high-purity crystals were obtained (Figure 10.23). In comparison, samples milled for a shorter time manifested the presence of small quantities of binary and ternary Cu–S phases. The obtained physico-chemical properties are comparable to colusite synthesized in a laboratory mill.

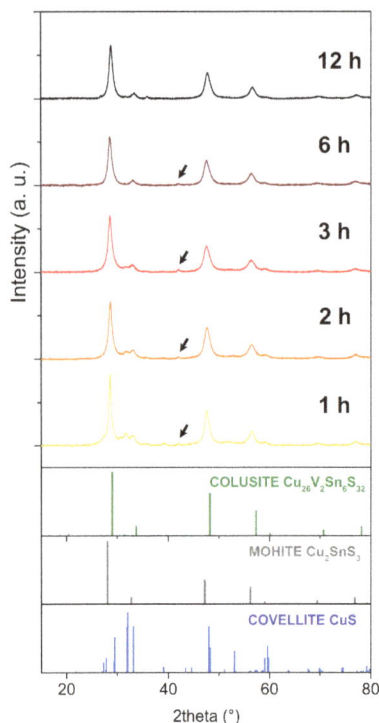

FIGURE 10.23   XRPD patterns of ball-milled elemental mixtures at milling times ranging from 1 h up to 12 h. Arrows indicate traces of vanadium metal. Reprinted with permission from ref. Hegedüs et al. (2020). Copyright 2020 Royal Society of Chemistry.

10.4 Large Scale of MeS (Me = Sn, Cu) Mechanochemical Synthesis in ESM-Commercial Approach

Two binary sulfides SnS and CuS were mechanochemically synthesized in an industrially acceptable manner.

10.4.1 Herzenbergite SnS

The results of the laboratory tests carried out in "satellite" mode showed that the milling load 447 g of Sn and S mixture with a molar ratio 1:1 is the most suitable for 60 min of milling and SnS product formation. By checking milling kinetics, it was found that only the orthorhombic phase of SnS-herzenbergite (JCPDS PDF 014-0620) is already generated during 60 min of milling.

Due to the exothermic nature of the reaction between tin and sulfur studied by Batsanov et al. (1994), and the subsequent heating of the milling tube, and the milling balls, a single module mill ESM 324-1ks (Siebtechnik, Germany) of 50 L volume with a cooling system was chosen for the semi-pilot experiments. Figure 10.24 shows the mill with hoses ensuring the inflow and outflow of water for cooling in the double casing of the mill.

The following operating conditions were implemented: 80% of the milling tube filled with the balls (35–40 mm diameter), the material of milling balls – WC; the mass of Sn and S powder – according to Sn:S molar ratio – 1:1; total throughput – 5, 7, 10, 13 and 15 kg, the milling atmosphere – air, the frequency of the motor – 1500 r.min^{-1}, the amplitude of the inhomogeneous vibrations – 8 mm, the cooling medium – water, the milling time – 13–60 min, and discharging of the product – pneumatic.

FIGURE 10.24 Eccentric vibratory mill ESM 324-1ks with the cooling system (photograph is a courtesy by the authors).

10.4.1.1 Parameter Studies

The mutual dependencies of the operational parameters were verified through parameter studies. The investigated parameters and their influence on the product quality were: batch size, process temperature, cooling, and milling media material. The dependence of the process temperature on the batch size is illustrated in Figure 10.25. The mechanochemical reaction between Sn and S has a characteristic sharp temperature increase after a period of mechanical activation of reaction precursors. With increasing batch size, the maximum temperature is reached. This points to a delay in reaction progress from 35 to 60 min of milling time. Therefore, further increasing the batch size would not be beneficial, as operating temperatures over 100°C should be avoided due to the melting of sulfur at 119°C.

Preliminary investigations showed that at the maximum batch size of 15 kg, it was possible to reach a temperature of approx. 55°C. At a 7 kg batch, which corresponds to a 50% decrease in the throughput, and with cooling, the maximum temperature of approx. 30°C was reached. Concerning a semi-continuous operation, it was recommended to limit the batch size to keep the maximum temperature between 30 and 50°C to allow the immediate filling of the next batch.

10.4.1.2 Quality of Product

The XRD patterns of produced SnS of various batches in the ESM 324-1ks with corresponding milling times are documented in Figure 10.26. The diffraction peaks of all three experiments match those of the SnS phase.

To compare the quality of the SnS products using WC and steel milling balls, a chemical analysis was performed. It can be seen from Table 10.3 that a relatively high content of Co and W compared to the reference material was found when WC balls were used.

Cobalt, which serves as a binding agent in WC balls, occurs together with W in the product due to wear. Surprisingly, when using steel balls, the amount of abraded Fe is about the same as when using WC balls. Only trace amounts of cobalt and tungsten were found. The used steel balls are alloyed with Co and Ni and are extremely wear-resistant. In terms of cost for WC balls and steel balls, the factor is 12 in favor of steel. In order to achieve the same throughput at the same reaction time as that of WC balls, the mill volume would have to be 30% higher. It means that, instead of the two-module mill (ESM 324-2ks), a three-module mill (ESM 324-3ks) should be used. The cost for the third module would eliminate the cost savings from using steel balls rather than WC balls. In terms of the sulfur content, dependencies on the product temperature during discharge from the mill have been established. Table

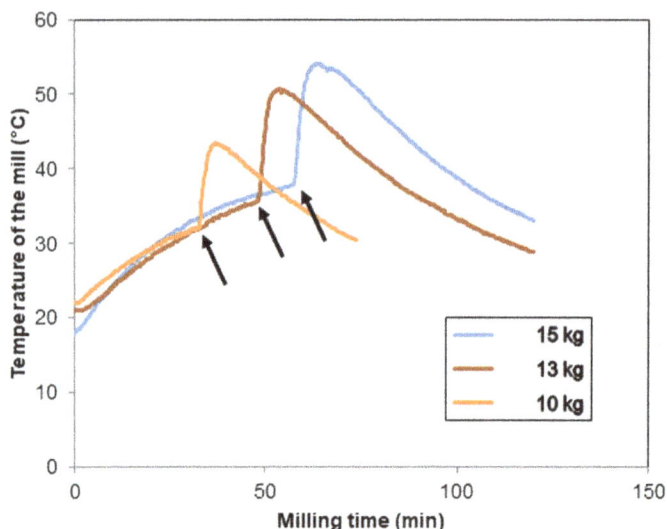

FIGURE 10.25   The dependence of the mill temperature on the batch size of precursors. Arrows indicate the onset of the mechanochemical reaction.

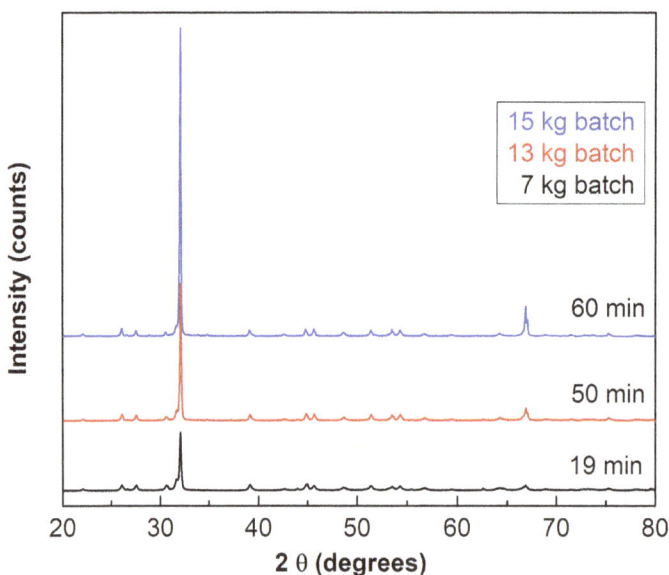

FIGURE 10.26   XRD patterns of SnS products mechanochemically synthesized at different milling times depending on the size of the precursors batch.

10.4 shows the sulfur content of the products as well as the free elemental sulfur content for throughputs between 5 and 15 kg.

The higher the throughput, the lower the elemental sulfur content. The reason for this is that at higher throughputs, the temperature of the mill causes the combustion of the elemental sulfur into SO_2.

The comparison of the grain size distribution parameters from X_{10} to X_{90}, and the specific surface S_v of the generated SnS product shows a good correlation with the reference material (see Table 10.5).

Of particular interest was the morphology of the generated SnS particles. The reference material indicates a platelet structure, which is even more evident in our product, regardless of the throughput. Figure 10.27(a–c) shows SEM micrographs of the reference material and the products with different batch sizes.

TABLE 10.3

Product Quality When Using WC Balls vs. Steel Balls in Comparison with a Reference Sample

SnS Batch Size/Milling Balls Material/ Milling Time	Co (ppm)	Fe (ppm)	Ni (ppm)	W (ppm)	Sn (%)
Reference sample	31	297	–	<10	72.0
7 kg/WC/19 min	1942	690	15	18713	75.5
7 kg/steel/45 min	11	603	8	112	75.4

TABLE 10.4

Elemental Sulfur of the SnS Products as a Function of the Throughput

SnS Batch Size/Milling Balls Material/Milling Time	S Total (%)	S Elemental (%)
Reference sample	18.71	10.4300
5 kg/WC/13 min	19.95	1.8085
7 kg/WC/19 min	19.36	1.0062
15 kg/WC/60 min	19.90	10.8181

TABLE 10.5

Grain Size Distribution Parameters of Reference Material and Product

SnS Batch Size/Milling Balls Material/Milling Time	X_{10} (μm)	X_{50} (μm)	X_{90} (μm)	S_v (m²/cm³)
Reference sample	3.6	23.3	58.5	10.71
7 kg/WC/19 min	3.5	28.7	77.3	10.74

All images show the platelet structure typical for lubricants. Perpendicular to the crystal axis, the material demonstrates eminent cleavage just like graphite, or molybdenite, so that during milling, mostly thin and very flexible platelets of SnS are generated, which are extremely resistant even under high compressive stress and thus exhibit the properties characteristic for lubricants (Gock 1977).

Considering the reaction heat defined by the throughput, an optimum batch size of 7 kg was identified. According to the experimental data, it is possible to produce 74 t (metric)/year of SnS using ESM 324-1ks for a three-shift operation. For a two-module machine, the annual throughput would amount to approx. 150 t (metric)/year of SnS. The average energy consumption per ton of product under the given circumstances is at least 350–400 kWh/t.

If steel balls were to be used instead of WC balls, the yearly production using a two-module ESM 324-2ks would be approximately 74 tons of SnS. The use of 40 mm steel balls could bear some optimization potential, thereby reducing the disadvantage of steel balls as compared to WC balls.

10.4.2 Covellite CuS

The kinetics of CuS mechanochemical synthesis was first investigated in a "satellite" mode with a milling load of 300 g of Cu and S (molar ratio Cu:S = 1:1). XRD patterns in Figure 10.28 confirmed that the reaction between Cu (manufactured electrolytically) and S was very fast and after 40 min of milling only covellite CuS (JCPDS 006-0464) with hexagonal structure is formed as the major phase (Achimovičová et al. 2019).

The dependence of the phase composition of CuS products on the milling time in Figure 10.29 showed 83% CuS phase content of the sample milled for 40 min, which indicates the completion of the mechanochemical reaction.

FIGURE 10.27   SEM micrographs of the reference material (a), the SnS product of 7 kg batch (b), and the SnS product of 15 kg batch (c).

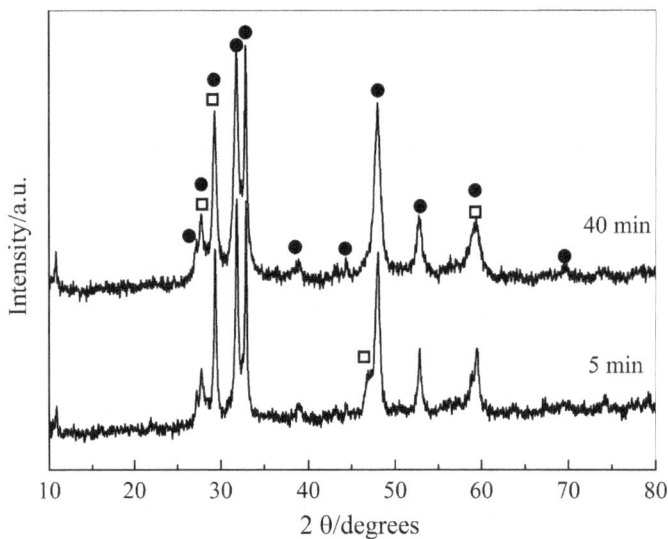

FIGURE 10.28   XRD patterns of Cu_e/S mixture milled for 5 and 40 min. ● – CuS, □ – Cu_2S. Reprinted with permission from ref. Achimovičová et al. (2019). Copyright 2018 Springer Nature.

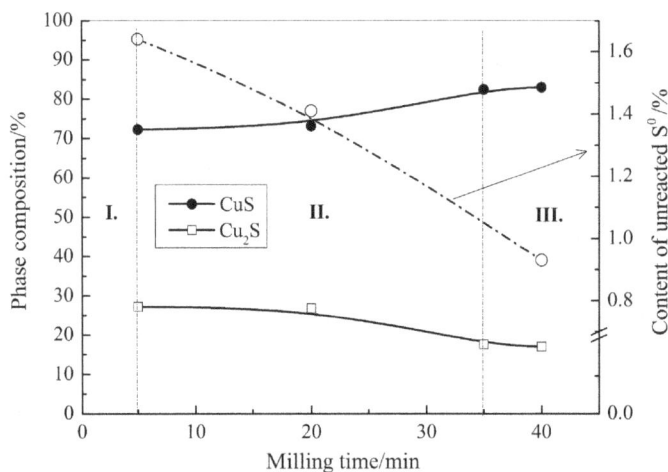

FIGURE 10.29   Dependence of the phase composition of CuS products and the content of unreacted S^0 on the milling time. Reprinted with permission from ref. Achimovičová et al. (2019). Copyright 2018 Springer Nature.

FIGURE 10.30   SEM image of CuS product (a), selected grain with the typical EDX spectrum (inset) (b). Reprinted with permission from ref. Achimovičová et al. (2019). Copyright 2018 Springer Nature.

The content of unreacted elemental sulfur in the CuS products determined by the Soxhlet extraction method decreased almost linearly with milling time (see Figure 10.29). CuS, mechanochemically synthesized for 40 min contains only 10.93% S^0, which confirmed 99% completion of the reaction.

Figure 10.30 shows the SEM images revealing the morphology of CuS product consisting of the larger agglomerated grains of homogeneous composition with sizes from at least 5 μm. The lamellar structure of the agglomerated grains formed from individual CuS nanoparticles is illustrated in Figure 10.30b.

The whole volume of the ESM mill was used for the proving of CuS large-scale production with the following operating conditions: the volume of the milling chamber– 50 L, the loading of the milling chamber – 150 kg balls (35 mm diameter), the material of the milling chamber and balls – ferroalloy, the mass of Cu powder (manufactured by atomization process) – 4.98 kg, the mass of S powder – 2.52 kg, Cu:S molar ratio – 1:1, the ball to powder ratio – 20:1, the milling atmosphere – air, the frequency of the motor – 1500 rpm, and the amplitude of the inhomogeneous vibrations – 20 mm. It was found that after 60 min of milling under ambient conditions, it is possible to produce a 7.5 kg batch of CuS product (Achimovičová et al. 2019).

10.5 Conclusion

It is established that high-energy milling produces powders with nanocrystalline structures. The unique properties of nanoscale chalcogenide particles lead to their use in a broad range of energy materials, such as photovoltaics and thermoelectrics. The main advantage of the mechanochemical approach to prepare advanced materials consists in the fact that it is a straightforward method, where the whole process is performed at ambient temperature, atmospheric pressure, short processing times, and in the absence of solvents. This chapter highlighted eccentric vibratory milling (ESM) as a feasible technique for the scale-up of mechanochemical reactions. It is a scalable process permitting the production of various materials in large amounts. The approach is economically feasible, environmentally friendly, and essentially waste-free. However, several issues still need to be solved for the solid-state synthesis of energy materials. Improving their productivity is vital for practical application.

Acknowledgments

This work was supported by the projects of the Slovak Research and Development Agency APVV (VV-18-0357), and the Slovak Grant Agency VEGA (2/0103/20, 2/0112/22).

REFERENCES

Achimovičová, M, M Baláž, V Girman, et al. 2020. Comparative study of nanostructured CuSe semiconductor synthesized in a planetary and vibratory mill. *Nanomaterials* 10(10):1386. doi: 10.3390/nano11061386

Achimovičová, M, P Baláž, J Ďurišin, et al. 2011. Mechanochemical synthesis of nanocrystalline lead selenide: Industrial approach. *International Journal of Materials Research* 102(4):441–445. doi: 10.3139/146.110496

Achimovičová, M, P Baláž, T Ohtani, et al. 2011. Characterization of mechanochemically synthesized ZnSe in a laboratory and an industrial mill. *Solid State Ionics* 192(1):632–637. doi: 10.1016/j.ssi.20110.07.009

Achimovičová, M, E Dutková, E Tóthová, Z Bujňáková, J Briančin, and S Kitazono. 2019. Structural and optical properties of nanostructured copper sulfide semiconductor synthesized in an industrial mill. *Frontiers of Chemical Science and Engineering* 13(1):164–170. doi: 10.1007/s11705-018-1755-2

Baláž, M, N Daneu, M Rajňák, et al. 2018. Rapid mechanochemical synthesis of nanostructured mohite Cu_2SnS_3 (CTS). *Journal of Materials Science* 53(19):13631–13642. doi: 10.1007/s10853-018-2499-6

Baláž, M, O Dobrozhan, M Tešinský, et al. 2021. Scalable and environmentally friendly mechanochemical synthesis of nanocrystalline rhodostannite ($Cu_2FeSn_3S_8$). *Powder Technology* 388:192–200. doi: 10.1016/j.powtec.2021.04.047

Baláž, P, M Achimovičová, M Baláž, et al. 2013. Hallmarks of mechanochemistry: From nanoparticles to technology. *Chemical Society Reviews* 42(18):7571–7637. doi: 10.1039/c3cs35468g

Baláž, P, M Achimovičová, M Baláž, et al. 2021. Thermoelectric Cu-S-based materials synthesized via a scalable mechanochemical process. *ACS Sustainable Chemistry and Engineering* 9(5):2003–2016. doi: 10.1021/acssuschemeng.0c05555

Baláž, P, M Baláž, M Achimovičová, Z Bujňáková, and E Dutková. 2017. Chalcogenide mechanochemistry in materials science: Insight into synthesis and applications (a review). *Journal of Materials Science* 52(20):11851–11890. doi: 10.1007/s10853-017-1174-7

Baláž, P, E Boldižarová, E Godočíková, and J Briančin. 2003. Mechanochemical route for sulphide nanoparticles preparation. *Materials Letters* 57(9–10):1585–1589. doi: 10.1016.S0167-577X(02)01037-6

Baláž, P, E Guilmeau, M Achimovičová, et al. 2021. Bismuth doping in nanostructured tetrahedrite: Scalable synthesis and thermoelectric performance. *Nanomaterials* 11(6). doi: 10.3390/nano11061386

Baláž, P, M Hegedüs, M Achimovičová, et al. 2018. Semi-industrial green mechanochemical syntheses of solar cell absorbers based on quaternary sulfides. *ACS Sustainable Chemistry and Engineering* 6(2):2132–2141. doi: 10.1021/acssuschemeng.7b03563

Baláž, P, M Hegedüs, M Reece, et al. 2019. Mechanochemistry for thermoelectrics: Nanobulk $Cu_6Fe_2SnS_8$/Cu_2FeSnS_4 composite synthesized in an industrial mill. *Journal of Electronic Materials* 48(4):1846–1856. doi: 10.1007/s11664-019-06972-7

Baláž, M. 2021. *Environmental Mechanochemistry*. Switzerland: Springer.

Baláž, P. 2000. *Extractive Metallurgy of Activated Minerals*. 1. Amsterdam: Elsevier.

Baláž, P. 2008. *Mechanochemistry in Nanoscience and Minerals Engineering*. Berlin: Springer.

Baláž, P, E Dutková, P Levinský, et al. 2020. Enhanced thermoelectric performance of chalcopyrite nanocomposite via co-milling of synthetic and natural minerals. *Materials Letters* 275:128107. doi: 10.1016/j.matlet.20210.128107

Baláž, P, E Guilmeau, N Daneu, et al. 2020. Tetrahedrites synthesized via scalable mechanochemical process and spark plasma sintering. *Journal of the European Ceramic Society* 40:1922–1930. doi: 10.1016/j.jeurceramsoc.20210.01.023

Baláž, P, M Hegedüs, M Baláž, et al. 2019. Photovoltaic materials: Cu_2ZnSnS_4 (CZTS) nanocrystals synthesized via industrially scalable, green, one-step mechanochemical process. *Progress in Photovoltaics: Research and Applications* 27(9):798–811. doi: 10.1002/pip.3152

Batsanov, SS, MF Gogulya, MA Brazhnikov, GV Simakovv, and II Masksimov. 1994. Behaviour of the reacting system Sn+S in shock-waves. *Combustion, Explosion, and Shock Waves* 30(3):361–365. doi: 10.1007/BF00789431

Cai, BW, HH Hu, HL Zhuang, and JF Li. 2019. Promising materials for thermoelectric applications. *Journal of Alloys and Compounds* 806:471–486. doi: 10.1016/j.jallcom.2019.07.147

Dutková, E, MJ Sayagues, M Fabián, M Baláž, and M Achimovičová. 2021. Mechanochemically synthesized ternary chalcogenide Cu_3SbS_4 powders in a laboratory and an industrial mill. *Materials Letters* 291. doi: 10.1016/j.matlet.2021.129566

Gock, E, and KE Kurrer. 1999. Eccentric vibratory mills - Theory and practice. *Powder Technology* 105(1–3):302–310. doi: 10.1016/S0032-5910(99)00152-7

Gock, E. 1977. *Beeinflussung des Löseverhalten Sulfidischer Rohstoffe Durch Festkörperreaktionen bei der Schwingmahlung*. TU Berlin.

Gock, E, and KE Kurrer. 1996. Eccentric vibratory mills-a new energy efficient way for pulverization. *Erzmetall* 49:434–442.

Godočíková, E, P Baláž, E Gock, WS Choi, and BS Kim. 2006. Mechanochemical synthesis of the nanocrystalline semiconductors in an industrial mill. *Powder Technology* 164(3):147–152. doi: 10.1016/j.powtec.2006.03.021

He, J, and TM Tritt. 2017. Advances in thermoelectric materials research: Looking back and moving forward. *Science* 357(6358). doi: 10.1126/science.aak9997

Hegedüs, M, M Achimovičová, HJ Hui, et al. 2020. Promoted crystallisation and cationic ordering in thermoelectric $Cu_{26}V_2Sn_6S_{32}$ colusite by eccentric vibratory ball milling. *Dalton Transactions* 49(44):15828–15836. doi: 10.1039/D0DT03368E

Hegedüs, M, M Baláž, M Tešinský, et al. 2018. Scalable synthesis of potential solar cell absorber Cu_2SnS_3 (CTS) from nanoprecursors. *Journal of Alloys and Compounds* 768:1006–1015. doi: 10.1016/j.jallcom.2018.07.284

Kavinchan, J, E Saksornchai, S Thongtem, and T Thongtem. 2018. One-step microwave assisted synthesis of copper antimony sulphide (Cu_3SbS_4) nanostructures: Optical property and formation mechanism study. *Chalcogenide Letters* 15(12):599–604.

Kumar, P, and K Singh. 2009. Wurtzite ZnSe quantum dots: Synthesis, characterization and PL properties. *Journal of Optoelectronic and Biomedical Materials* 1(1):59–69.

Liu, YC, Y Li, HY Kang, T Jin, and LF Jiao. 2016. Design, synthesis, and energy-related applications of metal sulfides. *Materials Horizons* 3(5):402–421. doi: 10.1039/C6MH00075D

Lokhande, AC, RBV Chalapathy, M He, et al. 2016. Development of Cu_2SnS_3 (CTS) thin film solar cells by physical techniques: A status review. *Solar Energy Materials and Solar Cells* 153:84–107. doi: 10.1016/j.solmat.2016.04.003

Pattrick, RAD, and AJ Hall. 1983. Silver substitution into synthetic zinc, cadmium, and iron tetrahedrites. *Mineralogical Magazine* 47(345):441–451. doi: 10.1180/minmag.1983.047.345.05

Ratz, T, G Brammertz, R Caballero, et al. 2019. Physical routes for the synthesis of kesterite. *Journal of Physics: Energy* 1(4). doi: 10.1088/2515-7655/ab281c

Ramasamy, K, H Sims, WH Butler, and A Gupta. 2014. Selective nanocrystal synthesis and calculated electronic structure of all four phases of copper-antimony sulfide. *Chemistry of Materials* 26(9):2891–2899. doi: 10.1021/cm5005642

Rui, XH, HT Tan, and QY Yan. 2014. Nanostructured metal sulfides for energy storage. *Nanoscale* 6(17):9889–9924. doi: 10.1039/C4NR03057E

Siebentritt, S, and S Schorr. 2012. Kesterites - A challenging material for solar cells. *Progress in Photovoltaics* 20(5):512–519. doi: 10.1002/pip.2156

Suekuni, K, K Tsuruta, T Ariga, and M Koyano. 2012. Thermoelectric properties of mineral tetrahedrites $Cu_{10}Tr_2Sb_4S_{13}$ with low thermal conductivity. *Applied Physics Express* 5(5). doi: 10.1143/APEX.5.051201

Vaidya, M, GM Muralikrishna, and BS Murty. 2019. High-entropy alloys by mechanical alloying: A review. *Journal of Materials Research* 34(5):664–686. doi: 10.1557/jmr.2019.37

van Embden, J, and Y Tachibana. 2012. Synthesis and characterisation of famatinite copper antimony sulfide nanocrystals. *Journal of Materials Chemistry* 22(23):11466–11469. doi: 10.1039/C2JM32094K

Vanalakar, SA, GL Agawane, SW Shin, et al. 2015. Non-vacuum mechanochemical route to the synthesis of Cu_2SnS_3 nano-ink for solar cell applications. *Acta Materialia* 85:314–321. doi: 10.1016/j.actamat.2014.11.043

Yu, WW, JC Falkner, BS Shih, and VL Colvin. 2004. Preparation and characterization of monodisperse PbSe semiconductor nanocrystals in a noncoordinating solvent. *Chemistry of Materials* 16(17):3318–3322. doi: 10.1021/cm049476y

Zhang, RZ, K Chen, B Du, and MJ Reece. 2017. Screening for Cu-S based thermoelectric materials using crystal structure features. *Journal of Materials Chemistry A* 5(10):5013–5019. doi: 10.1039/c6ta10607b

Zyoud, A, K Murtada, H Kwon, et al. 2018. Copper selenide film electrodes prepared by combined electrochemical/chemical bath depositions with high photo-electrochemical conversion efficiency and stability. *Solid State Sciences* 75:53–62. doi: 10.1016/j.solidstatesciences.2017.11.013

11

Scalable Solutions for Continuous Manufacturing by Mechanochemical-Assisted Synthesis

Jamie A. Leitch, Matthew T. J. Williams, and Duncan L. Browne

CONTENTS

11.1 Introduction

While mechanochemical techniques can offer potent sustainable improvements in reaction development and design, one of the major sticking points in the adoption in industrial settings is the ability to conduct protocols at scale.[1] This is due to the practical capabilities of ball-milling equipment – either the mixer or planetary mills – being largely limited at the milligram-to-gram scale.[2] While there are larger-scale ball-milling-derived alternatives available (such as drum mills), and other mechanochemical large-scale options such as jaw crushers and cutters, extruders have come to the fore as a potentially powerful enabling technology for increasing the viability of scalable mechanochemical-assisted methodology, as summarized in Figure 11.1.[3]

Already commonly used in the food and polymer industry,[4] continuous extrusion relies on the mechanical force applied by interwoven screws on materials either against the other screw or against the extruder barrel.[5] Importantly, extruders are most often a continuous technology, akin to solution-phase flow chemistry set-ups, which allows the, theoretically, never-ending processing of materials through a reaction barrel.[6] Resistant to scaling issues associated with traditional batch chemistry such as exotherm potential, slow reaction kinetics, and gas release, solution-phase continuous flow chemistry systems are becoming of increasing importance in the safe and sustainable manipulation of organic molecules at process scales.[7] For this reason, when studying metrics such as space–time yield (STY) which are used for reactor optimization and analysis, continuous reactors routinely often offer improvements and opportunities when optimizing manufacturing systems. Coupled with solvent minimization accessed through mechanochemical protocols, continuous extrusion has the potential for substantial process maximization.[8]

To date, three extruder sub-classes have been used in molecular organic synthesis, twin-screw extruders (TSE), micro-compounders, and a home-made single-screw extruder (SSE). We will highlight potential differences and distinct optimizable parameters; however, all three primarily operate through a similar standardized process. Materials are added to the extruder barrel either using a solid feeder or

DOI: 10.1201/9781003178187-14

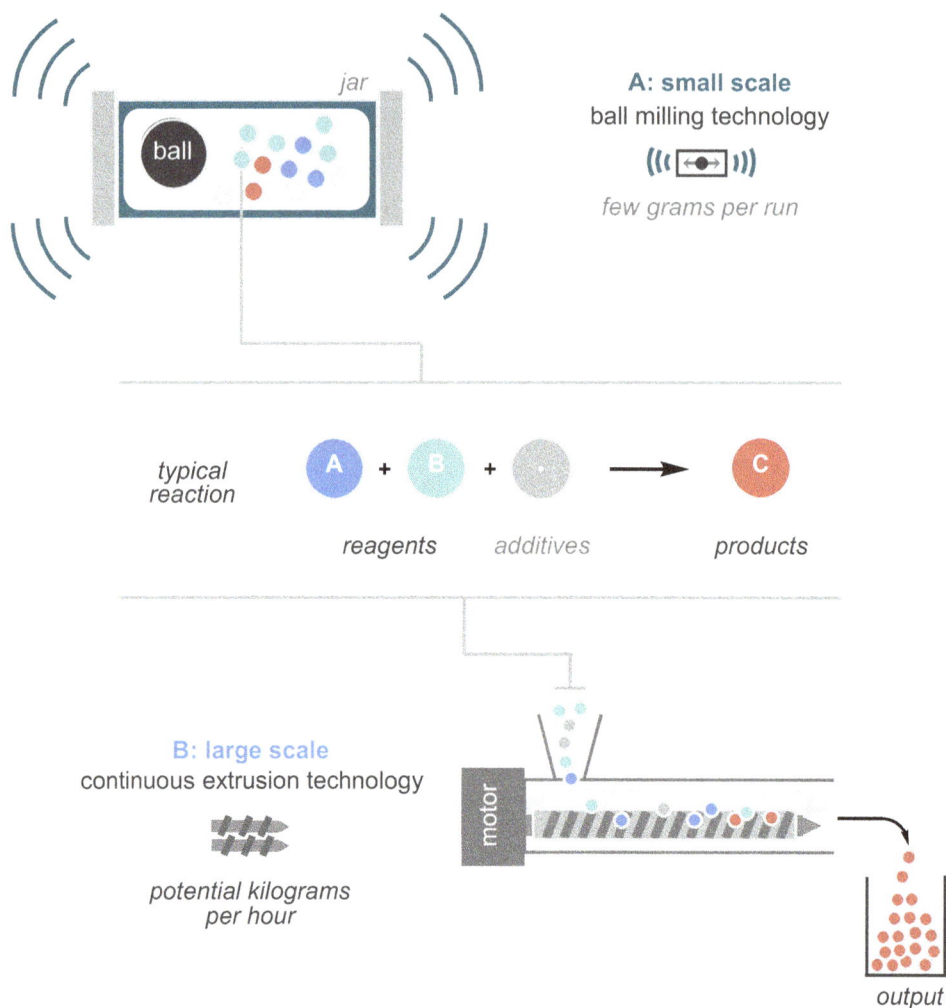

FIGURE 11.1   Scalable solutions to mechanochemistry: continuous extrusion.

via tubing/syringe for liquid feeds, or often a combination of the two. Powered by a motor, screws rotate together, processing the material through the extruder for a given time (known as the residence time). The material is then collected at the end where the – often – solid material can be submitted for further manipulation such as work-up and purification and/or analyzed by normal synthetic means such as NMR spectroscopy and TLC. Furthermore, in extrusion, powder X-ray diffraction (PXRD) or solid-state mass spectrometry can be employed to analyze the solid extrudate, also.

   This chapter will cover an introduction to different types of extruders, an overview of the synthetic chemistry which can be achieved using the reactors, and end with a few selected case studies to highlight what extrusion has thus far demonstrated to the community.

11.2 Introduction to Twin-Screw Extrusion

Twin-screw extruders are by far the most common type of extruders that have been employed in molecular organic synthesis, and an overview schematic is shown in Figure 11.2. They are based on a horizontal barrel containing two screws, which can accommodate a mixture of conveying, kneading, and sometimes reverse sections.[5, 8]

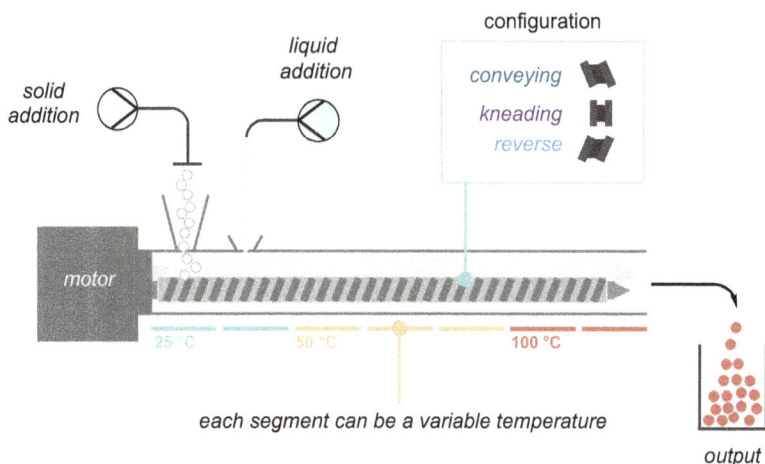

FIGURE 11.2 Twin-screw extruder overview.

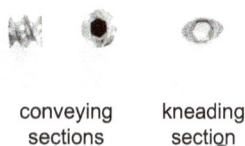

conveying   kneading
sections   section

FIGURE 11.3 Photos of a Thermofisher Process 11 twin-screw extruder.

Screw elements can be manually reconfigured allowing the user to fine-tune the reaction process. Further to this, elements within the kneading sections can be arranged at 30° or 60° orientations (in either forward or reverse orientation) or at 90° to maximize mixing and mechanical energy transfer. Within this context, it is largely regarded that the predominant amount of the mechanical forces solicited by an extruder on a given material is in the kneading sections.[9] Furthermore, reverse sections can be introduced to hold the material in these high-force kneading sections for longer periods of time. As discussed above, materials can either be added via solid additions through gravimetric/volumetric feeders, or via liquid additions from HPLC/syringe pumps, and often a combination of both. The barrel is generally split into multiple different sections which can be individually heated to independent temperatures.

11.3 Variables

Bringing in new technology and reactors to carry out molecular synthesis by mechanochemical means also introduces different variables/parameters. As detailed in Figure 11.4, there are seven key variables to consider with extrusion technology, all of which have a profound effect on the process: screw speed, feed rate, screw configuration, screw temperature, reaction morphology, residence time, and torque.

In Figure 11.4, the first five we can refer to as 'direct variables', those that we have direct control of when setting up an extrusion process. Notably in this case, within the context of mechanochemistry, four of the parameters (excluding reaction morphology) are largely unique to extrusion.[10] The latter two are 'indirect variables', as these parameters cannot be selected and result from the influence of the combination of the direct variables above. The direct variables can influence residence time and torque in the following ways:

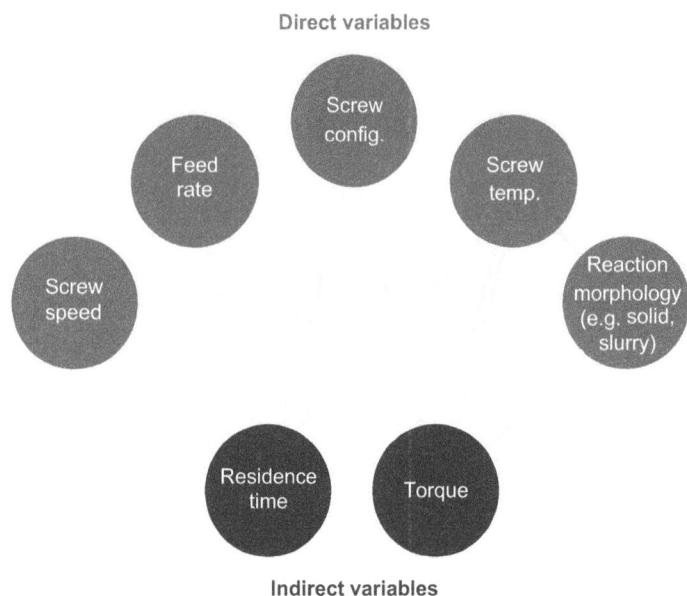

FIGURE 11.4 Extrusion variables with influencing opportunities shown with gray lines.

Screw speed: higher screw speeds generally lead to lower residence times and lower torques as the material is processed at faster speeds.

Feed rate: higher feed rates generally lower residence time but increase torque as there is more material to push through the kneading sections – but also risk higher chances of blockages.

Screw configuration: introduction of reverse-type sections slows down procession through the screw, increasing residence time and increasing torque.

Screw temperature: while increased temperature can augment/enable reactivity within the screw to take place, the temperature can also ease flow through the screw often reducing torque and residence time. Such increases are further exaggerated when phase changes take place at higher temperatures (e.g. a solid component becomes a liquid, or the product is in a different phase than the starting material(s)).

Reaction morphology: the overall morphology of the reaction mixture, i.e. more liquid or more solid will affect both residence time and torque. Here more liquid mixtures afford shorter residence times and lower torque values, and more solid mixtures the reverse. The overall reaction morphology can be simply adjusted by altering the equivalencies of reagents, but also through the addition of glidants (solid grinding auxiliaries), or liquid-assisted grinding (LAG) agents to the extrusion mixture.[11]

As shown in Figure 11.5, along with screw temperature, the two indirect variables make up the 'reaction-influencing variables', which govern reaction success. In short, how much mechanical energy is a material receiving, for how long, and at what temperature?

Residence time: as with solution-phase continuous flow technology, the residence time is incredibly important.[6] However, unlike in the solution phase where residence time is almost entirely determined by the pump flow rate and reactor volume, within an extruder, all the above variables can affect the overall residence time of the material within the extruder. The most profound difference in translating ball-milling studies to extrusion technology is the reaction time, translating from often a few hours in the ball mill to minutes and even seconds in the extruder.

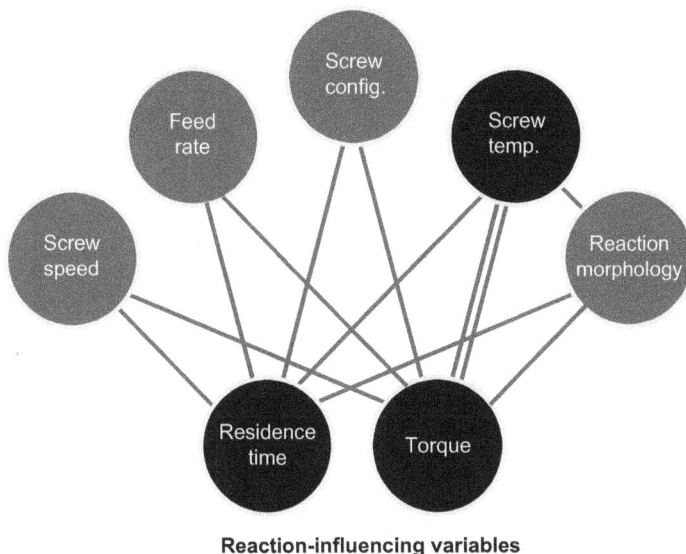

Reaction-influencing variables

FIGURE 11.5 Reaction-influencing extrusion variables.

Torque: measured in Nm or kg m^2 s^{-2}, torque is the force required for the motor to turn the screw within the barrel and is related to the resistance exhibited by the reaction materials. If the screw speed is not high enough (especially with the addition of reverse sections), the material can form blockages, leading to 'torqueing out' of the extruder and the shutdown of the reactor system. The value at which an extruder 'torques out' depends on the size and supplier of the extruder, as they are rated to different maximum torque values (generally given as a value per screw shaft). Despite this, torque is often beneficial to reaction performance. Torque (as well as screw speed) is proportional to specific mechanical energy (Equation 11.1).

$$\text{SME} = \frac{120 \cdot \pi \cdot n \cdot M}{\dot{m}} \quad \text{Mechanical energy} \propto \text{Torque} \tag{11.1}$$

where SME = specific mechanical energy, n = screw speed (rpm), M = torque (Nm), and $\dot{m}$ = throughput rate (kg h^{-1}).

Within this context, it could be perceived that increasing the screw speed (n) would have a direct impact on reaction efficiency by imparting more mechanical force; however, this in return will decrease the residence time (increasing throughput rate), potentially diminishing any gains made. It is worth noting that processes consistently running with high torque values can lead to increases in screw temperature through latent heating.

11.4 Overview of Twin-Screw Extrusion Protocols to Date

Building on pioneering work on using twin-screw extrusion for the continuous synthesis of co-crystals,[12] metal–organic frameworks,[13] and deep eutectic solvents,[9] since 2017 there has been an influx of reports using extruders for molecular organic synthesis. Despite this, the technique remains largely in its infancy, with lots of information from both synthetic and engineering aspects still yet to be well characterized.

In this section, we will be using a general schematic in Figure 11.6, with information split into seven sections including variables and resulting data (where available): (i) feed, (ii) screw speed, (iii)

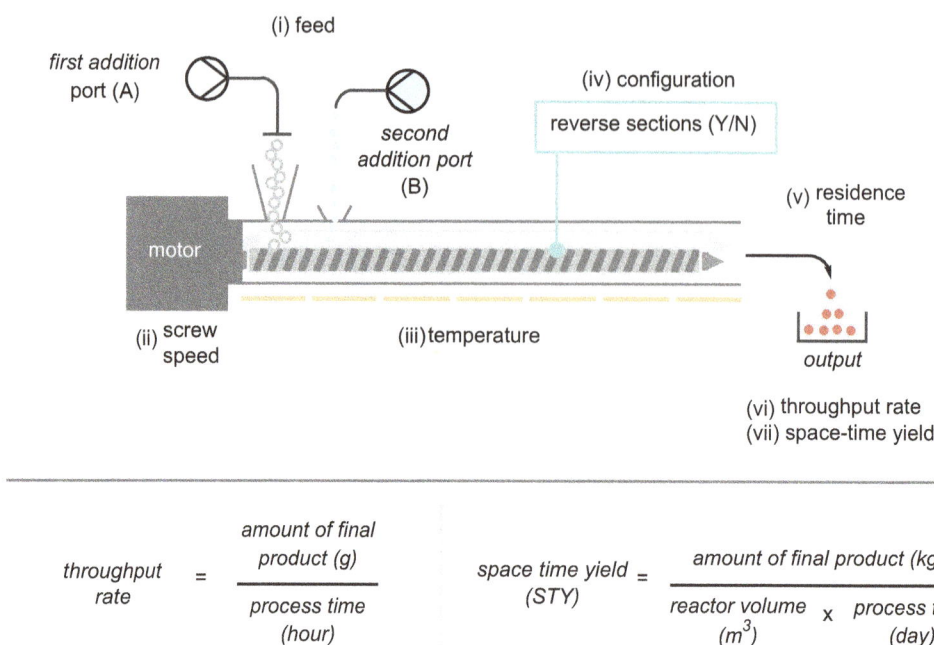

FIGURE 11.6 Overview schematic of extrusion protocols.

temperature, (iv) screw configuration, (v) residence time, (vi) throughput rate, and (vii) space–time yield. The relevant equations for calculating the latter two parameters are also given.

The first forays into the use of extrusion in organic synthesis were conducted by Crawford, James, and co-workers, with an overview detailed in Figure 11.7.[14] The authors studied an array of transformations beginning with Knoevenagel condensations (Figure 11.7a and b) using both barbituric acid and ethyl 2-cyanoacetate as pronucleophiles.

In the first of these cases (Figure 11.7a) they were able to add all solids into a feeder and achieved 100% conversion to the alkene product with impressive throughput rates of 520 g h^{-1} and space–time yield values an order of magnitude higher than many extrusion protocols to date (*vide infra*). In the case of the nitrile-substituted example (Figure 11.7b), this liquid was fed into a separate second addition port *via* a syringe pump. Notably, a base catalyst was required to suppress undesired polymerization of the pronucleophile and was not removed from the reaction mixture after extrusion.

Following this, when using dimedone as a pronucleophile, they uncovered that an *in situ* Michael addition took place (Figure 11.7c). After process optimization, including the addition of reverse sections, increase in screw speed, and mixing the reagents in a planetary mill prior to addition to the extruder, 100% conversion to the Michael adduct was achieved. Subsequently, as shown in Figure 11.7d and e, the authors also detailed further reaction examples of dimedone, including a double imine formation process with notable reaction efficiency providing 100% conversion less than 1 minute of residence time.

Figure 11.8 shows a collection of further reports using twin-screw extrusion for organic synthesis. Isoni and co-workers demonstrated an important advance in the use of extruders to process slurries rather than solids,[15] where in this case they were able to feed solid NaBH$_4$ dissolved in 1 M NaOH into an extruder fed with a solid benzaldehyde derivative. This process gave rise to an outstanding throughput of ~4.5 kg h^{-1}, and furthermore a full process design including pump feeds and purification was provided by the authors.

Alongside a telescoped cascade process (*vide infra*), Crawford, James, and co-workers reported the multi-component Petasis reaction from boronic acid, aldehyde, and amine feedstocks.[16] Making use of reverse sections to increase reaction efficiency and feeding the liquid amine *via* the second addition port, they achieved a route to valuable α-branched amine products with an excellent throughput rate of 185 g h^{-1}.

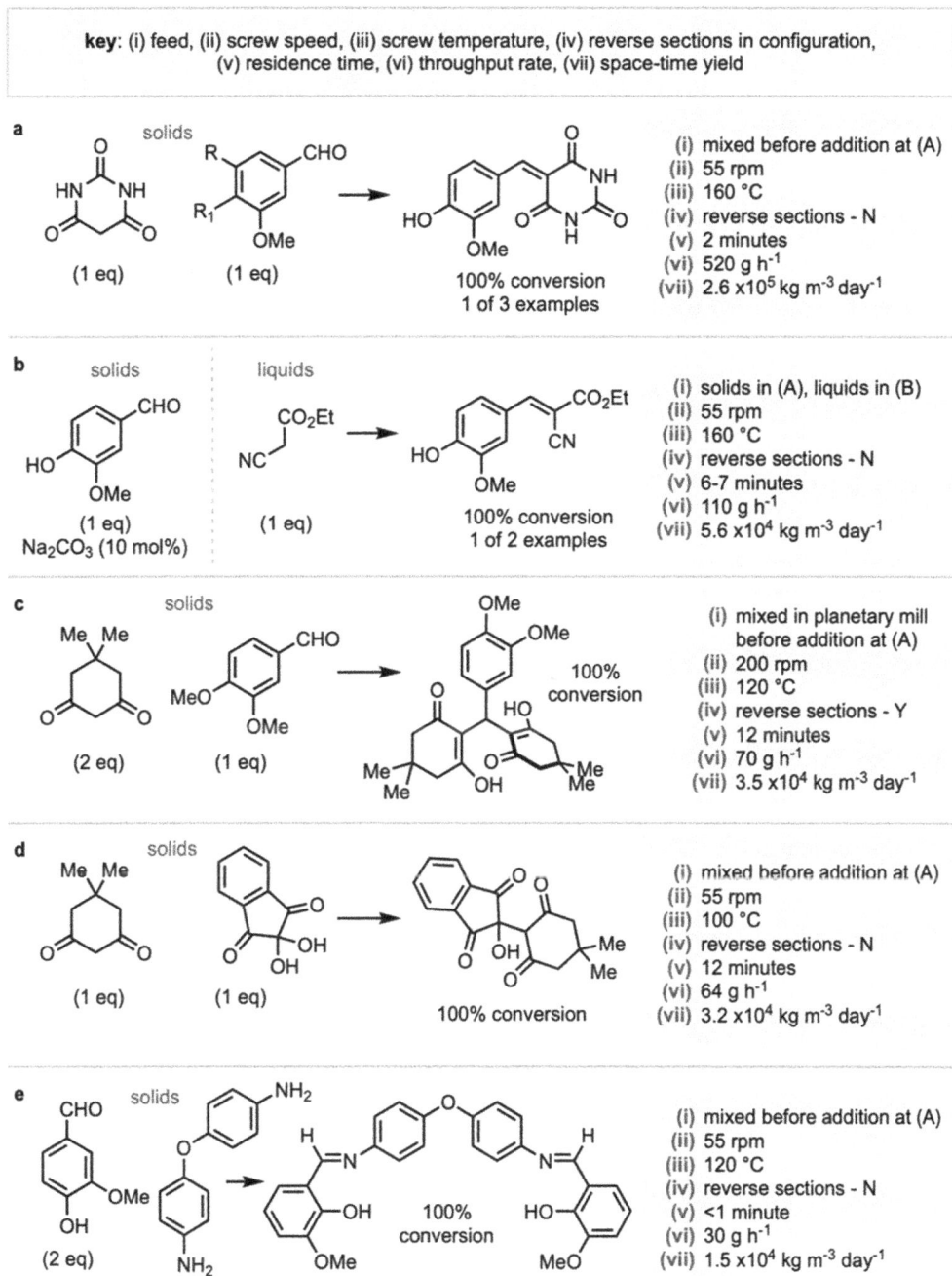

FIGURE 11.7 Overview of transformations covered in the initial report by Crawford, James, and co-workers.

Building on prior small-scale ball-milling work, Browne, James, and co-workers demonstrated their selective additive-dependent mono- and di-fluorination of 1,3-dicarbonyls.[17] Using solid Na$_2$CO$_3$, the di-fluorinated product predominated, whereas using MeCN as a liquid additive promoted the formation of the mono-fluorinated product, as shown in Figure 11.8c. By mixing all reaction components together and then manually feeding them into an extruder containing reverse sections, the designed protocols were able to afford both fluorinated species in a selective manner.

key: (i) feed, (ii) screw speed, (iii) screw temperature, (iv) reverse sections in configuration, (v) residence time, (vi) throughput rate, (vii) space-time yield

a: Isoni and co-workers

solids   liquids

NaBH$_4$ (s)
(0.26 eq)

in NaOH (1 M)

(1 eq)

89% conversion
1.41 kg in 17 min

(i) solid in (A), liquids in (B)
(ii) 200 rpm
(iii) 15 °C
(iv) reverse sections - N
(v) 16 seconds
(vi) 4,530 g h^{-1}
(vii) 1.3 x10^4 kg m^{-3} day^{-1}

b: Crawford, James and co-workers

solids   liquids

B(OH)$_2$ CHO

NHEt$_2$
(1 eq)

Me   OMe

(1 eq)   (1 eq)

100% conversion

(i) solid in (A), liquids in (B)
(ii) 150 rpm
(iii) 50 °C
(iv) reverse sections - Y
(v) 7 minutes
(vi) 185 g h^{-1}
(vii) 9.2 x10^4 kg m^{-3} day^{-1}

c: James, Browne and co-workers

(1 eq)   (2 eq)

NaCl (23.6 g), Na$_2$CO$_3$ (3 eq)

70% isolated yield

(i) mixed before addition at (A)
(ii) 280 rpm
(iii) 70 °C
(iv) reverse sections - Y
(v) 1.5 minutes
(vi) 6.8 g h^{-1}
(vii) 3.4 x10^3 kg m^{-3} day^{-1}

(1 eq)   (2 eq)

NaCl (25 g), MeCN (6 mL)

90% isolated yield
mono:di 6:1

(i) mixed before additon at (A)
(ii) 280 rpm
(iii) 60 °C
(iv) reverse sections - Y
(v) 3 minutes
(vi) 8.9 g h^{-1}
(vii) 4.4 x10^3 kg m^{-3} day^{-1}

d: James and co-workers

solid   liquid

H$_2$N

(2.4 eq)

Pigment Black 31
100% conversion
1 of 2 examples in automated synthesis

(i) solid in (A), liquids in (B)
(ii) 55 rpm
(iii) 110 °C
(iv) reverse sections - Y
(v) 7-8 minutes
(vi) 60 g h^{-1}
(vii) 3.0 x10^4 kg m^{-3} day^{-1}

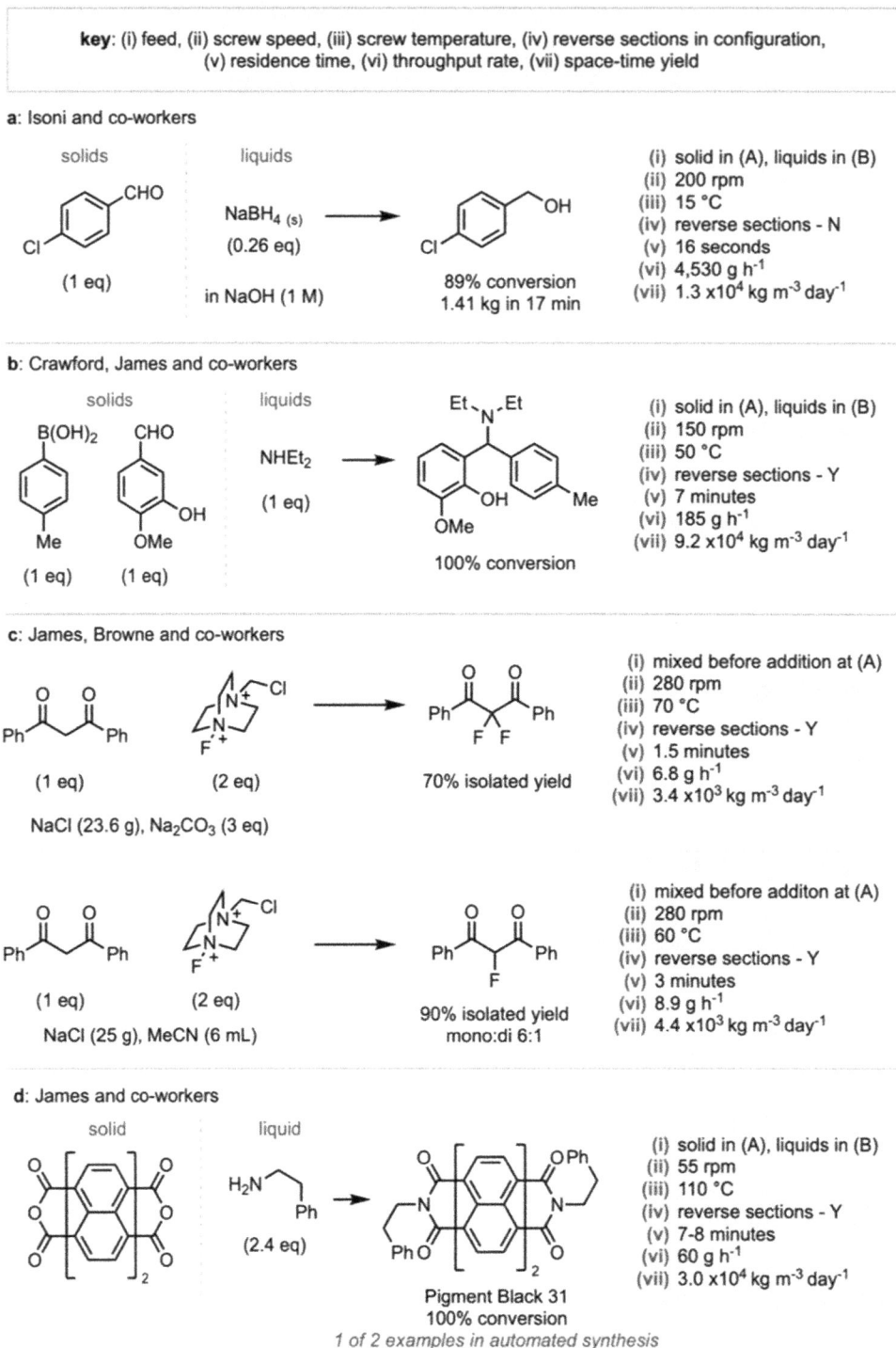

FIGURE 11.8 Overview of work on extrusion covered by multiple research groups.

As detailed in Figure 11.8d, James and co-workers reported the synthesis of naphthalene diimide (NDI) dyes using extrusion.[18] While the report contains more procedures using stirred manual addition, we wish to highlight the automated synthesis of Pigment Black 31. By adding the anhydride into the solid feeder and the amine through the liquid feeder, feeding along a screw containing reverse sections, the authors achieved an excellent throughput rate of 60 g h^{-1} of a compound with a notable lack of solubility, another benefit of using mechanochemical techniques to process materials and chemical reactions.

El-Remaily and co-workers have demonstrated the multi-component Ugi reaction using a twin-screw extruder, where the starting materials were first mixed, by hand, in a mortar and pestle before being added to an extruder, which was heated to 100 °C (Figure 11.9a).[19] In residence times of less than 20 minutes, they achieved the synthesis of a small selection of bis-amide fragments.

Within an in-depth study into the optimization of extrusion procedures, Andersen and co-workers explored the S$_N$Ar reaction between benzylamine and an electron-poor aryl fluoride (Figure 11.9b).[20] The authors utilized automated solid and liquid feeders and were able to demonstrate the influence of temperature on the effectiveness of their extrusion process. Using a screw system containing reverse sections, at 90 °C, they observed a 97% yield for the S$_N$Ar process. The authors also further explored the Knoevenagel process discussed previously (Figure 11.7a).

key: (i) feed, (ii) screw speed, (iii) screw temperature, (iv) reverse sections in configuration, (v) residence time, (vi) throughput rate, (vii) space-time yield

a: El-Remaily and co-workers

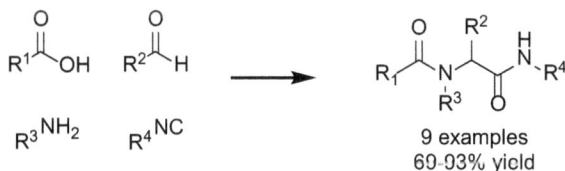

(i) mixed in pestle and mortar before addition at (A)
(ii) 50 rpm
(iii) 100 °C
(iv) reverse sections - n.d.
(v) <20 minutes
(vi) not given
(vii) not given

9 examples
69–93% yield

b: Andersen and co-workers

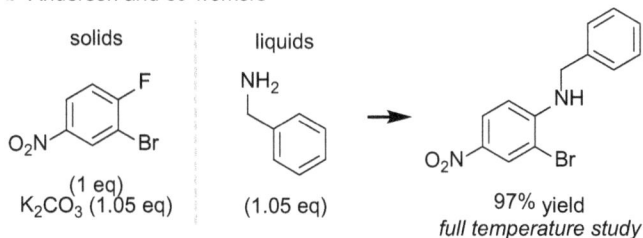

(i) solids in (A), liquids in (B)
(ii) 50 rpm
(iii) 90 °C
(iv) reverse sections - Y
(v) 15 minutes
(vi) not applicable
(vii) not applicable

97% yield
full temperature study

in depth study on screw config. and temperature

c: Halasz, Palčić, Hernández and co-workers

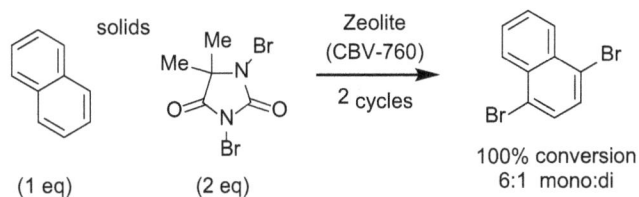

(i) mixed before addition at (A)
(ii) 20 Hz (1200 rpm)
(iii) room temperature
(iv) reverse sections - N
(v) not given
(vi) not given
(vii) not given

100% conversion
6:1 mono:di

FIGURE 11.9 Overview of work on extrusion covered by multiple research groups.

More recently, Halasz, Palčić, Hernández, and co-workers reported a small exploration into extrusion as part of a report on the ball-milling-enabled, zeolite-promoted bromination of naphthalene using 1,3-dibromo-5,5-dimethylhydantoin (DBDMH) as the brominating agent (Figure 11.9c.)[21] They demonstrated that at very high screw speeds (20 Hz, 1200 rpm), two cycles were required to enable 100% conversion to a mixture of 1-bromo- and 1,4-dibromo-napthalene.

11.5 Micro-Compounders

Micro-compounders are based on a vertical barrel containing two screws facing downwards with only conveying-type patterns on the screw system, as shown in Figure 11.10. Previously we have also termed this equipment 'recycling screw extrusion', as the premise is akin to twin-screw extrusion; however, the material can be recycled for a period of time before being ejected to deliver the final extrudate. Additionally, the whole system can be heated to high temperatures (~450 °C), while extrusion takes place.

In the plastics/formulation fields, micro-compounders are often seen as a 'stepping-stone' reactor towards extrusion processes as only a relatively small amount of material can be processed (5–15 g). With these reactors, the screw configuration cannot be altered and contain no kneading sections, and therefore mechanical force is only applied through shearing against the wall of the barrel.

To date, there has been only one report of organic synthesis using these micro-compounders by Métro and co-workers on one of the most important reactions in synthesis, amide bond formation.[22] Mixing an activated ester and an amine (such as a protected tryptophan and glycine in the examples below) along with NaHCO$_3$ and acetone and feeding into a micro-compounder, and allowing 10 minutes of recirculation time before ejection, afforded the amide product in 85% yield on a multi-gram scale. This methodology was applied to dipeptides and even tripeptides in programmable amide coupling, all in the absence of toxic coupling agents and bulk reaction solvents (Figure 11.11).

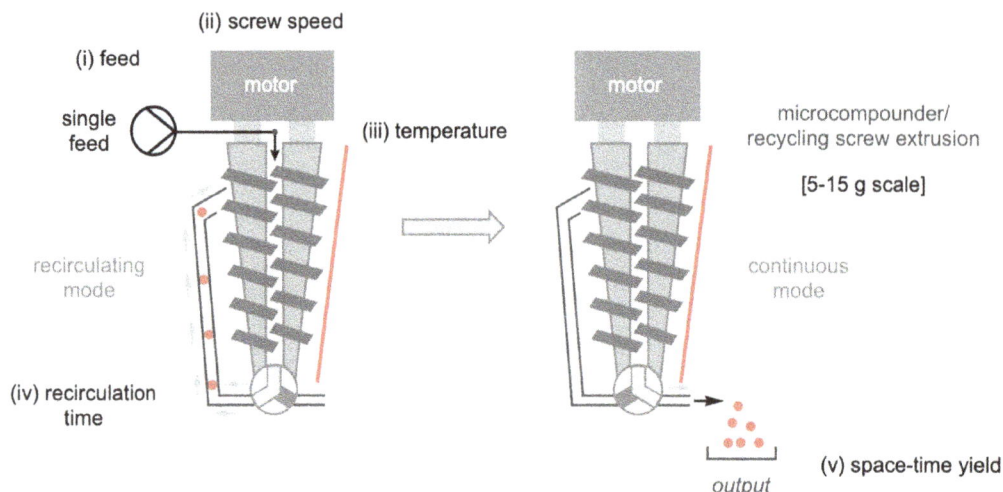

FIGURE 11.10   Micro-compounder/Recycling screw extrusion overview.

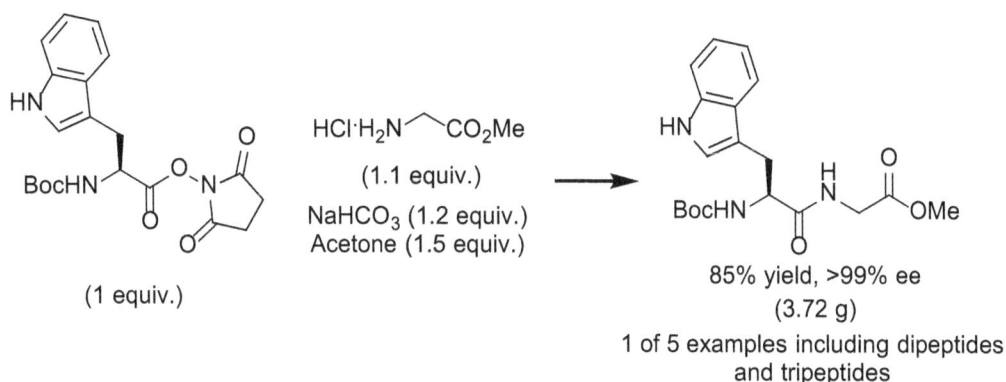

(i) feed: all mixed together before addition
(ii) screw speed: 150 rpm
(iii) screw temperature: 40 °C

(iv) recirculation time: 10 minutes
(v) STY: 4.8 x10³ kg m⁻³ day⁻¹

FIGURE 11.11 Micro-compounders and their use in recirculating screw extrusion.

11.6 Single-Screw Extruder

Aligned with other endeavors into the development of enabling technology techniques such as flow chemistry,[6] photochemistry,[23] and electrochemistry,[24] extrusion methodology has witnessed users creating home-made devices to study this concept.

There is one example of a single-screw extruder fabricated in-house using a single Teflon screw connected to a motor inside a standard laboratory condenser by Kulkarni and co-workers, as shown in Figure 11.12.[25] Fastened vertically the extruder is fed by two further Teflon screws on either side, where it is possible to feed two sets of materials from different ports.

The screw system also does not possess any kneading sections, only conveying, therefore, the mechanical force is transferred solely via the shear between the screws and barrel edge. Figures 11.13 and 11.14 show the array of transformations covered in this initial report by Kulkarni and co-workers with four pieces of data given: (i) feed information (one port or two ports), (ii) screw speed, (iii) temperature, and (iv) residence time. It must be noted that due to the low amounts of mechanical force elicited by this technique, coupled with very low residence times, it is likely that these reactions take place from high concentration and efficient mixing rather than any mechanical effects.

The authors covered a large amount of chemical reaction space, including diazo dye formation, binaphthyl coupling, alcohol oxidation, Pd-catalyzed cross coupling, aldol addition, acylation, and alcohol protection. All these transformations were conducted on a multi-gram scale (between 4 and 18 g), but remarkably the diazo dye reaction was scaled to the kilogram scale affording 983 g of product using a home-made extrusion device. This initial study provides hope and opportunity for the growing use of extrusion in organic synthesis, as the capital outlay for extrusion equipment will render their widespread use limited for years to come.

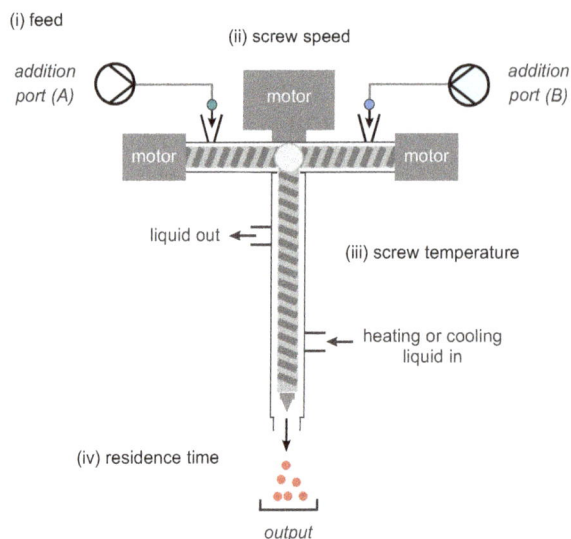

FIGURE 11.12   Home-made single-screw extrusion device.

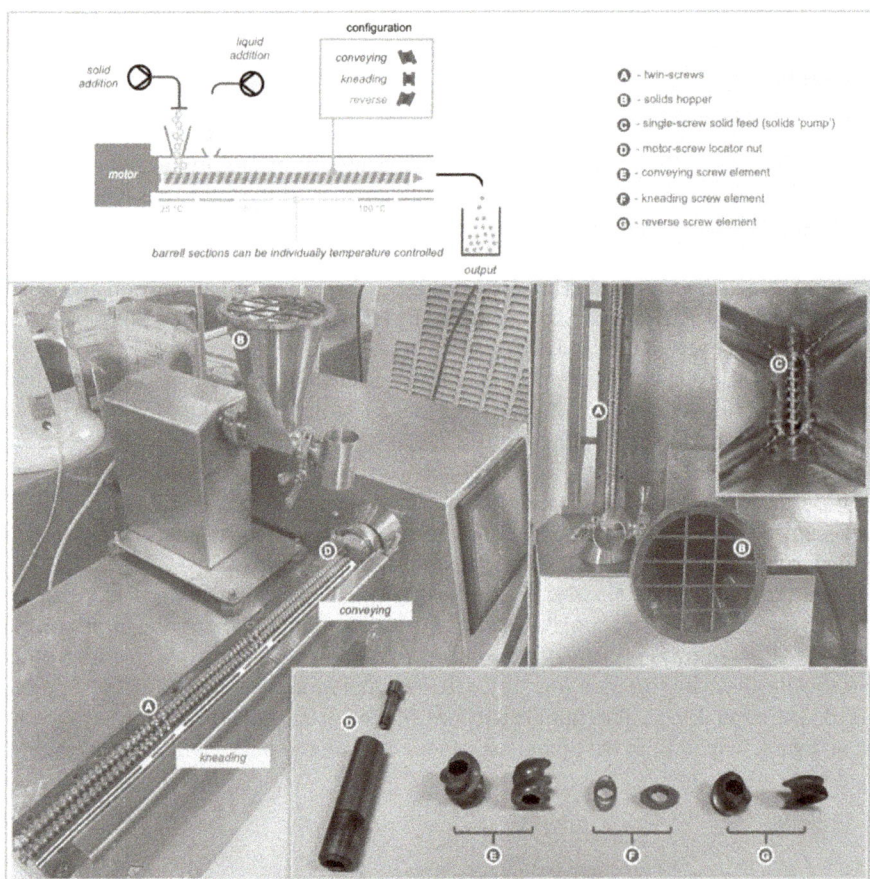

FIGURE 11.13   Overview of reactions carried out in a single-screw extruder.

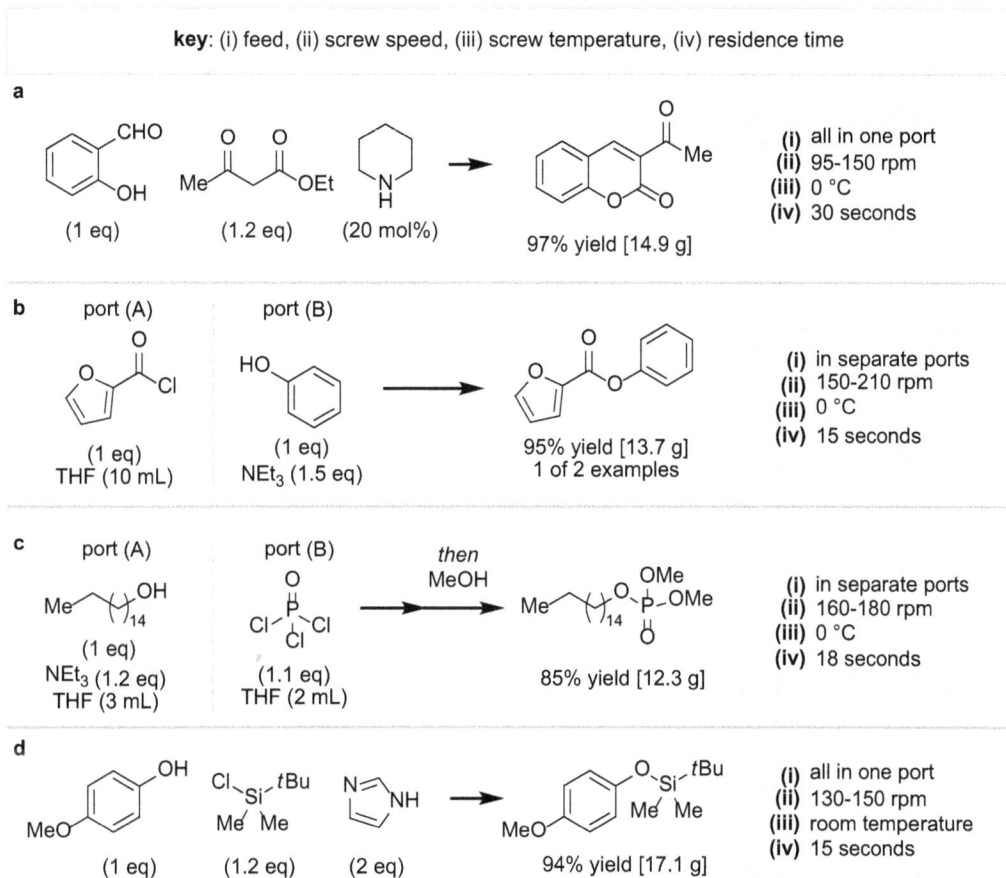

key: (i) feed, (ii) screw speed, (iii) screw temperature, (iv) residence time

FIGURE 11.14 Overview of reactions carried out in a single-screw extruder.

11.8 Twin-Screw Extrusion, What Can It Offer?

Having explored a range of extrusion techniques and reports to date, we have selected three reports to highlight what extrusion as an enabling technology can offer both the mechanochemistry community but also prospective 'extrusion-curious' industrial chemists.

In 2017, Crawford and James built on their previous work by developing a telescoped 'one-screw' reaction, summarized in Figure 11.15.[16] In this instance they fed vanillin, malononitrile, and catalytic quantities of sodium carbonate into the extruder using a solid feeding port. After zone three (just after the first kneading section), they then positioned a second solid addition port to add dimedone. This cascade Knoevenagel/Michael addition reaction was achieved in 100% conversion with an excellent throughput rate of 154 g h^{-1}. Notably in this example the authors initially worked on a two-stage heating protocol, with the first three sections at 120 °C and the last three at 160 °C; however, this led to 'bridging' problems around the feeding ports as the materials melted as they entered the extruder. This was circumvented using the extruder's fine temperature control with zone 1 at 25 °C, and zone 4 at 135 °C (dimedone melting point 147–150 °C). These alterations allowed a successful application towards the cascade product, making use of four different temperatures in a telescoped process.

Within this context, Browne, Leitch, and co-workers explored extrusion methodology to scale a nickel-catalyzed Suzuki–Miyaura reaction (Figure 11.16).[26] From small-scale ball-milling studies, the C–C coupling protocol was found to be very temperature dependent and a three-stage heating profile was found to be optimal (10 minutes at 25 °C, 10 minutes at midway point 63 °C, and 10 minutes at 100 °C).

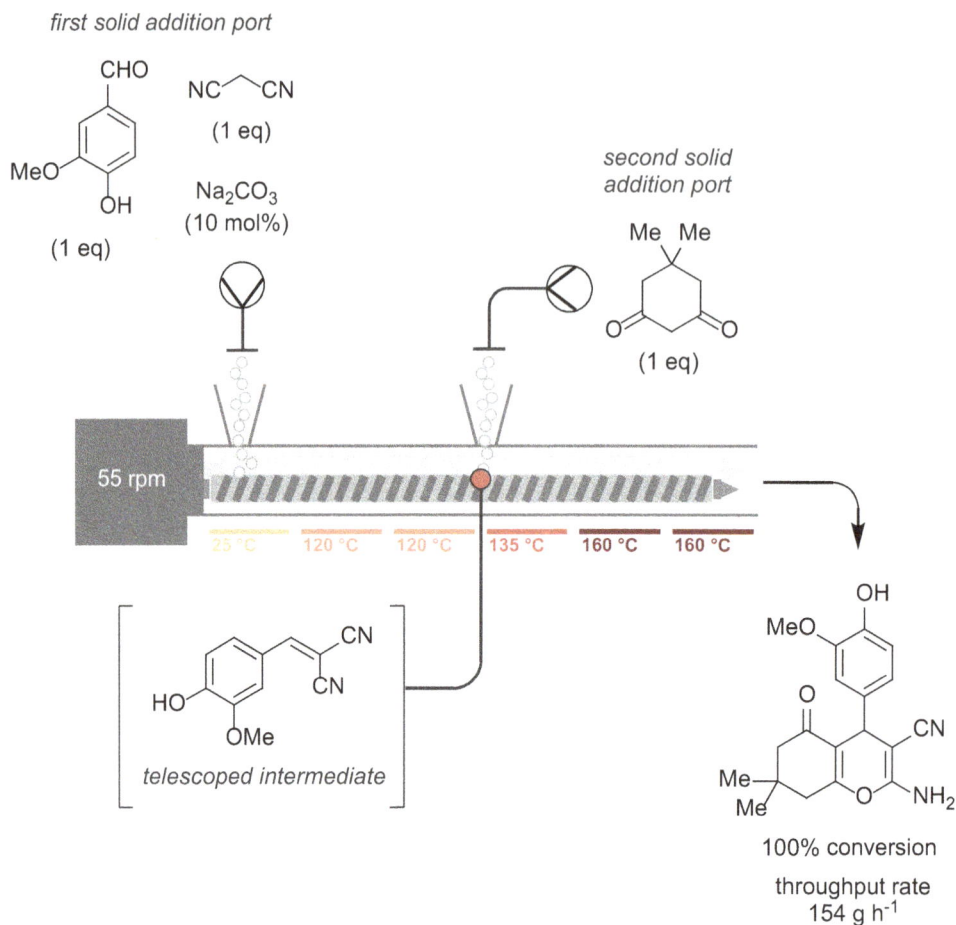

FIGURE 11.15   Telescoped multi-step reactions using a twin-screw extruder.

Translating this system to extrusion methodology – moving from 0.5 mmol scale to 100 mmol scale – the temperature change in the ball mill was mirrored (zones 1–2 at 25 °C, zones 3–4 at 63 °C, and zones 5–7 at 100 °C). Feeding the two reagents, catalyst, base, and grinding auxiliary into the solid feeder port and adding *n*-hexanol (liquid-assisted grinding agent) *via* syringe port at a screw speed of 50 rpm, the authors were able to isolate 13.76 g of the C–C coupled product. Further analysis at 10-minute timed intervals demonstrated an initiation phase as the reactor reached a steady state, also a feature noted by Andersen and co-workers,[20] and then a process-end drop-off. This highlights the potential for increasing the 'steady-state' section simply by running the reaction for longer in a continuous fashion. Considering the value of Suzuki–Miyaura-type methodology to pharmaceutical route design,[27] these studies serve to showcase the sustainable options extrusion can offer, all in the absence of bulk reaction solvent, with a base-metal catalyst, all under air atmospheres.

Within the context of the pharmaceutical industry, Crawford, Colacino, and co-workers reported the synthesis of two hydantoin active pharmaceutical ingredients (APIs) using extrusion methodology.[28] Based on the condensation reaction between 3-aminohydantoin hydrochloride and furfural derivatives, the authors demonstrated that hydrazone formation was possible in the absence of both base and solvent, without the need for purification. Here, the HCl, which is produced as a gas, can escape in gas vents from the open extruder (in comparison to a closed jar), and the water by-product may be present as 'crystallization' water in the final extrudate. The authors processed the mixture at 30 rpm using reverse sections,

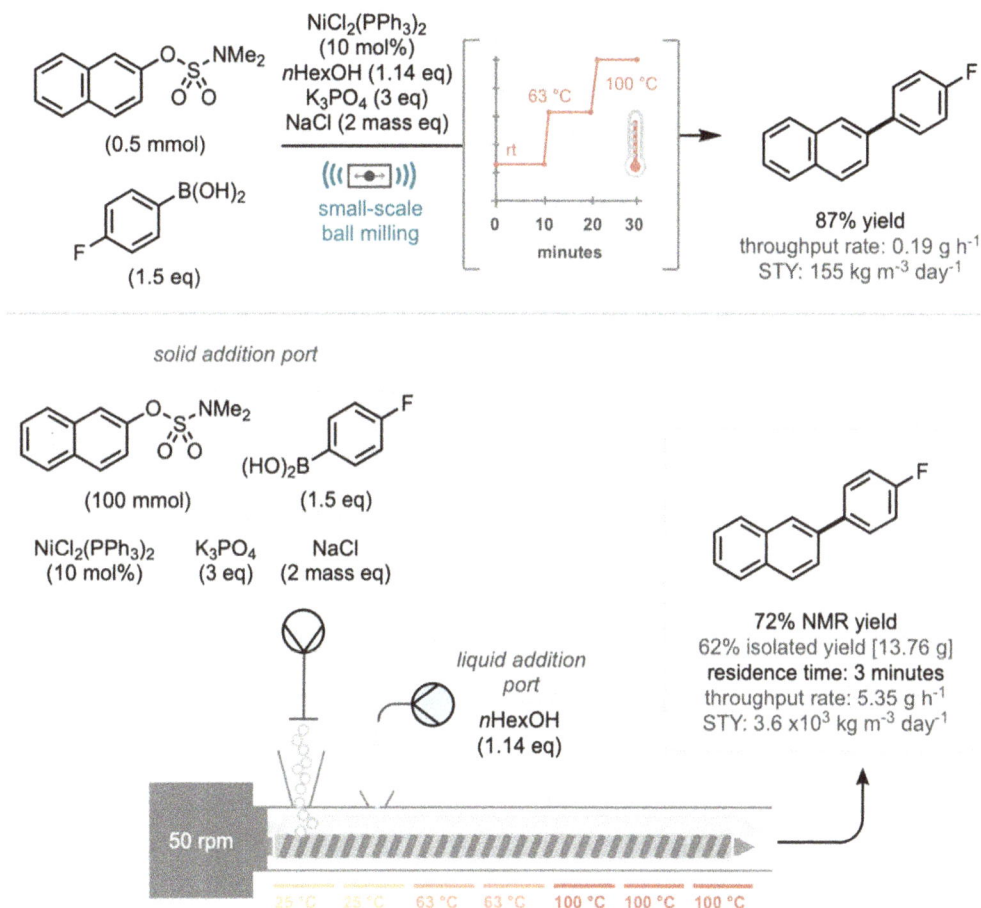

FIGURE 11.16 Heating profiles adapted from ball milling to extrusion.

enabling residence times of 40 minutes, the longest of any organic extrusion process to date, affording nitrofurantoin in 100% conversion and dantrolene in 87% conversion (Figure 11.17).

We direct readers to an interesting follow-up discussion on this chemistry including a full life cycle analysis of the environmental impacts of this extrusion methodology by Spatari and co-workers, with insights into global warming, terrestrial ecotoxicity, and operating costs.[29]

11.9 Conclusion

In conclusion we have presented extrusion as a new enabling technology in sustainable synthesis. The last 5 years have seen an influx of research interests from across the globe looking to extrusion as an answer to scale mechanochemical techniques. We have highlighted the types of extruders which have been used thus far in molecular organic synthesis: twin-screw extruders, micro-compounders (recirculating extruders), and a home-made single-screw extruder.

Extrusion has been shown to be able to process simple and more complex catalytic procedures on industrially relevant reaction scales of 4.5 kg h⁻¹. Three reports have been highlighted in more detail at the end of this chapter to try and demonstrate what extrusion methodology can offer any potential end user. Also, for those who are 'extrusion-curious', we point readers to our recent tutorial review with an extrusion 'how-to' reaction design and development.[8]

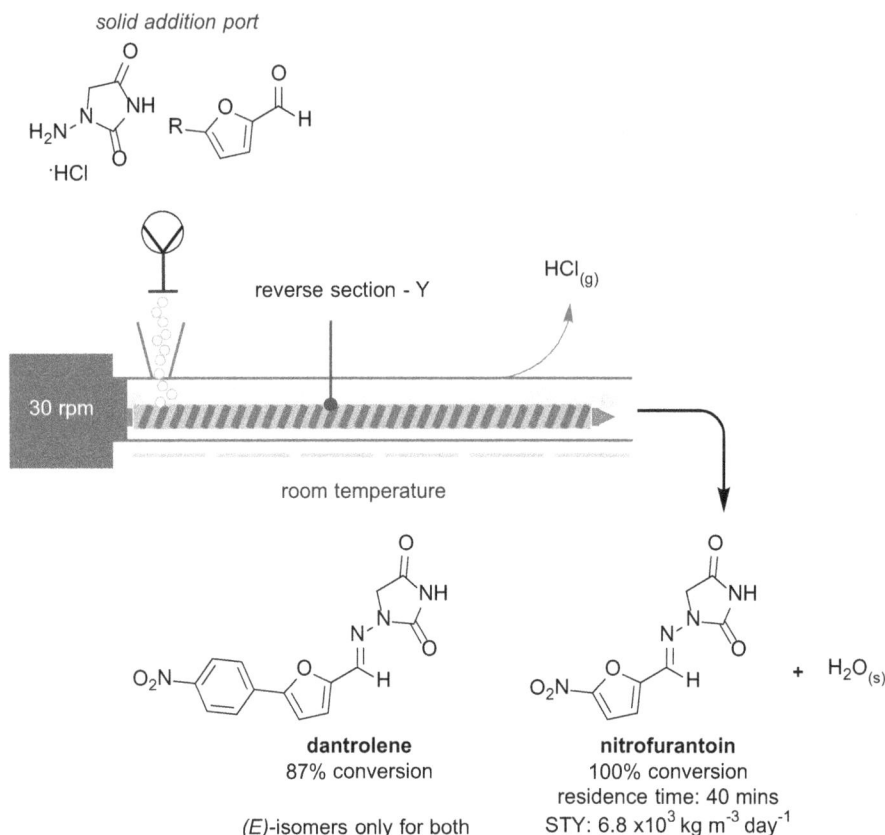

FIGURE 11.17   Synthesis of APIs using extrusion.

As pipeline lead candidates are becoming more and more complex, and in an ever-warming world, the need for sustainable developments in manufacturing has never been higher.[30] Even if one step in a synthetic route of an upcoming blockbuster active ingredient could be exchanged for extrusion, the implications in sustainable processing would be profound. We have no doubt continuous extrusion will play an important role in the projection of mechanochemistry as an unignorable sustainable technique in the future.

REFERENCES

1. E. Colacino, V. Isoni, D. Crawford, F. García, *Trends Chem.* 2021, *3*(5), 335–339.
2. (a) K. J. Ardila-Fierro, J. G. Hernández, *ChemSusChem* 2021, *14*(10), 2145–2162; (b) A. Stolle, T. Szuppa, S. E. S. Leonhardt, B. Ondruschka, *Chem. Soc. Rev.* 2011, *40*(5), 2317–2329; (c) J. L. Howard, Q. Cao, D. L. Browne, *Chem. Sci.* 2018, *9*(12), 3080–3094; (d) J. Andersen, J. Mack, *Green Chem.* 2018, *20*(7), 1435–1443; (e) I. N. Egorov, S. Santra, D. S. Kopchuk, I. S. Kovalev, G. V. Zyryanov, A. Majee, B. C. Ranu, V. L. Rusinov, O. N. Chupakhin, *Green Chem.* 2020, *22*(2), 302–315; (f) T. Friščić, C. Mottillo, H. M. Titi, *Angew. Chem. Int. Ed.* 2020, *59*(3), 1018–1029.
3. *Faraday Discuss.* 2014, *170*, 287–310.
4. (a) R. Cardinaud, T. McNally, *Eur. Polym. J.* 2013, *49*(6), 1287–1297; (b) H. Akdogan, *Int. J. Food Sci. Technol.* 1999, *34*(3), 195–207.
5. (a) D. E. Crawford, J. Casaban, *Adv. Mater.* 2016, *28*(27), 5747–5754; (b) D. E. Crawford, *Beilstein J. Org. Chem.* 2017, *13*, 65–75.
6. M. B. Plutschack, B. Pieber, K. Gilmore, P. H. Seeberger, *Chem. Rev.* 2017, *117*(18), 11796–11893.

7. L. Vaccaro, D. Lanari, A. Marrocchi, G. Strappaveccia, *Green Chem.* 2014, *16*(8), 3680–3704.
8. R. R. A. Bolt, J. A. Leitch, A. C. Jones, W. I. Nicholson, D. L. Browne, *Chem. Soc. Rev.* 2022, *51*(11), 4243–4260.
9. D. E. Crawford, L. A. Wright, S. L. James, A. P. Abbott, *Chem. Commun. (Camb)* 2016, 4215–4218.
10. Despite this, there have been select reports on temperature-controlled mechanochemistry, see for a selection of these: (a) S. Immohr, M. Felderhoff, C. Weidenthaler, F. Schüth, *Angew. Chem. Int. Ed.* 2013, *52*, 12688–12691; (b) R. Schmidt, C. F. Burmeister, M. Baláž, A. Kwade, A. Stolle, *Org. Process Res. Dev.* 2015, *19*(3), 427–436; (c) K. Užarević, V. Štrukil, C. Mottillo, P. A. Julien, A. Puškarić, T. Friščić, I. Halasz, *Cryst. Growth Des.* 2016, *16*(4), 2342–2347; (d) J. M. Andersen, J. Mack, *Chem. Sci.* 2017, *8*(8), 5447–5453; (e) T. Seo, N. Toyoshima, K. Kubota, H. Ito, *J. Am. Chem. Soc.* 2021, *143*(16), 6165–6175.
11. P. Ying, J. Yu, W. Su, *Adv. Synth. Catal.* 2021, *363*(5), 1246–1271.
12. (a) R. S. Dhumal, A. L. Kelly, P. York, P. D. Coates, A. Parakdar, *Pharm. Res.* 2010, *27*(12), 2725–2733; (b) C. Medina, D. Daurio, K. Nagapudi, F. Alvarez-Nunez, *J. Pharm. Sci.* 2010, *99*(4), 1693–1696.
13. D. E. Crawford, J. Casaban, R. Haydon, N. Giri, T. McNally, S. L. James, *Chem. Sci.* 2015, *6*(3), 1645–1649.
14. D. E. Crawford, C. K. G. Miskimmin, A. B. Albadarin, G. Walker, S. L. James, *Green Chem.* 2017, *19*(6), 1507–1518.
15. V. Isoni, K. Mendoza, E. Lim, S. K. Teoh, *Org. Process Res. Dev.* 2017, *21*(7), 992–1002.
16. D. E. Crawford, C. K. Miskimmin, J. Cahir, S. L. James, *Chem. Commun. (Camb)* 2017, *53*(97), 13067–13070.
17. Q. Cao, J. L. Howard, D. E. Crawford, S. L. James, D. L. Browne, *Green Chem.* 2018, *20*(19), 4443–4447.
18. Q. Cao, D. E. Crawford, C. Shi, S. L. James, *Angew. Chem. Int. Ed.* 2020, *59*(11), 4478–4483.
19. M. A. E. A. A. A. El-Remaily, A. M. M. Soliman, O. M. Elhady, *ACS Omega* 2020, *5*(11), 6194–6198.
20. J. Andersen, H. Starbuck, T. Current, S. Martin, J. Mack, *Green Chem.* 2021, *23*, 8501–8509.
21. K. J. Ardila-Fierro, L. Vugrin, I. Halasz, A. Palčić, J. G. Hernández, *Chemistry-Methods* 2022, e202200035.
22. Y. Yeboue, B. Gallard, N. Le Moigne, M. Jean, F. Lamaty, J. Martinez, T.-X. Métro, *ACS Sustain. Chem. Eng.* 2018, *6*(12), 16001–16004.
23. C. K. Prier, D. A. Rankic, D. W. C. MacMillan, *Chem. Rev.* 2013, *113*(7), 5322–5363.
24. C. Zhu, N. W. J. Ang, T. H. Meyer, Y. Qiu, L. Ackermann, *ACS Cent. Sci.* 2021, *7*(3), 415–431.
25. B. M. Sharma, R. S. Atapalkar, A. A. Kulkarni, *Green Chem.* 2019, *21*(20), 5639–5646.
26. R. R. A. Bolt, S. E. Raby-Buck, K. Ingram, J. A. Leitch, D. L. Browne, *Angew. Chem. Int. Ed.* 2022, e202210508.
27. S. D. Roughley, A. M. Jordan, *J. Med. Chem.* 2011, *54*(10), 3451–3479.
28. D. E. Crawford, A. Porcheddu, A. S. McCalmont, F. Delogu, S. L. James, E. Colacino, *ACS Sustain. Chem. Eng.* 2020, *8*(32), 12230–12238.
29. O. Galant, G. Cerfeda, A. S. McCalmont, S. L. James, A. Porcheddu, F. Delogu, D. E. Crawford, E. Colacino, S. Spatari, *ACS Sustain. Chem. Eng.* 2022, *10*, 1430–1439.
30. K. N. Ganesh, D. Zhang, S. J. Miller, K. Rossen, P. J. Chirik, M. C. Kozlowski, J. B. Zimmerman, B. W. Brooks, P. E. Savage, D. T. Allen, A. M. Voutchkova-Kostal, *Environ. Sci. Technol. Lett.* 2021, *8*(7), 487–491.

Part 4

Reduced Solvent Sustainable Technologies

12

Large-Scale Flow Chemistry

Nicole Neyt, Jaimee Jugmohan, Wessel Bonnet, Jenny-Lee Panayides and Darren Riley

CONTENTS

12.1 Introduction: Background and Driving Forces in Flow Chemistry

Historically, chemical manufacturing has to a large extent been reliant on so-called batch manufacturing wherein discrete batches of a chemical commodity are produced in a batch-by-batch manner.[1] This approach to manufacturing has been well refined over the last 150 years; however, due to its intrinsic nature, it suffers from several issues such as challenging dimensional upscaling, typical reliance on a large reactor footprint, safety concerns related to processing large volumes of material in single batches and reactor downtimes between runs.[1]

To overcome these and other shortcomings, focus has shifted in the past two decades to continuous manufacturing with a particular emphasis on the utilisation of flow chemistry.[1–3] Flow chemistry involves the continuous reacting/processing of chemical materials in flowing streams wherein reagents are combined at a mixing junction prior to undergoing reaction in a heat- and pressure-controlled reactor. Thereafter, the reagent stream will typically undergo an in-line quench prior to passage through a back-pressure regulator and collection as depicted in Figure 12.1. The modular nature of flow chemistry reactors further allows for the inclusion of optional in-line downstream processing (DSP) and purification modules as well as hyphenating the system with in-line, on-line and at-line analytical instrumentation.[2, 3]

Flow technology, although utilised to a limited degree by several pharmaceutical manufacturers like Merck Sharp Dohme, Pfizer and Novartis, is still regarded to be in its relative infancy, with most research and development on small scale occurring in the last 15 years. The technology at this stage is disruptive in nature and as a result buy-in from established pharmaceutical manufacturing sectors has been slow, despite the advantages afforded by the technology.[1]

Several of the advantages afforded by flow are:

- Solvents can be super-heated as flow systems are closed and can typically be pressurised through the use of appropriate back-pressure regulators.[3]
- Mixing can be achieved in seconds and with improved efficiency at the small scales used in flow chemistry.[3]
- Heat transfer is intensified as there is a significantly lower thermal mass of the fluid than the thermal mass of the system, which makes controlling the temperature of the media both faster and easier.[3]

DOI: 10.1201/9781003178187-16

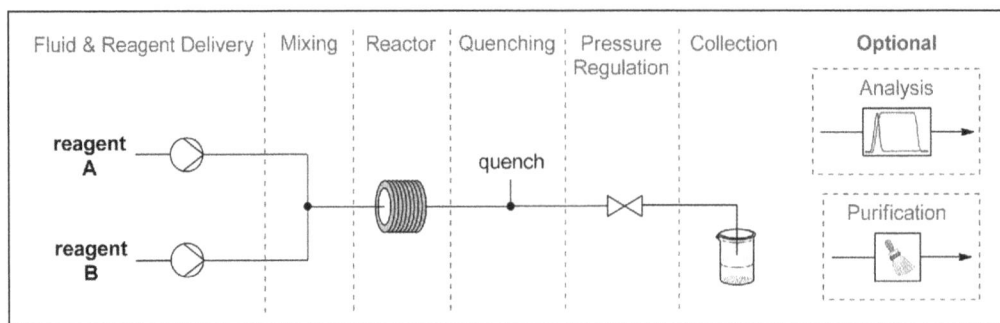

FIGURE 12.1   Schematic of a generic two-reagent flow setup.

- Safety is greatly improved as reactions are occurring on a micro-scale and the system operates under steady-state conditions.[3, 4]

- Flow systems can be more easily automated than batch processes, and it is possible through the use of external detectors to automate the investigation of several different parameters such as stoichiometry, temperature and residence time with little or no intervention.[3, 5]

- Multi-step reactions can be arranged in a continuous sequence, which can be beneficial if intermediates are unstable or hazardous, as they are reacted on-the-fly in small quantities.[1–3]

- Position along the flowing stream and reaction time point are directly related, which means that it is possible to arrange the system such that further reagents can be introduced into the flowing reaction stream at precisely the time point in the reaction that is desired.[3]

- Improved utility of solvents due to higher reagent concentrations and/or recycle loops as well as reagent usage facilitated by improved mixing and heating.[3, 6]

- In-line downstream processing and purification are possible.[2]

- Reactions that involve reagents containing dissolved gases are easily handled.[2, 3]

- Multi-phasic reactions can be performed in a straightforward way, with high reproducibility over a range of scales and conditions.[2]

- Scale-up of processes can be achieved with comparably less process development work.[3]

12.2 The Effect of Upscaling on Fluid Dynamics under Flow Conditions

The performance of a chemical reactor is underpinned by fluid dynamics, and an appreciation thereof is critical to understand the challenges associated with upscaling, regardless of whether batch or flow approaches are employed. Several theoretical models exist to describe the behaviour of fluids, and knowledge of the type of fluid flow is important in determining parameters related to mixing time (t_m), residence time distribution (RTD), mass transfer (MT) and heat transfer (HT).

The two most described types of fluid movement are laminar and turbulent flow. In the case of laminar flow, the fluid is viewed as moving in parallel sheets as depicted in Figure 12.2a.[7] When the fluid starts moving, the layers travel forward uniformly and there is little to no movement of fluid perpendicular to the direction of flow, with mixing only occurring due to kinetic diffusion. In the case of a tubular reactor, the layers in contact with the sides of the channel experience more friction and as a result move slower than those located towards the centre of the tube leading to dispersion effects.

As laminar flow increases in speed, it eventually transitions into the turbulent flow where molecules in the fluid line start travelling in random directions as they move down the length of the channel; see Figure 12.2b.[7] The flow nature can be described mathematically by Reynold's number (*Re*),[8] which for a circular pipe is defined as:

$$Re = \rho Q D_H / \mu A \tag{12.1}$$

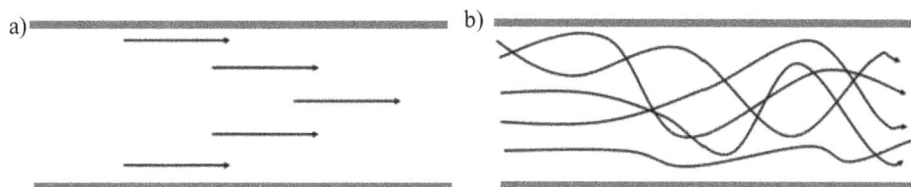

FIGURE 12.2 Generalised diagram of laminar flow (a) and turbulent flow (b) in a tube.

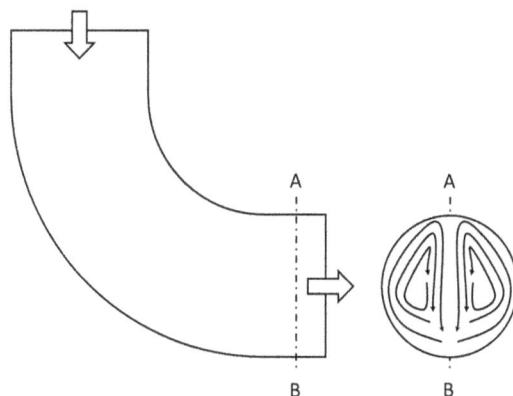

FIGURE 12.3 Secondary fluid flow forming Dean vortices.

where ρ is the fluid density, Q is the volumetric flow rate, D_H is the inside diameter of the pipe, μ is the dynamic viscosity of the fluid and A is the cross-sectional area of the pipe.[8]

Laminar flow is predicted for Re numbers <2300,[9] and turbulent flow for numbers >4000. In between these values, transitional flow occurs with laminar flow dominating near the centre of the pipe and turbulent flow at the inner surface of the pipe where friction is higher.

Consideration of the equation and its implication on fluid behaviour indicates that "direct" upscaling of flow reactors, although possible, still has associated challenges. In practice, scaling a reactor's channel dimension while retaining a constant flow rate results in the residence time increasing and the Re number decreasing. As the Re number is inversely proportional to the cross-sectional area of channel A, which decreases faster than the D_H value, modest increases in D_H results in dramatic decreases in Re. As a result, the flow becomes laminar on scaling, negatively affecting the t_m, MT and HT properties.

To counter this loss and retain a fixed residence time, the volumetric flow rate must be increased accordingly; however, as with the relationship between D_H and Re, the cross-sectional area A decreases faster than the increase in the volumetric flow rate Q. As a result, if residence time is maintained, fluid flow begins to transition to laminar flow as D_H increases. The situation is further complicated by secondary flow which arises when a fluid in motion encounters a curved section of a flow path.[8] As a volume of fluid enters the curved section, centripetal forces act on the flowing particles changing their main direction of travel by creating a transverse pressure gradient, resulting in the fluid in the centre being swept towards the outer side of the bend and the fluid near the channel surface being swept towards the centre. This creates a pair of counter-rotating cells known as Dean vortices that increase radial mixing shown in Figure 12.3.[8]

Added to this, stagnation zones can also arise, which are localised portions of fluid in a pipe where highly turbulent swirls form as spinning vortices.[8] These can be problematic leading to higher levels of friction and increasing backpressure and dispersion. Under microfluidic conditions, as channel dimensions are small, one can expect uniform mixing if transitional flow is achieved; however, as channel dimensions increase, the Re number decreases and fluid flow transitions to laminar flow.

An appreciation of fluid dynamics highlights that translation from the bench to production scales under flow conditions, which typically involves scaling factors that can range from 100 to 1000 times,[10] needs careful assessment and selection of an appropriate upscaling strategy to ensure that the control advantages afforded by the technology are not lost. In selecting an appropriate strategy, one needs to assess the impact the approach will have on the (i) pressure drop, (ii) mixing time, (iii) residence time distribution, (iv) mass transfer and (v) heat transfer. In this chapter, we provide a high-level overview of these criteria that does not require an extensive understanding of mathematics. Readers wishing to further explore these concepts are referred to the comprehensive review by Noël et al., which highlights recent studies and provides an in-depth account of the mathematical derivations of these terms.[10]

i) Pressure Drop (PD)

PD refers to the difference in total pressure between two points in a fluid-carrying network which arises as a result of the types of obstructions, the restrictions and the friction experienced by the moving fluid.[8] PD in a tubular channel is expressed as:

$$\Delta P = C/Re^n \cdot \rho \, L/2D_H \cdot u^2 \qquad (12.2)$$

where the C/Re^n term represents a friction factor f which is linked to the reactor geometry and flow conditions, ρ is the fluid density, L is the path length, u is the average fluid velocity and D_H is the channel diameter.[11] Large PDs are undesirable as they increase operational costs with regards to pumping units and are also associated with increased safety risks.

ii) Mixing Time (t_m)

Mixing takes place at several levels (macro, meso and micro) and significantly affects reactor performance with regards to axial dispersion, HT and MT.[10, 12, 13] The influence of mixing is most commonly quantified through the mixing time (t_m) term which refers to the time taken to achieve complete homogenisation at a molecular level.[12] Ideally, the t_m should be at least 10× faster than the rate of reaction to avoid issues associated with slow mixing.[10, 12] Under laminar flow, t_m can be defined as:

$$t_m = A(D^2/D_m) \qquad (12.3)$$

where A is a factor depending on the channel geometry, D is the channel diameter and D_m is the molecular diffusion coefficient.[10, 12] The equation highlights that under laminar flow t_m is only dependent on molecular diffusion, the inefficiency of which can be appreciated when typical values of D_m are considered. D_m is in the range of 10^{-9} m².s⁻¹ for liquid phase reactors, resulting in mixing times of 1–100 s for channel diameters of 0.1–1 mm and 1–100 min for channel sizes of 1–10 mm.[10]

Under turbulent flow, t_m is defined as:

$$t_m = 17.24\upsilon^{0.5}\varepsilon^{-0.5} \qquad (12.4)$$

where υ is the kinematic fluid viscosity and ε is the energy dissipation rate. Under fully turbulent flow, mixing is dependent primarily on the engulfment rate of turbulent eddies and mixing times in the range of 0.1–10 ms are achievable.[10, 14] At this rate, even fast reactions are unlikely to suffer from mixing issues; however, fully turbulent flow is rarely achieved under microfluidic conditions. Transitional flow on the other hand which is more readily accessible has the t_m terms defined as:

$$t_m = C\upsilon^{0.5}\varepsilon^{-0.5}\ln(Sc) \qquad (12.5)$$

where C is a coefficient dependent on the type of mixer employed, υ is the kinematic fluid viscosity and ε is the energy dissipation rate.[15]

iii) **Residence Time Distribution (RTD)**

RTD is a measure of how long a molecule resides within a reactor. In flow reactors RTD is described by a one-dimensional axial dispersion model rationalised in terms of the Bodenstein number (B_o):

$$B_o = \mu L / D_{ax} \tag{12.6}$$

where μ is the volumetric flow rate, L is the path length and D_{ax} is an axial dispersion coefficient. [10, 12]
  As B_o tends to zero, the axial dispersion increases, conversely as the B_o number increases, axial dispersion decreases and at values >100 a plug-flow regime is generally assumed.[10] The equation highlights that as the L/D ratio increases, the reactor will tend to plug-flow. Plug-flow is generally preferred affording a situation where reagents are uniformly distributed axially across the flow stream in the so-called "reagent plug".
  Under a laminar flow regime, axial dispersion remains constant and provided the B_o number is above 100 such reactors can be regarded as plug-flow. Importantly, as the B_o number scales linearly with the average residence time under laminar flow, slow reactions which require longer residence times are less affected by axial dispersion. Conversely, under turbulent flow, axial dispersion will decrease rapidly and if the L/D ratio is greater than 50, axial dispersion can be ignored, that being said, as noted previously, turbulent flow is rarely achieved under microfluidic conditions. Under more achievable transitional flow, the B_o number decreases rapidly with an increasing *Re* number, and as such, plug-flow will not be realised unless the L/D ratio is significantly high. As a result, additional strategies are needed to reduce axial dispersion, particularly when dealing with fast reactions. Axial dispersion is commonly reduced by introducing passive mixing structures to promote radial mixing.[10] Alternatively, helical tubular reactors are commonly employed to generate Dean vortices which improve radial mixing.[10, 16]

iv) **Heat Transfer (HT)**

HT is a measure of a reactor's capacity to transfer heat generated by chemical processes away from the reactor with the goal of achieving isothermal operation.[10] Under microfluidic conditions, if the HT capacity is not sufficient, one will tend to generate an axial temperature profile leading to uneven heating across the reaction plug.[10] The heat flux of a system q is defined as:

$$q = U.a(T_c - T) \tag{12.7}$$

where U is the overall heat transfer coefficient, a is the specific area of the channel wall, T_c is the temperature of the cooling medium and T is the temperature of the reaction mixture.[10, 12] The overall heat transfer coefficient is dependent on all resistances encountered, including the resistance between the channel wall and fluid stream, the channel wall itself and the cooling medium.[10, 12] The convective heat transfer in a fluid channel is described by the Nusselt number (*Nu*) defined as:

$$Nu = hD/k \tag{12.8}$$

where h is the heat transfer coefficient, D is the channel diameter and k is the heat conductivity of the fluid.[10, 12, 17] *Nu* can be further correlated to:

$$Nu = 1.615(Re.Pr.(D/L))^{0.33} \tag{12.9}$$

$$Nu = 0.023.Re^{0.8}.Pr^{0.33}(1+(D/L)^{0.67}) \tag{12.10}$$

for laminar and turbulent flow, respectively, where *Pr* is the Prandtl number which is determined by heat specificity, dynamic viscosity and thermal conductivity.[17] Although turbulent flow is not readily achieved, similar HT can be realised under transitional flow through the incorporation of passive mixing elements and/or through the use of helical reactors. Under these conditions, Nu can be defined as:

$$Nu = 0.912.Re^{0.82}.Pr^{0.25} \qquad (12.11)^{18,\ 19}$$

Equations (12.13) and (12.14) highlight a dependence on a similar Re term and show that similar HT can be achieved under transitional flow through the introduction of passive mixing elements.[10, 19] HT is more complex under biphasic conditions with toroidal circulation patterns leading to increased heat transfer. Ultimately, changes in a reactor's HT capacity on scaling can often be circumvented through the employment of appropriate reactor cooling systems.

v) **Mass Transfer (MT)**

Flow chemistry is ideally suited to handle multiphasic reactions, particularly liquid–liquid and gas–liquid as the small reactor dimensions afford a large well-controlled interfacial contact area.[10, 20, 21] In the case of mass transfer the specific mass flux m is defined as:

$$m = k_L a.(c^*-c) \qquad (12.12)$$

where k_L is the mass transfer coefficient and a in the interfacial surface area. Critically the interfacial surface area a is dependent on both the wetted surface area of the microreactor and the observed flow patterns (e.g. droplet, slug, annular).[22, 23] The mass transfer coefficient is usually expressed as the overall mass transfer coefficient $k_L a$ which incorporates the interfacial surface area and the following correlations:

$$K_L a = [0.44C_a^{-0.1}Re^{-0.65}(D/L)^{-0.1}]/\tau \qquad (12.13)$$

$$K_L a = 0.084Re_{gas}^{0.213}Re_{liq}^{0.937}Sc_{liq}^{0.5} (D_m/D^2) \qquad (12.14)$$

are obtained for liquid–liquid[24] and gas–liquid[25] mass transfer. As with heat transfer, the use of passive mixing elements is commonly employed to improve the mass transfer.[10]

12.3 Strategies for Upscaling Tubular Reactors under Continuous Flow Conditions

In general, a microfluidic or flow reactor is comprised of tubular channels, typically in the form of mixing chips or tubular (coil) reactors, the volume of which can be described as:

$$V = \pi/4.N.L.D^2 \qquad (12.15)$$

where N is the number of channels/flow paths, L is the reactor length and D is the channel diameter.[10] Reactors of different sizes can be related by a scale-up factor S defined as:

$$S = V_{large}/V_{small} = N_{large}L_{large}D^2_{large}/N_{small}L_{small}D^2_{small} \qquad (12.16)$$

where $S = S_L \times S_N \times S_D^2$.

Under flow conditions, general strategies employed for upscaling include (i) numbering up by increasing the number of channels (N), (ii) scaling out by increasing the flow path length (L) and (iii) sizing up by increasing the channel diameter (D) or a combination thereof.

i) *Numbering-Up*

Numbering-up increases the overall reactor volume (V) and as a result throughput through the use of multiple identical reactors arranged in a parallel fashion; this arrangement can involve the use of discrete microreactors or, alternatively, can involve numbering-up of the channels within a single reactor. The approach has the advantage that the mixing time, mass and heat transfer properties and residence time distribution remain identical across all the reactors and match those of a single reactor.[10] The approach in most of its iterations, however, suffers from fluid distribution issues wherein either reactor imperfections or the presence of precipitates or gas bubbles in the reaction mixture results in an uneven fluid distribution across the reactors, ultimately resulting in a loss of control and consistency.

The simplest approach to numbering-up, albeit the most costly, involves external numbering-up wherein each microreactor has its own pumping unit, temperature controller and downstream processing equipment. In such instances scale-up is direct, the approach does not suffer from fluid distribution issues and the approach adds redundancy in that a single reactor can be replaced without influencing or stopping the other reactors.

Unfortunately, the cost and complexity associated with external numbering-up are often undesirable on scale and, alternatively, an internal numbering-up approach is more commonly adopted wherein a single pumping unit is employed to feed the reagent stream into a distributing unit wherein the reagent stream is split into several lines which then feed into discrete reactors or channels. The approach, although more economical, suffers from the aforementioned fluid distribution issues requiring careful design.

Several external and internal flow distribution unit designs have been successfully employed, arguably the most encountered being bifurcation and chamber designs. In the case of external bifurcation distributors, the flow stream is split into two streams using an appropriate T- or Y-piece mixer, and each stream can thereafter be further split into 2^n channels prior to passage through a micro-reactor; see Figure 12.4a.[26] Under the same premise, commercial symmetrical manifolds and tree-type modular manifolds have been designed.[10] An alternative related design is a split and recombine design wherein a flow stream is split into N parallel channels with $N(N-1)/2$ bifurcation points but has $(N-2).(N-1)/2$ junction (recombine) points as shown in Figure 12.4b.[27, 28] The approach affords excellent fluid partitioning and through the use of sensors can detect blockages by comparing the ratio of flow rates between two bifurcation/recombine points.[28] Critically the approach requires that the geometric parameters of the various channels be carefully designed to allow uniform operation.

In the case of chamber designs, shown in Figure 12.5a the distributing unit and an analogous collecting unit take the form of a void chamber which in the case of the inlet unit needs to first be filled up before exiting into a series of parallel channels whose entry points are perpendicular to the flow direction.[10, 29–32] The approach is notable in that it reduces the flow resistance that occurs across the distribution unit channels; however, as flow rates increase, flow distribution begins to favour the centre of the chamber. Higher flow rates are made accessible through the insertion of a baffle into the distribution chamber to maintain laminar flow and create a fluidic damper prior to entry into the reactor channels.[10]

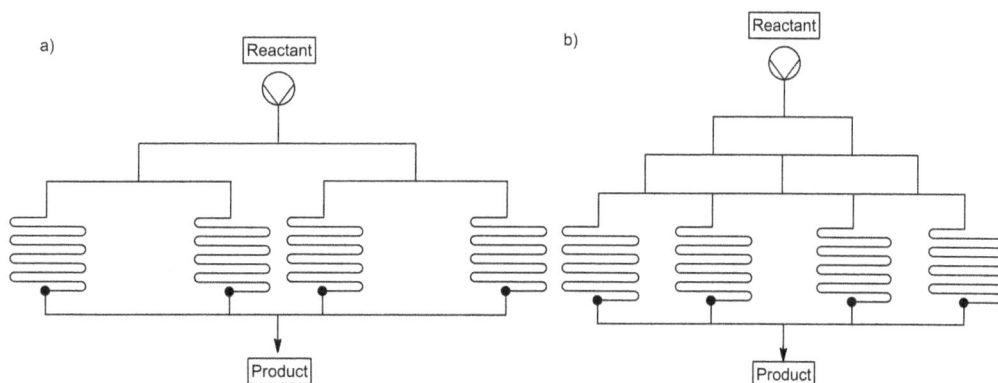

FIGURE 12.4 Numbering-up using a bifurcation approach (a) and split and recombine design (b).

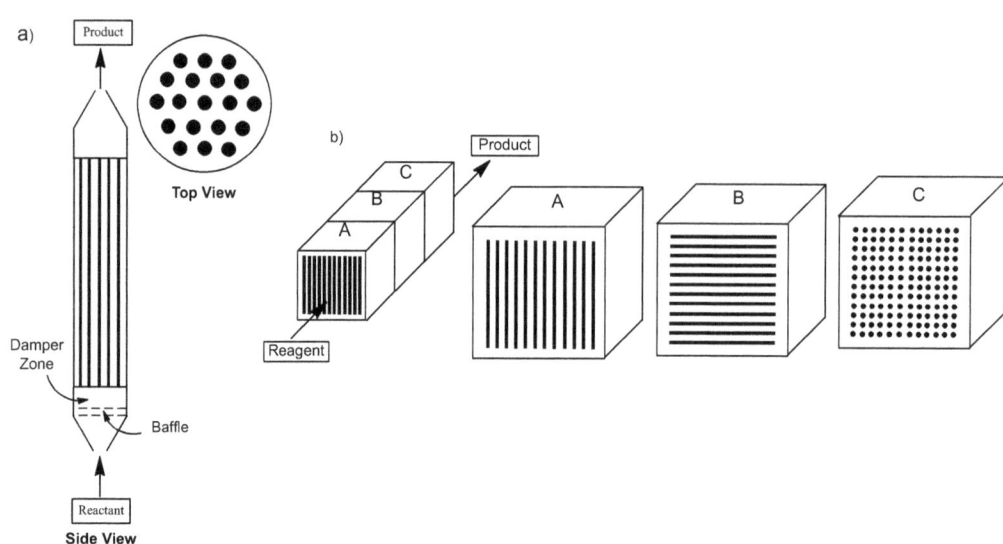

FIGURE 12.5   Numbering-up using a chamber design (a) and a thick-wall screen (b).

Thick-walled screen distributors, shown in Figure 12.5b, operate in a similar fashion to chamber distributors, allowing the distribution of flow streams into a grid of parallel channels.[33] The distributor operates by passing the flow stream through a cone diffusor prior to passage through a screen with parallel slits (A) wherein the slit thickness is significantly narrower than the thickness of the screen. Thereafter the flow stream is passed through an analogous screen (B) which has slits orientated at 90° relative to the slits in the first screen prior to passage into the reactor channels (C).

ii) *Scaling-Out*

Scaling-out involves one-dimensional scaling in the axial direction, wherein the flow path length L is increased while maintaining the channel diameter, typically achieved by connecting several reactors in series.[10, 17, 34] The approach allows one to increase productivity as an increased fluid flow rate is required to maintain a desired residence time. Under such instances, the number of channels and the channel diameters remain constant and as such, the reactor's scale-up factor can be represented as

$$S = S_L = V_{large}/V_{small} = N_{large}L_{large}D^2_{large}/ N_{snall}L_{small}D^2_{small}, \text{ where } S_D \text{ and } S_N = 1 \qquad (12.17)$$

If the residence time remains constant, S can further be expressed in terms of a ratio of Reynold's numbers (Re):

$$S_L = S_{Re} = Re_{large}/Re_{small} = u_{large}D_{Hlarge}/ u_{small}D_{Hsmall} \qquad (12.18)$$

where u = fluid velocity and D_H = channel diameter.[6]

Under these conditions, both the fluid velocity and the Reynolds number scale proportionately to S which facilitates a decrease in mixing time ($S^{-1.4}$).[10] As a result, desired transitional fluid flow is readily achieved, and the reactor tends to plug-flow on scaling, with the axial dispersion scaling as $S^{0.68}$ (assuming the use of a helical tubular reactor).[10] In terms of HT the Nu number decreases slightly ($S^{0.82}$), again assuming the use of a helical reactor.[10] MT for liquid–liquid systems decreases ($S^{-0.68}$) with increasing reactor volume, whereas gas–liquid systems increases ($S^{1.15}$).[10] In the case of liquid–liquid systems although the surface area of the reactor increases proportionally, the interfacial contact area between the phases decreases, resulting in an overall negative scaling of the MT coefficient. Conversely gas–liquid MT scales positively as the density difference between the phases is larger, resulting in the generation of stronger secondary flows.[10]

If Equation (12.2) which describes the PD is considered under conditions of laminar, transitional and turbulent flow n can be approximated to equal 1, 0.25 and 0, respectively,[35] leading to scale-up factors of S^2, $S^{2.75}$ and S^3.[10] As can be inferred, the strategy suffers from a steep increase in pressure drop which is limiting from an operational cost point of view.

iii) **Sizing-Up**

Sizing-up involves the scaling of channel geometries. The approach allows a higher throughput with a reduced pressure drop penalty when compared to scaling-out; however, the approach results in a steep drop-off of mass and heat transfer. By means of example, a simple one-dimensional increase in the channel diameter results in the reactor volume and the fluid flow rate scaling by S^2, and the Re number scaling by S^{-2}. The resulting PD scale-up factors are three orders of magnitude lower than those obtained from scaling-out.[10] In this instance, although the pressure drop penalty is less severe, the drop-off in Re number results in the reactor transitioning to laminar flow on scaling if a fixed residence time is maintained.

To overcome this issue, a two-dimensional scaling approach called geometric similarity screening (GSS) is employed, wherein the channel diameter and length are scaled proportionally.[35, 36] Under this approach, the reactor volume scales as S^3 and the Re number as $S^{0.67}$ affording comparable mixing to that achieved under scaling-out.[10] Overall, the PD is not as severe as with one-dimensional scaling-out, and the drop-off in Re number is not as severe as with one-dimensional sizing-up of the channel diameter. The approach readily allows access to transitional flow; however, it suffers as the surface area-to-volume ratio scales negatively, which, assuming the use of a helical tubular reactor, results in both the overall HT coefficient and the axial dispersion scaling negatively.[10] Overall MT for liquid–liquid systems scales negatively but to a lesser degree than under scaling-out.

Alternatively, constant pressure drop scaling (CPDS) can be employed, wherein the PD value is kept constant on scaling.[35, 36] Under laminar conditions, this approach is identical to geometric similarity scaling; however, under transitional flow conditions, the channel diameter scales to $S^{0.41}$ and the channel length to $S^{0.19}$, leading to a scaling of the channel diameter to channel length ratio of $S^{0.22}$. Under these conditions, the Re number scales as $S^{0.59}$ and t_m remains constant. As with GSS, the surface area-to-volume ratio scales negatively. As a result, the approach affords similar scaling of the Re number and the surface area to volume as geometric similarity scaling, but the scaling ratio leads to (i) increased axial dispersion with the B_o scaling negatively for transitional (helical tubular reactor) flow and (ii) decreased overall heat transfer. That being said, the approach is attractive as it keeps the t_m constant and maintains good MT while avoiding undesirable pressure drop.

As can be seen, there is no one–fits-all solution to upscaling under flow conditions. Instead, one needs to consider the relative advantages and disadvantages of the system in hand and the productivity rate required. In practice, most upscaling involves a combination of approaches with numbering-up combined with geometric similarity or constant pressure drop scaling. The reactor is scaled up physically to maximise productivity within cost and safety restrictions after which, numbering-up can increase the productivity further. The various scaling factors under a translational flow regime are summarised in Table 12.1.

TABLE 12.1

Summary of Scale-Up Factors for a Transitional Flow Regime under Different Scale-Up Strategies.

Scaling Factor	Numbering-Up	Scaling-Out	GSS	CPDS
Surface area-to-volume ratio	1	1	$S^{-0.33}$	$S^{-0.41}$
Reynolds number	1	S	$S^{0.67}$	$S^{0.59}$
Pressure drop	1	$S^{2.75}$	$S^{0.5}$	1
Mixing time (t_m)	1	$S^{-1.4}$	$S^{-1.4}$	1
Residence time distribution		$S^{0.68}$	$S^{-0.21}$	$S^{-0.41}$
Total heat transfer	1	$S^{0.82}$	$S^{-0.13}$	$S^{-0.33}$
Mass transfer (liquid–liquid)	1	$S^{-0.68}$	$S^{-0.47}$	$S^{-0.41}$

Adapted from Noël et al.[10]

12.4 Strategies for Upscaling Packed-Bed Reactors under Continuous Flow Conditions

Packed-bed reactors (PBRs) are a type of tubular reactor wherein a catalyst or solid/solid-supported matrix is packed, with reactants flowing over the solid matrix. In this instance, the contact surface area between the reactants and the solid matrix is imperative to determining the rate and extent of the reaction involved.

Particle shape and size require special consideration when scaling up a PBR system. Ideally, the design should maximise the exposed surface area of the solid without the particle size being so small as to cause blockages or dead zones within the reactor system.[37]

Two main scale-up methods can be employed for PBR systems: horizontal and vertical scale-up. Horizontal scale-up is considered the easiest of the two options, with the reactor system being multiplied in parallel to increase the throughput of the system. This approach is identical to the numbering-up explicated in the previous section.

Vertical scale-up, on the other hand, involves dimensional upsizing of the PBR to increase the volume of fluid that can be processed per unit time. This, however, changes the kinetic data, residence times, contact area and flow regimes throughout the PBR, making it hard to adapt and predict on scale-up.

One of the major concerns in the scale-up of tubular PBRs is the formation of hotspots along the length of the reactor.[38] Common in exothermic reactions, these temperature changes could pose a safety concern, promote side reactions or potentially denature the solid matrix that is present within the reactor system.

12.5 Downstream Processing under Continuous Flow Conditions

While the reactor is considered the core of any chemical process, careful attention needs to be placed on downstream processing (DSP) techniques, focusing on the separation and refinement of the components within the respective process streams. In-line DSP and purification under flow conditions remain challenging today with only liquid–liquid or liquid–gas separations being easily implemented on a scale.

i) *In-line Extraction*

One of the primary concerns raised with the separation of immiscible liquid–liquid or liquid–gas mixtures is the length of time required to achieve the gradient needed for separation – typically utilising gravity as the driving force. In a batch process, it is likely that the reactor will have to remain unagitated for lengthy periods of time to allow sufficient separation to occur. In response to this, Zaiput Flow Technologies has noted that at minimum, the use of batch reactors for separation processes can reduce productivity by approximately half. Zaiput has addressed the need for in-line extractions through the development of a separation system, that utilises the wetting properties of a liquid, in conjunction with a porous membrane system.[39] Wetting properties are defined as the ability of liquid molecules to be attracted to the surface of a solid. When the pores of the membrane have been filled, a pressure difference is created within the unit operation, allowing for the wettable liquid to be transported through the membrane with minimal impact on the secondary substance in that mixture. It is important to note that this technology makes use of the wettability of the liquid, and it can be utilised to separate liquids of similar densities. Additionally, this system can be utilised in processes with space constraints as a storage tank is not required and as with the case of Zaiput's commercial Z-FOX system, separation can occur continuously outside the reactor (see Figure 12.6 with operating capacities of up to 200 litres per hour). When employing an in-line extraction, the desired purity or recovery rate may not be achieved on an initial pass. This is especially prominent in cases with low partition coefficients. As a result, there is a need for multiple separation unit operations in series. Zaiput has also developed a multistage extraction unit operation which employs a countercurrent flow scheme.[40]

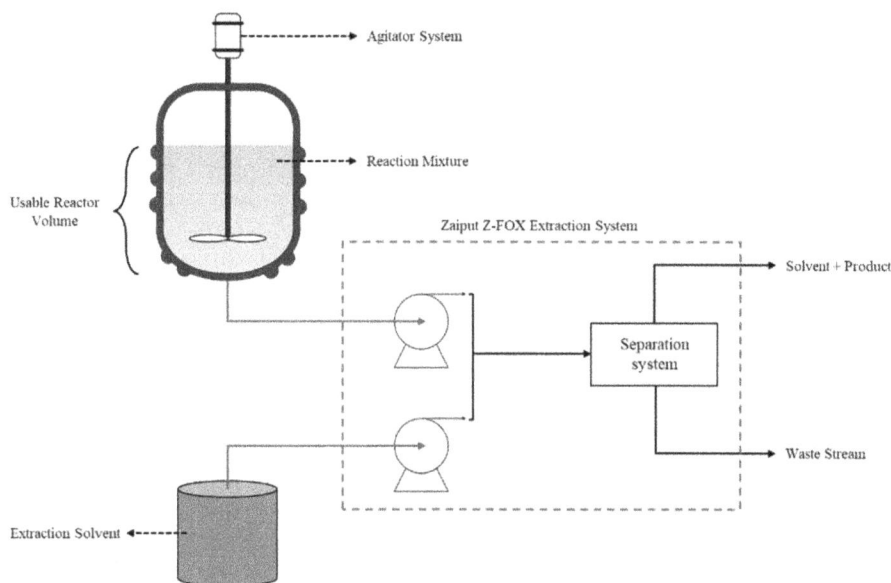

FIGURE 12.6 Zaiput Z-FOX extraction system.

The unit minimises the amount of extractant required with the raffinate being fed back into a previous stage in the extraction unit operation, with an organic solvent flowing in the reverse direction. Employing this approach, a higher solvent ratio increases efficiency, and a direct proportionality relationship exists between the number of stages present and the extraction efficiency. However, it is important to note that an increase in the number of stages will subsequently result in a higher cost for the unit operation, thereby serving as a limiting factor during process design. The reader is further directed to the review by Neyt and Riley for a detailed overview of DSP and purification techniques under flow conditions including approaches to crystallisation, distillation and chromatography.[2]

12.6 Sustainable Manufacturing in Flow

To conclude this chapter, we have included several examples of upscaling under continuous flow conditions highlighting firstly how cGMP production has been achieved and secondly how the scale-up of processes that are challenging, if not impossible, under batch conditions has been achieved under flow conditions.

i) *cGMP Production and Large-Scale API Manufacturing under Flow with In-line Analytics*

Today, the development of processes for robust commercialisation and the identification of scale-up strategy parameters remains a challenge. That being said, the number of patents published relating to continuous flow manufacturing has increased significantly over the past 15 years, with data suggesting an increase in the number of examples reported on a multi-kilogram scale. Furthermore, the complexity of flow chemistry reactions has shifted from the early years of single-step reactions to fully integrated multi-step processes, including advanced downstream processing and in-line analytics.[9, 41, 42] This drive towards implementing continuous manufacturing has seen major players advocating for its implementation. The Chinese State Administration of Work Safety, the US Food and Drug Administration (FDA) and the European Medicines Agency (EMA) have all strongly advocated for the implementation of flow technologies for manufacturing. This has resulted in several flow processes being developed under cGMP production capability: in 2015 lumacaftor and ivacaftor manufactured by Vertex, a combination medication used for treating cystic fibrosis, was approved along with the antiretroviral medication darunavir manufactured by Johnson and Johnson in 2016.[43] Eli Lilly and company has also developed several processes including a prexasertib monolactate monohydrate synthesis in 2017,[42] merestinib in 2019 [44, 45] and a novel aminopyridyloxypyrazole compound as a potential anticancer agent in 2020.[46] Ajinomoto Omnichem has also recently announced the acquisition of several continuous flow technologies specifically for the manufacturing of APIs in its pilot plant in Wetteren, Belgium.[43]

The World Health Organization (WHO) has established detailed guidelines for good manufacturing practices, and many countries have adopted these guidelines for their own requirements for GMP readiness. This mainly involves the assurance that products are consistently produced and controlled according to the quality standards (QC) appropriate to their intended use and as required by the marketing authorisation, clinical trial authorisation or product specification.

The Lilly research laboratories have demonstrated a kilogram-scale synthesis of prexasertib monolactate monohydrate **1**, a four-step synthesis, under cGMP flow conditions. The synthesis was successfully translated to flow as detailed in Figure 12.7, producing 24 kilograms of the targeted compound **1** suitable for use in human clinical trials. The flow synthesis consisted of eight continuous unit operations used to produce **1** at 3 kilograms per day. The operational units consisted of small continuous reactors, extractors, evaporators, crystallisers and filters in laboratory fume hoods. Under batch conditions, the first stage required excess hazardous hydrazine, the use of which could be mitigated utilising flow conditions.[42] Furthermore, online PAT (Process Analytical Technology) was heavily used during the development of the processes, it proved crucial during the cGMP production to demonstrate process health and process control. The PAT used during production included online HPLC, refractive index measurement, as well as temperature, pressure, and mass flow rates monitored by the distributed control system (DCS). DCS are computer-based software packages used to control hardware from a centralised human–machine interface. The online HPLC data generated from each step was not intended for cGMP decision-making; this was determined using manual, offline testing. The online PAT was intended to inform regarding minor adjustments to the process parameters to keep the process performance on target.

Further successes have been highlighted by Novartis/MIT demonstrating the flow process for aliskiren hemifumarate showing a production rate of 100 g/h (Figure 12.8). The synthesis reported consists of four stages, starting with the chemical intermediate **2**, which is melted and pumped into a tubular reactor 1 held at a 100°C, the material is then mixed with amide **3** (10 equiv) and acid catalyst **4** (1 equiv) to form compound **5**. The downstream processing includes a straightforward in-line extraction by adding water and ethyl acetate and passing this through a membrane extractor. The separated organic phase is fed into a two-phase crystallisation chamber to perform a mixed suspension, mixed product removal technique (MSMPR), thereby allowing the separation of catalyst **4** and compound **5**. From the crystalliser, the material is fed through a filter and passed onto the second stage whereby the concentration is accurately controlled.

The second reaction namely an acid-catalysed removal of the Boc protecting group is carried out in a tubular reactor (R2), where concentrated HCl is mixed with the slurry of **5** in ethyl acetate. The reaction is rapidly quenched online with NaOH (25 wt%), which is followed by a series of downstream processes to obtain compound **6**, which is reacted further with fumaric acid to form the fumarate salt **7** in an MSMPR reactor. All reactor components consist of stainless steel.[47]

Lastly MIT's pharmacy on-demand project has highlighted the synthesis of four APIs namely diphenhydramine hydrochloride, lidocaine hydrochloride, diazepam and fluoxetine hydrochloride that meet US Pharmacopoeia standards. These ingredients were synthesised in a modular reactor no bigger than the size of a refrigerator [1.0 metre (width) × 0.7 metres (length) × 1.8 metres (height)].

ii) *Photochemical Scale-Up*

Continuous flow photochemistry has probably seen the most growth over the past decade in both academia and industry. Flow technology has proven advantageous in photochemistry, showing excellence in safety, practicality and scale-up, which has been severely limited in the past due to the limitations of photon penetration in a fluid as described by the Beer–Lambert Law. The combination of continuous flow

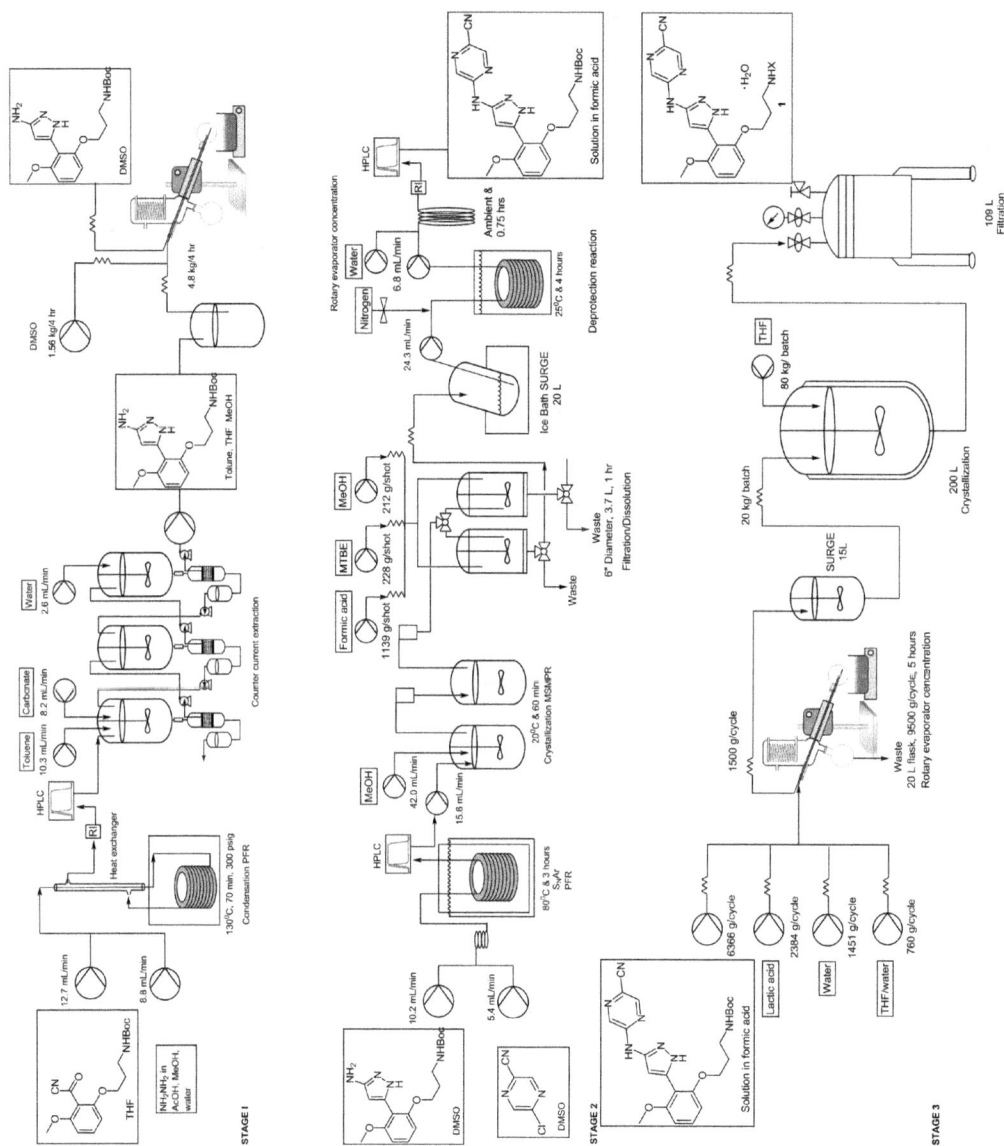

FIGURE 12.7   The synthesis of prexasertib monolactate monohydrate (adapted from Cole 2017).

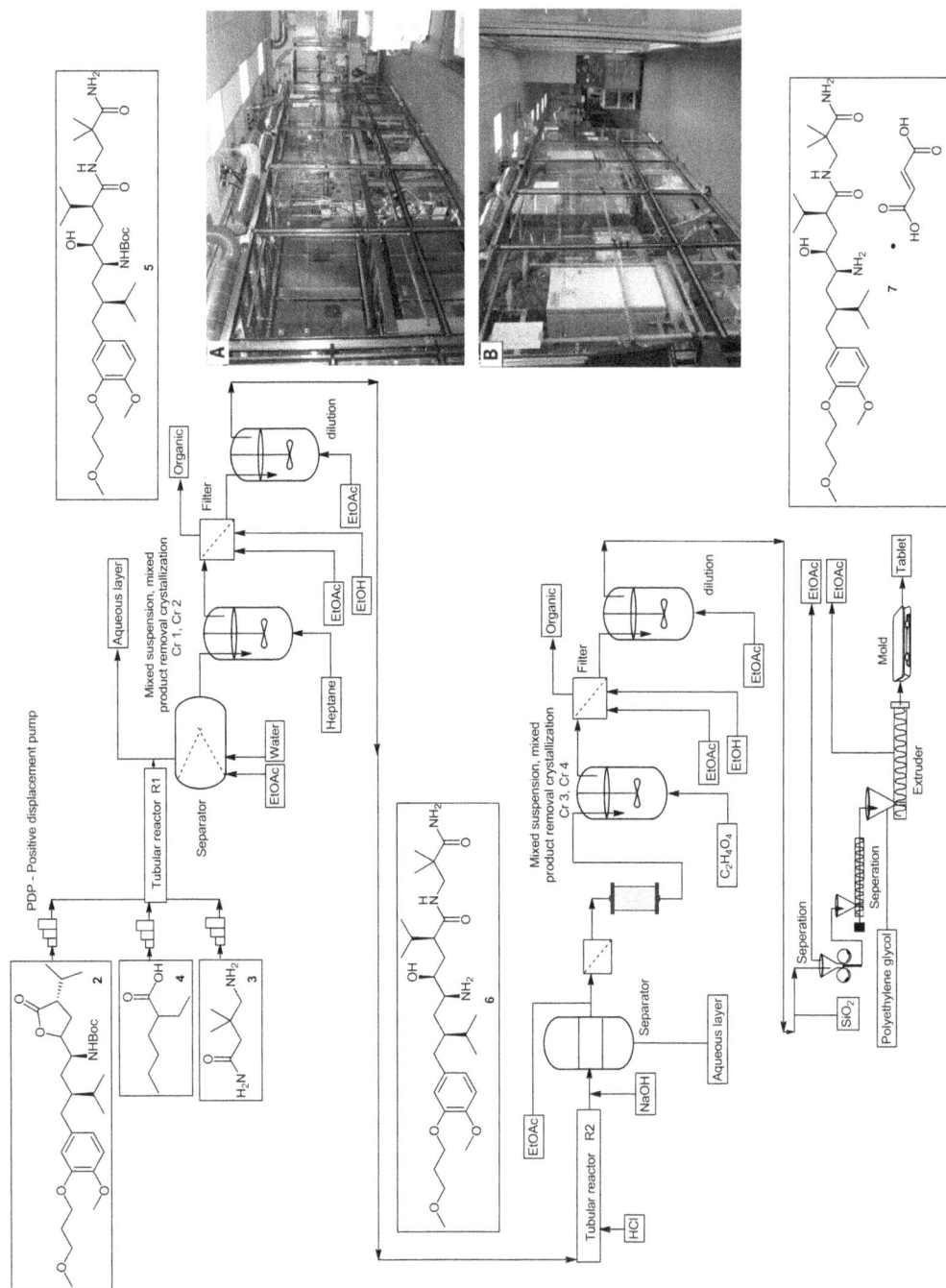

FIGURE 12.8 The synthesis of Aliskrien Hemifumarate: (A) front half of the process and (B) back half of the process (adapted from Trout 2017).

reactors along with energy-efficient LED lamps has seen the development of innovative reactor designs providing a powerful means to increase both the practicality and productivity of flow photochemistry.

More than a dozen papers have been published over the past 5 years depicting pilot scale (>100 g/day) and production scale (>1 kg/day) photochemical flow examples. These include a photochemical Negishi reaction, photochemical decarboxylation of trifluoroacetic acid anhydride, photooxidation of α-terpinene, photochemical [2 + 2]-cycloaddition reactions of ethylene, the photooxidation of citronellol, photocatalysed aryl amination reactions and a benzylic bromination.

Following the examples of photochemical reactions completed on a pilot scale, Alcazar and co-workers have on two separate occasions reported on a flow photochemical Negishi reaction demonstrating the formation of **8** on multi-gram scale (Figure 12.9). The process proceeds firstly by forming an organozinc reagent, facilitated by passing a primary halogen **9** through a packed-bed reactor containing zinc metal. Thereafter, the material proceeds to a second photoreactor facilitating the photochemical Negishi reaction, coupling **10** with an aryl halogen **11** in the presence of a nickel catalyst. The reaction was initially screened using a commercially available Vapourtec Photochem reactor (10 mL), with which they achieved a throughput of 800 mg/h. Translating this reaction to pilot scale, they opted to use the Corning G1 reactor (40 mL) increasing their productivity to 5.6 g/h, and using the larger reactor, they managed to increase their throughput seven-fold to 134 g/day.[48, 49] Kappe and co-workers have also demonstrated the utility of the Corning G1 photoreactor and optimised the reaction between a citraconic anhydride and ethylene gas to produce the corresponding cyclobutene. Initially the group focused on the importance of said chosen photosensitiser to facilitate the reaction and found thioxanthone to be ideal for this particular transformation. Optimised results were achieved with a 5.3 min residence time in a commercially available plate reactor of 2.77 mL, producing 2.1 g/h of the cyclobutene product. Further increasing the plate volume by adding five additional plates in series afforded a productivity rate of 10 g/h at a residence time of 5.5 min. This reaction proves hugely advantageous against the traditional batch counter-part due to the associated safety risks of using ethylene gas on large scale.[50]

The Stephenson group has published a multitude of photochemical trifluoromethylation reactions, demonstrating the synthesis of **12** on a kilogram scale. Initially the group developed a photochemical method to incorporate the trifluoromethyl groups into various arenes and heteroarenes, which involved the photochemical decarboxylation of trifluoroacetic anhydride facilitated by pyridine *N*-oxide in the presence of a photocatalyst. Various substrates were tested, and products were obtained with good yields. They then went on to demonstrate a scaled-up flow reaction based on a coil reactor design. The design consisted of PFA tubing coiled around a glass beaker housing several LED lights. The total volume of the reactor shown in Figure 12.10 was approximately 150 mL. The reactor was cooled by passing water through the cavities of the coiled reactor maintaining a temperature of 45 °C. Running the reactor continuously for 48 h provided a throughput of 0.95 kg of product **13** equating to 20 g/h of material produced.[51, 52]

Several groups have followed in these footsteps engineering efficient photochemical reactors with good results. George and co-workers have developed their own photochemical reactor based on a simple rotary evaporator. The photoreactor named the PhotoVap generates a thin film upon rotation for efficient radiation. The photoreactor comprises a rotary evaporator with banks of LED lights mounted above the surface of the rotating evaporation flask; see Figure 12.11. They demonstrated three different

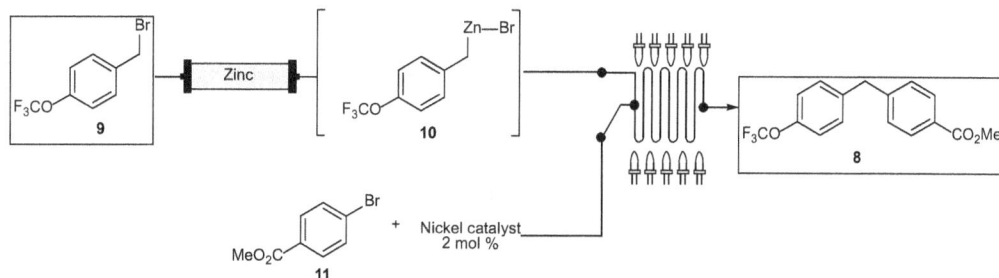

FIGURE 12.9 Photochemical Negishi reaction (adapted from Alcazar 2019).

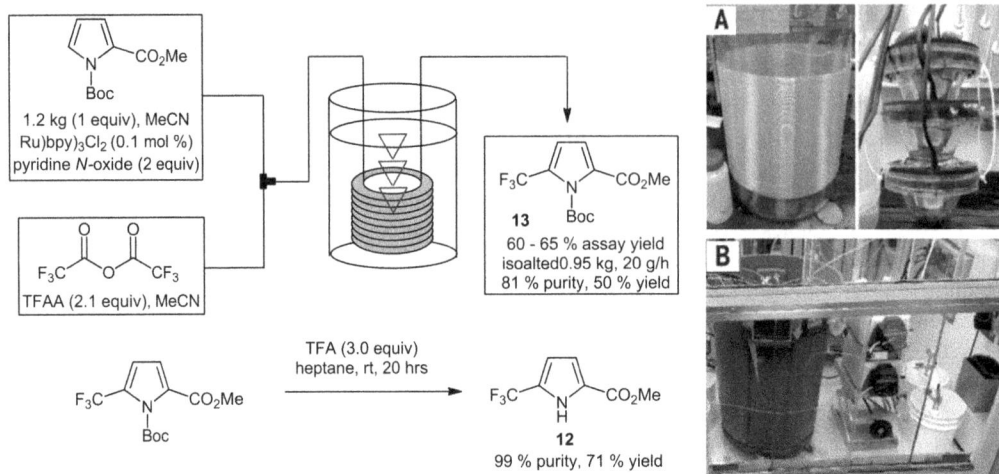

FIGURE 12.10 Photochemical trifluoromethylation reaction: (A) coiled reactor and (B) reactor immersed in water and cooled with a glycol chiller (adapted from Stephenson 2016).

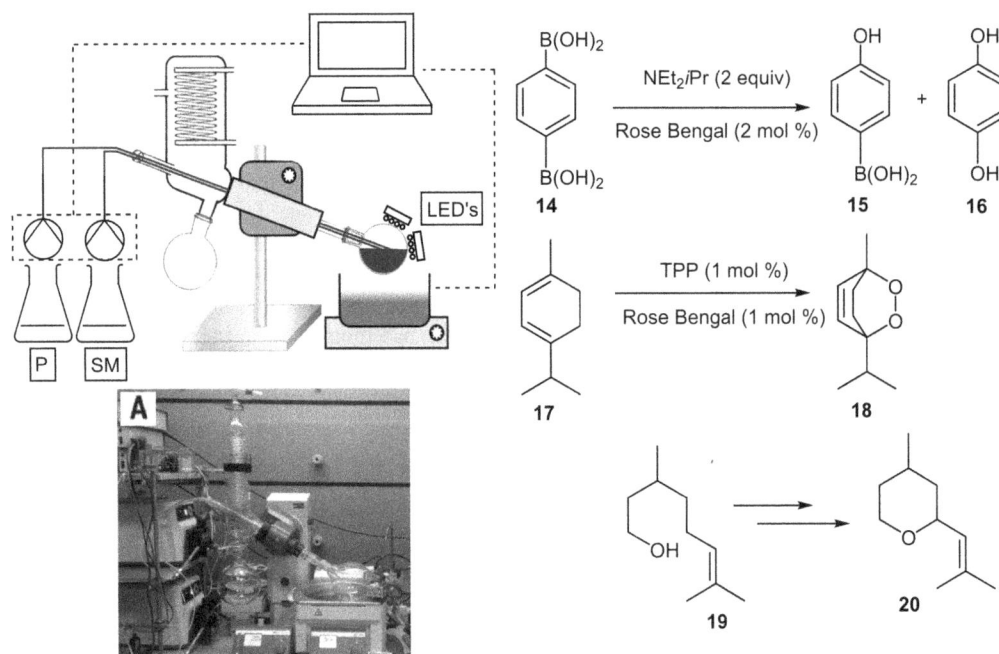

FIGURE 12.11 The PhotoVap (adapted from George 2016).

photooxidation reactions and investigated several parameters which included the flask size, volume, rotation speed and light intensity. The first reaction investigated was a photo desymmetrisation of benzene-1,4-diboronic acid **14** to the mono- and dihydroxylated products **15** and **16**. A 10 min residence time gave nearly quantitative conversion of the starting material **14**. The second reaction investigated was the photooxidation of α-terpinene **17** to form ascaridole **18** with the best example affording a yield of 87% in 60 seconds.

The final reaction investigated was the photooxidation of citronellol **19** to form hydroperoxides which are further reduced to form the corresponding diols followed by the dehydrative cyclisation to yield rose oxide **20**. The reaction proceeded to 99% conversion in 420 seconds. All of the reactions investigated

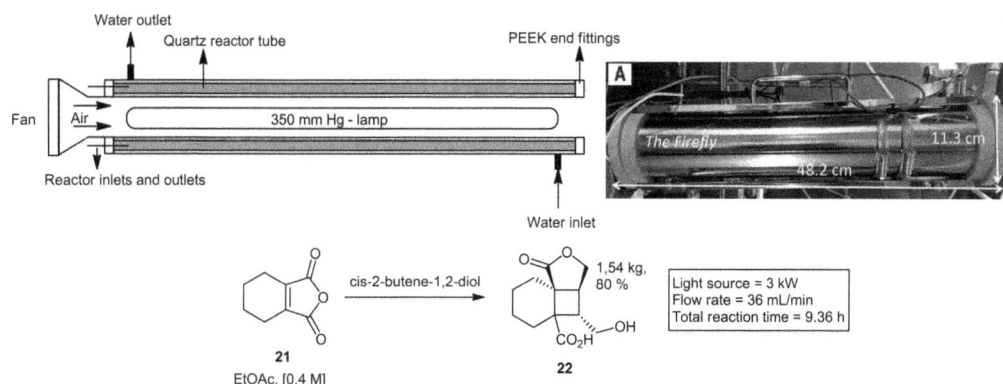

FIGURE 12.12 The firefly (adapted from Elliot 2016).

employed Rose Bengal as a photosensitiser, facilitating the generation of singlet oxygen required for efficient oxidation of the substrates. Using a 3 L flask, a productivity of 10 g/h could be achieved.[53]

Further examples of similar designs include the "Firefly" engineered by Elliot and co-workers. It consists of a tubular photo-reactor. Several organic transformations were investigated including the "Cookson's dione" synthesis. The 120 mL reactor consisted of a succession of quartz tubes connected together in series and arranged axially around a variable power mercury lamp. The reactor was housed in a jacketed metal frame to protect against any emitting UV light which further contributed to more radiating light from the jacket. The entire system was cooled with a fan displacing stagnant air within the reactor. They showed that 3,4,5,6-tetrahydrophthalic anhydride (THPA) **21** and cis-2-butene-1,4-diol can undergo an efficient [2 + 2] cycloaddition-lactonisation sequence to afford tricyclic lactone **22** upon direct UV irradiation with a staggering production of 4 kg in 24 h. They also demonstrated that doubling the concentration and subsequent power supply doubled the production rate to 8 kg/day. They further demonstrated a similar cyclisation with a productions rate of kilograms per day (Figure 12.12).[54]

The rotor stator spinning disc reactor (RS-SDR) is another alternate design by the Noël group showing the photocatalysed gas–liquid oxidation of α-terpinene to the drug ascaridole with Rose Bengal as a photosensitiser. Throughputs of over 1 kg/day (270 mmol/h) were achieved under visible light irradiation. The performance of the reactor is correlated to various process parameters such as rotation speed, liquid flow rate and catalyst concentration, among others. The reactor consisted of a rotor housed within a 64 mL reactor, a quartz window allowed for the efficient radiation of the material with an irradiated volume of 27 mL. White LEDs (120 W) were utilised to irradiate the mixture. An acute observation in the conversion and selectivity was seen with an increase from 37% to 97% and 75% to 90%, respectively, with an increase of rotation speed from 100 to 2000 RPM.[55]

Lasers have not traditionally been used to catalyse photochemical transformations due to potential safety hazards associated with the use thereof under large-scale conditions; however, Harper and co-workers have designed a photochemical reactor with the use of a fibre optic laser system to promote organic reactions with more selectivity and reactivity. The reactor was fitted with a laser beam focused onto a continuously stirred tank reactor (100 mL). The reactor operates in a semi-batch fashion, whereby materials are continuously fed into a stirred tank and removed as the reaction progresses. They investigated a C(sp²)-N(sp³) cross-coupling reaction between 1-bromo-4-(trifluoromethyl)benzene **23** and pyrrolidine **24** to form **25**, shown in Figure 12.13. Throughput of 1.2 kg/day of **25** was achieved.[56]

Scale-up in flow photochemistry has also been envisioned with a numbering-up approach, and Su and co-workers have developed a numbering-up strategy for the convenient scale-up of photochemical reactions, with minimal optimisations required from laboratory scale to pilot scale. The 8-capillary system was designed to allow gas–liquid photoreactions to occur in parallel. The aerobic oxidation of thiols **26** to disulfides **27** was used as a model reaction. The yield obtained for the numbering-up strategy was comparable to the yield obtained in a single device shown in Figure 12.14.

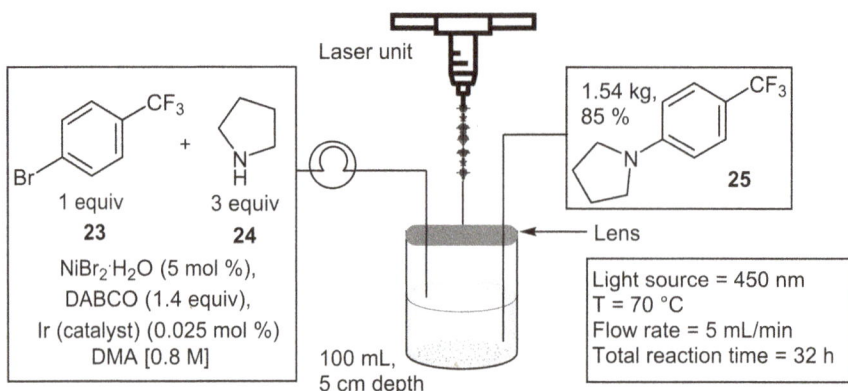

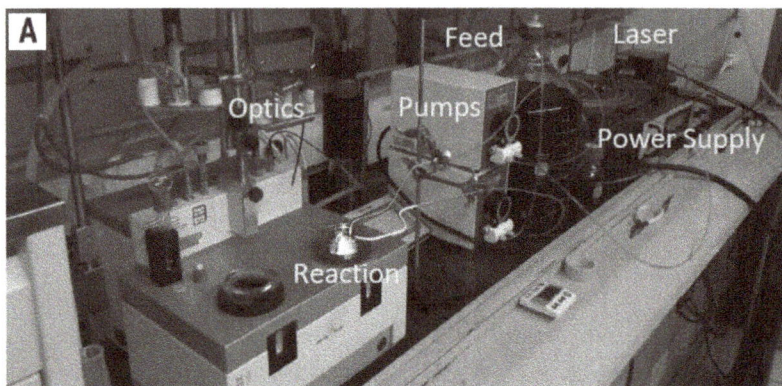

FIGURE 12.13   Laser-mediated photoreactor diagram: (A) photographic depiction of the laser-mediated flow assembly (adapted from Harper 2020).

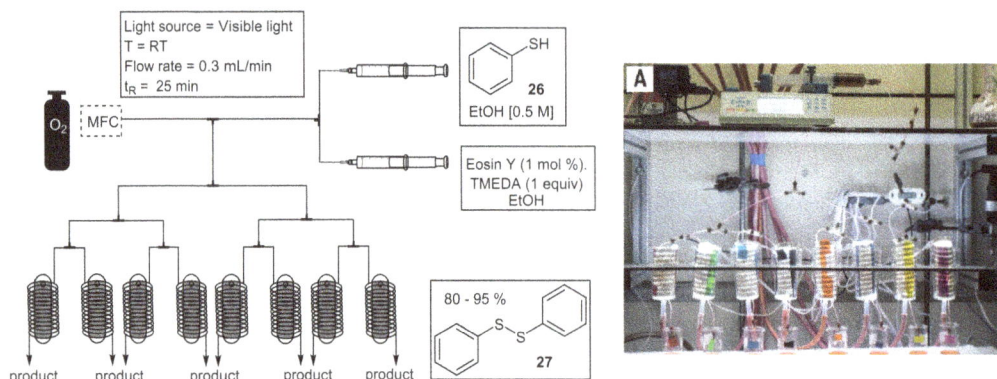

FIGURE 12.14   Photochemical scale-up under flow conditions: (A) photograph of the 8-capillary system (Adapted from Su 2016).

iii)   *Electrochemical Scale-Up*

Flow chemistry has improved the overall efficiency and reproducibility of electrochemical reactions. The technology has brought together chemists and engineers showing the scale-up potential of flow electrochemistry and the importance of small interelectrode gaps on pilot and industrial electrochemical processes. Classical scale-up of electrochemical reactions do not follow a dimensional scaling strategy as the cell voltage and energy costs would increase dramatically as the interelectrode gap increases.

FIGURE 12.15 Scalable Birch reduction under flow conditions: (A) 10 g scale, (B) 40 g scale and (C) 100 g scale with $250 total assembly cost.

Almost all electrochemical scale-up examples follow a numbering-up approach.[57] Peters and co-workers reported a Li-ion electrochemical Birch reduction in flow scalable to 100 g. They used a sacrificial anode (magnesium or aluminium), combined with a cheap, nontoxic and water-soluble proton source (dimethylurea), and a protectant: [tris(pyrrolidino)phosphoramide], all can allow for the multigram-scale synthesis of the key pharmaceutical building block **29** from **30** (Figure 12.15).[58]

iv) *Biphasic Reaction Scale-Up*

Most synthetic gases (CO, CO_2, O_2, H_2, C_2H_2, C_2H_4, CH_2O, F_2, O_3, Cl_2, CH_2N_2 and NH_3) have been utilised and employed using flow technologies.[59] Several gas modules and devices used to introduce and enhance gases in flow reactors have been developed, and many are now also available commercially. Such devices include tube-in-tube reactors and segmented flow reactors perfect for large-scale gaseous chemistry in flow. Hydrogenation reactions using heterogenous and homogenous catalysts have been well studied under flow conditions, and several review articles have been published describing these.[50, 60, 61] Commercially, Thales Nano has developed several flow reactors one of which is the well-known H-cube reactor specifically designed for the generation and introduction of hydrogen into flow.[62] Another commercially available reactor able to facilitate heterogenous and homogenous gaseous reactions is the FlowCAT system from the HEL group. It typically finds use in hydrogenation reactions, oxidations and carbonylations to name a few. Scaling is possible from 4 to 38 mL in reactor volume.

The Ley group has reported large-scale hydrogenation of ethyl nicotinate utilising a HEL FlowCAT trickle bed reactor shown in Figure 12.16. Several methods were investigated employing two different catalyst cartridges Pd/Al_2O_3 and Rh/Al_2O_3, and optimised results were achieved with a 12 mL reactor cartridge containing 13 g Pd/Al_2O_3. A 1.0 M stock solution of ethyl nicotinate was prepared and introduced into the FlowCAT at 7 mL/min along with 0.6 mL/min H_2 resulting in a throughput of 64.5 g/day of reduced ethyl nicotinate. Utilising a different cartridge packed with 4 g of Rh/Al_2O_3 and a stock solution concentration of 3.0 M ethyl nicotinate equating to a flow rate of 3 mL/min, they managed to obtain a throughput of 81.6 g/h.[63]

Some steps towards scaled use of ozone have already been conducted. Roth and co-workers utilising a Corning G1 reactor have demonstrated the ozonolysis β-pinene **31** to nopinone **32** producing 16 g of product per hour (Figure 12.17). The original report on this work showed a productivity rate of 0.24 g/h, and following this paper, the productivity was increased by 67-fold, showing a successful scale-up of the formation of nopinone utilising flow. Optimal results showed that 0.1 M β-pinene **31** and an ozone concentration of 6.95 (wt%) performed best equating to 16.2 g/h productivity and a 12 s residence time.[64]

In another example, The Lonza Group a Swiss company specialising in biotechnology demonstrated a tonne-scale ozonolysis reaction of chrysanthemic acid **32**. The continuous flow process boasted a production rate capable of processing over 0.5 tonnes of product per day. They further showed a scaled-down bench reactor, which is an exact scale-down of the plant system and uses a gas–liquid loop reactor equipped with a micro-Venturi injector. The loop reactor system consisted of four main parts, an injector for ozone, a residence time module, a gas–liquid separation unit, a pump and an ozone analyser (Figure 12.18).[65]

Several more examples utilising ozone in CSTRs and film shear flow reactors in a continuous fashion have been highlighted by many groups, producing in some cases up to 285 g/h of material.[66–68]

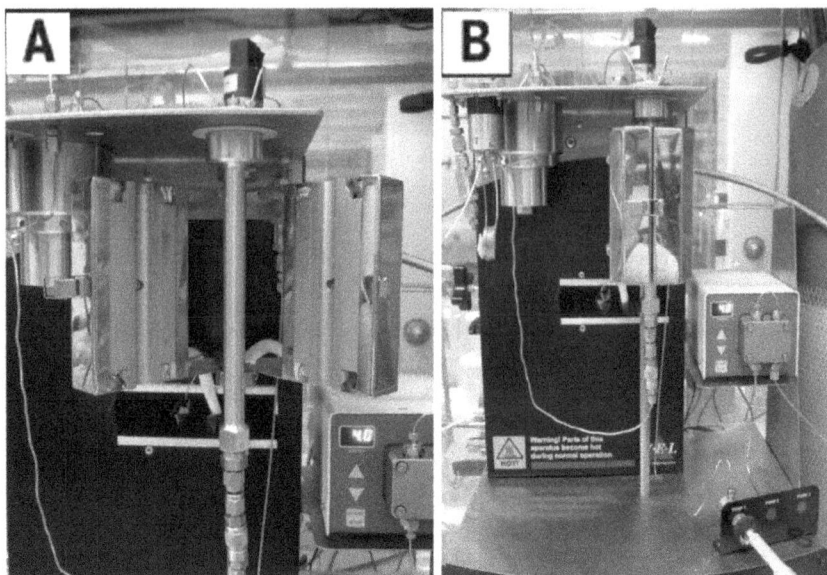

FIGURE 12.16   Large-scale hydrogenation of ethyl nicotinate utilising a HEL FlowCAT trickle bed reactor: (A) column reactor and (B) entire system (adapted from Ley 2014)

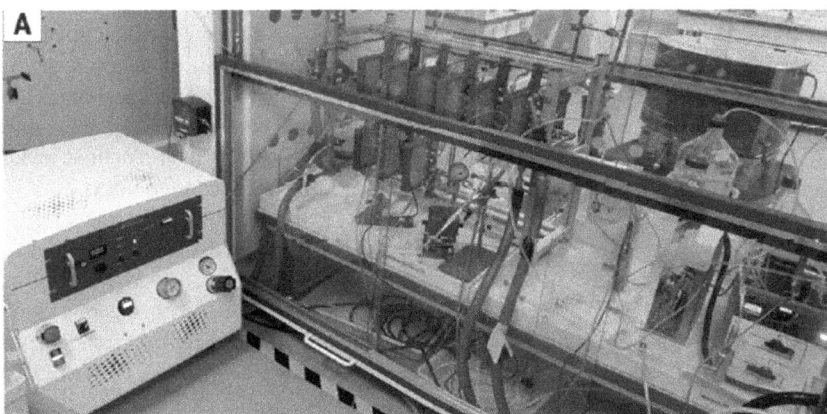

FIGURE 12.17   Corning G1 reactor demonstrating the ozonolysis of β-pinene to nopinone

v) Scale-Up of Hazardous Chemistries

The use of flow chemistry has unlocked the safe scale-up of many processes deemed too hazardous or which posed significant risks under normal batch conditions. The Naber group along with Merck has developed a flow process towards a chiral ketimine compound **33** via the addition of organolithium **34** and an aryl halide intermediate **35** at the pilot-plant scale. Critical conditions such as the flow rate, temperature and mixing performance were examined. Online process analytical technology was used to collect real-time data on batches to demonstrate steady-state operation over multiple hours and also to confirm the robustness of the continuous process against fouling. These experiments at the pilot-plant scale provided improved operating parameters that enabled the implementation of this chemistry for the commercial manufacture of verubecestat, currently being evaluated in Phase III trials for the treatment of Alzheimer's disease. Three different ko-flow mixers were tested, with optimised results showing the formation of compound **5** in more than >100 kg with a conversion of 88%; see Figure 12.19.[69]

FIGURE 12.18 The Lonza group downscaled loop reactor.

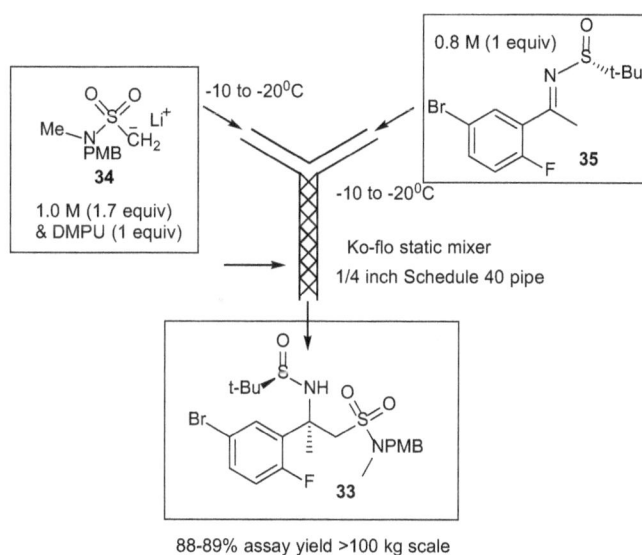

88-89% assay yield >100 kg scale

FIGURE 12.19 Koflo static mixer.

The next reaction that we have decided to highlight is a large-scale Matteson reaction, a key reaction towards the synthesis of a β-lactamase inhibitor vaborbactum. Two concept designs were visualised and implemented. The first concept was based on a loop reactor, whereby a small volume of material was circulated with a high pump flow rate. The temperature of the reaction was accurately controlled with a heat exchanger. The second concept consisted of a small volume CSTR, and temperature was controlled via a cooling jacket. With both concepts, a second larger CSTR was employed for downstream processing. A stream of the boronate adduct mixture exiting reactor R3 was combined with a stream of the zinc chloride stock solution in THF. The stable product is then directed to a buffer tank for further continuous downstream processing. The main points of difference between the two processes are that the first concept is a continuous loop reactor where the material is fed by means of a centrifugal pump and the temperature of the reactor is controlled though a heat exchanger. In the second concept the feed from R3 enters the CSTR along with the zinc chloride in THF. The temperature is controlled by a cooling jacket via a temperature sensor. The best results were achieved utilising the loop reactor, which was installed in the production facility and attached to the already existing tube reactor for the continuous Matteson

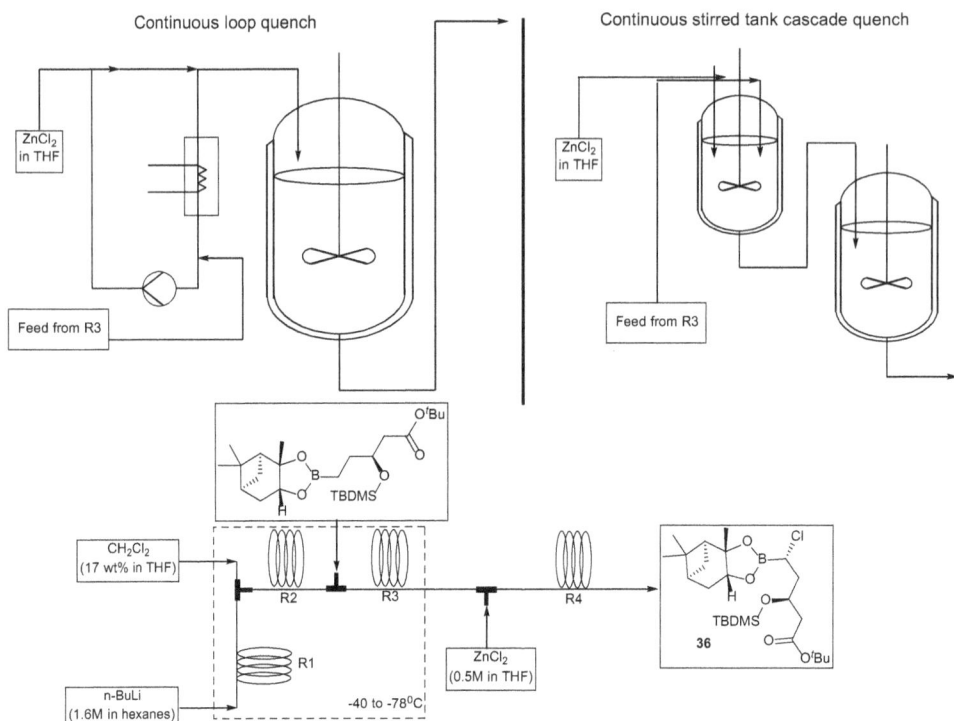

FIGURE 12.20   A continuous loop reactor and a continuous tank cascade reactor designed for a large-scale Matteson reaction.

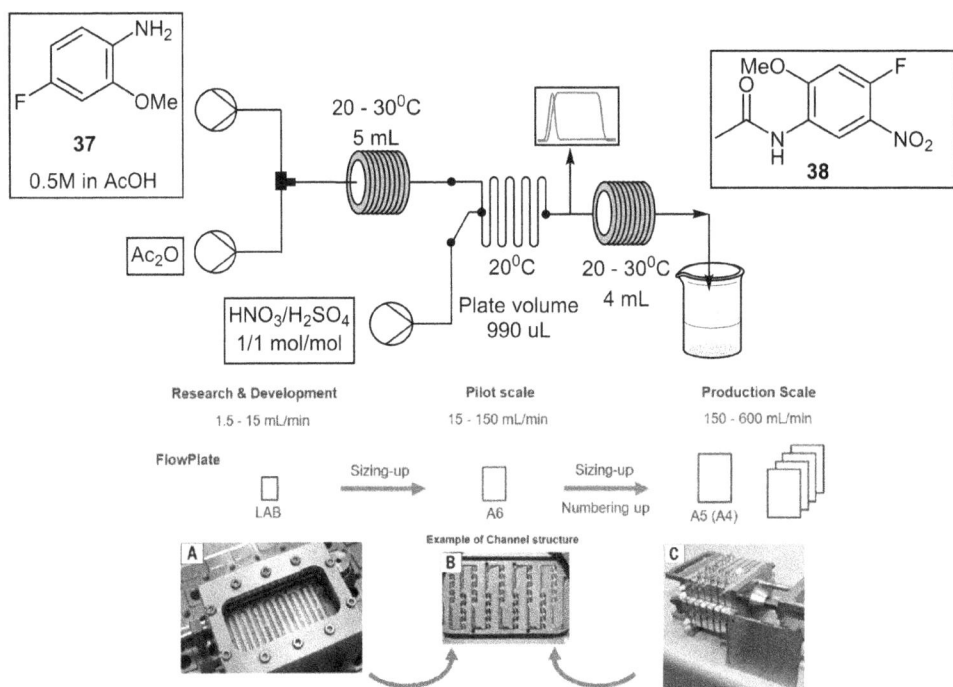

FIGURE 12.21   Flow nitration: (A) single flow plate used during R&D, (B) channel structure and (C) scale-up of flow plates.

reaction shown in Figure 12.20. Several hundreds of kilograms of intermediate **36** were produced with high chemical purity (>98% area by HPLC) and yield (97%).[70]

The development of a scalable telescoped continuous flow procedure for the acetylation and nitration of 4-fluoro-2-methoxyaniline **37** was described by Kappe and co-workers. Following the acetylation and nitration, the process proceeds via a batch deprotection affording 4-fluoro-2-methoxy-5-nitroaniline **38**, a key building block in the synthesis of osimertinib, a third-generation epidermal growth factor receptor tyrosine kinase inhibitor (EGFR-TKI). The hazards associated with the nitration of organic compounds make the scale-up under traditional batch chemistry difficult. The group investigated an acetic acid/aqueous nitric acid mixture as a predominantly kinetically controlled nitration regime and a water-free mixture of acetic acid, fuming nitric acid and fuming sulphuric acid (oleum) as a mass-transfer-limited nitration regime. The group utilised a modular microreactor platform to facilitate the nitration which was also monitored by an in-line temperature regime. They also identified that it was vital to protect the amine functionality through acetylation to avoid any side reactions. The process parameters and equipment configuration were optimised at the laboratory scale for acetylation and nitration to improve the product yield and purity. The two steps could be successfully telescoped, and the laboratory-scale flow process was operated for 80 min to afford the target molecule in 82% isolated yield over two steps, corresponding to a throughput of 25 mmol/h. The process was then transferred to an industrial partner for commercial implementation and scaled up by the use of higher flow rates, sizing-up and numbering-up

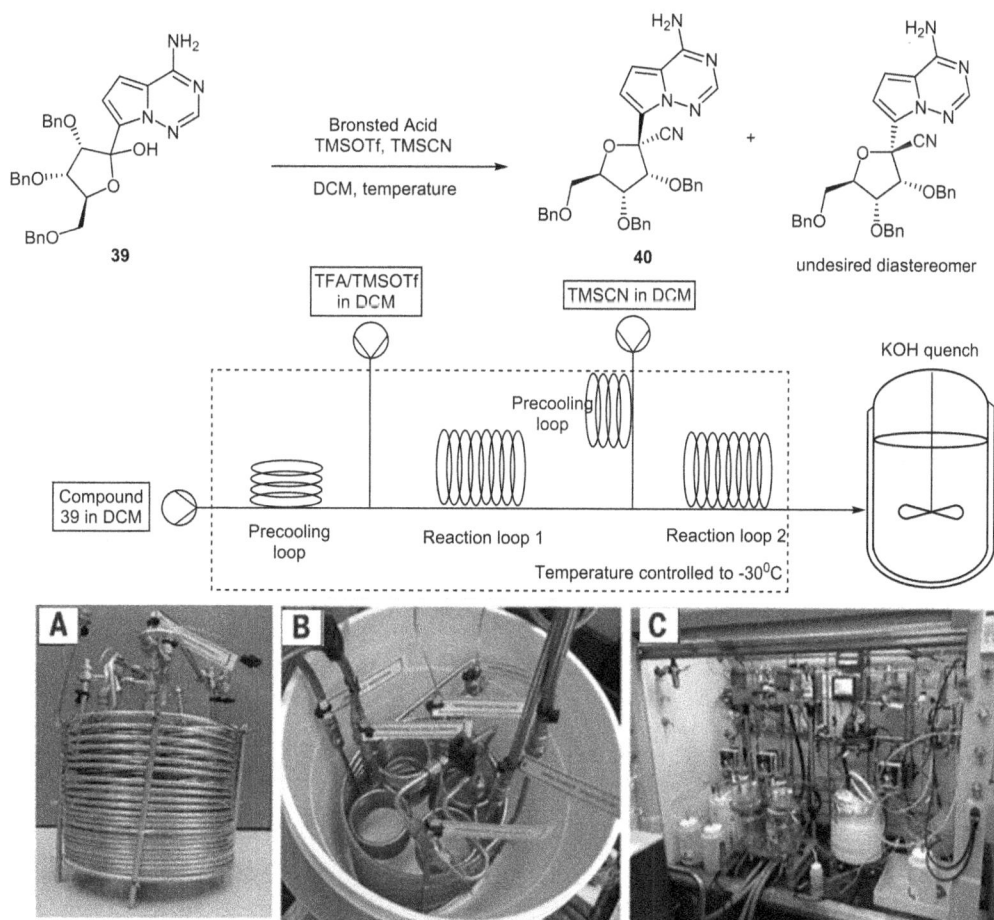

FIGURE 12.22 Cyanation of a glycoside compound: (A) and (B) stainless steel coiled flow reactor and (C) kilo-lab flow reactor assembly (adapted from Heumann 2020).

of the microreactor platform to pilot scale to afford the product in 83% isolated yield, corresponding to a throughput of 2 mol/h (0.46 kg/h) (Figure 12.21).[71]

Heuman and co-workers implemented flow technology describing the cyanation of a glycoside compound, a key intermediate in the synthesis of remdesivir. Cyanation chemistry at manufacturing scales using batch equipment is known to be challenging due to the hazardous nature of the reagents used, often requiring tight control of reaction parameters including cryogenic temperature ranges. The process was first investigated utilising a Vapourtec R series system consisting of four pumps used to introduce the four different reagents shown in Figure 12.22.

Compound **39** was sequentially combined with TFA, followed by TMSOTf and then TMSCN. Initial studies suggested that the use of Bronsted Acid was negated. They processed a 100 g of material **39** which equated to an 84% yield of compound **40**. Following the successful demonstration of the 100 g scale, they carried on to kilo-lab-scale preparation of compound **40** which was then successfully demonstrated in a 78% yield based on a 250 kg run.[72] Further large-scale reactions demonstrated on flow include an aldol reaction and a method developed for the synthesis of pyrroles, both processes able to demonstrate processing of large volumes of material.[73, 74]

REFERENCES

1. Riley, D. L.; Strydom, I.; Chikwamba, R.; Panayides, J. L., Landscape and opportunities for active pharmaceutical ingredient manufacturing in developing African economies. *React Chem Eng* 2019, *4*(3), 457–489.
2. Neyt, N. C.; Riley, D. L., Application of reactor engineering concepts in continuous flow chemistry: A review. *React Chem Eng* 2021, *6*(8), 1295–1326.
3. Plutschack, M. B.; Pieber, B.; Gilmore, K.; Seeberger, P. H., The hitchhiker's guide to flow chemistry(II). *Chem Rev* 2017, *117*(18), 11796–11893.
4. Gutmann, B.; Cantillo, D.; Kappe, C. O., Continuous-flow technology—A tool for the safe manufacturing of active pharmaceutical ingredients. *Angew Chem Int Edit Engl* 2015, *54*(23), 6688–6728.
5. Jas, G.; Kirschning, A., Continuous flow techniques in organic synthesis. *Chem Eur J* 2003, *9*(23), 5708–5723.
6. Sagandira, C. R.; Nqeketo, S.; Mhlana, K.; Sonti, T.; Gaqa, S.; Watts, P., Towards 4th industrial revolution efficient and sustainable continuous flow manufacturing of active pharmaceutical ingredients. *React Chem Eng* 2022, *7*(2), 214–244.
7. Fusi, L., An Introduction to fluid mechanics. *MAT Ser A* 2016, *21*, 0–111.
8. White, F. M., *Fluid Mechanics.* 6th edition. McGraw-Hill, 2010.
9. Cole, K. P.; Groh, J. M.; Johnson, M. D.; Burcham, C. L.; Campbell, B. M.; Diseroad, W. D.; Heller, M. R.; Howell, J. R.; Kallman, N. J.; Koenig, T. M.; May, S. A.; Miller, R. D.; Mitchell, D.; Myers, D. P.; Myers, S. S.; Phillips, J. L.; Polster, C. S.; White, T. D.; Cashman, J.; Hurley, D.; Moylan, R.; Sheehan, P.; Spencer, R. D.; Desmond, K.; Desmond, P.; Gowran, O., Kilogram-scale prexasertib monolactate monohydrate synthesis under continuous-flow cGMP conditions. *Science* 2017, *356*(6343), 1144–1150.
10. Dong, Z.; Wen, Z.; Fang, Z.; Kuhn, S.; Noel, T., Scale-up of micro- and milli-reactors: An overview of strategies, design principles and applications. *Chem Eng Sci: X* 2021, *10*, 100097.
11. Kockmann, N., Pressure loss and transfer rates in microstructured devices with chemical reactions. *Chem Eng Technol* 2008, *31*(8), 1188–1195.
12. Kockmann, N., Transport phenomena in micro process engineering. *Heat Mass Transf* 2008, 1–365.
13. Ghanem, A.; Lemenand, T.; Della Valle, D.; Peerhossaini, H., Static mixers: Mechanisms, applications, and characterization methods - A review. *Chem Eng Res Des* 2014, *92*(2), 205–228.
14. Zhang, F.; Marre, S.; Erriguible, A., Mixing intensification under turbulent conditions in a high pressure microreactor. *Chem Eng J* 2020, *382*.
15. Reichmann, F.; Vennemann, K.; Frede, T. A.; Kockmann, N., Mixing time scale determination in microchannels using reaction calorimetry. *Chem Ing Tech* 2019, *91*(5), 622–631.
16. Klutz, S.; Kurt, S. K.; Lobedann, M.; Kockmann, N., Narrow residence time distribution in tubular reactor concept for Reynolds number range of 10–100. *Chem Eng Res Des* 2015, *95*, 22–33.
17. Kockmann, N.; Roberge, D. M., Scale-up concept for modular microstructured reactors based on mixing, heat transfer, and reactor safety. *Chem Eng Process* 2011, *50*(10), 1017–1026.

18. Baehr, H. D.; Stephan, K. *Heat Mass Transf.* Springer, Berlin Heidelberg, 2006.
19. Kockmann, N., Transport phenomena and chemical reactions in modular microstructured devices. *Heat Transf Eng* 2017, *38*(14–15), 1316–1330.
20. Raimondi, N. D.; Prat, L.; Gourdon, C.; Tasselli, J., Experiments of mass transfer with liquid-liquid slug flow in square microchannels. *Chem Eng Sci* 2014, *105*, 169–178.
21. Sattari-Najafabadi, M.; Esfahany, M. N.; Wu, Z.; Sunden, B., Mass transfer between phases in micro-channels: A review. *Chem Eng Process* 2018, *127*, 213–237.
22. Sobieszuk, P.; Aubin, J.; Pohorecki, R., Hydrodynamics and mass transfer in gas-liquid flows in micro-reactors. *Chem Eng Technol* 2012, *35*(8), 1346–1358.
23. Kashid, M.; Kiwi-Minsker, L., Quantitative prediction of flow patterns in liquid-liquid flow in micro-capillaries. *Chem Eng Process* 2011, *50*(10), 972–978.
24. Kashid, M. N.; Gupta, A.; Renken, A.; Kiwi-Minsker, L., Numbering-up and mass transfer studies of liquid-liquid two-phase microstructured reactors. *Chem Eng J* 2010, *158*(2), 233–240.
25. Yue, J.; Chen, G. W.; Yuan, Q.; Luo, L. G.; Gonthier, Y., Hydrodynamics and mass transfer characteristics in gas-liquid flow through a rectangular microchannel. *Chem Eng Sci* 2007, *62*(7), 2096–2108.
26. Su, Y. H.; Kuijpers, K.; Hessel, V.; Noel, T., A convenient numbering-up strategy for the scale-up of gas-liquid photoredox catalysis in flow. *React Chem Eng* 2016, *1*(1), 73–81.
27. Nagaki, A.; Hirose, K.; Tonomura, O.; Taniguchi, S.; Taga, T.; Hasebe, S.; Ishizuka, N.; Yoshida, J., Design of a numbering-up system of monolithic microreactors and its application to synthesis of a key intermediate of valsartan. *Org Process Res Dev* 2016, *20*(3), 687–691.
28. Tanaka, Y.; Tonomura, O.; Isozaki, K.; Hasebe, S., Detection and diagnosis of blockage in parallelized microreactors. *Chem Eng J* 2011, *167*(2–3), 483–489.
29. Togashi, S.; Miyamoto, T.; Asano, Y.; Endo, Y., Yield improvement of chemical reactions by using a microreactor and development of a pilot plant using the numbering-up of microreactors. *J Chem Eng Jpn* 2009, *42*(7), 512–519.
30. Jang, S.; Vidyacharan, S.; Ramanjaneyulu, B. T.; Gyak, K. W.; Kim, D. P., Photocatalysis in a multi-capillary assembly microreactor: toward up-scaling the synthesis of 2H-indazoles as drug scaffolds. *React Chem Eng* 2019, *4*(8), 1466–1471.
31. Park, Y. J.; Yu, T.; Yim, S. J.; You, D.; Kim, D. P., A 3D-printed flow distributor with uniform flow rate control for multi-stacked microfluidic systems. *Lab Chip* 2018, *18*(8), 1250–1258.
32. Wei, M.; Fan, Y. L.; Luo, L. G.; Flamant, G., Design and optimization of baffled fluid distributor for realizing target flow distribution in a tubular solar receiver. *Energy* 2017, *136*, 126–134.
33. Rebrov, E. V.; Ekatpure, R. P.; de Croon, M. H. J. M.; Schouten, J. C., Design of a thick-walled screen for flow equalization in microstructured reactors. *J Micromech Microeng* 2007, *17*(3), 633–641.
34. Zhang, J. S.; Wang, K.; Teixeira, A. R.; Jensen, K. F.; Luo, G. S., Design and scaling up of microchemical systems: A review. *Annu Rev Chem Biomol* 2017, *8*, 285–305.
35. Thakur, R. K.; Vial, C.; Nigam, K. D. P.; Nauman, E. B.; Djelveh, G., Static mixers in the process industries - A review. *Chem Eng Res Des* 2003, *81*(A7), 787–826.
36. Nauman, K. D., *Chemical Reactor Design, Optimization, and Scale-Up.* McGraw-Hill, 2008.
37. Afandizadeh, S.; Foumeny, E. A., Design of packed bed reactors: Guides to catalyst shape, size, and loading selection. *Appl Therm Eng* 2001, *21*(6), 669–682.
38. Jakobsen, H. A. *Applied Thermal Engineering.* Springer, Berlin, Heidelberg, 2021.
39. Zaiput Flow Technologies, Liquid-liquid/gas-liquid separators, n.d. Retrieved January 5, from Zaiput Flow Technologies: https://www.zaiput.com/product/liquid-liquid-gas-separators/.
40. Zaiput Flow Technologies, Zaiput Flow Technologies. (n.d.). Multistage Extraction. Retrieved January 5, from Zaiput Flow Technologies: https://www.zaiput.com/product/multistage-extraction/.
41. Zhang, P.; Weeranoppanant, N.; Thomas, D. A.; Tahara, K.; Stelzer, T.; Russell, M. G.; O'mahony, M.; Myerson, A. S.; Lin, H.; Kelly, L. P.; Jensen, K. F.; Jamison, T. F.; Dai, C.; Cui, Y.; Briggs, N.; Beingessner, R. L.; Adamo, A., Advanced continuous flow platform for on-demand pharmaceutical manufacturing. *Chem Eur J* 2018, *24*(11), 2776–2784.
42. Adamo, A.; Beingessner, R. L.; Behnam, M.; Chen, J.; Jamison, T. F.; Jensen, K. F.; Monbaliu, J.-C. M.; Myerson, A. S.; Revalor, E. M.; Snead, D. R.; Stelzer, T.; Weeranoppanant, N.; Wong, S. Y.; Zhang, P., On-demand continuous-flow production of pharmaceuticals in a compact, reconfigurable system. *Science* 2016, *352*(6281), 61–67.

43. Gérardy, R.; Emmanuel, N.; Toupy, T.; Kassin, V.-E.; Tshibalonza, N. N.; Schmitz, M.; Monbaliu, J.-C. M., Continuous flow organic chemistry: Successes and pitfalls at the interface with current societal challenges. *Eur J Org Chem* 2018, *2018*(20–21), 2301–2351.

44. Cole, K. P.; Reizman, B. J.; Hess, M.; Groh, J. M.; Laurila, M. E.; Cope, R. F.; Campbell, B. M.; Forst, M. B.; Burt, J. L.; Maloney, T. D.; Johnson, M. D.; Mitchell, D.; Polster, C. S.; Mitra, A. W.; Boukerche, M.; Conder, E. W.; Braden, T. M.; Miller, R. D.; Heller, M. R.; Phillips, J. L.; Howell, J. R., Small-volume continuous manufacturing of Merestinib. Part 1. Process development and demonstration. *Org Process Res Dev* 2019, *23*(5), 858–869.

45. Reizman, B. J.; Cole, K. P.; Hess, M.; Burt, J. L.; Maloney, T. D.; Johnson, M. D.; Laurila, M. E.; Cope, R. F.; Luciani, C. V.; Buser, J. Y.; Campbell, B. M.; Forst, M. B.; Mitchell, D.; Braden, T. M.; Lippelt, C. K.; Boukerche, M.; Starkey, D. R.; Miller, R. D.; Chen, J.; Sun, B.; Kwok, M.; Zhang, X.; Tadayon, S.; Huang, P., Small-volume continuous manufacturing of Merestinib. Part 2. Technology transfer and cGMP manufacturing. *Org Process Res Dev* 2019, *23*(5), 870–881.

46. Brewer, A. C.; Hoffman, P. C.; White, T. D.; Lu, Y.; McKee, L.; Boukerche, M.; Kobierski, M. E.; Mullane, N.; Pietz, M.; Alt, C. A., Design, development, and execution of a continuous-flow-enabled API Manufacturing Route. *Organomet Chem Ind Pract Approach* 2020, 23–60.

47. Mascia, S.; Heider, P. L.; Zhang, H.; Lakerveld, R.; Benyahia, B.; Barton, P. I.; Braatz, R. D.; Cooney, C. L.; Evans, J. M.; Jamison, T. F.; Jensen, K. F.; Myerson, A. S.; Trout, B. L., End-to-end continuous manufacturing of pharmaceuticals: Integrated synthesis, purification, and final dosage formation. *Angew Chem Int Ed Engl* 2013, *52*(47), 12359–12363.

48. Abdiaj, I.; Fontana, A.; Gomez, M. V.; de la Hoz, A.; Alcazar, J., Visible-light-induced nickel-catalyzed Negishi cross-couplings by exogenous-photosensitizer-free photocatalysis. *Angew Chem Int Ed Engl* 2018, *57*(28), 8473–8477.

49. Abdiaj, I.; Horn, C. R.; Alcazar, J., Scalability of visible-light-induced nickel negishi reactions: A combination of flow photochemistry, use of solid reagents, and in-line nmr monitoring. *J Org Chem* 2019, *84*(8), 4748–4753.

50. Irfan, M.; Glasnov, T. N.; Kappe, C. O., Heterogeneous catalytic hydrogenation reactions in continuous-flow reactors. *ChemSusChem* 2011, *4*(3), 300–316.

51. Beatty, J. W.; Douglas, J. J.; Cole, K. P.; Stephenson, C. R., A scalable and operationally simple radical trifluoromethylation. *Nat Commun* 2015, *6*, 79112.

52. Beatty, J. W.; Douglas, J. J.; Miller, R.; McAtee, R. C.; Cole, K. P.; Stephenson, C. R. J., Photochemical perfluoroalkylation with pyridine N-oxides: Mechanistic insights and performance on a kilogram scale. *Chem* 2016, *1*(3), 456–472.

53. Clark, C. A.; Lee, D. S.; Pickering, S. J.; Poliakoff, M.; George, M. W., A simple and versatile reactor for photochemistry. *Org Process Res Dev* 2016, *20*(10), 1792–1798.

54. Elliott, L. D.; Berry, M.; Harji, B.; Klauber, D.; Leonard, J.; Booker-Milburn, K. I., A small-footprint, high-capacity flow reactor for UV photochemical synthesis on the kilogram scale. *Org Process Res Dev* 2016, *20*(10), 1806–1811.

55. Chaudhuri, A.; Kuijpers, K. P. L.; Hendrix, R. B. J.; Shivaprasad, P.; Hacking, J. A.; Emanuelsson, E. A. C.; Noël, T.; van der Schaaf, J., Process intensification of a photochemical oxidation reaction using a Rotor-Stator Spinning Disk Reactor: A strategy for scale up. *Chem Eng J* 2020, *400*.

56. Harper, K. C.; Moschetta, E. G.; Bordawekar, S. V.; Wittenberger, S. J., A laser driven flow chemistry platform for scaling photochemical reactions with visible light. *ACS Cent Sci* 2019, *5*(1), 109–115.

57. Noel, T.; Cao, Y.; Laudadio, G., The fundamentals behind the use of flow reactors in electrochemistry. *Acc Chem Res* 2019, *52*(10), 2858–2869.

58. Peters, B. K.; Rodriguez, K. X.; Reisberg, S. H.; Beil, S. B.; Hickey, D. P.; Kawamata, Y.; Collins, M.; Starr, J.; Chen, L.; Udyavara, S.; Klunder, K.; Gorey, T. J.; Anderson, S. L.; Neurock, M.; Minteer, S. D.; Baran, P. S., Scalable and safe synthetic organic electroreduction inspired by Li-ion battery chemistry. *Science* 2019, *363*(6429), 838–845.

59. Mallia, C. J.; Baxendale, J. R., The use of gases in flow synthesis. *Org Process Res Dev* 2015, *20*(2), 327–360.

60. Cossar, P. J.; Hizartzidis, L.; Simone, M. I.; McCluskey, A.; Gordon, C. P., The expanding utility of continuous flow hydrogenation. *Org Biomol Chem* 2015, *13*(26), 7119–7130.

61. Riley, D. L.; Neyt, N. C., Approaches for performing reductions under continuous-flow conditions. *Synthesis-Stuttgart* 2018, *50*(14), 2707–2720.

62. Mini, H.-C. ThalesNano Nanotechnology, Inc, Budapest, Hungary. http://thalesnano.com/products/h-cube-series/H-Cube_Mini.

63. Ouchi, T.; Battilocchio, C.; Hawkins, J. M.; Ley, S. V., Process intensification for the continuous flow hydrogenation of ethyl nicotinate. *Org Process Res Dev* 2014, *18*(11), 1560–1566.

64. Vaz, M.; Courboin, D.; Winter, M.; Roth, P. M. C., Scale-up of ozonolysis using inherently safer technology in continuous flow under pressure: Case study on β-pinene. *Org Process Res Dev* 2021, *25*(7), 1589–1597.

65. Nobis, M.; Roberge, D. M., Mastering ozonolysis: Production from laboratory to ton scale in continuous flow. *Chim Oggi* 2011, *29*, 56–58.

66. Allian, A. D.; Richter, S. M.; Kallemeyn, J. M.; Robbins, T. A.; Kishore, V., The development of continuous process for alkene ozonolysis based on combined in situ FTIR, calorimetry, and computational chemistry. *Org Process Res Dev* 2011, *15*(1), 91–97.

67. Cochran, B. M.; Corbett, M. T.; Correll, T. L.; Fang, Y. Q.; Flick, T. G.; Jones, S. C.; Silva Elipe, M. V.; Smith, A. G.; Tucker, J. L.; Vounatsos, F.; Wells, G.; Yeung, D.; Walker, S. D.; Bio, M. M.; Caille, S., Development of a commercial process to prepare AMG 232 using a green ozonolysis-Pinnick tandem transformation. *J Org Chem* 2019, *84*(8), 4763–4779.

68. Kendall, A. J.; Barry, J. T.; Seidenkranz, D. T.; Ryerson, A.; Hiatt, C.; Salazar, C. A.; Bryant, D. J.; Tyler, D. R., Highly efficient biphasic ozonolysis of alkenes using a high-throughput film-shear flow reactor. *Tetrahedron Lett* 2016, *57*(12), 1342–1345.

69. Thaisrivongs, D. A.; Naber, J. R.; Rogus, N. J.; Spencer, G., Development of an organometallic flow chemistry reaction at pilot-plant scale for the manufacture of Verubecestat. *Org Process Res Dev* 2018, *22*(3), 403–408.

70. Stueckler, C.; Hermsen, P.; Ritzen, B.; Vasiloiu, M.; Poechlauer, P.; Steinhofer, S.; Pelz, A.; Zinganell, C.; Felfer, U.; Boyer, S.; Goldbach, M.; de Vries, A.; Pabst, T.; Winkler, G.; LaVopa, V.; Hecker, S.; Schuster, C., Development of a continuous flow process for a Matteson reaction: From lab scale to full-scale production of a pharmaceutical intermediate. *Org Process Res Dev* 2019, *23*(5), 1069–1077.

71. Köckinger, M.; Wyler, B.; Aellig, C.; Roberge, D. M.; Hone, C. A.; Kappe, C. O., Optimization and scale-up of the continuous flow acetylation and nitration of 4-Fluoro-2-methoxyaniline to prepare a key building block of osimertinib. *Org Process Res Dev* 2020, *24*(10), 2217–2227.

72. Vieira, T.; Stevens, A. C.; Chtchemelinine, A.; Gao, D.; Badalov, P.; Heumann, L., Development of a large-scale cyanation process using continuous flow chemistry en route to the synthesis of remdesivir. *Org Process Res Dev* 2020, *24*(10), 2113–2121.

73. McMullen, J. P.; Marton, C. H.; Sherry, B. D.; Spencer, G.; Kukura, J.; Eyke, N. S., Development and scale-up of a continuous reaction for production of an active pharmaceutical ingredient intermediate. *Org Process Res Dev* 2018, *22*(9), 1208–1213.

74. Nieuwland, P. J.; Segers, R.; Koch, K.; van Hest, J. C. M.; Rutjes, F. P. J. T., Fast scale-up using micro-reactors: Pyrrole synthesis from micro to production scale. *Org Process Res Dev* 2011, *15*(4), 783–787.

13

Sustainable Synthesis of API in Water in Industrial Setting

Ning Ye and Fabrice Gallou

CONTENTS

13.1 Introduction

The last couple of decades have witnessed substantial advances in the field of green and sustainable chemistry with the framework of green chemistry being formulated and to a large extent accepted [1], the advent of circular economy and more recently significant pressure from public opinion and governmental bodies [2]. And yet, the continued global growth and demand makes the gap to a sustainable economy still wider.

Conceptually, design thinking is the basis for doing good and sustainability. One should carefully assess which product we intend to make, to not just produce and manufacture it but also to ensure that it bears the additional function of eventually being traceless and harmless to society as a whole. This notion has found limited adoption for a long time and only now, starts being nurtured with for example the IMI initiative on biodegradable medicines [3]. Needless to say that significant efforts and time will be required to achieve this task. More realistically, design applied to less conceptual and more applied chemistry aspects, whether around synthetic strategies or process-related, has started to demonstrate great impact. Remarkable tools have been established to measure the performance of processes and syntheses, with for example to name just a few the notion of E-factor [4], PMI [5], or iGAL [6]. Very useful guides around the selection of solvents and reagents have been critically compiled in collective efforts (ACS Roundtable and Chem21 guides) [7]. New biorenewable chemical entities have been discovered and started coming to market as sustainable alternatives to the more widely encountered petroleum-derived chemicals [8]. Computer-aided (retro)-synthetic tools have flourished and significantly strengthened the predictive power of computers [9]. While all these tools and subsequent significant advances are required to cope with the current urgency of the situation, linking all these pieces together has appeared critical and a pre-requisite for success. For example, proper use of metrics should be utilized to advantageously drive down the footprint of processes. This holistic notion of the big picture is intrinsically built into the concept of circular economy and the overall Life-Cycle Analysis and should be an integral part of a good design and execution of processes and syntheses of any material. In this following account, we

DOI: 10.1201/9781003178187-17

would like to offer a perspective, will try to educate the readers on the benefits specific to the minimal and optimal use of organic solvents in syntheses and will discuss how this can impact the overall environmental performance and what are the current limitations and future perspectives, more specifically in the pharmaceutical industry.

13.2 Water as a Sustainable Bulk Medium

Solvents are recognized to be responsible for almost two-thirds of the overall footprint of chemical processes within the pharmaceutical industry [10]. To a large extent, these chemicals come out of the petroleum industry and thus do not bear inherently the notion of sustainability. While recycling is always technically feasible and optional, it is still energy-intensive and a significant root cause of high carbon release. Besides, such recycling is at best only sporadically conducted to the relatively low-volume projects encountered within the pharmaceutical industry that do not facilitate building in and reusing recovered solvents, such that they more than often end up being incinerated, thus contributing further to the overall carbon release. The rationale for their usage tends to be that one needs a medium to enable the expected chemistry and such compounds to create our diverse and value-added products. Nature on the other hand does not utilize organic solvents and yet gives rise to a wealth of compounds of incredible complexity. The trick Nature consists of building compartments with the proper affinity of the compounds to make and all other critical components, relying on soft matter objects, whether vesicles, micelles or other supramolecular arrangements. This has allowed Nature to give rise to a huge diversity of compounds, functionalities or physical properties. Even more striking is the ability of Nature to conduct multiple transformations in a row to result in hugely sophisticated constructs. With this in mind, does the justification of organic solvents to create our precious medicines remain credible? Certainly less. Since a little more than a decade, such a status quo has been hugely challenged, and the concept of running chemistry the way Nature does has been tremendously pushed by a couple of academic investigators, namely professors Bruce Lipshutz and Sachin Handa, and ourselves. From the onset, the overarching objective was simple: contribute to a better world! This meant concretely finding out a way to impact substantially the way to produce better chemicals, in the water! Under our lead, this was obviously mostly focused on making pharmaceutically relevant compounds, but the concept would apply and actually does too many other fields.

13.2.1 Benefits and Scope of Transformations

One of the pre-requisites to offering such an opportunity to streamline synthetic sequences and have the ability to run multiple steps in a row, in cascade, one pot or by whichever appropriate mode is to have a large toolbox of enabled chemistry from a single medium. The more useful and common/widespread the chemistry covered, the more impactful the concept will be obviously. Polar aprotic solvents such as NMP or DMF exhibit such a broad scope but have proven reprotoxic, requiring their rapid replacement [11]. While there has been a tremendous amount of effort in the last several years to identify sustainable alternative solvents, whether biorenewable or not, most if not all identified hits still suffer from a narrow field of applicability. Of particular interest when one focuses on the typical requirements of the pharmaceutical industry, amide bond formation or cross-coupling is rarely all covered by a common medium, thus leading to the need to switch solvents in synthetic sequences, inducing costly, whether financially or productively and environmentally, processes. An attractive option has emerged in the last 15 years with the advent of chemistry in water. With this medium, the ability to run chemistry on water, that is to say at the interface of the medium and the insoluble reaction partners as it is most often the case, had been recognized many years ago [12]. The scope was however rather limited not just to specific transformations but also to specific substrates that would behave well enough in this medium and would not suffer from mass transfer issues. More recently, enabled by surfactants or even certain polymers, the scope of applicability dramatically increased [13]. Remarkable selectivities also tend to be observed in such transformations, whether chemo- or regio-selectivity, that further enable much simpler processes in virtue of the mild conditions that especially come from micellar chemistry [14]. While in traditional solvents elevated temperatures are required for a given transformation, for example in a cross-coupling event, the compartmentalization and compression effect resulting from the soft matter created in bulk water can give rise to exquisite selectivities and unique reactivity even with minimal activation energy, that is to say low to moderate temperature. Such an effect also enables the use of minimal excess of catalysts, and reaction partners, as they tend not to over-react or decompose. For example, in a standard Suzuki–Miyaura cross-coupling (SMCC) transformation, protodeboronation tends to be a standard manifold of decomposition, further accelerated at elevated temperature, that is anyhow required to offer sufficient reactivity, leaving as an only option the need to use (large) excess of the boronate species. Running the reaction in water under micellar conditions allows to minimize such decomposition, and thus to minimize or even suppress post-reactional operations that are normally necessary to remove undesired components. Following up on the example of the Suzuki–Miyaura cross-coupling, we have routinely been reducing excess of the boronate species from 1.5–2 equivalents down to 1.05–1.2 equivalents [15]. Embedded into the bigger picture, this also translates into the ability to go directly into the next step, as carry-over reactions from impurities can be reduced to the bare minimum. It is such that one very rapidly has the opportunity to either proceed directly within the same pot or interrupt the sequence *via* simple isolation, e.g. precipitation, without purifying thoroughly the product of the first transformation (see Figure 13.1). One must now understand and be aware that on average reactions account for no more than 50% of the overall mass utilization and cycle time. This means that one has the ability to now drastically reduce the mass efficiency, and sometimes even cut by more than 50% both the mass efficiency, translating into a much better environmental footprint, and productivity. This can be repeated until the right control and/or end point of a sequence where then a classical and more extensive work-up and purification set of operations has to be utilized.

Within the span of a few years, a significantly large number of protocols for important transformations were developed, for the most part, applicable to a wide scope of substrates [16]. We had indeed focused to make the chemistry impactful for our standard pharmaceutical portfolio, bearing a significant number of functionality, of heteroatoms and overall high complexity. Besides the structural diversity, we aimed at addressing significant differences in the physical behavior of the synthetized compounds and aimed at developing robust protocols with the widest scope. Amide bond formation [17], almost all kind of cross-couplings [18], nucleophilic aromatic substitutions [19], reductions of various functionalities [20], oxidations [21], halogenation [22], metathesis [23], biocatalytic transformations [24], photochemistry and photo-redox chemistry [25], are some of the most common transformations for which a practical procedure was developed. So that when considering a sequence, it now becomes possible to envision a fully streamlined approach. For example, Scheme 13.1 depicts one of the standard synthetic sequences we are encountering. Structures have been blinded, but practitioners will recognize the commonality with many target compounds within the pharmaceutical or even agrochemical industries, along with the type of chemistry encountered. A classical approach relies on the use of the best solvent for every individual step and requires purification at certain points to enable a sufficiently high quality ultimately.

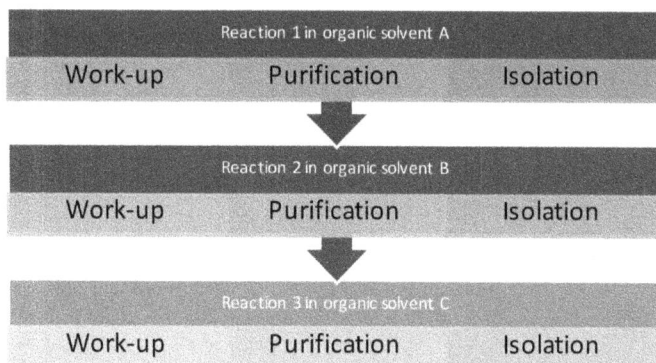

FIGURE 13.1 Traditional sequence in organic solvents *vs* water.

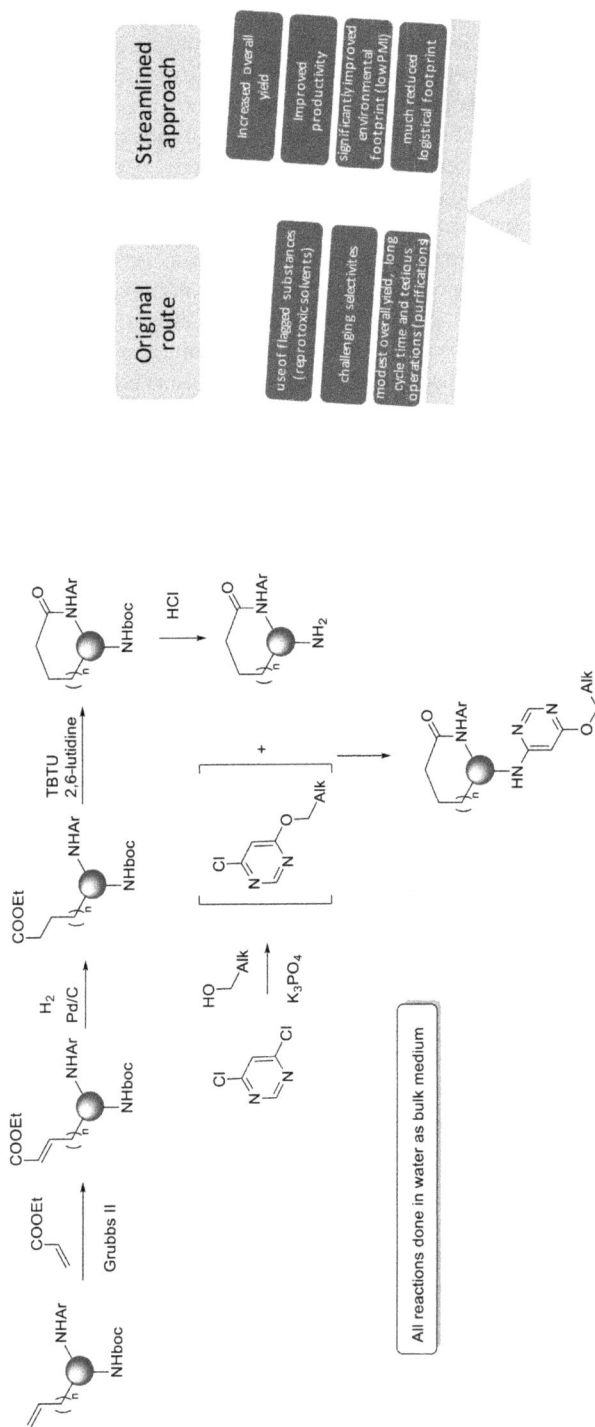

SCHEME 13.1   Standard streamlined protocol.

Thus, extensive work-up, isolation and purification contribute to significant mass utilization. Translated into a carbon dioxide release, and even with solvent recycling, which in any case rarely happens until there is sufficiently high-volume requirement, this rapidly adds up to high overall carbon dioxide that ends up being released (see Figure 13.1). The opportunity that has been created here is twofold: first, when properly executed under micellar conditions especially, high selectivity and minimal excess are utilized as mentioned above; second, with high purity of even crude product of each transformation and the ability to conduct the next transformation in the same medium comes the opportunity to cut down on operations. Practically, this means omitting a work-up or limiting it to the bare minimum, typically to remove inorganic salts and fully control the solubilities, isolating crude products of high enough purity without extensive purification(s) or using the product in solution as such, or even conducting sequences within the same pot. The impact of each of these scenarios of course can differ, but one must recognize that in all cases, significant environmental benefits are taking place. Besides, with water still as the bulk medium and the ability to pre-treat efficiently the waste water for optimal disposal in standard wastewater treatment plants [26], substantial carbon dioxide savings can be generated.

13.2.2 Streamlined Synthetic Sequences

Such a streamlined sequence as discussed and drawn in Scheme 13.1 is a typical opportunity encountered when working in water under micellar conditions, and obviously has a profound impact. A careful analysis of processes within our own portfolio indeed indicates that on average, an individual process in water under micellar conditions leads to a reduced PMI by 20–30%, and a larger CO_2 release reduction when water is properly pre-treated prior to disposal, based on the analysis of about a hundred process steps and dozens of synthetic sequences carried out within our internal portfolio. When utilized properly on multiple steps in a row, even more synergies become apparent and allow to further cut on the number of operations, classically washings and crystallizations *via* direct isolation of a sufficiently pure crude product, or use as such in solution, giving rise to drastic PMI reduction, typically well above 50% compared to the baseline in organic solvents. There is of course no easy rule as this concept of running chemistry in water under micellar conditions especially is highly specific, depending to a large extent on the physical properties of every chemical entity involved in a process. Nevertheless, it has by now been amply demonstrated that complex individual processes and sequences of processes could be run to result in spectacular benefits. Within the next section, we will try to share some of the most striking and remarkable examples.

Albeit a relatively nascent area of applied chemistry, some spectacular applications were already made in the industry. We first reported in a blinded way the assembly of one of our API candidates in 2016 [15c], and later disclosed the structures along the sequence for the program (see Scheme 13.2) [13, 27]. The main features of the work here relied on a series of compatible transformations that rendered amenable a streamlined sequence in water. The sequence was however originally designed for organic solvents, hence limited benefits, which nevertheless appeared as substantial (PMI reduced by more than 30%, cycle time divided by almost 2). The sequence consisted of nucleophilic aromatic substitution, Suzuki–Miyaura cross-coupling, hydrolysis and amide bond formation (S_NAr + SMCC + Hydrolysis + Amidation). We had been particularly interested in such a sequence as it encompassed the most prominent transformations from within the pharmaceutical portfolio. The most prominent impact was the drastic reduction of a consecutive acylation side reaction that would give rise to an undesired impurity, proven difficult to remove. The micellar approach that required low temperature here offered an elegant solution to this problem, which suppressed the need for extensive purifications and loss of yield. It must also be noted that the sequence had no particular tailoring for the medium.

A few years later, we disclosed once again in a blinded manner a novel series, so as to as rapidly as possible share our learnings and accelerate a change towards more sustainable practices (see Scheme 13.3) [15b]. The exciting feature here was the demonstration of the design of the sequence of the exact physical properties of the intermediates along the sequence specifically for the medium, in this case the lipophilic colloidal objects in bulk water. While we had rapidly shown that we could significantly improve the most challenging Suzuki–Miyaura cross-coupling step, we were able to show that further

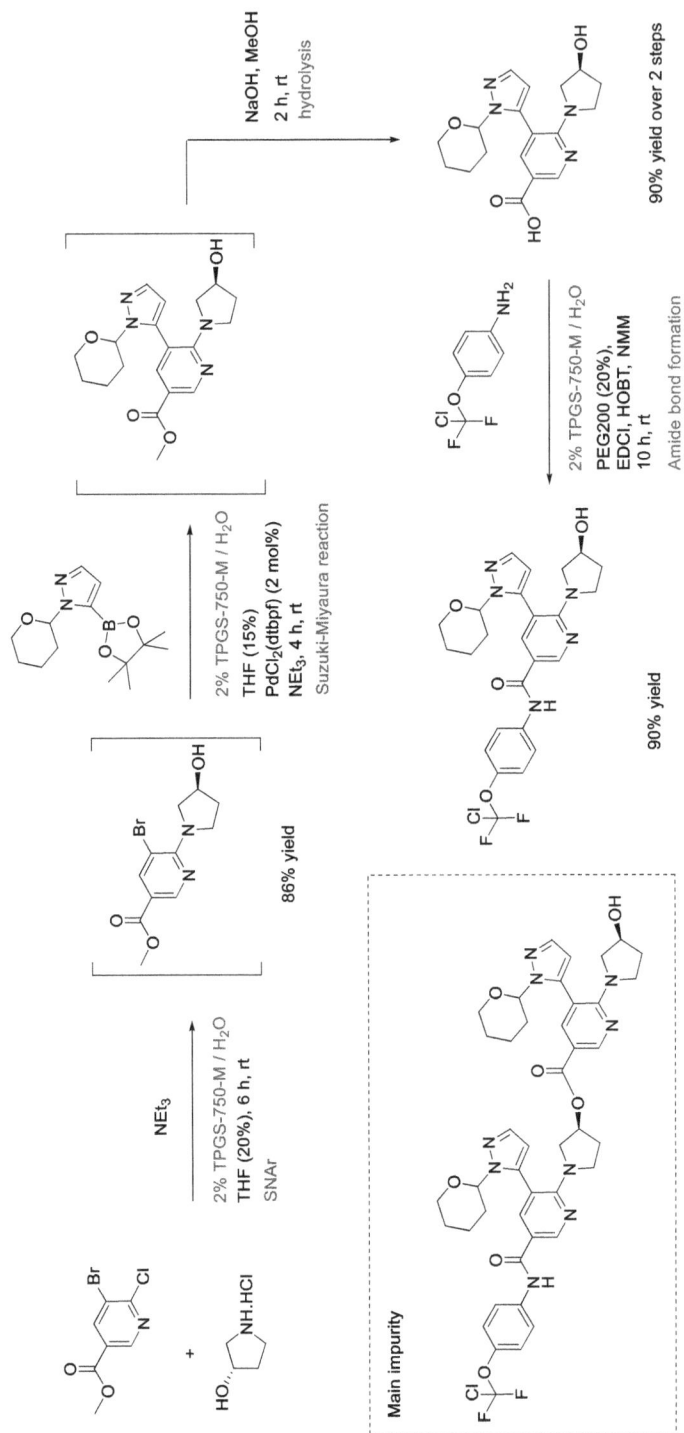

SCHEME 13.2   Our first streamlined sequence in bulk water.

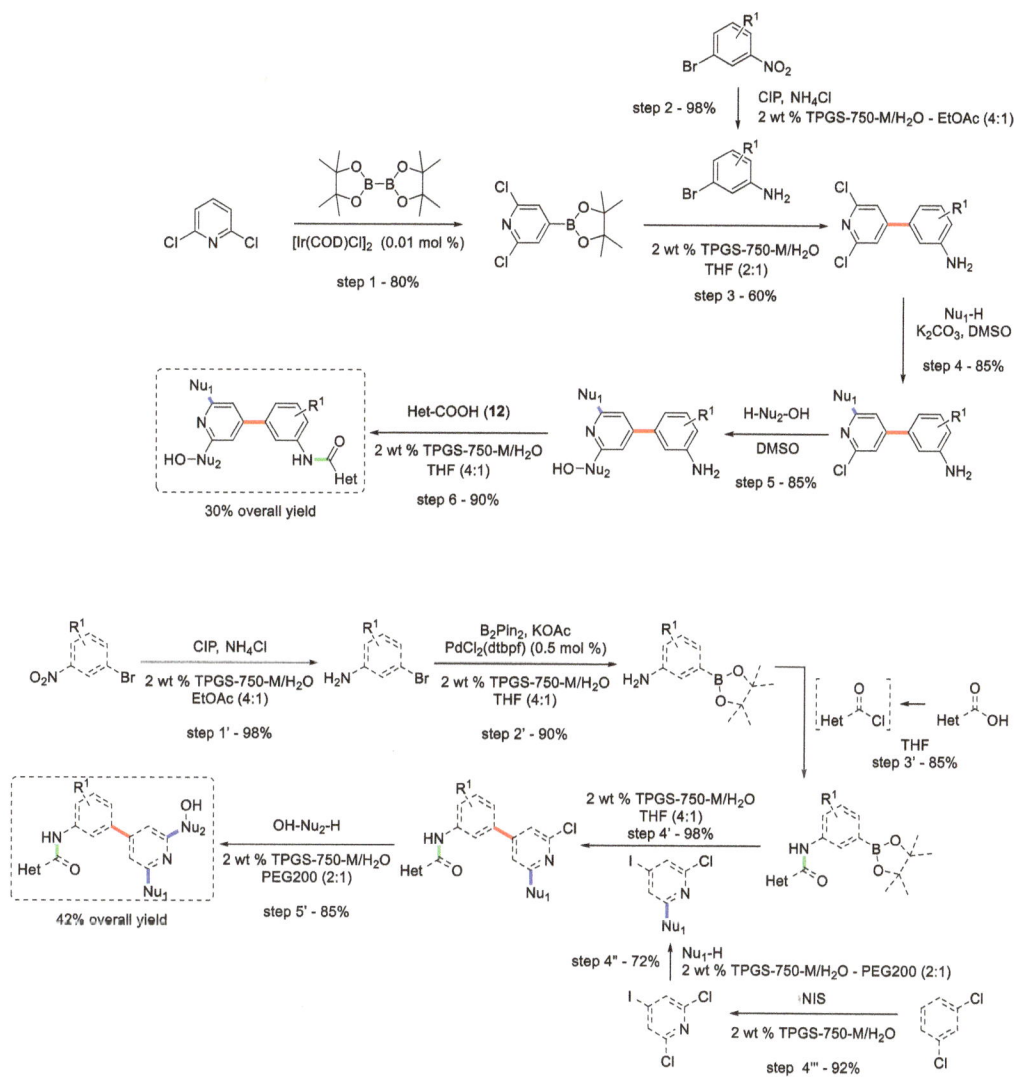

SCHEME 13.3 How tailoring the physical properties of the intermediates enabled a superior process and synthetic sequence.

tuning of the physical properties of the intermediates could leverage even better the medium and lipophilic micelles resulting in our case. The sequences were here even richer and more diverse, covering a sequence of halogenation, nitro reduction, Suzuki–Miyaura cross-coupling, two times a nucleophilic aromatic substitution and an amide bond formation (Halogenation + Nitro Reduction + SMCC + S$_N$Ar + S$_N$Ar + Amidation). Most exciting once again was the exquisite selectivity obtained in the course of the sensitive Suzuki–Miyaura cross-coupling, amidation and nucleophilic aromatic substitution that could give rise to over-reaction. Once again, the mild temperature conditions associated with the micellar phenomena especially enabled outstanding selectivity. Further leveraging of the compartmentalization effect would come from the appendage of more greasy moieties and substituents that would favor the segregation of the reactive components into or in the vicinity of the micellar environment, *via* the choice of optimal ancillary groups (for example favoring pinacol boronate instead of boronic acid, or introducing lipophilic moieties prior to the key steps).

Concomitantly, chemists at Takeda reported an outstanding application of chemistry in water, pushing the boundaries of science in this arena with an elegant full assembly of one of their APIs (see Scheme 13.4)

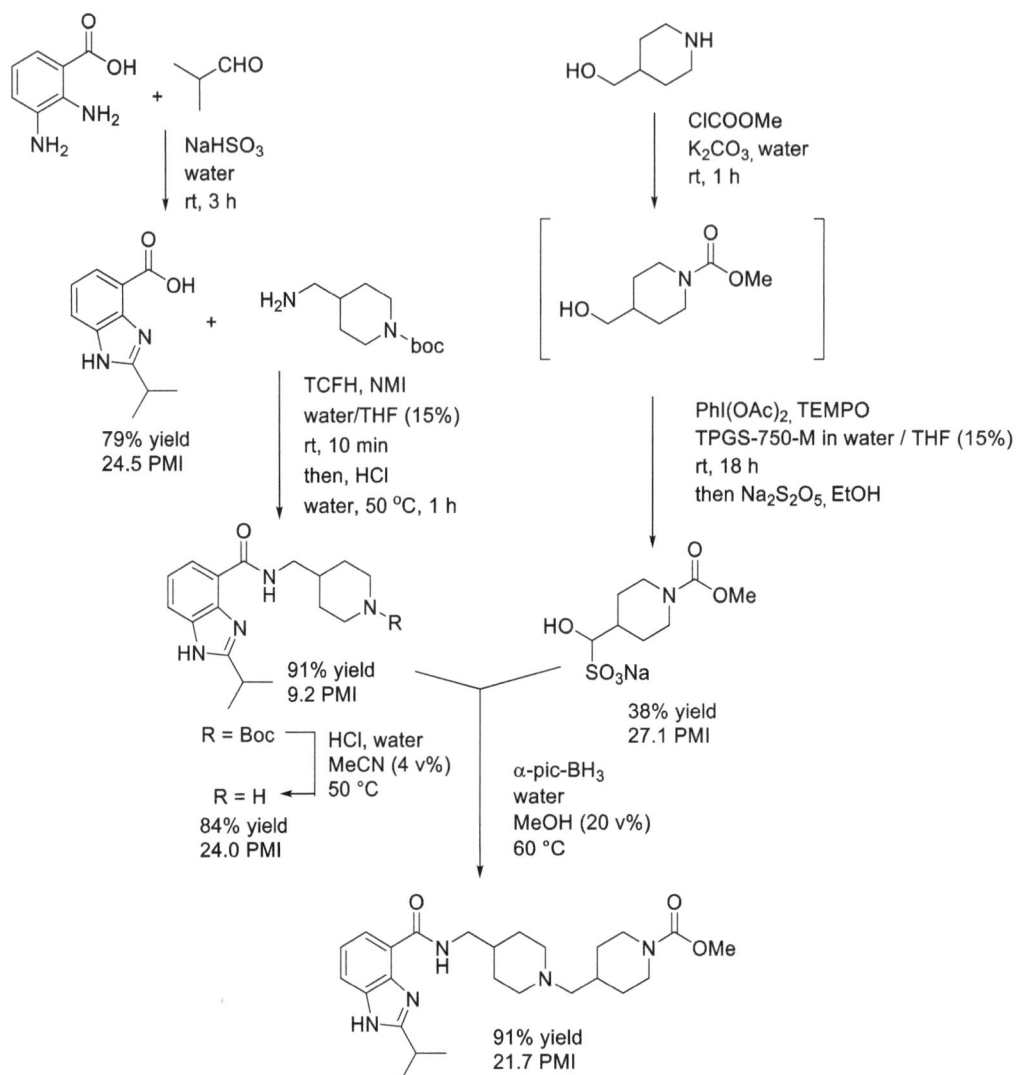

SCHEME 13.4 Example of chemistry in water from Takeda.

[28]. In this work, Bailey et al. nicely streamlined the strategy to rely on the optimal medium, whether water only, or in addition to an aid, for example a surfactant that could act as a micelle-forming agent, or a polymer. The sequence here relied on condensation, amidation, acylation, oxidation, hydrolysis and reductive amination (Condensation + Amidation + Acylation + Oxidation + Reductive Amination) and ultimately led to a very streamlined and efficient synthesis and process, deprived of undesirable solvents especially (dichloromethane and other polar aprotic reprotoxic solvents utilized originally).

Once again here, the immediate benefits of such a streamlined solvent strategy resulted in excellent mass efficiency and operational simplicity. This for the most part came from highly selective transformations that required a minimal number of post-reactional operations. Novel protocols had to be invented, for example for a highly innovative reductive amination procedure that relied on the use of the more readily purified sodium bisulfite adduct that underwent smooth reductive amination with the piperidine fragment in water [20b]. Careful control of the pH in the various operations was essential to ensure the proper solubilization of the various reaction components and showed the importance of

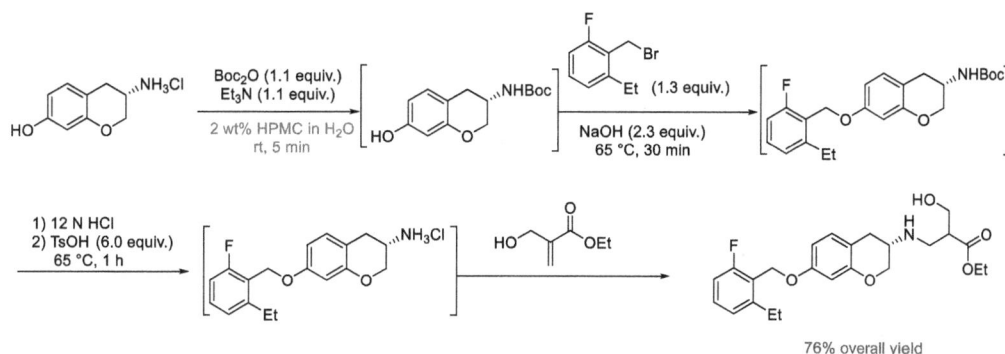

SCHEME 13.5 Example of chemistry in water from Abbvie.

physico-chemical phenomena in the outcome of a chemical transformation, also as a means to facilitate direct isolation. This point is one often not considered by chemists who tend to rely on readily dissolving all reaction components and thus missing opportunities created by dynamic systems. Overall, the water route allowed to boost the yield from 35% to 56%, to reduce the cumulative PMI from 350 to 79, respectively, 106 and 55 for PMI water while preserving high purity above 99.8%.

Chemists at Abbvie also set the stage for such opportunities, in their case relying on a polymer, HPMC, to enable rapid and selective transformations, in virtue of hydrophobic effects (see Scheme 13.5) [29]. They hypothesized that HPMC in water would create in virtue of the hydrophobicity of some of its regions' reactive pockets that would function similarly to colloidal objects and micelles in particular. Even more striking was the kinetic effect on transformations that seem to occur orders of magnitude more rapidly than under a micellar regimen. The technology has been demonstrated on a wide scope of transformations and was exemplified in a sequence for a series of nucleophilic substitution events. It resulted in a pharma-like compound in a highly streamlined manner and a high overall yield of 76%. Even more striking is the operational simplicity as the whole sequence was demonstrated in essentially a single pot, each of the new reaction components being added once the reaction is over. This approach is again unique to the concept of compatible chemistry in a streamlined medium, water here, that truly enables such opportunities.

13.3 On-Going Academic Efforts and Their Potential

13.3.1 Demonstrated Sequences

In the course of these industrial efforts, academic findings, oftentimes conducted in collaboration with industrial partners, showed the potential of synthetic sequences to generate elegantly and efficiently pharmaceuticals or agrochemicals. Boscalid, Rinskor, Indomethacin, Celecoxib, Ponatinib, Axitinib, Venetoclax, Sonidegib or advanced intermediates along their synthesis were demonstrated to be generated in a convenient and effective manner using such an approach of chemistry in water (see overview in Figure 13.2) [30–39]. To a large extent, the success of such approaches relied on a sufficiently large toolbox of chemistry developed in the course of a short period of time, that allowed picking the suitable strategy for the efficient assembly, en route, to the desired API targets. Of primary importance is the diversity and scope of chemistry covered that gave the unique ability to tackle a rich pattern of chemistry.

13.3.2 Promising Future Developments

While the latter examples constitute fairly advanced precedents, there is still ample potential from a growing toolbox. Multiple other opportunities have indeed been created and will undoubtedly be

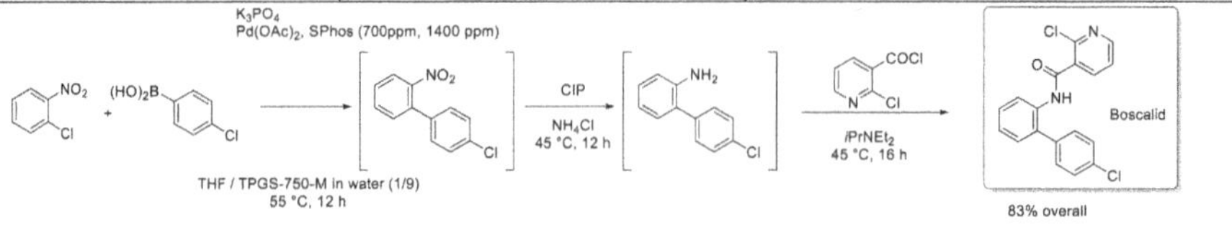

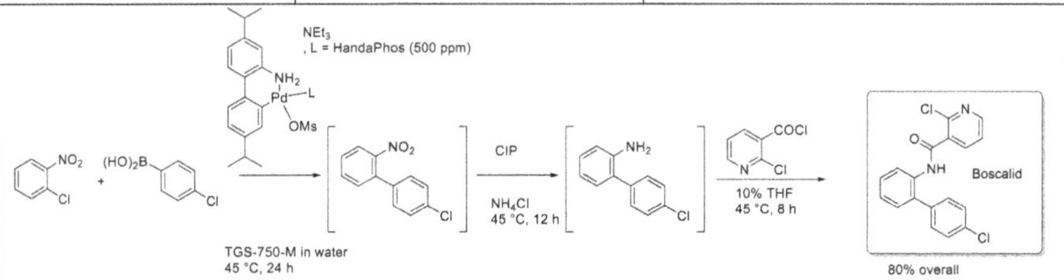

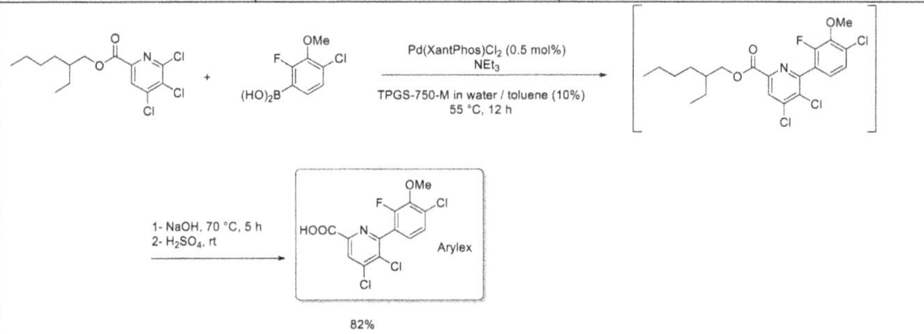

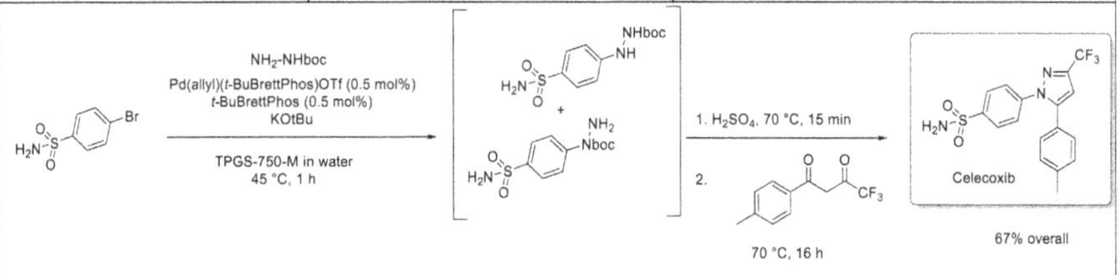

Entry	Reference	Target compound	Sequence
1	[30]	Boscalid	SMCC + Nitro Reduction + Amidation
		K$_3$PO$_4$, Pd(OAc)$_2$, SPhos (700ppm, 1400 ppm); THF / TPGS-750-M in water (1/9), 55 °C, 12 h → CIP, NH$_4$Cl, 45 °C, 12 h → iPrNEt$_2$, 45 °C, 16 h. **Boscalid, 83% overall**		
2	[31]	Boscalid	SMCC + Nitro Reduction + Amidation
		NEt$_3$, L = HandaPhos (500 ppm); TGS-750-M in water, 45 °C, 24 h → CIP, NH$_4$Cl, 45 °C, 12 h → 10% THF, 45 °C, 8 h. **Boscalid, 80% overall**		
3	[32]	Arylex	SMC + Hydrolysis
		Pd(XantPhos)Cl$_2$ (0.5 mol%), NEt$_3$, TPGS-750-M in water / toluene (10%), 55 °C, 12 h → 1- NaOH, 70 °C, 5 h; 2- H$_2$SO$_4$, rt. **Arylex, 82%**		
4	[33]	Celecoxib	C-N CC + Condensation + Condensation
		NH$_2$-NHboc, Pd(allyl)(t-BuBrettPhos)OTf (0.5 mol%), t-BuBrettPhos (0.5 mol%), KOtBu, TPGS-750-M in water, 45 °C, 1 h → 1. H$_2$SO$_4$, 70 °C, 15 min; 2. 70 °C, 16 h. **Celecoxib, 67% overall**		
5	[34]	Hedgehog (Hh) signaling antagonist	SMCC + SMCC + Reduction + Amidation

FIGURE 13.2   Some recent synthetic sequences in water.

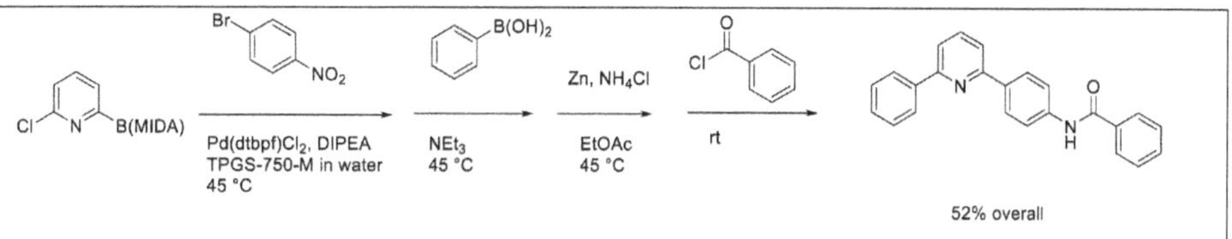

6	[35]	Ponatinib	SMCC + Hydrolysis + SCC

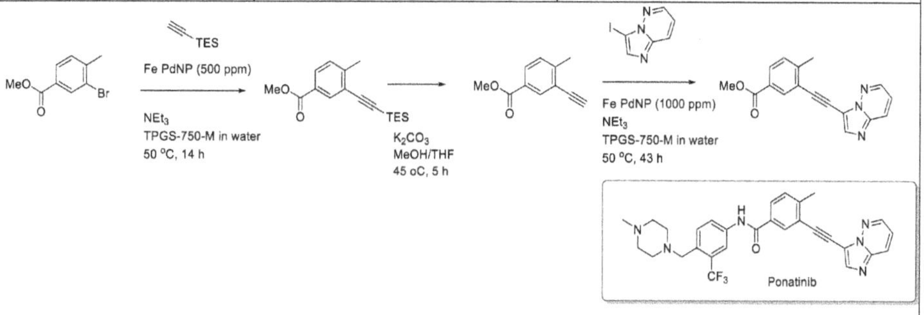

7	[36]	Binol ligand	SMCC + Hydrolysis + Cycloaddition + Reduction + SN

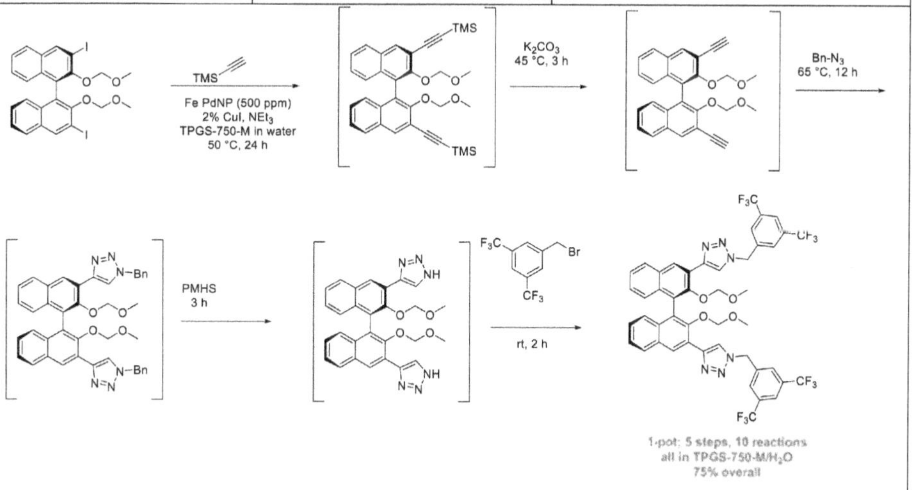

8	[37]	Axitinib	Reduction + C-N CC + Hydrolysis

FIGURE 13.2 Contd.

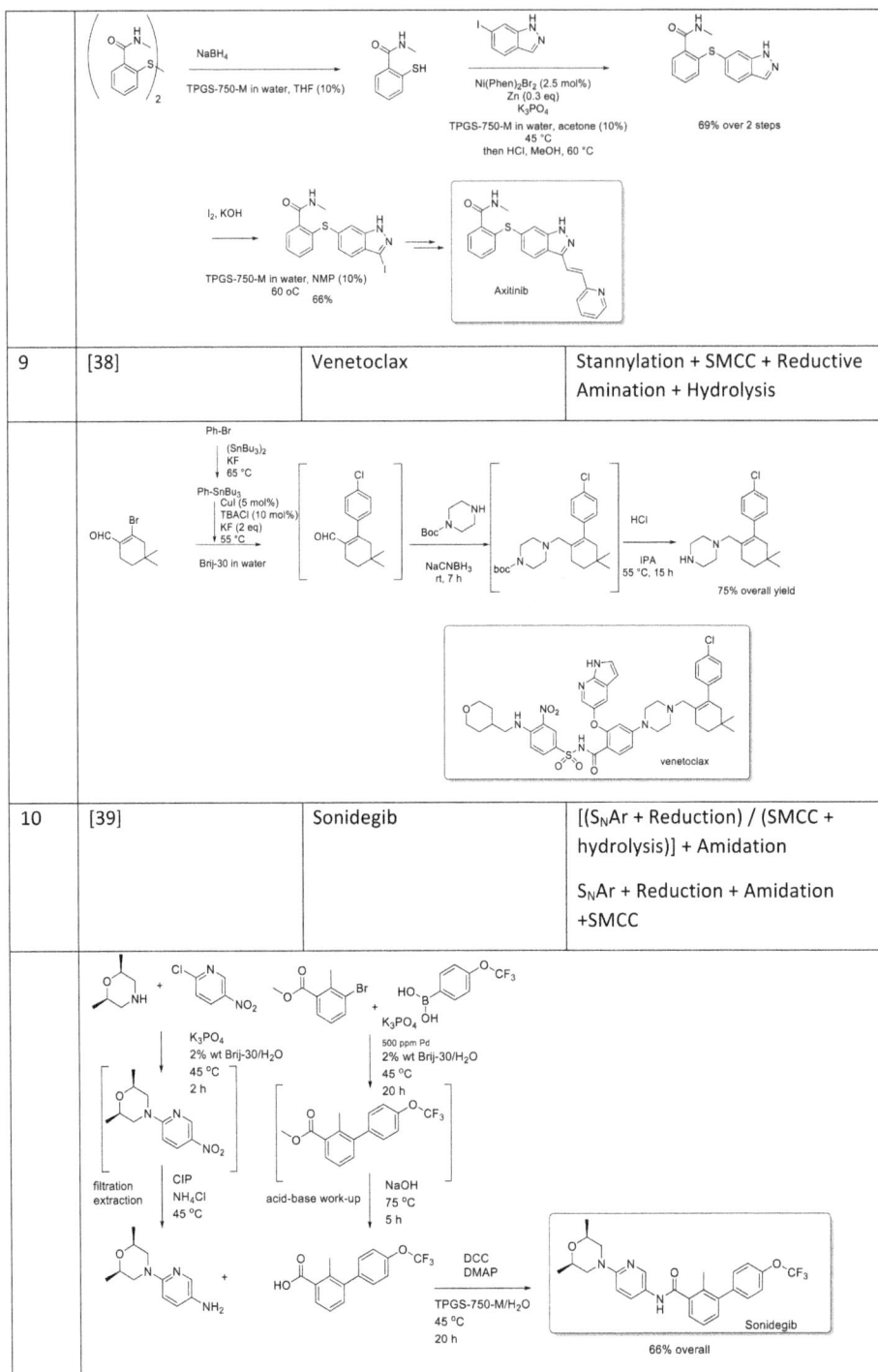

9	[38]	Venetoclax	Stannylation + SMCC + Reductive Amination + Hydrolysis
10	[39]	Sonidegib	[(S_NAr + Reduction) / (SMCC + hydrolysis)] + Amidation S_NAr + Reduction + Amidation + SMCC

FIGURE 13.2 Contd.

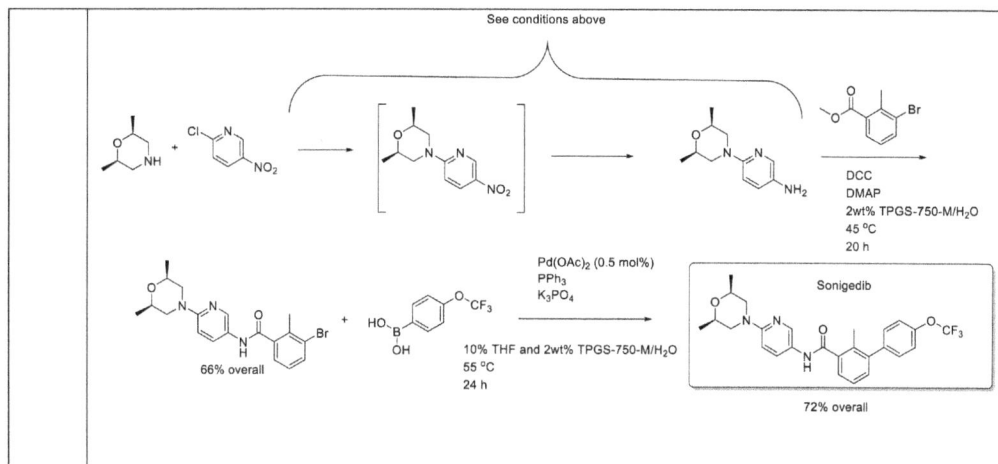

FIGURE 13.2 Contd.

appearing soon in the literature to further demonstrate the impact of such a strategy. We would like to highlight below the few ones that we believe should have a profound impact.

- Lipshutz laid the ground for effective peptide chemistry in water, synonymous with phasing out of the reprotoxic polar aprotic solvents, and opening up tremendous hope in this always impactful class of compounds relying for the most part on the iterative type of chemistry [40]. While it has not (yet) been utilized in an iterative manner, the foundations are now here to fully exploit this opportunity (see Figure 13.3a).

- Further extension of the many Suzuki–Miyaura cross-coupling for C-(sp^2)-C-(sp^2) now enables the more demanding C-(sp^2)-C-(sp^3) cross-coupling, embedded into complex synthetic sequences [41] (see Figure 13.3,b).

- Increased understanding and the development of a robust protocol for the *in situ* generation of palladium nanoparticles allowed us to develop and embed effective Mizoroki-Heck cross-couplings into complex synthetic sequences [42] (see Figure 13.3c).

- More recently, we also reported a highly promising protocol for cross-electrophilic coupling [43], which can further facilitate the assembly of complex fragments when combined with the already existing toolbox of chemistry (see Figure 13.3d).

13.3.3 The Case for Biocatalysis: On the Way to Merging Chemo- and Biocatalysis

Even more exciting in our mind is the potential to merge bio- and chemo-catalysis, as started being reported. While each individual step conducted in bulk water can offer significant benefits whether from selectivity (chemo-, regio-, stereo-selectivity), productivity, environmental and financial, it is when concatenated that the real impact multiplies and becomes prominent. It is indeed building onto the inherent selectivity of micellar chemistry and the large toolbox of reachable transformations that one can even more drastically cut on the number of ancillary operations previously required in the traditional way to conduct chemistry, that is to say in organic solvents. More impressively has been the ability to link previously incompatible words such as that of bio- and chemical transformations, and thus enable a merged field of catalysis, rather than dogmatizing several communities, that of biocatalysis, of chemo-catalysis, or organo-catalysis. We are still at the "friction" stage for merged catalysis in water, but the feasibility is established (see Figure 13.4 [24b, 44, 45, 46]. There is a path to operationally simple and potentially truly sustainable chemistry ahead of us. One must also

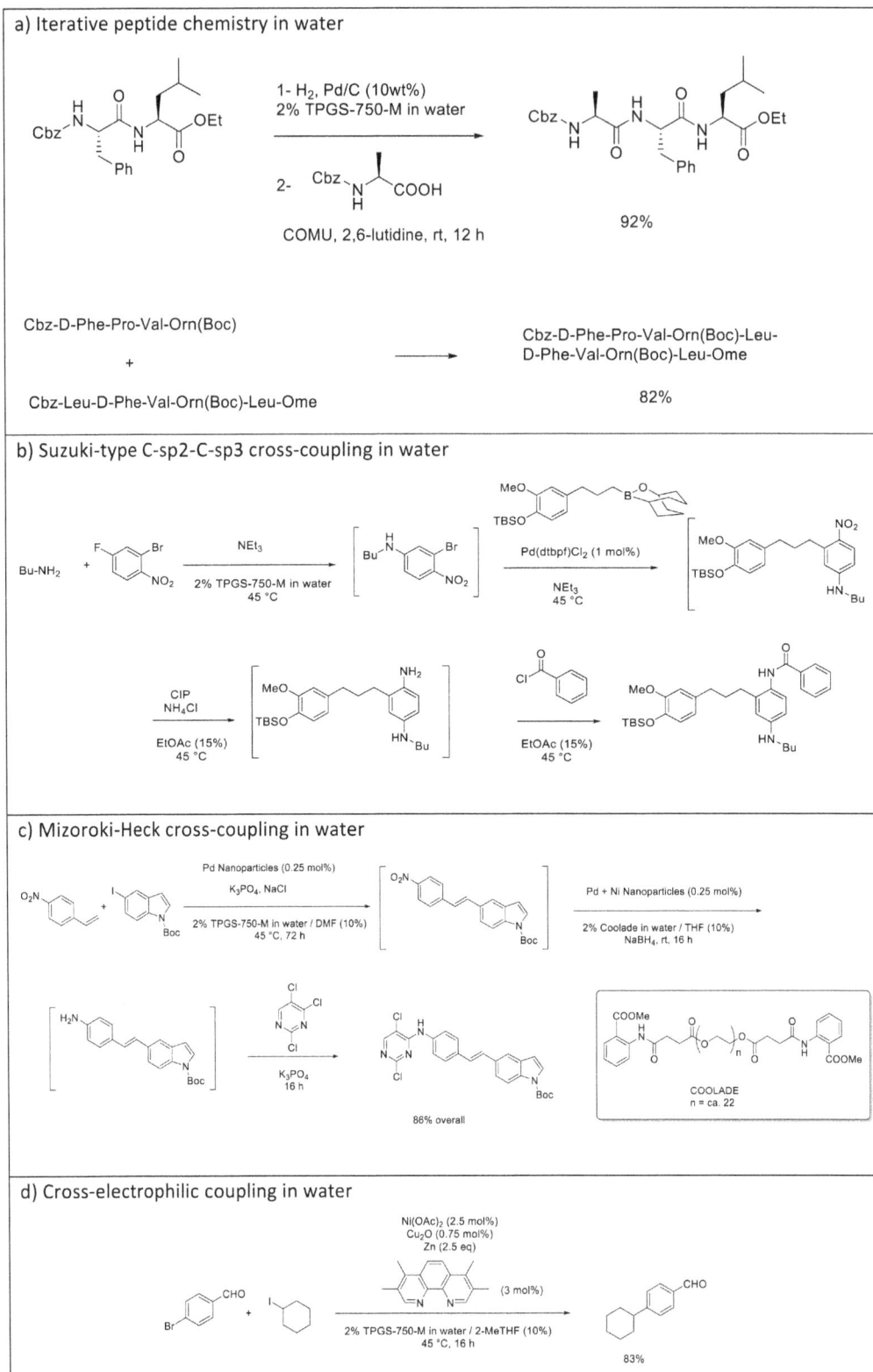

FIGURE 13.3   Promising leads with potential future impact.

Entry	Reference	Scheme
1	[24b]	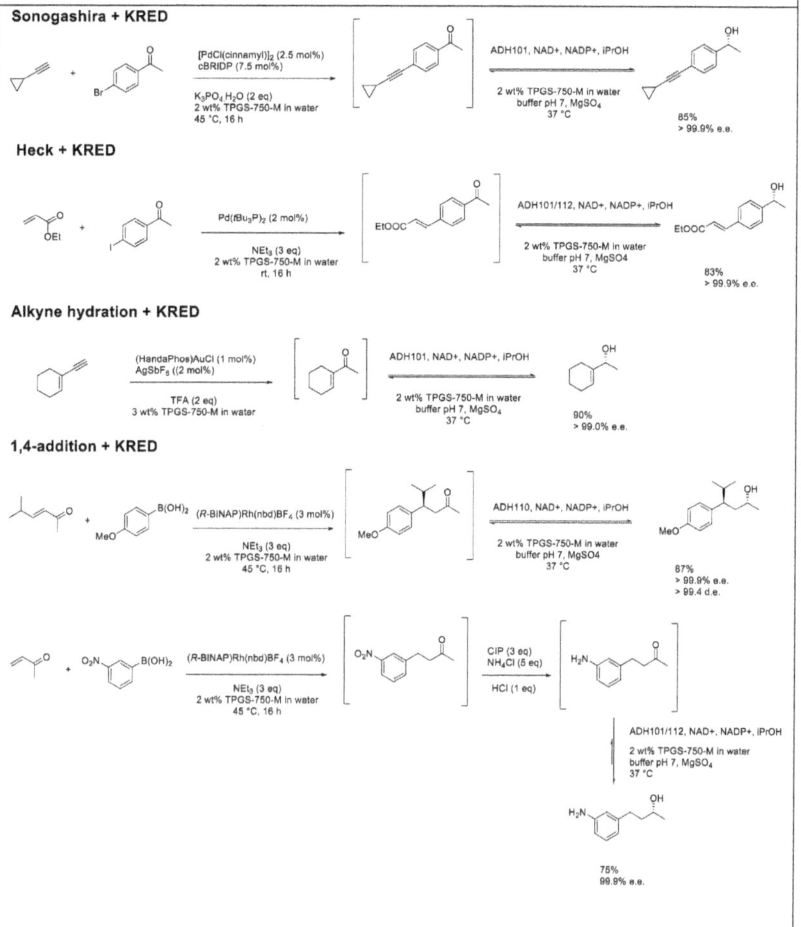 **Sonogashira + KRED** **Heck + KRED** **Alkyne hydration + KRED** **1,4-addition + KRED**
		Sonogashira + KRED
		Heck + KRED
		Hydration + KRED
		Conjugate addition + KRED

FIGURE 13.4 Recent examples of sequences using biotransformations.

<table>
<tr><td>2</td><td>[44]</td><td>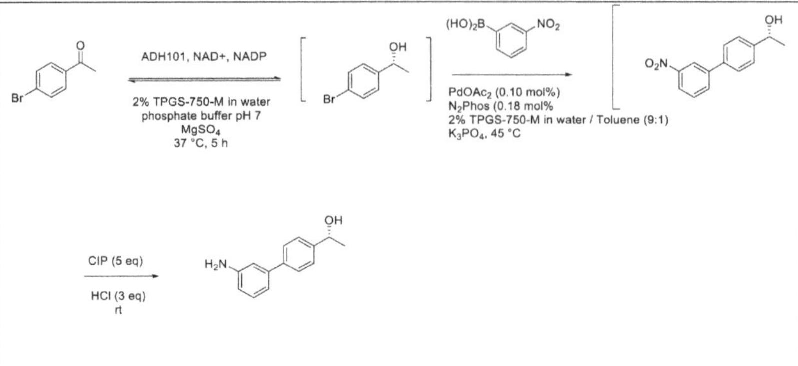
ADH101, NAD+, NADP
2% TPGS-750-M in water
phosphate buffer pH 7
$MgSO_4$
37 °C, 5 h
$(HO)_2B$
NO_2
$PdOAc_2$ (0.10 mol%)
N_2Phos (0.18 mol%
2% TPGS-750-M in water / Toluene (9:1)
K_3PO_4, 45 °C
O_2N
OH
Br
CIP (5 eq)
HCl (3 eq)
rt
H_2N</td></tr>
<tr><td></td><td></td><td>KRED + SMCC + reduction</td></tr>
<tr><td>3</td><td>[45]</td><td>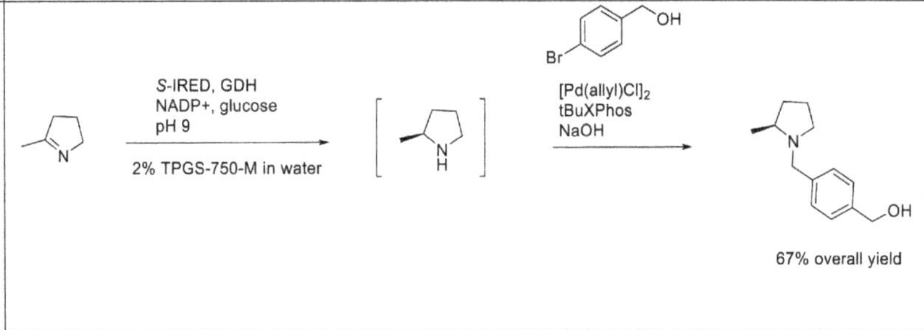
S-IRED, GDH
NADP+, glucose
pH 9
2% TPGS-750-M in water
Br
OH
$[Pd(allyl)Cl]_2$
tBuXPhos
NaOH
N
H
67% overall yield</td></tr>
<tr><td></td><td></td><td>IRED + C-N CC</td></tr>
<tr><td>4</td><td>[46]</td><td>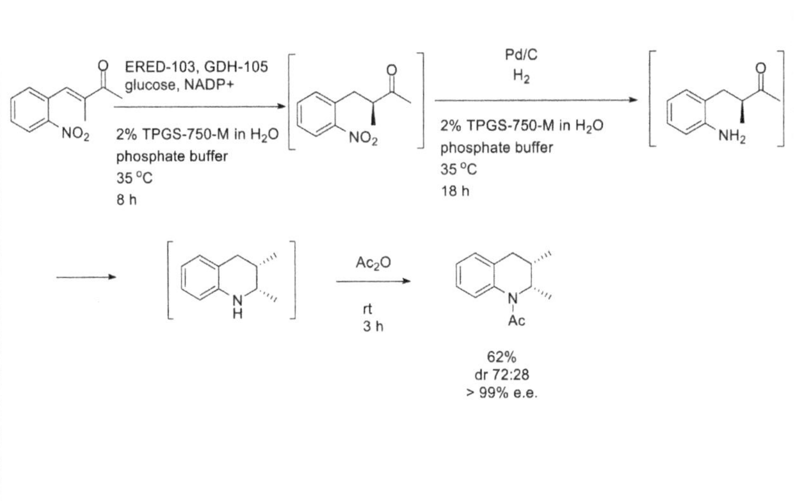
ERED-103, GDH-105
glucose, NADP+
2% TPGS-750-M in H_2O
phosphate buffer
35 °C
8 h
NO_2
Pd/C
H_2
2% TPGS-750-M in H_2O
phosphate buffer
35 °C
18 h
NH_2
Ac_2O
rt
3 h
N
H
Ac
62%
dr 72:28
> 99% e.e.</td></tr>
<tr><td></td><td></td><td>ERED + reduction + reductive amination + acylation</td></tr>
</table>

FIGURE 13.4 Contd.

recognize that sticking to one or the other technologies without openness is not the proper attitude and therefore leave the door open to any good practices that will contribute to making the world better. Significant advances in the field of mechanochemistry, which can potentially even suppress the need for any medium, can for example further strengthen our toolbox of options and further accelerate the transition to a sustainable way of doing chemistry. More important is to embrace scientific curiosity and cultural acceptance of novel practices without dogma, provided they are well funded and substantiated by data. It must be clearly recognized as well that we are still in the infancy of the

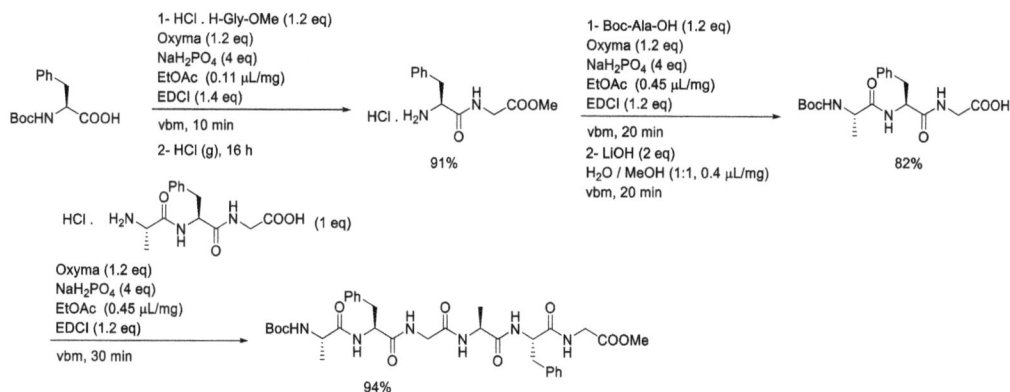

SCHEME 13.6 Liquid-Assisted Grinding iterative assembly of a hexapeptide.

chemistry in water with or without surfactant, on-water chemistry still being a valid option in several cases. Water, when not used properly, can indeed be an even worse medium and is becoming in any case a future precious chemical entity that one should not take for granted.

13.3.3.1 Outlook

Conceptually similar is the use and development of mechanochemical sequences, with potentially even higher impact on our environmental footprint. Individual synthetic protocols have indeed been built over the years to little by little enlarge the toolbox of chemistry [47]. The field of iterative chemistry and that of amide bonds especially has developed rapidly, with key milestones. Lamaty, Jerome et al. demonstrated the assembly of a hexapeptide using a ball mill in the promising gram scale. In 77% overall yield over five linear steps, using a minimal quantity of EtOAc for a more effective Liquid-Assisted Grinding effect (see Scheme 13.6) [48]. Combined with the more recent work from the authors on epimerization-free peptide bond formation [49], it makes us envision a profound impact of the technology in the development of sustainable peptide synthesis. Highly interestingly, a similar concept of mechanochemistry was proposed to explain prebiotic peptide bond formation [50].

One could therefore envision continued efforts in the field and hope that it will further accelerate our change towards better practices.

We hope that this account will serve as a source of inspiration and encouragement for anyone involved in sustainability efforts and strongly encourage practitioners to join forces in the collective efforts to more rapidly change the practices.

REFERENCES

1. (a) Anastas, P. T.; Warner, J. C. *Green Chemistry: Theory and Practice*. Oxford University Press, New York, 1998. (b) Anastas, P. T.; Zimmerman, J. B. *Environ. Sci. Technol.* 2003, *37*(5), 94A–101A. (c) Abraham, M. A.; Nguyen, N. *Environ. Prog.* 2003, *22*(4), 233–236.
2. A European Green Deal. European Commission (europa.eu); accessed on November 1, 2021.
3. PREMIER | IMI Innovative Medicines Initiative (europa.eu); accessed on November 1, 2021.
4. Sheldon, R. A. *Chem. Ind. (Lond.)* 1992, 903–906.
5. Jimenez-Gonzalez, C.; Ponder, C. S.; Broxterman, Q. B.; Manley, J.-B. *Org. Process Res. Dev.* 2011, *15*(4), 912–917.
6. Roschangar, F.; Sheldon, R. A.; Senanayake, C. H. *Green Chem.* 2015, *17*(2), 752–768.
7. (a) Prat, D.; Pardigon, O.; Flemming, H.-W.; Letestu, S.; Ducandas, V.; Isnard, P.; Guntrum, E.; Senac, T.; Ruisseau, S.; Cruciani, P.; Hosek, P. *Org. Process Res. Dev.* 2013, *17*(12), 1517–1525. (b) ACS Green Chemistry Institute Pharmaceutical Roundtable *Reagent Guides*. https://reagents.acsgcipr.org/; accessed on November 1, 2021.

8. (a) Marathianos, A.; Liarou, E.; Hancox, E.; Grace, J. L.; Lestera, D. W.; Haddleton, D. M. *Green Chem.* 2020, *22*(17), 5833–5837. (b) Luterbacher, J. S.; Komarova, A. O.; Dick, G. R. *Green Chem.* 2021, *23*(9), 3459–3467. (c)Aycock, D. F. *Org. Process Res. Dev.* 2007, *11*(1), 156–159.

9. (a) Coley, C. W.; Rogers, L.; Green, W. H.; Jensen, K. F. *ACS Cent. Sci.* 2017, *3*(12), 1237–1245. (b) Szymkuc, S.; Gajewska, E. P.; Klucznik, T.; Molga, K.; Dittwal, P.; Startek, M.; Bajczyk, M.; Grzybowski, B. A. *Angew. Chem. Int. Ed.* 2016, *55*(20), 5904–5937. (c) Segler, M. H. S.; Preuss, M.; Waller, M. P. *Nature* 2018, *555*(7698), 604–610. (d) https://www.bing.com/search?q=ASKCOS&FORM =ANAB01&PC=U531 SYNTHIA™ Retrosynthesis Software, Powerful AI Software Accelerating Drug Discovery. Merck (merckgroup.com); Allchemy – Resource-aware AI for drug discovery; Accelerating synthetic chemistry with a predictive retrosynthesis solution (elsevier.com); Homepage (ibm.com) all; accessed on November 1, 2021.

10. Ashcroft, C.; Dunn, P.; Hayler, J.; Wells, A. *Org. Process Res. Dev.* 2015, *19*(7), 740–747.

11. Substance Information - ECHA (europa.eu); accessed on November 1, 2021.

12. Rideout, D. C.; Breslow, R. *J. Am. Chem. Soc.* 1980, *102*(26), 7816–7817.

13. Cortes-Clerget, M.; Yu, J.; Kincaid, J. R. A.; Walde, P.; Gallou, F.; Lipshutz, B. H. *Chem. Sci.* 2021, *12*(12), 4237–4266.

14. Hauk, P.; Wencel-Delord, J.; Ackermann, L.; Walde, P.; Gallou, F. *Curr. Opin. Colloid Interface Sci.* 2021, *56*, 101506.

15. (a) Parmentier, M.; Wagner, M.; Wickendick, R.; Gallou, F. *Org. Process Res. Dev.* 2020, *24*(8), 1536–1542. (b) Lippincott, D. J.; Landstrom, E.; Cortes-Clerget, M.; Lipshutz, B. H.; Buescher, K.; Schreiber, R.; Durano, C.; Parmentier, M.; Ye, N.; Wu, B.; Shi, M.; Yang, H.; Andersson, M.; Gallou, F. *Org. Process Res. Dev.* 2020, *24*(5), 841–849. (c) Gallou, F.; Isley, N. A.; Ganic, A.; Onken, U.; Parmentier, A. *Green Chem.* 2016, *18*(1), 14–19.

16. (a) Gallou, F. *Chimia* 2020, *74*(7), 538–548. (b) Parmentier, M.; Gabriel, C. M.; Guo, P.; Isley, N. A.; Zhou, J.; Gallou, F. Curr. Opin. Green Sustain. Chem. 2017, *7*, 13–17. (c) Lipshutz, B. H.; Gallou, F.; Handa, S. *ACS Sustain. Chem. Eng.* 2016, *4*(11), 5838–5849.

17. (a) Shi, M.; Ye, N.; Wang, H.; Chen, W.; Yang, H.; Wang, J.; Cheung, C.; Parmentier, M.; Gallou, F.; Wu, B. *Org. Process Res. Dev.* 2020, *24*(8), 1543–1548. (b) Gallou, F.; Guo, P.; Parmentier, M.; Zhou, J. *Org. Process Res. Dev.* 2016, *20*(7), 1388–1391. (c) Parmentier, M.; Wagner, M. K.; Magra, K.; Gallou, F. *Org. Process Res. Dev.* 2016, *20*(6), 1104–1107. (d) Gabriel, C. M.; Keener, M.; Gallou, F.; Lipshutz, B. H. *Org. Lett.* 2015, *17*(16), 3968–3971.

18. (a) Ansari, T. N.; Gallou, F.; Handa, S. *Organometallic Chemistry in Industry: A Practical Approach*, First Edition. © 2020 Wiley-VCH Verlag GmbH & Co. KGaA. (b) Ye, N.; Wu, B.; Zhao, K.; Ge, X.; Zheng, Y.; Shen, X.; Shi, L.; Cortes-Clerget, M.; Regnier, M. L.; Parmentier, M.; Gallou, F. *Chem. Comm. (Camb)* 2021, *57*(62), 7629–7632. (c)Handa, S.; Ibrahim, F.; Ansari, T.; Gallou, F. *ChemCatChem* 2019, *9*, 4229–4233. (d) Brocklehurst, C. E.; Gallou, F.; Hartwieg, J. C. D.; Palmieri, M.; Rufle, D. *Org. Process Res. Dev.* 2018, *22*(10), 1453–1457. (e) Adenot, A.; Landstrom, E. B.; Gallou, F.; Lipshutz, B. H. *Green Chem.* 2017, *19*(11), 2506–2509. (f) Sharma, S.; Buchbinder, N. W.; Braje, W.; Handa, S. *Org. Lett.* 2020, *22*(15), 5737–5740. (g) Ansari, T.; Taussat, A.; Clark, A.; Nachtegaal, M.; Plummer, S.; Gallou, F.; Handa, S. *ACS Catal.* 2019, *9*(11), 10389–10397. (h) Bihani, M.; Bora, P. P.; Nachtegaal, M.; Gallou, F.; Handa, S. *ACS Catal.* 2019, *9*(8), 7520–7526. (i) Jakobi, M.; Gallou, F.; Sparr, C.; Parmentier, M. *Helv. Chim. Acta* 2019, *102*(3), e1900024. (j) Pang, H.; Hu, Y.; Yu, J.; Gallou, F.; Lipshutz, B. H. *J. Am. Chem. Soc.* 2021, *143*(9), 3373–3382. (k) Takale, B. S.; Thakore, R. R.; Casotti, G.; Li, X.; Gallou, F.; Lipshutz, B. H. *Angew. Chem. Int. Ed.* 2020, *60*, 4158–4163. (l) Yu, T.; Pang, H.; Cao, Y.; Gallou, F.; Lipshutz, B. H. *Angew. Chem. Int. Ed.* 2020, *60*, 3708–3713. (m) Lee, N. R.; Moghadam, F. A.; Braga, F. C.; Lippincott, D. J.; Zhu, B.; Gallou, F.; Lipshutz, B. H. *Org. Lett.* 2020, *22*(13), 4949–4954. (n) Akporji, N.; Thakore, R. R.; Cortes-Clerget, M.; Anderson, J.; Landstrom, E.; Aue, D. H.; Gallou, F.; Lipshutz, B. H. *Chem. Sci.* 2020, *11*(20), 5205–5212. (o) Landstrom, E. B.; Handa, S.; Aue, D. H.; Gallou, F.; Lipshutz, B. H. *Green Chem.* 2018, *20*(15), 3436–3443. (p) Handa, S.; Andersson, M. P.; Gallou, F.; Reilly, J.; Lipshutz, B. H. *Angew. Chem. Int. Ed.* 2016, *55*(16), 4914–4918. (q) Handa, S.; Wang, Ye; Gallou, F.; Lipshutz, B. H. *Science* 2015, *349*(6252), 1087–1091. (r) Zhang, Y.; Takale, B. S.; Gallou, F.; Reilly, J.; Lipshutz, B. H. *Chem. Sci.* 2019, *10*(45), 10556–10561. (s) Handa, S.; Bo, J.; Bora, P. P.; Wang, Y.; Zhang, X.; Gallou, F.; Reilly, J.; Lipshutz, B. H. *ACS Catal.* 2019, *9*(3), 2423–2431. (t) Bhattacharjya, A.; Klumphu, P.; Lipshutz, B. H. *Nat. Commun.* 2015, *6*(1), 7401–7406. (u) Cortes-Clerget, M.; Akporji,

N.; Takale, B. S.; Wood, A.; Landstrom, E.; Lipshutz, B. H. Earth-Abundant and Precious Metal Nanoparticle Catalysis. In Kobayashi S. (ed.) *Nanoparticles in Catalysis. Topics in Organometallic Chemistry*, vol. 66. Springer, Cham, 2020. (v) Borlinghaus, N.; Kaschel, J.; Klee, J.; Haller, V.; Schetterl, J.; Heitz, S.; Lindner, T.; Dietrich, J. D.; Braje, W. M.; Jolit, A. *J. Org. Chem.* 2021, *86*(2), 1357–1370.

19. (a) Lee, N. R.; Gallou, F.; Lipshutz, B. H. *Org. Process Res. Dev.* 2017, *21*(2), 218–221. (b) Isley, N. A.; Linstadt, R. T. H.; Kelly, S. M.; Gallou, F.; Lipshutz, B. H. *Org. Lett.* 2015, *17*(19), 4734–4737. (c) Borlinghaus, N.; Ansari, T.; von Garrel, L. H.; Ogulu, D.; Handa, S.; Wittmann, V.; Braje, W. *Green Chem.* 2021, *23*(11), 3955–3962.

20. (a) Li, X.; Thakore, R. R.; Takale, B. S.; Gallou, F.; Lipshutz, B. H. *Org. Lett.* 2021, *23*(20), 8114–8118. (b) Li, X.; Iyer, K. S.; Thakore, R. R.; Bailey, J. D.; Leahy, D. K.; Lipshutz, B. H. *Org. Lett.* 2021, *23*(18), 7205–7208. (c) Takale, B. S.; Thakore, R. R.; Gao, E. S.; Gallou, F.; Lipshutz, B. H. *Green Chem.* 2020, *22*(18), 6055–6061. (d) Wood, A. B.; Cortes-Clerget, M.; Kincaid, J. R. A.; Akkachairin, B.; Singhania, V.; Gallou, F.; Lipshutz, B. H. *Angew. Chem. Int. Ed.* 2020, *59*(40), 17587–17593. (e) Akporji, N.; Lieberman, J.; Maser, M.; Yoshimura, M.; Boskovic, Z.; Lipshutz, B. H. *Chem* 2019, *11*, 5743–5747. (f) Fennewald, J. C.; Landstrom, E. B.; Lipshutz, B. H. *Tetrahedron Lett.* 2015, *56*(23), 3608–3611. (g) Isley, N. A.; Hageman, M. S.; Lipshutz, B. H. *Green Chem.* 2015, *17*(2), 893–897. (h) Slack, E. D.; Gabriel, C. M.; Lipshutz, B. H. *Angew. Chem. Int. Ed.* 2014, *53*(51), 14051–14054. (i) Lee, N. R.; Bikovtseva, A. A.; Cortes-Clerget, M.; Gallou, F.; Lipshutz, B. H. *Org. Lett.* 2017, *19*(24), 6518–6521. (j) Pang, H.; Gallou, F.; Sohn, H.; Camacho-Bunquin, J.; Delferro, M.; Lipshutz, B. H. *Green Chem.* 2018, *20*(1), 130–136. (k) Feng, J.; Handa, S.; Gallou, F.; Lipshutz, B. H. *Angew. Chem. Int. Ed.* 2016, *55*(31), 8979–8983. (l) Gabriel, C. M.; Parmentier, M.; Riegert, C.; Lanz, M.; Handa, S.; Lipshutz, B. H.; Gallou, F. *Org. Process Res. Dev.* 2017, *21*(2), 247–252. (m) Kelly, S. M.; Lipshutz, B. H. *Org. Lett.* 2014, *16*(1), 98–101.

21. Lippincott, D. J.; Trejo-Soto, P. J.; Gallou, F.; Lipshutz, B. H. *Org. Lett.* 2018, *20*(17), 5094–5097.

22. Bora, P. P.; Bihani, M. B.; Plummer, S.; Gallou, F.; Handa, S. *ChemSusChem* 2019, *12*(13), 3037–3042.

23. Voigtritter, K.; Ghorai, S.; Lipshutz, B. H. *J. Org. Chem.* 2011, *76*(11), 4697–4702.

24. (a) Zhou, J.; Guo, P.; Gai, Y.; Parmentier, M.; Gallou, F.; Kong, W. *"Improved enzymatic catalysis using buffer medium containing surfactants "* WO2018134710. (b) Cortes-Clerget, M.; Akporji, N.; Zhou, J.; Gao, F.; Guo, P.; Parmentier, M.; Gallou, F.; Berthon, J.-Y.; Lipshutz, B. H. *Nat. Commun.* 2019, *10*(1), 2169–2177. (c) Singhania, V.; Cortes-Clerget, M.; Dussart-Gautheret, J.; Akkachairin, B.; Yu, J.; Akporji, N.; Gallou, F.; Lipshutz, B. H. Lipase-catalyzed esterification in water enabled by nanomicelles. Applications to 1-pot multi-step sequences. *Chem. Sci.*, 2022, *13*(5), 1440–1445. (d) Akporji, N.; Dussart-Gautheret, J.; Singhania, V.; Gallou, F.; Lipshutz, B. H. *Chem. Comm. (Camb)* 2021, *57*(89), 11847–11850.

25. (a) Bu, M.-J.; Cai, C.; Gallou, F.; Lipshutz, B. H. *Green Chem.* 2018, *20*(6), 1233–1237. (b) Ogulu, D.; Bora, P. P.; Ansari, T. N.; Sharma, S.; Parmar, S.; Gallou, F.; Kozlowski, P. M.; Handa, S. *ChemSusChem* 2021, *14*(13), 2704–2709. (c) Finck, L.; Brals, J.; Pavuluri, B.; Gallou, F.; Handa, S. *J. Org. Chem.* 2018, *83*(14), 7366–7372.

26. Krell, C.; Parmentier, M.; Padeste, L.; Buser, M.; Hellsttern, J.; Glowienke, S.; Schreiber, R.; Wickendick, R.; Hueber, L.; Haenggi, R.; Sciascera, L.; Zheng, X.; Clarke, A.; Gallou, F. *Org. Process Res. Dev.* 2021, *25*(4), 900–915.

27. Serrano-Luginbühl, S.; Ruiz-Mirazo, K.; Ostaszewski, R.; Gallou, F.; Walde, P. *Nat. Rev. Chem.* 2018, *2*(10), 306–327.

28. Bailey, J. D.; Helbling, E.; Mankar, A.; Stirling, M.; Hicks, F.; Leahy, D. K. *Green Chem.* 2021, *23*(2), 788–795.

29. Petkova, D.; Borlinghaus, N.; Sharma, S.; Kaschel, J.; Lindner, T.; Klee, J.; Jolit, A.; Haller, V.; Heitz, S.; Britze, K.; Dietrich, J.; Braje, W. M.; Handa, S. *ACS Sustain. Chem. Eng.* 2020, *8*(33), 12612–12617.

30. Takale, B. S.; Thakore, R. R.; Mallarapu, R.; Gallou, F.; Lipshutz, B. H. *Org. Proc. Res. Dev.* 2019, *24*, 101–105.

31. Takale, B. S.; Thakore, R. R.; Handa, S.; Gallou, F.; Reilly, J.; Lipshutz, B. H. *Chem.Sci.* 2019, *10*(38), 8825–8831.

32. Takale, B. S.; Thakore, R. R.; Irvine, N. M.; Schuitman, A. D.; Li, X.; Lipshutz, B. H. *Org. Lett.* 2020, *22*(12), 4823–4827.

33. Landstrom, E. B.; Akporji, N.; Lee, N. R.; Gabriel, C. M.; Braga, F. C.; Lipshutz, B. H. *Org. Lett.* 2020, *22*(16), 6543–6546.

34. Isley, N. A.; Wang, Y.; Gallou, F.; Handa, S.; Aue, D. H.; Lipshutz, B. H. *ACS Catal.* 2017, *7*(12), 8331–8337.
35. Handa, S.; Jin, B.; Bora, P. P.; Wang, Y.; Zhang, X.; Gallou, F.; Reilly, J.; Lipshutz, B. H. *ACS Catal.* 2019, *9*(35), 2423–2431.
36. unpublished results.
37. Yu, T.; Pang, H.; Cao, Y.; Gallou, F.; Lipshutz, B. H. *Angew. Chem. Int. Ed.* 2021, *60*(7), 3708–3713.
38. Takale, B. S.; Thakore, R. R.; Casotti, G.; Li, X.; Gallou, F.; Lipshutz, B. H. *Angew. Chem. Int. Ed.* 2021, *60*(8), 4158–4163.
39. Bo, J.; Gallou, F.; Reilly, J.; Lipshutz, B. H. *Chem. Sci.* 2019, *10*(12), 3481–3485.
40. Cortes-Clerget, M.; Berthon, J.-Y.; Krolikewicz-Renimel, I.; Chaisemartin, L.; Lipshutz, B. H. *Green Chem.* 2017, *19*(18), 4263–4267.
41. Lee, N. R.; Linstadt, R. T. H.; Gloisten, D. J.; Gallou, F.; Lipshutz, B. H. *Org. Lett.* 2018, *20*(10), 2902–2905.
42. Pang, H.; Hu, Y.; Yu, J.; Gallou, F.; Lipshutz, B. H. *J. Am. Chem. Soc.* 2021, *143*(9), 3373–3382.
43. Ye, N.; Wu, B.; Zhao, K.; Ge, X.; Zheng, Y.; Shen, X.; Shi, L.; Cortes-Clerget, M.; Regnier, M. L.; Parmentier, M.; Gallou, F. Chem. Commun. 2021, *57*(62), 7629–7632.
44. Akporji, N.; Thakore, R. R.; Cortes-Clerget, M.; Anderson, J.; Landstrom, E.; Aue, D. H.; Gallou, F.; Lipshutz, B. H. *Chem. Sci.* 2020, *11*(20), 5205–5212.
45. Cosgrove, S. C.; Thompson, M. P.; Ahmed, S. T.; Parmeggiani, F.; Turner, N. J. *Angew. Chem. Int. Ed.* 2020, *59*(41), 18156–18160.
46. Akporji, N.; Singhania, V.; Dussart-Gautheret, J.; Gallou, F.; Lipshutz, B. H. *Chem. Commun.* 2021, *57*(46), 11847–11850.
47. (a) Espro, C.; Rodriguez-Padron, D. *Curr. Opin. Green Sustain. Chem.* 2021, *30*, 100478. (b) Ying, P.; Yu, J.; Su, W. *Adv. Synth. Catal.* 2021, *363*(5), 1246–1271. (c) Colacino, E.; Isoni, V.; Crawford, D.; García, F. *Trends Chem.* 2021, *3*(5), 335–339. (d) Kappe, O. C.; James, M.; Bolm, C. *J. Org. Chem.* 2021, *86*(21), 14242–14244.
48. Yeboue, Y.; Rguioueg, N.; Subra, G.; Martinez, J.; Lamaty, F.; Métro, T.-X. *Eur. J. Org. Chem.* 2021, *30*, 631–635.
49. Yeboue, Y.; Jean, M.; Subra, G.; Martinez, J.; Lamaty, F.; Métro, T.-X. *Org. Lett.* 2021, *23*(3), 631–635.
50. Stolar, T.; Grubešić, S.; Cindro, N.; Meštrović, E.; Užarević, K.; Hernández, J. G. *Angew. Chem. Int. Ed.* 2021, *60*(23), 12727–12731.

Annex

Terminology and Nomenclature in Mechanochemistry

Mamoru SENNA

CONTENTS

1. Introduction

As the interests in science and technology around the concept *mechanochemistry* are mounting, we often encounter a linguistic gap during discussions in the interdisciplinary mechanochemical community comprising people with widely varying backgrounds. Organic synthesis and mineral processing, for instance, have both deep interest in modern mechanochemistry. However, background and mentality in these societies are fundamentally different.[1, 2] It is therefore timely to reconsider the nomenclature used in the ever- growing interdisciplinary area of scientific and technological importance for the sake of smoother mutual understanding among mechanochemists in different felds.

There are already a number of comprehensive reviews on the topic.[1, 3–5] In addition, different research groups started proposing symbols to describe ball- milling activated reactions.[6] which have been recently summarised in a review article.[2] Moreover, the history of mechanochemistry has been comprehensively described by Takacs including changes in the nomenclature with time.[5]

It is now clear that a simple dictionary-type alphabetical glossary does not seem to be the best solution for the present purpose. Therefore, this section discusses the topic in a broad sense. Starting from the definition of the central keyword, mechanochemistry or mechanical activation, based on some legendary books and articles. Then, the concordance of related nomenclature, relevance to more general basic science-rooted terminology and a short list of related symbols will follow. Some additional remarks are added to the conclusion.

2. Origin, definition and interpretation of "mechanochemistry" and "mechanical activation"

It seems appropriate to begin the discussion with the most important nomenclature, which is the word *mechanochemistry*. This compound word comprises mechanics and chemistry so illustrates

DOI: 10.1201/9781003178187-18

the synergy between these two scientific genres. There are other branches of chemistry where the energy required to "excite" the starting materials is not (only) thermal vibrational energy. Electro-, photo-, radiation-, magneto-, microwave- or sonochemistry are other representative of this category with well-established nomenclature. They are often generalized as *non-conventional chemical processes*.[4, 7–9]

One of the origins of the modern concept, *mechanochemical reaction*, is solid- state chemistry. V.V. Boldyrev and K Mayer complied a book, "Festkörperchemie-Beiträge aus Forschung und Praxis" with 28 chapters in German language, and published in 1973 in German Democratic Republic (at that time).[10] In this book, the last 3 chapters are dedicated to mechanochemistry. They are entitled: §26. chemical transformation during mechanical stressing (by P.A. Thiessen, G. Heinicke and K. Meyer); §27. Change in the reactivity of solids by preliminary mechanical treatment (by R. Schrader and B. Hoffmann), and §28. Mechanochemical decomposition of inorganic solids (by G.S. Chodakov and E.G. Avvakumov). The almost forgotten nomenclature, *tribo-plasma reaction*, or *impact reaction*, *topography* or *tribochemical deformation*, appeared in chapter 26. The statements in these chapters, published in half a century ago, are supported with a series of experimental and analytical evidences based on then the top-mode tools. They might be worth reexamining with modern analytical tools to date.

In the book "Tribochemistry" authored by Gerhardt Heinicke,[11] the author explained the book title as "chemical consequences of solids under mechanical stressing stochastic and hierarchical" by starting from the foregoing work of Ostwald. [12] While this original reference of Ostwald is hard to access, there are several follow up interpretations with explicit application examples recently published.[13, 14] Heinicke has therefore very cautiously defined mechanochemistry as "*a branch of chemistry dealing with the chemical and physicochemical changes of substances of all states of aggregation due to the influence of mechanical energy*".[11] Here, aggregation means that of ions, atoms or atomic clusters toward molecules, crystals, supramolecules or polymers, distinct from the grouping of small particles used in colloid chemistry or powder technology.

As the concept of mechanochemistry has been extended to organic chemistry mainly in the interests of organic synthesis aided by mechanical stressing, nomenclature has also been drastically extended. The issue was systematically summarized by Boldyreva in a recent review.[1] The principal focus on mechanically aided organic synthesis are the manipulation of C-C covalent bonds, which are not involved in previous discussions involving inorganic crystals.[15] This made for a new theoretical interpretation of the mechanochemistry in terms of the *potential energy surface*,[16] which describes the potential energy of atomic aggregates as a function of the configuration of the constituent atomic species. Care should be taken that organic synthesis *via* a mechanochemical routes often passes via a solid/liquid *interfacial intermediates*, so that a traditional understanding that mechanochemistry is a part of *solid-state chemistry* may need to be reconsidered.

Theoretical basis of mechanochemistry was explored in depth by Butyagin and coworkers as summarized in his monograph.[17] Starting from the crystalline as a ground state, the study focused on the *mechanical disordering* and the *reactivity of solids*. The latter concept was inspired by the *solid-state reaction*, discussed and compiled initially by Hedvall *et al*.[18] Butyagin further interpreted the *mechanochemical processes* in terms of *energy yield*, G, (mol/MJ), which is defined as the number of moles of the substance used on consuming 1 MJ of mechanical energy.[19] The concept is based on the process of grinding or similar machinery, since the bridging basic science with industrial processes needs consciousness of the energy to be invested in a given process. Energy yield is of particular importance right now, where mechanochemistry is understood as a promising tool for green chemical process to mass production *via* improved process engineering.

It is also to be noted that many new insights of organic mechanochemistry have been referred in the same monograph as a separated chapter, "*Mechanochemical transformation in organic substances*" authored by A.M. Dubinskaya.[17] The author already discussed in a systematic manner the two different categories of bond breakage in an organic crystal, *i.e.*, that of strong covalent bonds within a molecule and much weaker non-covalent bonds aggregating organic molecules within a crystal.[17]

3. Comparison and concordance of nomenclature

As discussed in ref[2], "*tribochemistry*" has significant concordance with mechanochemistry.[11] The former came from more mechanical and engineering perspective, while the letter is more popular across the wider scientific community – including organic synthesis or pharmaceutics.[20] The keyword "*mechanical alloying*" has started in an attempt to obtain a technique for the extension of solid solubility in systems where the equilibrium solid solubility is limited.[21, 22] Therefore, mechanical alloying was limited to metallic systems. Later, its interpretation has also been extended to metal oxides, halides, and other compounds; as well as merged with the extended concept of *mechanochemical synthesis* or *mechanosynthesis*.[23] These two keywords are compatible and used in wide range of technological application without a clear inorganic / organic barrier.

Not less important is the concept of *mechanical activation*.[24, 25] Mechanosynthesis is inevitably preceded by excitation, and hence activation, of the starting materials. However, mechanical activation also implies activation by mechanical stressing, for subsequent chemical processes like dissolution or *leaching* in hydrometallurgy,[26] administration of pharmaceutic products[27], or post heating for ceramics,[28] *inter alia*. Here, the *relaxation* process of mechanically activated states plays an important role.[29, 30] Similar concept is also expressed as *mechanochemical activation*.[31] However, the latter nomenclature is not always appropriate, since the action is mechanical stressing and the consequence is the induced chemical reaction.

Care should be taken that the apparently identical nomenclature, mechanochemistry, is used in *mechanical biology*. Mechanical strain is known to stimulates the synthesis of collagen in vascular smooth muscle cells, associated with stretch induced collagen and total protein synthesis.[32, 33] At a glance, mechanical biology seems to be far from the mechanochemistry associated with materials science. However, careful study may reveal some scientific resemblance.[34, 35] Another resembling nomenclature is *chemomechanics*. This is another interplay between mechanical structure and chemistry. e.g., degradation of oxide films induced by environmental water[36], or removal of caries by *N*- monochloroglycine.[37] Mechanochemistry is also used in pharmaceutical field and very recently, the term *medicinal mechanochemistry* has been introduced to identify "*...mechanochemical procedures for generating pharmaceutically relevant fragments and functionalities,,for the assembly of API solid forms, as well as for the synthesis of APIs themselves*".[38]

4. Relevance to more general nomenclature

4.1 Stress-strain related chemical changes

While phenomenological description of related terminology seems to be endless, here we try to understand mechanochemistry in terms of established, common basic nomenclature without ambiguity. Here, the anatomy of the nomenclature requires combining elements, or components, of the entire phenomena on the very basics of physics and chemistry. Basically, it is *inelastic deformation* of the molecule or solid mass, which leaves chemical consequences triggering mechanochemical effects at all.[39] Stress-strain relationships become much more complicated with *molecular crystals*. In this context, deformation and rupture of C-C bonds in molecular crystals under mechanical force will be discussed below. The topic is of primary interests in the mechanical strength of polymeric products[40].

In the framework of *crystal engineering*,[41] *cocrystal* synthesis by mechanical stressing is increasingly becaming under the spotlight.[13, 15, 27, 42, 43]. The basic principle in this genre stands upon the concept, that "molecular crystals are *supermolecule* par excellence".[44] What is happening in the milling jar with a mixture of organic compounds, is based on the principles of *molecular recognition* and *self-assembly*.[15]

4.2 Excited states

Regardless of whether the reacting system is purely *solvent-free solid-state* of with the participation of liquid phase, *excitation of electronic energy* under *mechanical stressing* is one of the unique fundamentals. Conventional chemical kinetics is based on the concept of *Arrhenius activation energy*, with a few exceptions.[45, 46] Temperature dependence of mechanochemical phenomena is in most cases peculiar *non-Arrhenius* character, as Butyagin explored in detail.[47, 48] Their discussion is based on the *atomic processes at local interfaces*, and direct excitation of electronic energy state. They introduced the concept of *dose of absorbed mechanical energy*. Interplay between non-conventional excitation processes, e.g., mechanochemistry and photochemistry, is increasingly studied, e.g., mechanochemically gated photo-switching phenomena.[49] There appeared a number of mechanochemistry-related combined nomenclature, such as sono-mechanochemistry;[50] stress-induced photochemistry,[51, 52] or direction-dependent properties.[53]

Despite the non-Arrhenius character of mechanochemical reactions, the role of temperature and related *hot-spot* model will be briefly referred. As far as we operate milling, or similar, action for mechanochemical effects at room temperature, and above, the role of the thermal vibration has always to be take into account. The role of temperature on mechanochemical reactions has historically been discussed under the nomenclature, *hot-spot*.[5] Closer examination revealed however, that it cannot play a main role due to its locality and short life time.[46]

5. Symbols

Graphical symbols are convenient to quickly understand common items in any professional community. Three attached circles, for instance, are used as a symbol of mechanochemical activation.[6] Some preliminary set of icon-type symbols for representative actions, materials or conditions for mechanical activation or mechanochemical reactions were later proposed.[2] Letter symbols have also been used as listed by Gutman in his book entitled Mechanochemistry of materials.[54] While most of them are common to wider general mathematics or physical chemistry. some unique letter symbols specific to mechanochemistry are also included, e.g., $\tilde{f}$: mechanochemical activity coefficient; H_μ; microhardness; U_f: specific energy of shape changing deformation. Such symbols are not universally recognized, and hence to be defined at the beginning of the article or presentation.

6. Concluding remarks

In the present section, glossaries associated with the interplay between mechanics or mechanical properties of the substances and chemical species are displayed and compared. Important keywords appeared are summarized in **Fig. 1**. Since they cannot be ordered in a single scale, they are displayed in a random manner.

Established interpretation is comfortable to use among experts but not always for those from peripheral genres. For better understanding among those with widespread backgrounds, nothing is more important than to start from common basics. It is also very important to explain some of the important keywords at the beginning of an article of presentation, like checking the convention at the beginning of the contract bridge game with a wider society. By the same token, flexible interpretation of similar nomenclature with quite different implication is also important.

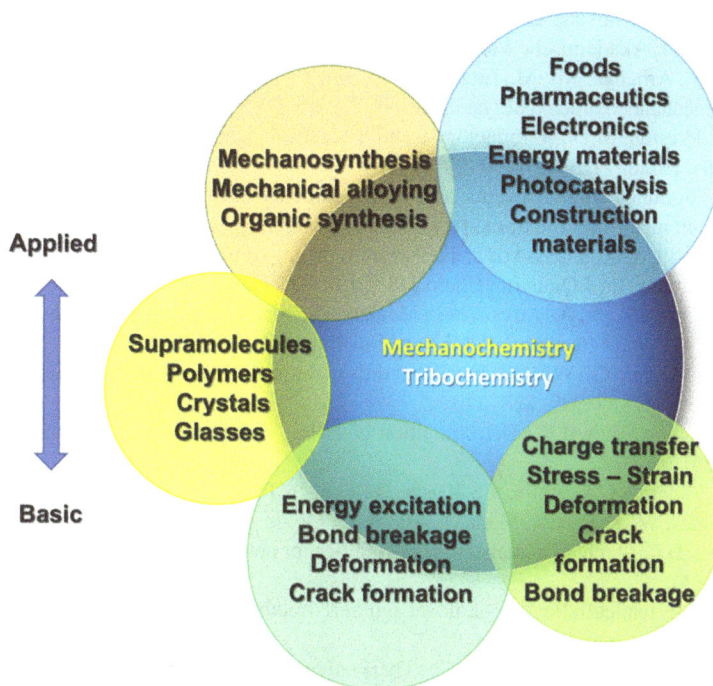

FIGURE 1 Crossover of the glossaries around mechanochemistry.

REFERENCES

1. Boldyreva, E., Mechanochemistry of inorganic and organic systems: what is similar, what is different?, Chem Soc Rev, 42 (2013) 7719–7738.
2. Michalchuk A. A. L., Boldyreva E.V., Belenguer A.M., Emmerling, F., Boldyrev V.V., Tribochemistry, Mechanical Alloying, Mechanochemistry: What is in a Name?, Front Chem, 9 (2021) 685789.
3. Balaz, P., Achimovicova, M., Balaz, M., Billik, P., Cherkezova-Zheleva, Z., Criado, J. M., Delogu, F., Dutkova, E., Gaffet, E., Gotor, F. J., Kumar, R., Mitov, I., Rojac, T., Senna, M., Streletskii, A., Wieczorek-Ciurowa, K., Hallmarks of mechanochemistry: from nanoparticles to technology, Chem Soc Rev, 42 (2013) 7571–7637.
4. Sepelak, V., Duvel, A., Wilkening, M., Becker, K. D., Heitjans, P., Mechanochemical reactions and syntheses of oxides, Chem Soc Rev, 42 (2013) 7507–7520.
5. Takacs, L., The historical development of mechanochemistry, Chem Soc Rev, 42 (2013) 7649–7659.
6. Rightmire, N.R., Hanusa, T.P., Advances in organometallic synthesis with mechanochemical methods, Dalton Trans, 45 (2016) 2352–2362.
7. Senna, M., Preparation of functional materials via non-convewntional routes, Ann Chem Sci Mater, 27 (2002) 3–14.
8. Musa, C., Locci, A. M., Licheri, R., Orrù, R., Cao, G., Vallauri, D., Deorsola, F.A., Tresso, E., Suffner, J., Hahn, H., Klimczyk, P., Jaworska, L., Spark plasma sintering of self-propagating high-temperature synthesized TiC0.7/TiB2 powders and detailed characterization of dense product, Ceramics Inter, 35 (2009) 2587–2599.
9. Suslick, K.S., Mechanochemistry and sonochemistry: concluding remarks, Faraday Discuss, 170 (2014) 411–422.
10. Boldyrev, V.V., Meyer, K. Festkörperchemie - Beiträge aus Forshung und Praxis, VEB Deutcher Verlag für Grundstoffindustrie, Leipzig, 1973.
11. Heinicke, G. Tribochemistry, Akademie-Verlag, Berlin, 1984.

12. Ostwald, W. Die chemische Literatur und die Organisation der Wissenschaft, Handbuch der allgemeinen Chemie 1, Akademische Verlagsgesellschaf, Leipzig,, (1919).
13. Germann, L.S., Arhangelskis, M., Etter, M., Dinnebier, R. E., Friscic, T., Challenging the Ostwald rule of stages in mechanochemical cocrystallisation, Chem Sci, 11 (2020) 10092–10100.
14. Jones, W., Eddleston, M.D., Introductory lecture: Mechanochemistry, a versatile synthesis strategy for new materials, Faraday Discuss, 170 (2014) 9–34.
15. Friscic, T., Supramolecular concepts and new techniques in mechanochemistry: cocrystals, cages, rotaxanes, open metal-organic frameworks, Chem Soc Rev, 41 (2012) 3493–3510.
16. Quapp, W., Bofill, J.M., Ribas-Ariño, J., Toward a theory of mechanochemistry: Simple models from the very beginnings, Int J Quantum Chem 118 (2018) e25775.
17. Butyagin, P., Dubinskaya, A., Advances in Mechanochemistry: Physical and Chemical Processes under Deformation, Harwood Academic Publishers, Reading, UK, 1998.
18. Hedvall, J.A., Cohn, G. Wissenschaftliche und technische Sammelreferate. Reaktionen im festen Zustand, Kolloid-Z, 88 (1939) 224–240.
19. Butyagin, P.V., Pavlichev, I.K., Determination of energy yield of mechanochemical reactions, React Solid, 1 (1986) 361–372.
20. James, S.L., Adams, C.J., Bolm, C., Braga, D., Collier, P., Friščić, T., Grepioni, F., Harris, K.D.M., Hyett, G., Jones, W., Krebs, A., Mack, J., Maini, L., Orpen, A.G., Parkin, I.P., Shearouse, W.C., Steed, J.W., Waddell, D.C. Mechanochemistry: opportunities for new and cleaner synthesis, Chem Soc Rev, 41 (2012), 413–447.
21. Murty, B.S., Mechanical alloying - a novel synthesis route for amorphous phases, Bull Mater Sci, 16 (1993) 1–17.
22. Ma, E., Atzmon, M., Pinkerton, F.E., Thermodynamic and magnetic properties of metastable FexCu100–x solid solutions formed by mechanical alloying, J Appl Phys, 74 (1993) 955.
23. Tsuzuki T., McCormick P.G., Mechanochemical synthesis of nanoparticles, J Mater Sci, 39 (2004).
24. Boldyrev, V.V., Mechanical activation and its application in technology, Mater Sci Forum, 269–272 (1998) 227–234.
25. Boldyrev, V.V. Mechanochemical processes with the reaction-induced mechanical activation. Chemo-mechanochemical effect, Russ Chem Bull Int Ed, 67 (2018) 933- 948.
26. Baláž, P., Aláčová, A., Achimovičová, M., Ficeriová, J., Godočíková, E., Mechanochemistry in hydro-metallurgy of sulphide minerals, Hydrometallurgy, 77 (2005) 9–17.
27. Batzdorf, L., Zientek, N., Rump, D., Fischer, F., Maiwald, M., Emmerling, F., Make and break - Facile synthesis of cocrystals and comprehensive dissolution studies, J Mol Struct, 1133 (2017) 18–23.
28. Vojisavljević, K., Malič, B., Senna, M., Drnovšek, S., Kosec, M., Solid state synthesis of nano-boehmite-derived CuAlO2 powder and processing of the ceramics, J Eur Ceramic Soc, 33 (2013) 3231–3241.
29. Boldyrev, V.V., Tkáčová, K., Mechanochemistry of Solids_Past, Present, and Prospects, J Mater Syn Process, 8 (2000) 121–132.
30. Bychkov, A., Matveeva, A., Introducing Students to Energy-Efficient Mechanochemistry of Biopolymers, J. Chem. Educ., 99 (2022) 2630–2635.
31. Prokof'ev V.Y., Gordina N.E., Comminution and mechanochemical activation, Glass Ceram, 69 (2012) 65–70.
32. Li Q., Muragaki Y., Hatamura I., Ueno H., Ooshima A., Stretch-Induced Collagen Synthesis in Cultured Smooth Muscle Cells from Rabbit Aortic Media and a Possible Involvement of Angiotensin II and Transforming Growth Factor-ß, J Vasc Res, (1998) -.
33. Murtada, S.I., Kroon, M., Holzapfel, G.A., A calcium-driven mechanochemical model for prediction of force generation in smooth muscle, Biomech Model Mechanobiol, 9 (2010) 749–762.
34. Zhang, L., Richardson, M., Mendis, P., Role of chemical and mechanical stimuli in mediating bone fracture healing, Clin Exp Pharmacol Physiol, 39 (2012) 706–710.
35. Okuda, S., Inoue, Y., Watanabe, T., Adachi, T., Coupling intercellular molecular signalling with multi-cellular deformation for simulating three-dimensional tissue morphogenesis, Interface Focus, 5 (2015) 20140095.
36. Belde, K.J., Bull, S.J., Chemomechanical effects in optical coating systems, Thin Solid Films, 515 (2006) 859–865.
37. Maragakis, G.M., Hahn, P., Hellwig, E., Chemomechanical caries removal: a comprehensive review of the literature, Int Dent J, 51 (2001) 291–299.

38. Tan, D., Loots, L., Friščić, T. Towards medicinal mechanochemistry: evolution of milling from pharmaceutical solid form screening to the synthesis of active pharmaceutical ingredients (APIs), Chem Commun 52 (2016) 7760–7781.

39. Senna M., Schönert K., Direct observation of inelastic deformation and mechanochemical activation of indented quartz single crystal, Powder Technol, 32 (1982) 217–221.

40. Nevejans, S., Ballard, N., Fernández, M., Reck, B., García, S.J., Asua, J.M., The challenges of obtaining mechanical strength in self-healing polymers containing dynamic covalent bonds, Polymer, 179 (2019) 121670.

41. Desiraju, G.R., Crystal engineering: a holistic view, Angew Chem Int Ed Engl, 46 (2007) 8342–8356.

42. Trask, A.V., Jones, W., Crystal enginnering of organic cocrystals by the solid-state grinding approach, Top Curr Chem, 254 (2005) 41–70.

43. Jones, W., Mechanochemistry and Its Role in Novel Crystal Form Discovery, in: K. Roberts, R. Docherty, R. Tamura (Eds.) NATO Science for Peace and Security Series A-Chemistry and Biology, Springer, Dordrecht, 2017.

44. Dunitz, J.D., Phase transitions in molecular crystals from a chemical viewpoint, Pure Appl Chem, 63 (1991) 177–185.

45. Gonsalvez M.A., Nieminen R.M., Sato K., Arrhenius and Non-Arrhenius Behaviour During Anisotropic Etching, Sensor Mater, 17 (2005) 187–197.

46. Užarević, K., Štrukil, V., Mottillo, C., Julien, P.A., Puškarić, A., Friščić, T., Halasz, I., Exploring the Effect of Temperature on a Mechanochemical Reaction by in Situ Synchrotron Powder X-ray Diffraction, Cryst Growth Des, 16 (2016) 2342–2347.

47. Butyagin, P.V., The role of interfaces in the reactions of low-temperature mechanochemical synthesis, Colloid J, 59 (1997) 425–431.

48. Borunova, A.B., Streletskii, A.N., Mudretsova, S.N., Leonov, A.V., Butyagin, P.Y., Low-temperature mechanochemical synthesis of nanosized silicon carbide, Colloid J, 73 (2011) 605–613.

49. Hu, X., McFadden, M.E., Barber, R.W., Robb, M.J., Mechanochemical Regulation of a Photochemical Reaction, J Am Chem Soc, 140 (2018) 14073–14077.

50. Sogabe K., Senna M., Preparation and properties of hydroxyapatite silk fibroin nanostructured dispersion by simultaneous ultrasonication and mechiancal agitation, J Meastable Nanocryst Magterl, 15–16 (2003) 289–294.

51. Ivanov, A.G., Velitchkova, M.Y., Allakhverdiev, S.I., Huner, N.P.A., Heat stress- induced effects of photosystem I: an overview of structural and functional responses, Photosynth Res, 133 (2017) 17–30.

52. Khatri, K., Rathore, M.S., Salt and osmotic stress-induced changes in physio- chemical responses, PSII photochemistry and chlorophyll a fluorescence in peanut, Plant Stress, 3 (2022) 100063.

53. Yoo, S.-H., Na, Y., Hwang, W., Jang, W., Soon, A., First-Principles Calculations of Heteroanionic Monochalcogenide Alloy Nanosheets with Direction-dependent Properties for Anisotropic Optoelectronics, ACS Appl Nano Mat, 4 (2021) 5912- 5920.

54. Gutman, E.M., Mechanochemistry of Materials, Cambridge International Science Publishing, Camridge, 1998.

Index

For Product Safety Concerns and Information please contact our EU
representative GPSR@taylorandfrancis.com
Taylor & Francis Verlag GmbH, Kaufingerstraße 24, 80331 München, Germany

www.ingramcontent.com/pod-product-compliance
Lightning Source LLC
Chambersburg PA
CBHW080915220326
41598CB00034B/5576

9 781032 013220